Microphysical Processes in Clouds

Microphysical Processes in Clouds

K. C. Young
University of Arizona

New York Oxford
OXFORD UNIVERSITY PRESS
1993

Oxford University Press

Oxford New York Toronto
Delhi Bombay Calcutta Madras Karachi
Kuala Lumpur Singapore Hong Kong Tokyo
Nairobi Dar es Salaam Cape Town
Melbourne Auckland Madrid

and associated companies in
Berlin Ibadan

Published by Oxford University Press, Inc.,
200 Madison Avenue, New York, New York 10016

Library of Congress Cataloging-in-Publication Data
Young, Kenneth C., 1941–
Microphysical processes in clouds/Kenneth C. Young.
p. cm. Includes bibliographical references and index.
ISBN 0-19-507563-3
1. Cloud physics. I. Title.
QC921.5.Y68 1993
551.57′6–dc20 93-7839

9 8 7 6 5 4 3 2 1
Printed in the United States of America
on acid-free paper

PREFACE

This book derives from a two-semester course on cloud microphysics and precipitation processes that has been taught at the University of Arizona on a biennial basis since 1975. There is a clear emphasis on fundamental concepts: a firm grasp of the fundamentals will enable students to assimilate specialized topics more readily and will prepare them for innovative research in cloud physics. The landmark texts of Mason (*The Physics of Clouds*, 1957) and Fletcher (*The Physics of Rainclouds*, 1962) serve as models for this text, both in focus and in depth of coverage.

Students are expected to have a basic understanding of physical meteorology and an introduction to thermodynamics. This text is directed to upper-division undergraduate and graduate students in the physical sciences. Researchers in atmospheric physics may find the updates of basic theories and concepts to be beneficial. Exercises are graded from one star (*) problems designed for undergraduate students, to two star (**) problems for graduate students, to three star (***) problems designed to challenge everyone.

There is a reliance on numerical models, particularly in the last six chapters. This reflects the author's bias. The text is not intended to be a "cookbook" for developing numerical models of microphysical processes, although many of the requisite equations and examples are provided. Rather, the introduction of modeling results is intended to provide insight into the interactions between the various microphysical processes, and is not intended to give "answers." Often a simple numerical model will reveal unsuspected synergisms that, once uncovered, seem intuitively obvious.

This book is purposely limited to the microphysical aspects of clouds and precipitation processes. Dynamical aspects of clouds and their influence on the development of precipitation have been examined in the recent book by Cotton and Anthes, *Storm and Cloud Dynamics* (1989). Cloud electricity and radar meteorology are not within the intended focus of this text.

The first eight chapters follow a traditional approach to cloud microphysics, discussing the processes of nucleation, diffusional growth, and growth by collection for isolated particles. The basic equations for each topic are developed for water drops and then extended to ice particles in the chapter immediately following.

Chapters 9 and 10 build on the development of the first eight chapters and show how these processes act synergistically to produce precipitation. Although emphasis is on the behavior of ensembles of particles, an effort has been made to place these discussions in the context of real cloud systems. Simple numerical models are used to illustrate the interactions between nucleation, diffusion, and collection processes, and to provide the student with insight into the synergistic nature of the development of precipitation through the coalescence and ice crystal mechanisms.

Additional insight into the development of precipitation is given by examining efforts to modify the precipitation process by "cloud seeding." Glaciogenic seeding for both static and dynamic effects, as well as hygroscopic seeding, are examined in Chapter 11. The growth of hail and its modification are examined in Chapter 12.

The final chapter on cirriform clouds has been included because cirrus clouds are important in the radiative budget of the earth. Unlike low clouds that tend to produce cooling at the surface, thin cirriform clouds tend to produce warming at the surface. The low temperatures and high altitudes at which cirrus clouds form provide a markedly different microphysical environment and consequently involve some interesting physics.

In the interest of notational simplicity, excessive use of superscripts, subscripts, primes, overbars, underbars, and weird symbols has been minimized: r means radius; m means mass. As a result, the same symbol may be used in different ways in different chapters, although the meaning should be clear from the context. Within a given chapter, there is only one use for a given symbol. In virtually all references to particle sizes, the radius or its equivalent for ice crystals is used. The author believes that the CGS system of units is the natural system for cloud microphysics, and that micrometers or grams should be used as appropriate to avoid excessive use of exponential notation.

A book is never produced in isolation. I am particularly indebted to the following people: my former thesis advisor, Roscoe Braham, who undoubtedly cringes at my use of the van't Hoff factor rather than the molal osmotic coefficient; my office mate in graduate school, Andy Heymsfield, for many stimulating arguments; and my good friends, Charlie and Nancy Knight, for making me realize that real clouds are more complicated than computer models. I am also deeply in debt to Toshiko Nakamatsu, who made sure that I would finish graduate school.

I would also like to express my appreciation to the late Lou Battan for his encouragement in the early stages of this book and to Phil Krider for his continued support, in spite of my iconoclastic tendencies. I also thank Jim Abbott for making artwork out of my sketchings.

Finally, I would like to acknowledge the contributions of all the overworked students, too numerous to mention, for questions and comments on the lecture notes. To all the students who will use this text in the future, I ask you to scratch beneath the surface and think in new ways. The joy of science lies in conceiving ideas heretofore unthought.

Arizona K.C.Y.
September, 1992

CONTENTS

Microphysical Processes in Clouds

1

THE FORMATION OF CLOUDS AND PRECIPITATION

1.1. Introduction

The formation of clouds and the subsequent development of precipitation result from the three basic processes of nucleation, diffusion, and collection. One should realize that these processes are highly interactive and quite dependent on the cloud's dynamical structure. The solution of these processes in a real framework is often synergistic; that is, the result is more than a simple summation of the individual processes. The synergistic nature of these processes has only recently come to be appreciated. The purpose of this chapter is to place them in context as a prelude to a detailed examination of the individual processes.

1.2. How air may become saturated

It is possible to identify three ways in which air may become supersaturated. These are by the removal of heat (diabatic cooling), by mixing, and by adiabatic processes.

1.2.1. *Diabatic cooling*

Heat may be removed from an air parcel by any of the three basic heat transfer modes: radiation, conduction, or convection. In many cases, it is not possible to distinguish between molecular diffusion (conduction) and turbulent diffusion (convection), and, for the purposes of this discussion, these will be grouped together.

Heat loss by radiation may occur within the atmosphere where gradients of absorbing (emitting) gases occur or where layers of particles exist. The principal absorbing gas in this regard is water vapor, since most of the other absorbing gases are fairly well mixed. The particles (aerosols) may consist of water (or ice), or any of a wide variety of chemical species, both soluble and insoluble, solid and liquid. The importance of radiational cooling within the atmosphere in producing clouds is not known, although it has been shown to have a marked effect on cloud particle growth under certain conditions.

Diabatic cooling resulting from conduction and convection occurs when warm air overlies a colder surface. This situation may result from horizontal transfer (advection) of the air, for example warm air moving over cold land, or it may result from nocturnal cooling of the surface due to emission of infrared radiation. What is commonly referred to as 'radiation' fog represents cooling of the surface by radiation and the subsequent transfer of heat from the air to the surface by conduction and convection. 'Advection' fog results from the horizontal movement of warm air over a cold surface.

1.2.2. *Mixing*

The attainment of supersaturated conditions by mixing of two parcels of air is a result of the positive curvature of the water vapor–liquid water phase transition line in PT-space (see Appendix 1). Any mixture of two saturated parcels having different temperatures must become supersaturated because the average vapor density of the mixture is greater than the saturation vapor density at the average temperature. It is also possible to mix two subsaturated parcels and attain a supersaturated condition, provided their difference in temperature is great enough.

In nature, mixing to produce supersaturated conditions may occur in two ways. The obvious way is when two saturated or nearly saturated air masses come together from different source regions (with different temperatures). The formation of fog off the eastern coast of Canada due to mixing between the cold moist air overlying the Labrador current and the warm moist air overlying the Gulf Stream is an example of this process. Fog formed in this manner is generally termed 'mixing' fog.

The second way in which this process operates in nature is when moist air overlies water of a different temperature. Consider the case when cold air overlies warmer water. Even if the air is at saturation, a vapor density gradient will exist, resulting in an upward flux of water vapor. A similar temperature gradient is present, resulting in an upward flux of sensible heat. The transport of water vapor and heat is by turbulent diffusion, leading to comparable transfer rates for both water vapor and heat. This may be viewed as a variant of the mixing fog process, i.e., mixing unmodified cold moist air aloft with air warmed (and humidified) in proximity to the warm water surface. Fog formed in this manner is generally termed 'steam' fog.

Warm moist air overlying colder water will produce supersaturated conditions by the same mechanism. The cooling of the air from below increases its stability and reduces the eddy size, limiting the fog to a relatively shallow layer. Fog formed in this manner is generally termed 'sea breeze' fog. Steam fogs are usually deeper than sea breeze fogs since the steam fog layer is much less stable, having been warmed from below.

1.2.3. *Adiabatic cooling*

As air expands, it cools, tending to approach saturation. However, the vapor density decreases as the pressure decreases, tending to decrease the saturation ratio. In other words, for condensation to occur in an updraft, the saturation vapor pressure must decrease more rapidly than the total pressure (or the water vapor partial pressure, assuming the mixing ratio is constant).

The differential change in temperature of the air may be expressed by the differential form of the Poisson relationship (see Appendix 1, (A.1.7)) and by the differential change in saturation vapor pressure from the Clausius–Clapeyron equation, assuming that the latent heat is not a function of temperature. In differential form, this is

$$dT = \frac{R\ T\ d\ (\ln\ P)}{M_{\mathrm{d}}\ c_{\mathrm{p}}} = \frac{R\ T^2\ d(\ln\ e_s)}{L_{\mathrm{v}}\ M_{\mathrm{v}}}. \tag{1.1}$$

Integrating (1.1) from some initial state $[e_{\mathrm{s}}(T_{\mathrm{o}}), P]$ to a final state $[e_{\mathrm{s}}(T), P]$, this becomes

$$\frac{e_{\mathrm{s}}\ (T)}{e_{\mathrm{s}}\ (T_{\mathrm{o}})} = \left(\frac{P}{P_{\mathrm{o}}}\right)^X = \left(\frac{e}{e_{\mathrm{o}}}\right)^X, \tag{1.2}$$

where

$$X = \frac{L_{\mathrm{v}}\ M_{\mathrm{v}}}{c_{\mathrm{p}}\ T\ M_{\mathrm{d}}}.$$

Note that this treatment assumes the condensing gas is a minor constituent of the atmosphere (see Appendix 1).

Examination of (1.2) reveals that X must be greater than 1 for condensation to occur in an updraft. The latent heat of vaporization ($L_{\mathrm{v}}M_{\mathrm{v}}$) for water is 45.06 kJ mol^{-1}; the specific heat capacity for dry air is 29.098 J mol^{-1} K^{-1}. Assuming a typical temperature for earth's troposphere of 273 K, X is 5.67, i.e., condensation occurs in updrafts.

For condensation in downdrafts, the condensing gas must have a low molar latent heat and the carrier gas must have a high molar specific heat. McDonald (1964) discusses several consequences that would result from condensation in downdrafts. The release of latent heat in downdrafts would tend to decelerate the downdraft and suppress convection, whereas updraft condensation tends to accelerate updrafts and promote convection. Condensate amounts would tend to be low and precipitation rare, since the condensate vapor density would likely decrease with height, assuming sources (evaporation) of condensate are at the surface.

Bohren (1986) presents a more rigorous examination of this problem, recognizing the temperature dependence of the latent heat. It may be noted that the latent heat of vaporization decreases with increasing temperature, reaching zero at the critical temperature.[1] This implies that any condensate, under the

[1] This is the definition of the critical temperature.

appropriate conditions, could condense in either updrafts or downdrafts, although condensation in downdrafts is probably exceedingly rare.

1.3. The formation of cloud particles

In § 1.2 a number of situations that may result in supersaturated conditions were noted. Of these, the most common and most important in the troposphere is the adiabatic process, condensation in updrafts. Depending on the stability of the atmosphere and the mechanism forcing vertical motion, different types of clouds may arise, including cumulus, cirrus, nimbostratus or altocumulus lenticularis. However, all of these types of clouds are formed as a result of condensation in upward moving air.

As air rises, its saturation ratio increases. Once saturation is achieved, it is possible in principle to produce a condensed water phase. As will be shown, this takes places preferentially on small aerosol particles of soluble material at slight supersaturations with respect to water. These nuclei are termed *cloud condensation nuclei* (CCN). The number of these particles which are 'activated,' i.e., become droplets, increases as the supersaturation increases in the updraft.

As these droplets grow, they tend to deplete the water vapor excess above water saturation. After some time, the depletion rate will equal and then exceed the rate at which vapor is being 'made available' in the updraft, and no further activation of CCN occurs. The numbers of droplets formed range from 30 to 3000 per cm^3, depending largely on the CCN characteristics and the updraft velocity.

Things are never this simple in nature, although this picture is useful under some conditions such as near the base of newly formed cumulus clouds. Updrafts are rarely very steady, either spatially or temporally, and the supersaturation may be expected to fluctuate accordingly. Mixing processes tend to bring cloud-free air into the cloud, along with more CCN, and tend to reduce the supersaturation. The presence of larger drops, formed earlier in the cloud's history and falling into the nucleation zone, helps deplete the available water vapor, tending to reduce the supersaturation. However, these large drops also serve to remove the smaller droplets by collecting them, resulting in a tendency to increase the supersaturation.

Cloud droplets grow rapidly to sizes on the order of 5 μm radius by a diffusive transfer of water vapor to the drop surface. The condensation of water vapor releases latent heat, which must be conducted away from the drop. The growth rate is determined by the balance between these two processes. The temperature of the drop differs from the environment, and depends on the drop size and the saturation ratio as well as on the ambient temperature and pressure.

As a droplet grows by diffusion, the rate of change of its radius (dr/dt) decreases as $1/r$. The growth of droplets by diffusion to sizes large enough for the drops to fall as rain would require many hours (even days) under typical supersaturations believed to exist within clouds.

1.4. The formation of precipitation

There are two mechanisms by which a cloud consisting of small droplets can produce significant numbers of large particles (precipitation). The mutual collection of water drops through collision and subsequent coalescence is commonly referred to as the 'warm rain' or *coalescence process*. The growth of ice particles in a predominantly supercooled water cloud due to the differences between water and ice saturation vapor densities is commonly referred to as the 'cold rain' or *ice crystal process*. Langmuir (1948) is generally credited with the first quantitative formulation of the coalescence process; Bergeron (1935) is often credited with advancing the concepts of the ice crystal process, although much of the groundwork was laid earlier.

It should be borne in mind that both of these processes may act concurrently within the same cloud. Often it is difficult to assess which of these mechanisms is responsible for producing the observed precipitation. The determination of which mechanism is dominant is less important than realizing that both mechanisms may be operative, and may interact synergistically. Still, it should be noted that a large fraction of the world's precipitation results under conditions in which the ice crystal process is operative.

1.4.1. *The coalescence mechanism*

The coalescence mechanism refers to the production of large drops by collisions and coalescences (collections) between the smaller drops. By virtue of the random distribution of the droplets in space, a very small fraction of the cloud droplets will be favored in the sense that they will undergo an inordinate number of collection events and grow much larger than their fellows. Each collection event tends to increase the disparity in fall velocities, making further collection events more likely. Once a droplet has attained a radius of 30 to 50 μm, it can grow rapidly to raindrop sizes (millimeters in radius).

The $1/r$ dependence of the diffusion growth rate implies that a population of droplets will tend to become more nearly equal in radius. One might assume that this narrowing of the spectrum would make it less likely for two droplets to collide and coalesce. However, in a population of droplets growing by diffusion, the rate of drop–drop interactions increases rapidly as the population grows. Since the fall velocity of small droplets is proportional to r^2, the fall velocity spectrum broadens as the droplets grow. The drop–drop interaction rate increases since the collection efficiencies, the drop cross-sections, and the difference in fall velocities all increase with increasing liquid water content. Thus, as droplets grow by diffusion in the updraft, the drop–drop interaction rate increases, forcing the coalescence process.

Twomey (1976) points out that the production rate for large drops by coalescence is so sensitive to the liquid water content that even small regions of relatively high liquid water content may dominate the coalescence process in the cloud. Thus, even though conditions in the greater portion of a cloud may not appear to favor the initiation of the coalescence process, precipitation may be initiated in relatively limited regions and spread throughout the cloud.

The ultimate size of raindrops is determined by breakup processes. Several modes of drop breakup have been observed in laboratories, most of these resulting from collisions between drops. Thus, the process of the collection of droplets, which allows a drop to grow, is also the process that ultimately limits the drop's growth.

Three different modes of drop breakup have been quantified by Low and List (1982). The most frequent drop breakup mode is filament breakup, in which the smaller drop contacts the larger drop near its edge. Temporary coalescence and subsequent gravitational separation pulls out a filament, which disrupts into several fragments. Sheet breakup is next most frequent, and results when the smaller drop collides closer to the larger drop's center, pulling out a sheet which disrupts into fragments. Disc breakup results from near-center contact and is a major source of small droplets.

The interaction between the coalescence and breakup processes results in an approximately exponential drop size spectrum. Raindrop spectra of this type are often referred to as Marshall–Palmer distributions after the researchers who first described this feature (1948). Numerical modeling studies (Valdez and Young, 1985) have reproduced the exponential nature of the raindrop size spectrum and have suggested a trimodal spectrum with peaks near 0.12, 0.4, and 1.0 mm radius.

1.4.2. *The ice crystal mechanism*

The ice crystal mechanism relies on the difference between water and ice saturation vapor densities at temperatures below 0°C. Obviously, this mechanism can only operate at temperatures below freezing and should be most efficient at temperatures near −15°C, where the difference between water and ice saturation vapor densities is the greatest. The introduction of ice particles into a supercooled water cloud allows the ice particles to grow as the water drops evaporate.

In order to produce precipitation particles, the mass of the largest particles must be increased by a factor of 10^4 to 10^5. Since cloud droplet concentrations are on the order of 100 to 1000 cm^{-3}, this implies that ice particle concentrations on the order of 1 to 10 ℓ^{-1} are required for producing precipitation. If one ice crystal per liter were introduced into a cloud containing 500 drops cm^{-3}, when the process is complete, the ice crystal would have a mass of 5×10^5 droplets and would have an appreciable fall velocity, allowing it to fall out of the cloud and reach the ground. If the number of ice crystals introduced were on the order of 10 cm^{-3}, each crystal would have a mass of only 50 droplets and a fall velocity not much greater than the droplets in the original cloud.

The number of ice particles in a cloud is largely determined by the ice nucleation processes. Although these processes are poorly understood, not only in clouds but also in the laboratory, it is known that the rate of ice particle production depends on the temperature, the character of the nuclei, the concentrations and sizes of water drops present, the updraft velocity, and the prior history of the cloud.

It is fortuitous that the characteristics which render a nucleus effective in forming ice are relatively rare, in contrast to the characteristics which render a nucleus an effective CCN. Whereas hygroscopic substances which form CCN are relatively common, substances that are epitaxial[2] with ice, and which are therefore effective ice nuclei, are relatively rare. Thus, natural clouds tend to produce low concentrations of ice particles, which then are able to grow to sizes large enough to reach the ground as precipitation. If this were not the case, it would be difficult to form precipitation by the ice crystal mechanism.

Ice particles can grow by collecting water drops, a process termed *accretion* (or riming), or by collecting other ice particles, a process termed *aggregation*. Unrimed or lightly rimed ice crystals are generally referred to as 'ice crystals.' As the crystal becomes more heavily rimed and loses its characteristic shape (habit), it is referred to as a graupel (pl. graupeln). If the latent heat released as the collected water drops freeze is not sufficient to raise the particle temperature to 0°C, the particle density remains fairly low, generally 0.6 g cm^{-3} or less. If the rate of heat loss by the particle is not sufficient to keep its temperature below 0°C, the unfrozen water may be retained within the ice structure, leading to densities closer to the density of bulk ice (0.9 g cm^{-3}). Such particles are termed 'hail.' The mutual collection of ice crystals leads to very low density aggregates of ice crystals, which are termed 'snowflakes.'

It should be noted that these definitions are based on how the particles are formed rather than how the particles appear, the latter being the common practice in meteorology. However, it is more meaningful to classify particles by how they form when studying cloud microphysics, although it should be recognized that these definitions may cause confusion when used outside the field.

1.5. Modification of precipitation processes

Much of the interest in cloud physics has been stimulated by the desire to modify precipitation processes. The object of any modification effort is to produce fairly substantial changes with minimal input. The variety of such proposals covers almost all aspects of cloud microphysics, and also extends to direct alteration of the macrophysical structure of the cloud or cloud system. In this text, attention will be focused on techniques which act on the microphysical processes, with the primary intent of altering of the microphysical structure of the cloud.

The first serious and scientifically sound efforts to modify clouds were initiated by the discoveries that dry ice (Schaefer, 1946) or silver iodide (Vonnegut, 1947) could cause the conversion of a supercooled water cloud to ice. Since these substances could produce high concentrations of ice particles at temperatures for which there are few active natural ice nuclei, the ice crystal mechanism could be induced in many clouds having temperatures between 0°C and −15°C. Field trials using aircraft to drop dry ice pellets into supercooled

[2] Epitaxy refers to having a similar molecular structure.

stratus clouds clearly showed that such clouds could be converted to ice particles with subsequent precipitation. However, stratus clouds have relatively low water contents and the precipitation that can be derived from them is not significant, although these techniques have proven highly useful in clearing supercooled fogs from airports.

The extension of these techniques to clouds more capable of producing significant amounts of precipitation has been largely inconclusive. Some experiments have produced statistically significant increases in precipitation, a few have demonstrated decreases, while most have been inconclusive. The use of ground-based silver iodide generators in many of these experiments has led to uncertainties about the particular clouds or volumes which have been treated and about the concentrations of silver iodide reaching the target clouds.

A major problem appears to be in producing a near optimal concentration of ice particles. It was noted above that ice particle concentrations on the order of 1 to 10 ℓ^{-1} are favorable for precipitation. Concentrations much less than this would produce inconsequential amounts of precipitation; concentrations much larger would fail to produce large particles due to competition.

The number of ice particles per gram of silver iodide that would be produced in a specific cloud (given its macrostructure and microstructure) probably cannot be determined to within an order of magnitude at the present time. The variations in the concentration of silver iodide within the cloud may account for another order of magnitude uncertainty. These uncertainties often cover the range from a deficit to a surfeit of ice crystals.

A second promising technique is to introduce large collection centers into the cloud to promote the coalescence process. Salt particles and water drops have both been used in attempts to stimulate the coalescence process in cumulus clouds. However, the much larger mass requirements plus the need to deliver the material directly to the cloud to be treated have argued against this technique, in comparison to seeding with silver iodide. Advantages of attempts to enhance the coalescence process are that the microphysical processes are better understood, there appears to be no upper limit on the amount of material that can be used, and the techniques can be applied to a wider range of clouds.

Once a cloud has been 'seeded,' it is not possible to quantify the amount of precipitation it would have produced if it had not been seeded. Since no two clouds (or storms) are alike, it is necessary to resort to carefully designed, statistical experiments in order to assess the efficacy of the modification. Thus, a properly designed modification experiment should have both physical and statistical components.

1.6. Classification of microphysical processes

In §§ 1.3 and 1.4, a number of microphysical processes were introduced which describe the initial formation of cloud particles and how these particles may grow to large sizes so as to result in precipitation. There are three basic categories of processes, namely, (i) phase nucleation, (ii) growth or evaporation by diffusion, and (iii) interparticle collection. These form a temporal sequence in which nucleation precedes diffusion growth, which precedes growth by collection.

The equations that describe processes within a given category are usually quite similar, whether the system is water or ice. Generally, the treatment for water is simpler, and each process will be introduced with reference to the liquid water phase. Subsequently, these treatments will be extended to handle the ice phase.

Three distinct phase nucleation processes are of interest, namely, nucleation of water from vapor, nucleation of ice from vapor, and nucleation of ice from water. These nucleation processes are termed *condensation*, *deposition*, and *freezing*, respectively. These nucleation processes may proceed within a pure system (homogeneous) or within a system containing 'impurities' (heterogeneous).

Growth and evaporation of particles by diffusion processes are complicated by kinetic effects for small particles, ventilation effects for large particles, and competition effects for particles which reside in a population of particles. Solution of the diffusion equation for ice crystals is complicated by the variety of shapes (habits) that ice crystals may assume.

Four types of interparticle collection process are possible. Large or more rapidly falling particles are thought of as collecting smaller or more slowly falling particles. The collection of a small water drop by a larger one is termed coalescence; the collection of a small ice particle by a larger one is termed aggregation; the collection of a small ice particle by a larger water drop is termed freezing; the collection of a small water drop by a large ice particle is termed accretion.

Problems

1.1* Consider a saturated air mass, initially at 20°C. Assume it is cooled isobarically to 15°C without condensation. What value of supersaturation would it attain?

1.2** Consider two adjoining, saturated air masses (A and B), at 15°C and 20°C, respectively. Assume they are both at the same level in the atmosphere and mix along their vertical interface. What is the maximum supersaturation achieved within the mixing zone?

Hint: Plot the supersaturation as a function of mass fraction of A (or B).

1.3** Determine whether or not you would expect condensation to occur in updrafts or downdrafts for each of the following atmospheres. Compare the dry and pseudo-adiabatic lapse rates for each atmosphere. What do your results indicate in terms of convection in each of these cases?

(a) H_2O in CO_2 at 0°C, total pressure of 1 atm.
(b) SO_2 in CO_2 at 100°C, total pressure of 2 atm.
(c) NH_3 in a mixture of 70% H_2 and 30% He (by volume) at -180°C, total pressure of 10 atm.

Computer problem

1.4** Consider an air mass, initially at 5°C and 98% relative humidity, which overlies an ocean current whose surface temperature is maintained at 10°C. Assume this air mass becomes well mixed to a pressure (height) of 950 mbar so that the temperature lapse rate is dry adiabatic and the water vapor mixing ratio decreases linearly from the surface (at water saturation) to its original value at 950 mbar (98% at 5°C).

Determine the profile of supersaturation with height. Estimate the heights at which clouds would be expected to form.

Repeat for humidities of 96% and 94%.

Hint: Calculate temperature, densities for dry air and water vapor, actual and saturation mixing ratios, and height in meters as functions of pressure from 1013 mbar to 950 mbar.

REFERENCES

Bergeron, T, 1935, 'On the physics of clouds and precipitation,' *Proc. 5th Assembly UGGI, Lisbon* **2**, 156.

Bohren, C.F., 1986, 'Cloud formation on descent revisited,' *J. Atmos. Sci.,* **43**, 3035.

Langmuir, I., 1948, 'The production of rain by a chain reaction in cumulus clouds at temperatures above freezing,' *J. Meteorol.* **5**, 175.

Low, T.B., and List, R., 1982, 'Collision, coalescence and breakup of raindrops. Part I: Experimentally established coalescence efficiencies and fragment size distributions in breakup, *J. Atmos. Sci.* **39**, 1591.

Marshall, J.S., and Palmer, W.McK., 1948, 'The distribution of raindrops with size,' *J. Meteorol.* **5**, 165.

McDonald, J.E., 1964, 'On a criterion governing the mode of cloud formation in planetary atmospheres,' *J. Atmos. Sci.* **21**, 76.

Schaefer, V.J., 1946, 'The production of ice crystals in a cloud of supercooled water droplets,' *Science* **104**, 457.

Twomey, S., 1976, 'The effects of fluctuations in liquid water content on the evolution of large drops by coalescence,' *J. Atmos. Sci.* **33**, 720.

Valdez, M.P., and Young, K.C., 1985, 'Number fluxes in equilibrium raindrop populations: A Markov chain analysis,' *J. Atmos. Sci.* **42**, 1024.

Vonnegut, B., 1947, 'The nucleation of ice formation by silver iodide, *J. Appl. Phys.* **18**, 593.

2

THE STRUCTURE OF WATER AND ICE

2.1. Introduction

An examination of the water molecule and how its properties affect the microscopic structure and the macroscopic form of ice, the internal structure of water, and the surfaces of water and ice is a useful prelude to the topics in the following chapters. The treatment in this chapter is intended to be brief and basic. The reader is referred to Fletcher (1970) or Hobbs (1974) for much more detail on these subjects.

2.2. The water molecule

Chemical analysis of water indicates that the water molecule is comprised of two atoms of hydrogen attached to a single atom of oxygen. The fact that the water molecule possesses a permanent dipole moment (6.10×10^{-30} C-m) leads one to the conclusion that the water molecule cannot be linear. This is supported by the observation that the specific heat of water vapor (c_v) at 100°C is ~3.3 k, where k is the Boltzmann constant (see Appendix 2). Since 3/2 k are contributed by the three translational modes, the remaining 3/2 k must result from three rotational modes (since vibrational modes will not be excited at this temperature). A linear molecule may only have two rotational modes (see Fig. 2.1), hence the water molecule cannot be linear.

The oxygen atom has eight orbital electrons, two in the 1s shell, two in the 2s shell, and four in the 2p shell. For bonding purposes, the six electrons in the second shell may be considered hybridized, i.e., they form four equivalent orbitals (sp^3), directed towards the corners of a regular tetrahedron. Two of these orbitals are filled (lone-pair orbitals) and two contain an unpaired electron, available for bonding.

In the water molecule, the two unfilled orbitals participate in bonding with hydrogen atoms. Because of the repulsion between the unfilled orbitals, the repulsion between the hydrogens/bond orbitals, and the attraction between the hydrogens/bond orbitals and the lone-pair orbitals, the bond angles are deformed from the tetrahedral configuration with bond angles of 109°28'. The observed angle between the bond orbitals is roughly 105°, the angle between the lone-pair orbitals is roughly 120°, and the bond angle between the lone-pair orbitals and the bond orbitals is roughly 108°. A simplified representation of the water molecule is shown in Fig. 2.2.

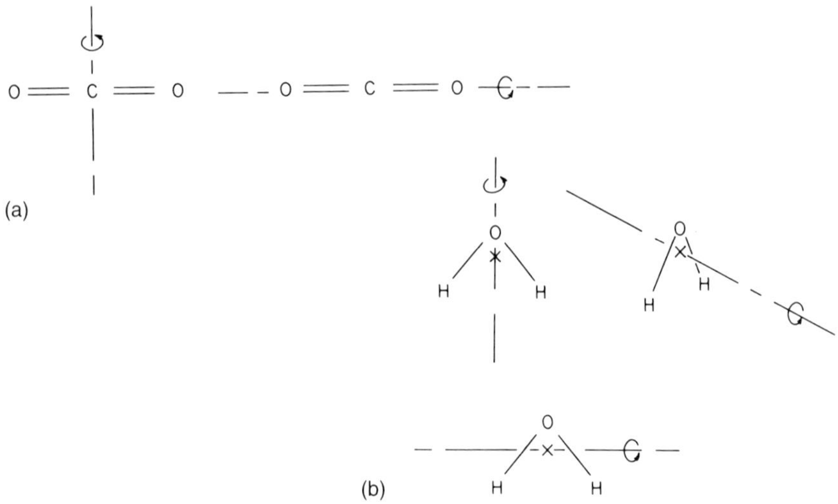

FIG. 2.1. Rotational modes for triatomic molecules. (a) The two modes for a linear molecule such as CO_2. (b) The three modes for a nonlinear molecule such as water. Note that the axes of rotation pass through the center of mass indicated by a cross.

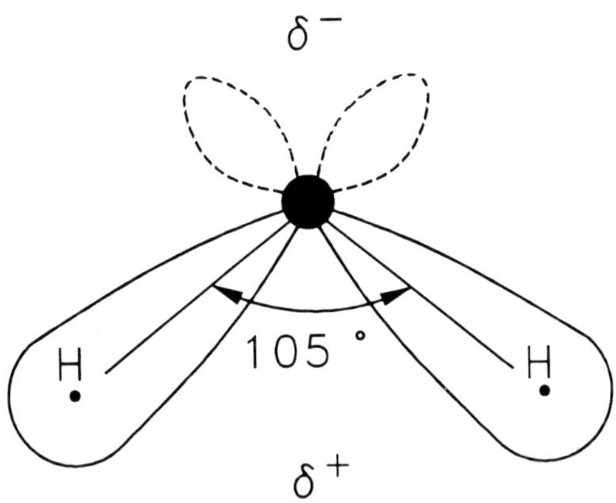

FIG. 2.2. Schematic diagram of a water molecule. The two covalent bonds form an angle of 105° in the plane of the diagram. The two lone-pair orbitals are directed behind and in front of the plane of the diagram.

2.3. The structure and form of ice

There are several ways in which water molecules link together to form three-dimensional structures. The common form of ice which is the stable form over the range of pressure and temperature normally encountered in the atmosphere is called Ice Ih. This distinguishes it from seven other known stable phases of ice, and from the metastable, cubic form of Ice I, called Ice Ic. Ice Ih is the only form of ice that will be considered here.

2.3.1. *The crystal structure*

Water molecules are held together in the ice lattice structure by a type of bonding called *hydrogen bonding*. A hydrogen bond is an intermolecular bond, formed between a hydrogen atom of one molecule and a highly electronegative atom of another molecule. Since hydrogen bonds arc formed only with highly electronegative elements such as oxygen, fluorine, and nitrogen, such bonds are considered to be mostly electrostatic, in contrast to the covalent H–O bond within the water molecule. The hydrogen bond is highly directional, with the hydrogen of one molecule being associated with a lone-pair orbital of another molecule.

Since the water molecule possesses two covalently bonded hydrogens and two lone-pair orbitals, an interlocking, three-dimensional lattice structure such as that shown in Fig. 2.3 may be built up. The intermolecular bonds are directed towards the corners of a regular tetrahedron with the hydrogens lying very nearly on the bond axes. The O–O bond distance is 0.276 nm with the hydrogen situated roughly 0.09 nm from its parent oxygen, retaining the integrity of the water molecule within the ice lattice.

Since the bonding in the ice structure is tetrahedral, it is possible to construct six-membered rings without deforming the bond angles. Such rings are not planar but are bent as shown in Fig. 2.4. These rings may be joined to form a sheet or layer as shown in Fig. 2.5.

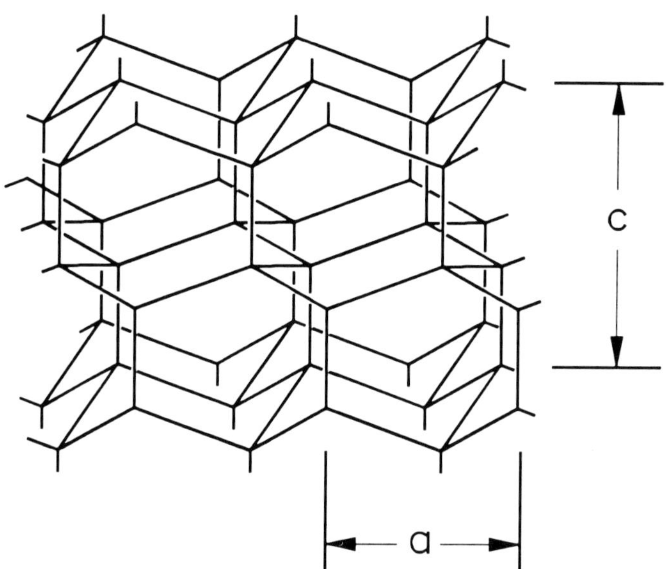

Fig. 2.3. Diagram of the three-dimensional lattice structure of Ice Ih in perspective. Intersections represent positions of oxygen atoms; lines represent hydrogen bonds. Dimensions a and c of the unit cell are indicated.

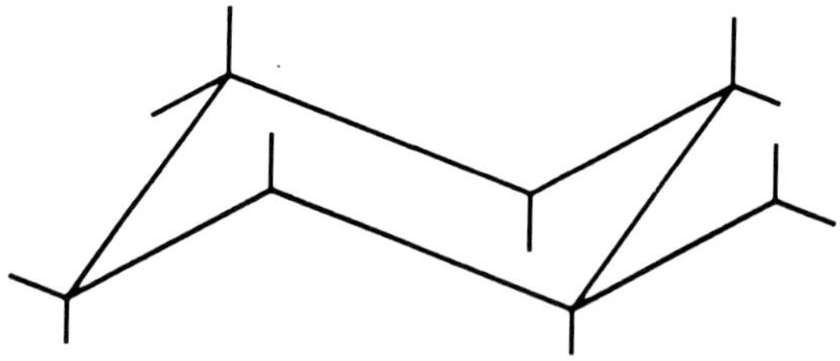

FIG. 2.4. An unstrained, six-membered ring of water molecules.

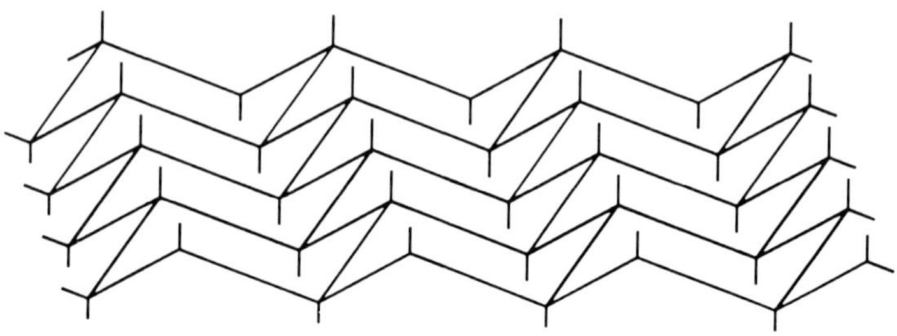

FIG. 2.5. A layer of rings such as that depicted in Fig. 2.4. Such sheets form the basal planes of the macroscopic ice crystal.

The downward-directed bonds of one layer must connect with the upward-directed bonds of the underlying layer in order to form the three-dimensional structure shown in Fig. 2.3. It is clear that alternate layers must be mirror images of each other and that the Ice Ih lattice structure consists of the two types of layers stacked alternately (ABABA . . .).

Viewing the lattice structure in Fig. 2.3 from above, i.e., perpendicular to the layers which form the basal plane of the crystal, a pattern of regular hexagons may be observed, as shown in Fig. 2.6. This pattern may be built up from individual unit cells, shown by the broken lines. A unit cell is the smallest unit of the lattice which contains all the information necessary to describe the entire structure. The unit cell base is shown in Fig. 2.6 as a parallelogram with sides a and b. The vertical extent of the unit cell is shown in Fig. 2.3 with a height c, forming a right parallelepiped. It may be shown that each unit cell contains four water molecules (see Problem 2.1).

The dimensions of the unit cell are a = b = 0.452 nm, and c = 0.736 nm (for temperatures within the range of interest in meteorology). The angles α and β (between sides a and c, and b and c, respectively) are 90°; the angle γ between sides a and b is 120°. The density of ice as calculated from these lattice parameters is $0.919\,\mathrm{g\,cm^{-3}}$ (see Problem 2.2), in good agreement with macroscopic measurements of 0.916 to $0.918\,\mathrm{g\,cm^{-3}}$.

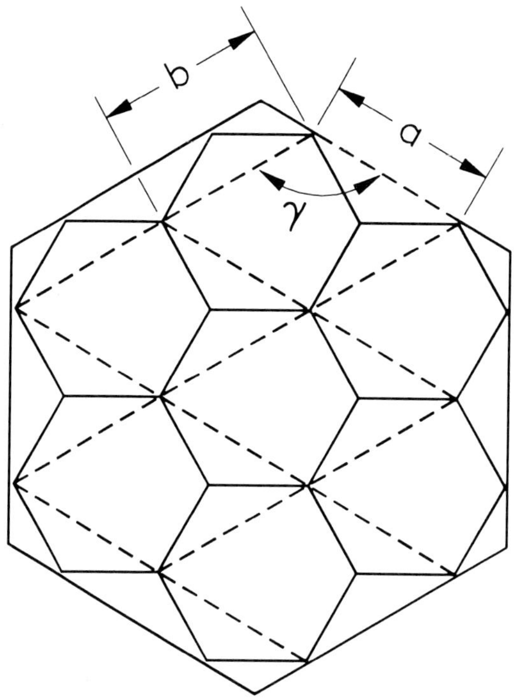

Fig. 2.6. A top view of the ice lattice shown in Fig. 2.3, looking perpendicular to the basal face. The six-membered rings shown in Fig. 2.4 appear as hexagons and join together to form a macroscopic hexagon (circumscribed). The bases of five unit cells are shown with sides a and b and angle γ between them.

2.3.2. *The crystal shape*

A rapidly growing crystal takes the form of a right hexagonal prism as shown in Fig. 2.7. Such hexagonal symmetry in the crystal shape is a consequence of the hexagonal symmetry in the crystal lattice. Figure 2.6 shows an arrangement of unit cells which produces hexagonal symmetry in the macroscopic crystal. The top and base of the hexagonal prism are termed the basal faces; the six remaining sides are termed the prism faces.

Crystal axes are defined perpendicular to the faces of the macroscopic crystal and pass through the center of the crystal. The c-axis is perpendicular to the basal faces and is parallel to the c-dimension of the unit cell. There are three a-axes, spaced 60° apart, directed to the centers of opposite prism faces, as shown in Fig. 2.7. For a symmetrical crystal, the three a-axes are equivalent.

The crystal dimensions are defined in terms of the a- and c-axes. In this text, the a-axis dimension is defined as half of the distance between the prism face centers (equal to the radius of an inscribed circle), and the c-axis dimension is

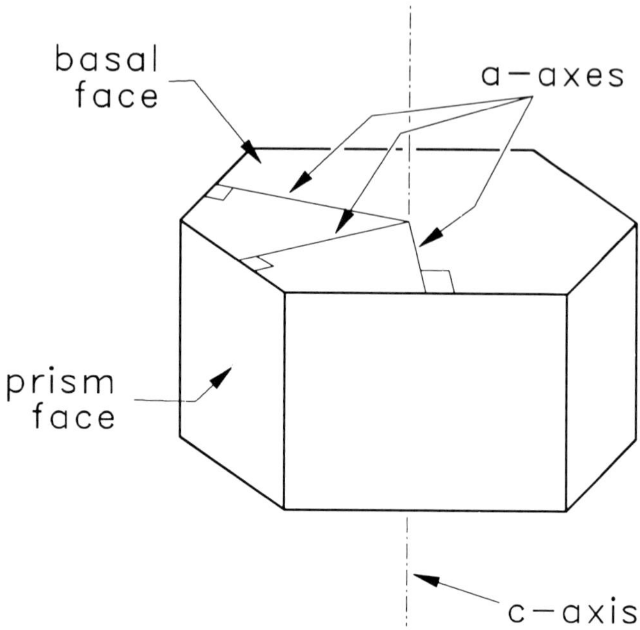

FIG. 2.7. The macroscopic structure of an ice crystal. Basal and prism faces of the right hexagonal prism and the a- and c-axes are shown.

half of the perpendicular distance between the basal faces. The surface area and volume of such a hexagonal prism can be shown to be

$$A = 4\,(3)^{1/2}\,a^2 + 8\,(3)^{1/2}\,ac \quad \text{and} \quad V = 4\,(3)^{1/2}\,a^2c.$$

To a good approximation, the crystal may be approximated as a right circular cylinder (see Problem 2.3).

It is necessary to distinguish between the equilibrium crystal shape and the nonequilibrium form a crystal assumes when it is growing rapidly. An ice crystal which is allowed to remain at ice saturation for a long period of time (days), will tend to become spherical in shape. Sublimating crystals may also tend toward a spherical shape if they are motionless although falling columns may tend toward a more extreme aspect ratio (c/a) as they sublimate.

A rapidly growing ice crystal will resemble a hexagonal prism with an aspect ratio ranging from that of a thin plate to that of a long column (or needle). Ice crystals may grow without completely filling in the hexagonal prism defined by the maximum extension of their a- and c-axes. Such skeletal growth structures reflect conditions of very rapid growth, resulting in crystal forms such as needles, hollow columns, dendrites, and sector plates. These ice crystal forms will be treated in more detail in Chapter 6.

2.4. The structure of liquid water

Although liquid water cannot possess a structure such as that visualized for ice, the influence of hydrogen bonding and approximately tetrahedral coordination serve to produce a short range order within the liquid. This tendency toward order, or coherence, extends several molecular diameters and involves tens of molecules. The extent or degree of coherence decreases as the temperature is increased.

One may visualize the melting point of ice as that point where the number of broken hydrogen bonds is so large that the ice lattice structure can no longer remain rigid. The ice lattice breaks down into clusters of various sizes plus unbonded molecules and ions (H_3O^+, HO^-, etc.). Since hydrogen bonds are continually breaking and re-forming, the structure is fluid, with a tendency toward short range order.

It is tempting to picture the clusters of water molecules as having an ice-like (Ice Ih) structure. The molecular dynamics model presented by Stillinger and Rahman (1972), however, does not support this picture, although they note that polygonal structures with four to seven sides are common. Models have been advanced which suggest that unbonded water molecules are incorporated within the relatively open structure of the clusters, thereby explaining why liquid water has a greater density than ice. However, the model of Stillinger and Rahman does not support the distinction between molecules bonded together in clusters and unbonded molecules contained within the clusters. A more detailed discussion of the structure of liquid water is contained in chapter 3 of Pruppacher and Klett (1978).

The fraction of hydrogen bonds that are broken increases with increasing temperature. This implies that the average cluster size decreases with increasing temperature. This leads to an increase in the bulk density as the temperature is increased. This tendency is opposed by the natural thermal expansion which dominates at temperatures warmer than 4°C, serving to explain the maximum density for liquid water near 4°C.

2.5. The water, ice, and air interfaces

It is more correct to identify water–air, ice–air, and water–ice boundaries as *interfaces* although the former two are usually termed *surfaces*. This is generally acceptable since the effect of air on the structure of water or ice at the interface may be ignored due to its low density and its limited solubility in water or ice. Here, the water–air and ice–air interfaces will be referred to as *surfaces*.

The surface energy of a system such as water or ice can be lowered if the outer surface contains the more polarizable ions or the more polarizable parts of the molecules. Weyl (1951) illustrates this in terms of the changes occurring on a nascent surface of NaCl (see Fig. 2.8). Here, the ions polarize immediately to reduce the electric field at the surface, with the more polarizable anions being more effective at this. The surface field may be reduced still further as the large anions migrate outward and tend to shield the cation field. Thus, the surface free

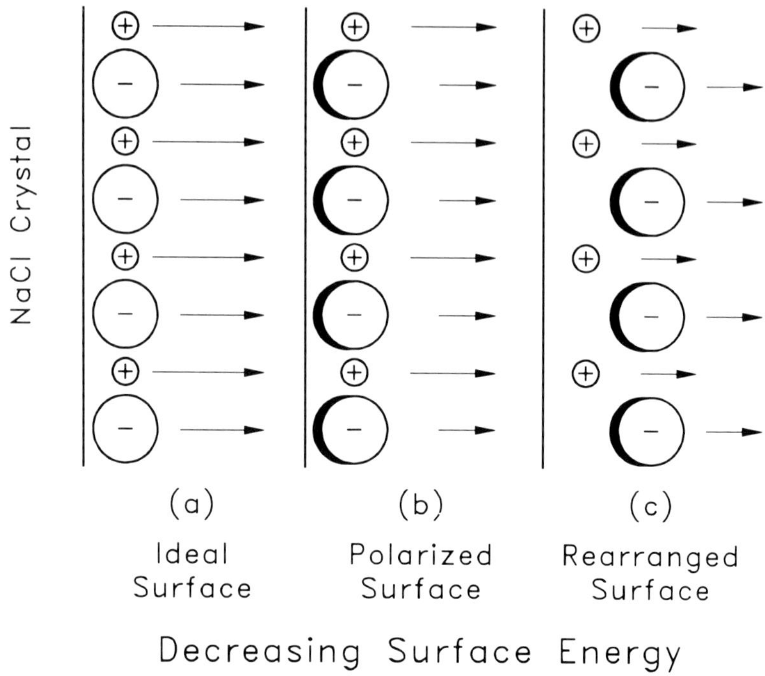

FIG. 2.8. Conceptual model of electronic deformation and ionic rearrangement in a sodium chloride surface showing the polarization of the ions and their migration in order to reduce the surface electric field. After Weyl (1951). Reprinted by permission of Academic Press.

energy is lowered as the more polarizable ions occupy a larger fraction of the surface.

Treatment of the surface of liquid water is more complex than this since it is not clear which portion of the molecule is more polarizable and would be directed outward in the lowest free energy configuration. In fact, this simple view is incomplete, and a treatment of the dipole and quadrupole interactions of the water molecules is necessary.

Fletcher (1968) presents such a model and concludes that the low surface energy orientation is with the hydrogens directed outward. Due to hydrogen bonding interactions between the water molecules, this surface orientation is carried into the bulk liquid with a characteristic length on the same order as the coherence distance (tens of molecules). Such ordering increases the free energy of the system, tending to offset the reduction in free energy associated with reducing the electric field at the surface. Fletcher deduces that 74% of the molecules are oriented so that their hydrogens are directed outward.

Molecular re-orientation of the ice lattice surface can only occur through defects in the lattice structure. Fletcher (1968) points out that the energy penalty for such surface re-orientation is so severe as to preclude re-ordering. However, re-ordering may occur if the ice lattice is covered by a thin, highly disordered layer, similar to the surface layer of liquid water. Such a layer would not have the same structure as liquid water and would be expected to merge gradually with the ice lattice.

Fletcher assumes that this quasi-liquid layer on the surface has a uniform thickness, and produces a free energy penalty equivalent to melting that volume of ice. However, the assumption of a well-defined surface layer leads to a discontinuity (in his treatment) at the ice–'water' interface, leading to uncertainties in his results. Fletcher estimates the thickness of this quasi-liquid layer on the ice surface decreases from 3 nm at −1°C to 1 nm at −5°C. However, he points out that impurities (such as dissolved CO_2) could result in retention of such a quasi-liquid layer to much colder temperatures.

The water–ice interface is much more difficult to define since the internal structures of water and ice are so similar. One may expect an increase in the number of broken hydrogen bonds as one proceeds from the ice to the water phase. However, hydrogen bonds are continually being formed and broken and the definition of an interface is vague.

For further discussions of the properties of the ice surface and its associated phenomena, the reader is referred to chapter 6 of Hobbs (1974).

Problems

2.1* Show that the unit cell of Ice Ih contains a total of four molecules.

2.2* Calculate the density of Ice Ih using the lattice parameters a = b = 0.452 nm and c = 0.736 nm.

2.3** What are the percent errors incurred in calculating the total surface area and volume of an ice crystal if one assumes that ice crystals are right circular cylinders rather than right hexagonal prisms?

2.4*** Methane, ammonia, water, hydrogen fluoride, and neon have the following melting and boiling points and molecular weights:

	Molecular weight	Boiling point (°C)	Melting point (°C)
CH_4	16	−161.5	−184.0
NH_3	17	−33.4	−77.7
H_2O	18	+100.0	0.0
HF	20	+19.4	−92.3
Ne	20	−245.9	−248.7

Considering what you know about bonding and molecular structure, how do you account for this behavior?

REFERENCES

Fletcher, N.H., 1968, 'Surface structure of water and ice. II. A revised model,' *Phil. Mag.* **18**, 1287.

Fletcher, N.H., 1970, *The chemical physics of ice*. Cambridge: Cambridge University Press.

Hobbs, P.V., 1974, *Ice physics*. Oxford: Clarendon Press.

Pruppacher, H.R., and Klett, J.D., 1978, *Microphysics of clouds and precipitation*. Dordrecht: Reidel.

Stillinger, F.H., and Rahman, A., 1972, 'Molecular dynamics study of temperature effects on water structure and kinetics,' *J. Chem. Phys.* **57**, 1281.

Weyl, W.A., 1951, 'Surface structure of water and some of its physical and chemical manifestations,' *J. Colloid Sci.* **6**, 389.

3

NUCLEATION OF THE LIQUID PHASE

3.1. Introduction

Even when the formation of a new phase within a parent phase may seem to be favored on the basis of the relative stabilities of the two phases, the requirement of forming an interface often retards or prevents nucleation. This restriction is often referred to as an *energy barrier* or a *barrier to nucleation*, and is quantified in terms of an *activation energy*. A study of nucleation processes, then, is a study of how this barrier may be surmounted, a study involving thermodynamical and statistical mechanical concepts.

The treatment of homogeneous condensation will serve to introduce a general treatment of nucleation problems. The procedure to be followed is to define the thermodynamic potential for the system of interest and find its extrema. This defines equilibrium embryonic drop sizes as a function of the saturation ratio. If the embryos are large and the effects of statistical fluctuations in the embryo size are not significant, a definition of the equilibrium size as a function of saturation ratio serves to define the behavior of the system. If the embryos are small and their sizes strongly influenced by statistical fluctuations, a kinetic treatment is required to derive a nucleation rate which defines the rate at which *supercritical* embryos are produced.

3.2. Homogeneous condensation

Homogeneous condensation refers to the direct condensation of vapor to liquid in a system originally consisting only of vapor. The random clustering of molecules within the vapor may produce supercritical clusters, i.e., clusters that are larger than the equilibrium size, provided the supersaturation (in the vapor) is high enough. Such supercritical clusters may then continue to grow, bounded only by the supply of water vapor. It will be shown that the critical cluster size decreases with increasing supersaturation (the Kelvin equation), leading to an increase in the production of supercritical clusters as the supersaturation is increased.

3.2.1. *Saturation vapor pressure over a droplet*

The equilibrium vapor pressure over a convex water surface such as a drop is higher than the vapor pressure over a plane surface of water at the same temperature. This is a consequence of the work which must be performed on the system in order to increase the drop's surface area, *i.e.*, to allow it to grow. The derivation of the Laplace relation between the internal and external pressures for a spherical drop in equilibrium will lead to the derivation of the Kelvin equation which describes the equilibrium vapor pressure over a curved surface.

(i) The Laplace relationship. The Laplace relationship gives the difference between the inside and outside pressures of a spherical drop in equilibrium with its vapor. This relationship can be derived from the definition of surface tension, using only the concept of work.

The reversible work which must be performed on a system in order to increase its surface area A by some increment dA is $\sigma\, dA$, thereby defining σ, the surface tension. Here, the surface is considered to be an infinitesimal layer belonging to the drop, serving to separate the condensed phase from its vapor. This is not strictly true, as the discussion of the surface of water in the previous chapter showed.

The apparatus depicted in Fig. 3.1 could, in principle, be used to form a drop as follows. The piston applies a pressure (P_2) to the bulk liquid, which is extruded through an infinitely fine syringe to form the drop. The vapor pressure (P_1) surrounding the drop is always the equilibrium vapor pressure of the drop (which is not constant as it will be shown to be a function of the drop radius). The applied pressure P_2 is just sufficient to cause the drop to grow so that the process may be conducted quasi-statically. Thus, at any given instant, the system can be considered to be in equilibrium.

The differential work done by the piston on the bulk liquid to form the drop is $P_2 dV_2$ while the work done by the drop on the vapor is $P_1 dV_2$. Clearly, the work required to increase the surface area is their difference,

$$\sigma\, dA = (P_2 - P_1)\, dV_2\,, \tag{3.1}$$

following the definition of surface tension given above.

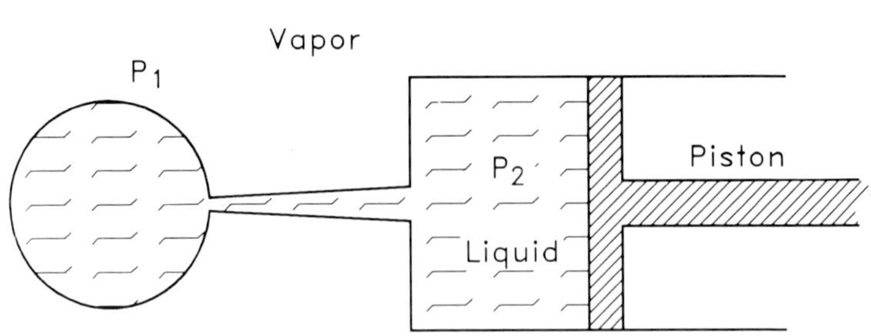

Fig. 3.1. An apparatus for forming a water drop. The piston applies pressure at P_2 to the water, which is squeezed out through an infinitely fine orifice into the drop. The drop expands against the vapor at pressure P_1.

If the drop is assumed to be spherical, it follows that $dA = 8\pi r\, dr$ and $dV_2 = 4\pi r^2\, dr$. Introducing these relationships into (3.1), the Laplace relation follows as

$$P_2 - P_1 = \frac{2\,\sigma}{r}\,. \tag{3.2}$$

The elevation of the meniscus in a narrow tube extending above the flat surface of a liquid serves to illustrate the difference in pressure between the bulk liquid and that contained by a curved surface. In this case, the curvature is negative, so the pressure within the tube is lower than the pressure in the bulk liquid, raising the meniscus. This observation implies that the surface tension is greater than zero.

(ii) A thermodynamic potential for the drop/vapor system. The concept of a thermodynamic 'potential' rests on the second law of thermodynamics, which states that the entropy S must increase for a system which undergoes an irreversible change over a closed path

$$\oint_{\text{irrev}} dS > 0\,.$$

For a reversible path, the change in entropy is zero

$$\oint_{\text{rev}} dS = 0\,.$$

Consider a system, initially in equilibrium, which is perturbed very slightly to a new equilibrium state, both changes taking place reversibly. The change in entropy of the system is zero.

Consider a second system which is not at equilibrium initially. Any slight perturbation must be irreversible and the entropy of the system must increase. Thus, if a system can be perturbed slightly and returned such that $\delta S = 0$, the system must have been in equilibrium initially.

Thermodynamic potentials Y have two common usages. They can be used as indicators of equilibrium as in the foregoing example, using entropy as a thermodynamic potential. The criterion for equilibrium is that $\delta Y = 0$.

Thermodynamic potentials can also be used to determine the direction in which a system tends to change. Consider the first system discussed above. In order to displace this system to a new equilibrium state, work must be done on the system (at constant temperature) which leads to a decrease in entropy. To keep the system from returning to its original state, the system must be constrained. The entropy in the constrained state is lower than in the initial state. Thus, if the constraint were removed, the system would spontaneously revert to the state of higher entropy.

One difficulty encountered in applying the concept of thermodynamic potential is that certain thermodynamic quantities are defined only for equilibrium states. In the above treatment, this difficulty has been circumvented by perturbing the system slightly by doing work on it so as to displace it to a new equilibrium position for which the entropy is defined. In essence, this places additional constraints on the system, allowing the entropy to be calculated even though the exact nature of the constraint is not specified. Such perturbations are called 'virtual variations' (δ). The reader is referred to the excellent treatment by Reiss (1965) for a more detailed discussion.

A generalized equilibrium condition for an isothermal, closed system may be expressed as

$$\delta U + DW_{\text{rev}} - T\delta S = 0\,, \tag{3.3}$$

where DW_{rev} is the total reversible work performed by the system on the environment and U represents the internal energy. This is a feature of equilibrium since it is true only for systems at equilibrium, following the arguments given above. Thus, for the water vapor/water drop system depicted in Fig. 3.2, a thermodynamic potential may be defined as

$$Y = U + P_1V_1 + P_1V_2 - TS\,, \tag{3.4}$$

where the internal energy and entropy of the surface are included as part of the drop.

For such a system at constant temperature and vapor pressure P_1, the equilibrium condition may be written in terms of virtual variations as

$$[\delta Y]_{T,P_1} = \delta U + P_1\delta V_1 + P_1\delta V_2 - T\delta S\,. \tag{3.5}$$

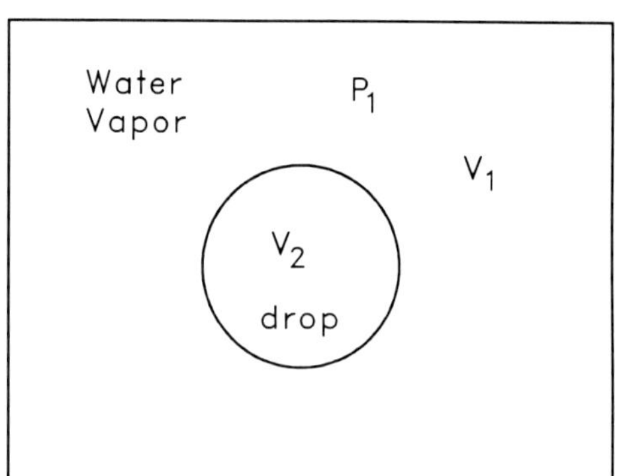

Fig. 3.2. The water drop/water vapor system. Note that the drop does work on the environment at P_1 and not at the drop pressure P_2.

Here, the work performed by the system has two components, representing the total reversible work. Thus, Y is a suitable thermodynamic potential and $[\delta Y]_{T,P} = 0$ represents an equilibrium condition.

The internal energy and entropy of the system may be expressed in terms of the contributions from each of the two phases, i.e.,

$$S = S_1 + S_2 \qquad \text{and} \qquad U = U_1 + U_2 ,$$

where phase 2 is understood to include the drop and its surface. Substituting these relationships into (3.4), the thermodynamic potential may be expressed as

$$Y = (U_1 + P_1V_1 - TS_1) + (U_2 + P_1V_2 - TS_2) = G_1 + R_2 . \tag{3.6}$$

The first term ,dealing with the vapor phase, is a Gibbs free energy (G). However, the thermodynamic potential for this system is *not* a Gibbs free energy since the contribution from the drop (R_2) contains the term P_1V_2 and not P_2V_2. This arises since the drop performs work on the environment through the vapor at pressure P_1.

It can be demonstrated that the contribution from the drop consists of a Gibbs free energy plus a surface term, i.e.,

$$R_2(T,P_2) = [G_2(T,P_1)]_{\text{bulk}} + \sigma A . \tag{3.7}$$

Note that the drop does work at P_1 but the drop itself is at P_2. Here, $[G_2(T,P_1)]_{\text{bulk}}$ refers to the Gibbs free energy which the material of the drop would possess if it were part of a bulk phase at (T,P_1). This may be expressed as

$$[G_2(T,P_1)]_{\text{bulk}} = [U_2(T,P_1)]_{\text{bulk}} + P_1[V_2(T,P_1)]_{\text{bulk}} - T[S_2(T,P_1)]_{\text{bulk}} .$$

Combining this with the definition of R_2 from (3.6) and making the assumption that the drop is incompressible, i.e.,

$$V_2(T,P_2) = [V_2(T,P_1)]_{\text{bulk}} ,$$

it follows that

$$\begin{aligned} R_2(T,P_2) - [G_2(T,P_1)]_{\text{bulk}} &= U_2(T,P_2) - [U_2(T,P_1)]_{\text{bulk}} \\ &\quad - T(S_2(T,P_2) - [S_2(T,P_1)]_{\text{bulk}} , \end{aligned}$$

where the bulk terms do not include a surface contribution. This may be written as

$$R_2(T,P_2) - [G_2(T,P_1)]_{\text{bulk}} = \Delta U - T\Delta S = -\Delta W ,$$

for the system in equilibrium. Here $-\delta W$ refers to the reversible work done on the system to form the drop from bulk liquid at P_1. It is necessary to show that $-\delta W = \sigma A$ for (3.7).

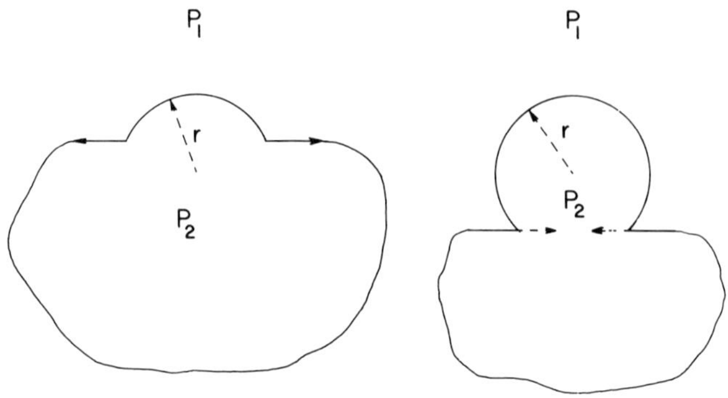

Fig. 3.3. Schematic diagram illustrating the work required to form the surface of a drop. The membrane enclosing the reservoir of liquid is squeezed, doing work at P_2. The drop does work on the vapor at P_1. The opening (a) and closing (b) of the orifice both require work to be done since the surface tension acts against both the opening and the closing.

Consider a system consisting of bulk liquid at P_2 and vapor at P_1, separated by a membrane which maintains the pressure difference. Since the bulk liquid is assumed to be incompressible, no work is required to increase the pressure of a volume V of the liquid from P_1 to P_2.

Now consider forming a drop by opening an orifice in the membrane and extruding the volume V as shown in Fig. 3.3. This requires P_2V work to be done on the membrane while the extruded drop does P_1 work on the vapor. The net work done in extruding the drop is $(P_2 - P_1)V$.

In order to maintain the pressure difference $(P_2 - P_1)$, the radius of curvature of the drop must equal its final radius as the drop is being extruded. The orifice opens and closes to permit this, reaching a maximum opening when exactly half of the drop has been formed. The opening and closing of the orifice is resisted by surface tension. The work required can be shown to equal $V(P_2 - P_1)/2$ (see Problem 3.8).

The total work required to form the drop is $3V(P_2 - P_1)/2$. From the Laplace relation (3.2), $P_2 - P_1 = 2\,\sigma/r$. The total work then is $4\pi r^2\sigma$ or σA. Thus, the thermodynamic potential may be written as

$$Y = G_1(T,P_1) + [G_2(T,P_1)]_{\text{bulk}} + \sigma A\,. \tag{3.8}$$

(iii) Derivation of the Kelvin equation. Since G_1 and $[G_2]_{\text{bulk}}$ depend on P_1, T, and the number of molecules in each phase (g_1 and g_2), applying the condition for equilibrium with the temperature and vapor pressure P_1 constant allows (3.8) to be expressed as

$$\delta Y = 0 = \left[\frac{\partial G_1}{\partial g_1}\right]_{T,P_1} \delta g_1 + \left[\frac{\partial([G_2]_{\text{bulk}} + \sigma A)}{\partial g_2}\right]_{T,P_1} \delta g_2\,. \tag{3.9}$$

Since the system is closed, changes in the numbers of molecules must satisfy $\delta g_1 = -\delta g_2$. Equation (3.9) then reduces to

$$\left(\mu_1(T,P_1) - \mu_2(T,P_1 - \left[\frac{\partial(\sigma A)}{\partial g_2} \right]_{T,P_1} \right) \delta g_1 = 0 \, ,$$

where the chemical potential μ at a constant temperature and pressure is

$$\mu = \frac{\partial G}{\partial g} \, . \tag{3.10}$$

For equilibrium, this implies that

$$\mu_1(T,P_1) = \mu_2(T,P_1) + \left[\frac{\partial(\sigma A)}{\partial g_2} \right]_{T,P_1} \, . \tag{3.11}$$

The equilibrium vapor pressure over the droplet will be expressed with reference to the equilibrium (saturation) vapor pressure over the bulk liquid P_1^∞ where the superscript ∞ refers to a plane surface ($r = \infty$). For the bulk phase in equilibrium with the vapor

$$\mu_1(T,P_1^\infty) = \mu_2(T,P_1^\infty) \, .$$

This allows (3.11) to be written as

$$\mu_1(T,P_1) - \mu_1(T,P_1^\infty) = \mu_2(T,P_1) - \mu_2(T,P_1^\infty) + \left[\frac{\partial(\sigma A)}{\partial g_2} \right]_{T,P_1} \, . \tag{3.12}$$

The difference in chemical potential per molecule between P_1 and P_1^∞ follows from the definition of Gibbs free energy. For a closed system at constant temperature

$$d\mu = \bar{v} dP$$

where $\bar{v}$ is the partial molecular volume. For the vapor, the ideal gas relation may be used to solve for this difference as

$$\mu_1(T,P_1) - \mu_1(T,P_1^\infty) = \int_{P_1^\infty}^{P_1} \bar{v} dP = kT \int_{P_1^\infty}^{P_1} d(\ln P) \, .$$

For the liquid, the assumption of incompressibility allows the integration

$$\mu_2(T,P_1) - \mu_2(T,P_1^\infty) = \bar{v}_2 \int_{P_1^\infty}^{P_1} dP = \bar{v}_2 \, (P_1 - P_1^\infty) \, .$$

For the condensed phase, the concentration of molecules N_L is preferred to the partial molecular volume $\bar{v}_2$. This reduces the equilibrium condition (3.12) to

$$kT \ln\left(\frac{P_1}{P_1^\infty}\right) = \frac{1}{N_L}(P_1 - P_1^\infty) + \sigma\left[\frac{\partial A}{\partial g_2}\right]_{T,P_1},$$

with the assumption that the surface tension is independent of the drop radius.

The total volume of the drop V is

$$V = \frac{g}{N_L} = \frac{4\pi r^3}{3}$$

where g (unsubscripted) refers to the number of molecules in the condensed phase. Solving for the area,

$$A = (4\pi)^{1/3}\left(\frac{3g}{N_L}\right)^{2/3}, \tag{3.13}$$

and differentiating with respect to g,

$$\frac{\partial A}{\partial g} = \frac{2}{N_L r}.$$

The equilibrium condition then reduces to

$$kT \ln\left(\frac{P_1}{P_1^\infty}\right) = \frac{1}{N_L}(P_1 - P_1^\infty) + \frac{2\sigma}{N_L r}. \tag{3.14}$$

Since $(P_1 - P_1^\infty) \ll 2\sigma/r$ (see Problem 3.1), (3.14) may be simplified to obtain the familiar form of the Kelvin equation,

$$\ln\left(\frac{P_1}{P_1^\infty}\right) = \frac{2\sigma}{N_L kTr} = \ln\left(\frac{e}{e_{s,w}}\right). \tag{3.15}$$

Note that e and $e_{s,w}$ have been adopted to represent the ambient and water saturation vapor pressure, respectively. An excellent alternate derivation of the Kelvin equation is given by Long (1969).

(iv) Interpretation of the Kelvin equation. Figure 3.4 is a plot of the equilibrium saturation ratio (P_1/P_1^∞) over a pure water drop as a function of its radius. A drop which finds itself above the curve will grow indefinitely since its equilibrium vapor pressure would be lower than the vapor pressure of its environment and would remain so as it grows. A drop which finds itself below the curve will evaporate since the ambient vapor pressure is less than that over the drop's

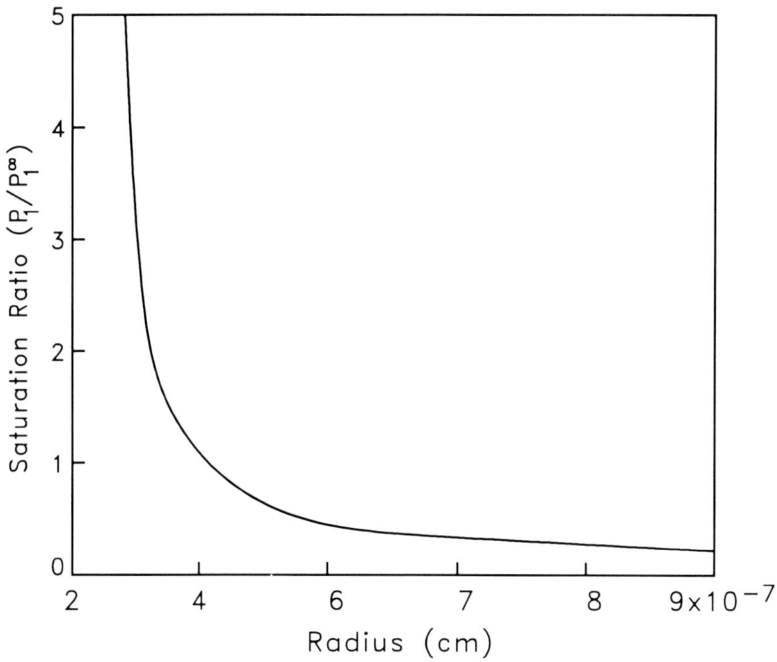

FIG. 3.4. Equilibrium saturation ratio for pure water drops as a function of drop radius. Values are calculated from the Kelvin equation (3.15). The curve represents an unstable equilibrium condition for which a drop will either grow or evaporate.

surface. It is clear that the Kelvin equation represents an unstable equilibrium which, in practice, does not represent a condition that can be maintained.[1]

The high equilibrium saturation ratios for very small drops imply that a significant barrier to homogeneous condensation exists. The number of molecules contained within a 2×10^{-7} cm radius drop is about 1100. It is not likely that droplets of this size would arise by random clustering of molecules in the vapor even given a saturation ratio of 1.8 (the equilibrium saturation ratio for a drop of this size). It is intuitively reasonable that homogeneous condensation requires saturation ratios in excess of this value, as will be shown later.

3.2.2. *Homogeneous nucleation rate*

The homogeneous nucleation rate refers to the rate at which randomly formed clusters of water molecules exceed their critical size (as defined by the Kelvin equation) so that they can continue to grow indefinitely. First, the probability that molecular clusters (embryos) of a given size may exist will be determined on the

[1] It is conceivable that a drop could be enclosed within a very small box such that the growth of the drop would so deplete the vapor phase that the drop would tend to evaporate, returning to the 'unstable' equilibrium radius.

basis of the theory of fluctuations (e.g., Long, 1969). Then, the rate at which clusters larger than the critical size are formed will be determined following the treatment given by Farley (1952).

(i) The population of embryos. The Boltzmann distribution (3.16) has been used to describe the expected concentration of randomly-formed clusters containing g molecules (N_g) given the concentration of unassociated molecules (N_1).

$$N_g = N_1 \exp\left[\frac{-(\Delta Y)_g}{kT}\right] \tag{3.16}$$

The difference in thermodynamic potential $(\Delta Y)_g$ refers to the difference between unassociated vapor molecules and a cluster containing g molecules. One difficulty in applying the Boltzmann distribution is that the thermodynamic potential (Y_g) may only be defined for equilibrium states. Thus, $(\Delta Y)_g$ has not been defined for any g not equal to 1 or to g^*, corresponding to the critical (Kelvin) radius. Following the treatment of Reiss (1965), the concept of a reversible constraint which reversibly displaces the system to a new equilibrium position can be employed to circumvent this problem.

Assume that the initial state with initial constraints χ_0 is in stable equilibrium. Now apply additional constraints χ_1 in a reversible manner so that the system is displaced to a new equilibrium position. If the additional constraints are removed, the system will return to its initial state. Thus, the thermodynamic potential of a system containing an embryo of g molecules where $g < g^*$ may be defined.

Now visualize a system in which there are embryos of many sizes, adding or losing molecules in a random fashion. Constraints cannot be applied in such a fashion that *all* of these embryos are in equilibrium. Consequently, the concept of a thermodynamic potential cannot be used to strictly describe the concentration of each cluster size. However, it may be assumed that this is a good approximation over a narrow range of g, and in particular, for g close to g^*, which is the principal range of interest.

The Boltzmann distribution pertains to systems where the concentration of embryos of any size is constant, and there is no net flux of molecules from one size to another. This is referred to as a nontransient distribution.

The Boltzmann distribution may be used to predict the concentrations of clusters of water molecules in subsaturated vapor. The concentrations of clusters decreases with increasing cluster size, as shown in Fig. 3.5. This follows from (3.16) since $(\Delta Y)_g$ increases with increasing g. Physically, clusters of any size will tend to evaporate in subsaturated vapor.

Application of the Boltzmann distribution to supersaturated vapor leads to difficulties. Since the Kelvin equation represents an unstable equilibrium, embryos larger than the Kelvin radius (critical size) tend to grow, suggesting that $(\Delta Y)_g$ decreases with increasing cluster size for $g > g^*$. Thus, (3.16) would predict large numbers of supercritical embryos, with the number of such embryos

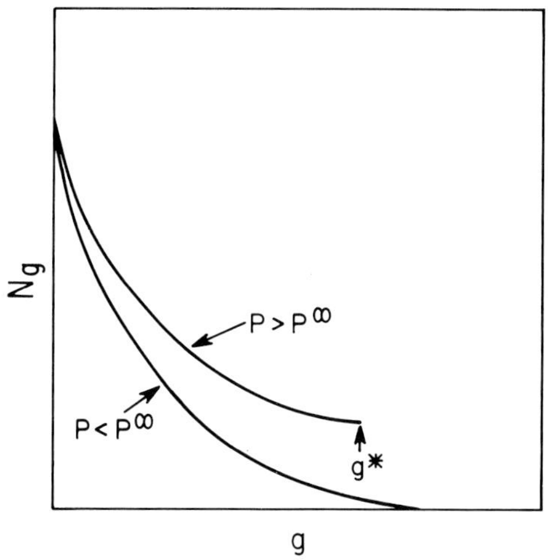

FIG. 3.5. The concentration of embryos containing g molecules (N_g) as a function of g. The two curves depict subsaturated and supersaturated conditions. Only the supersaturated case exhibits a critical size (g^*).

increasing with cluster size. Eventually this requirement would exhaust the total number of water molecules available. Physically, this may be interpreted as the unlimited growth of a randomly formed, supercritical embryo until it incorporates all the molecules. Given sufficient time, such an embryo must form. Clearly, this violates the assumption of equilibrium conditions.

This difficulty is commonly circumvented by restricting embryo sizes to $g \leq g^*$. Embryos which grow from g^* to $(g^* + 1)$ molecules are assumed to be broken up into single molecules and returned to the reservoir of unassociated molecules. The removal of critical clusters must be balanced by a net flux of clusters to the critical size. Thus the distribution of clusters is transient and differs from the nontransient Boltzmann distribution in that the ratio between the transient and nontransient concentrations must decrease with increasing g.

The following treatment will assume a steady state, transient distribution which yields a flux of embryos through the critical size. This flux is what is meant by the term *nucleation rate*. The transient distribution will be related to the nontransient Boltzmann distribution to facilitate determination of the nucleation rate.

(ii) Derivation of the nucleation rate. Consider homogeneous condensation within a large volume at constant temperature and vapor pressure P_1. Within this volume, there is a changing population of embryonic clusters such that, on average, a given cluster tends to grow with time. Individual embryos may exchange molecules with the vapor phase, gaining or losing molecules such that there is a flux of embryos in both directions. The flux of embryos growing from

$(g - 1)$ to g will be expressed as $\vec{J}_g$ and the flux of embryos evaporating from g to $(g - 1)$ as $\overleftarrow{J}_g$. The net flux of embryos from $(g - 1)$ to g is then

$$J_g = \vec{J}_g - \overleftarrow{J}_g .$$

This assumes that all changes in embryo size are due to the gain or loss of single molecules, probably a very good assumption.

The net flux of embryos from $(g - 1)$ to g may be written in terms of the number of embryos N, the surface area per embryo A, and the rates at which molecules are gained $\vec{\omega}$ or lost $\overleftarrow{\omega}$ per unit surface area. For the transient population of embryos N',

$$J_g = A_{g-1}\, \vec{\omega}_{g-1}\, N'_{g-1} - A_g\, \overleftarrow{\omega}_g\, N'_g) . \tag{3.17}$$

Note that this reduces to

$$A_{g-1}\, \vec{\omega}_{g-1}\; N_{g-1} = A_g\, \overleftarrow{\omega}_g\, N_g , \tag{3.18}$$

for the nontransient population of embryos ($J_g = 0$).

The net flux (3.17) may be related to the nontransient distribution as follows

$$\begin{aligned} J_g &= A_{g-1}\, \vec{\omega}_{g-1}\, N_{g-1} \left(\frac{N'_{g-1}}{N_{g-1}} \right) - A_g\, \overleftarrow{\omega}_g\, N_g \left(\frac{N'_g}{N_g} \right) \\ &= A_{g-1} \vec{\omega}_{g-1}\, N_{g-1} \left[\frac{N'_{g-1}}{N_{g-1}} - \frac{N'_g}{N_g} \right] \\ &= A_{g-1}\, \vec{\omega}_{g-1}\, N_{g-1} \left[-\Delta \left(\frac{N'_g}{N_g} \right) \right] . \end{aligned}$$

Expressing this in differential form,

$$J_g = -A_g\, \vec{\omega}_g\, N_g \frac{d}{dg} \left(\frac{N'_g}{N_g} \right) . \tag{3.19}$$

Now, to evaluate the change in N'_g/N_g with g, rewrite (3.19) as

$$\frac{d}{dg} \left(\frac{N'_g}{N_g} \right) = \frac{-J_g}{\vec{\omega}_g} \frac{1}{A_g\; N_g} ,$$

and integrate to

$$\left. \frac{N'_g}{N_g} \right|_{g=\infty} - \left. \frac{N'_g}{N_g} \right|_{g=1} = \frac{-J_g}{\vec{\omega}_g} \int_1^\infty \frac{1}{A_g\; N_g} dg , \tag{3.20}$$

assuming that the flux of embryos through any size g is constant. Since this integral will be approximated by evaluating it near g^* where $1/N_g$ is a maximum, all that is required of J_g is that it be roughly constant in the vicinity of g^*. Note that this then becomes the definition of the nucleation rate J.

The left-hand side of (3.20) may be evaluated as follows. At $g = 1$, $N_g'/N_g \approx 1$ since the reservoir of unassociated molecules must be very large (see Problem 3.6). At $g = \infty$, $N_g'/N_g = 0$ since the net flux toward larger sizes requires that the $(g+1)$th embryo must be present in a lower concentration relative to its nontransient concentration than the gth embryo, i.e.,

$$\frac{N'_{g+1}}{N_{g+1}} < \frac{N'_g}{N_g}.$$

Hence, the left-hand side of (3.20) reduces to -1 and the nucleation rate may be expressed as

$$J = \frac{\vec{\omega}_{g^*}}{\displaystyle\int_1^{\infty} \frac{dg}{A_g\ N_g}} \tag{3.21}$$

The integral in (3.21) may be approximated by considering only the region in the vicinity of g^*. Rewriting the Boltzmann distribution (3.16) as

$$N_g = N_{g^*}\ exp\left[\frac{(\Delta Y)_{g^*} - (\Delta Y)_g}{kT}\right]$$

reduces the integral to the form

$$\int_0^{\infty} \exp\left(-a^2\ y^2\right)\ dy\,,$$

the solution of which is $\pi^{1/2}/2a$. First,

$$\frac{(\Delta Y)_g - (\Delta Y)_{g^*}}{kT} \approx \chi\ (g - g^*)^2,$$

where χ is the third term of the Taylor's expansion,

$$\chi = -\frac{1}{2}\frac{d^2}{dg^2}\left[\frac{(\Delta Y)_g}{kT}\right]. \tag{3.22}$$

Second, A_g may be assumed to be roughly constant in the vicinity of g^*. Now the solution to the integral follows as

$$\int_0^\infty \frac{dg}{N_g} = \frac{1}{N_{g^*}} \int_0^\infty \exp[-\chi\,(g - g^*)^2]dg = \frac{1}{2N_{g^*}} \left[\frac{\pi}{\chi}\right]^{1/2} . \tag{3.23}$$

In order to evaluate χ, $(\Delta Y)_g$ must be expressed as a function of g. By differentiating the thermodynamic potential (3.8) and making use of the chemical potential (3.10),

$$dY = \mu_1(T,P_1)\,dg_1 + \mu_2\,(T,P_1)\,dg_2 + \sigma\,dA\,. \tag{3.24}$$

Since the system is closed, $dg_1 = -dg_2$. Integrating (3.24),

$$\Delta Y = g_2[\mu_2\,(T,P_1) - \mu_1\,(T,P_1)] + \sigma A\,. \tag{3.25}$$

From part (iii) in § 3.2.1 the solution to

$$\mu_2\,(T,P_1) - \mu_1\,(T,P_1)$$

is

$$\frac{1}{N_L}\,(P_1 - P_l^\infty) - kT\,\ln\left(\frac{P_1}{P_l^\infty}\right),$$

which may be approximated as

$$-kT\,\ln\left(\frac{e}{e_{s,w}}\right).$$

The change in thermodynamic potential then reduces to

$$(\Delta Y)_g = -gkT\,\ln\left(\frac{e}{e_{s,w}}\right) + \sigma A\,. \tag{3.26}$$

In order to evaluate (3.22), embryos are assumed to be spherical, and the area A from (3.13) is substituted in (3.26) as

$$(\Delta Y)_g = -gkT\,\ln\left(\frac{e}{e_{s,w}}\right) + 4\pi\sigma\left(\frac{3g}{4\pi N_L}\right)^{2/3} . \tag{3.27}$$

Now solving for χ at $g = g^*$ yields

$$\chi = \left(\frac{36\pi}{N_L^2}\right)^{1/3} \frac{\sigma}{9\ kT} g^{*-4/3} .$$

The nucleation rate J follows from (3.21) as

$$J = \frac{2A_{g^*}\ \vec{\omega}_g\ N_{g^*}}{3} \left(\frac{36\pi}{N_L^2}\right)^{1/6} \left(\frac{\sigma}{\pi kT}\right)^{1/2} g^{*-2/3}. \tag{3.28}$$

Once again replacing A_{g^*} from (3.13), the nucleation rate reduces to

$$J = \frac{4\vec{\omega}_{g^*}\ N_{g^*}}{N_L} \left(\frac{\sigma}{kT}\right)^{1/2} . \tag{3.29}$$

The nucleation rate is expressed in terms of the concentration of critical embryos for the nontransient Boltzmann distribution. A very simple approach might suggest a nucleation rate equal to the nontransient concentration of critical embryos multiplied by the probability of an embryo becoming supercritical, i.e.,

$$J = A_{g^*}\ \vec{\omega}_{g^*}\ N_{g^*} . \tag{3.30}$$

Evaluation of (3.28) for $T = 0°C$ and a supersaturation sufficient to allow a nucleation rate of 10^3 events per second shows that this simple approach overestimates the nucleation rate by a factor of 14.

It should not be construed that the actual nucleation rate is simply (3.30) with N'_{g^*} substituted for N_{g^*}. Such simple treatments neglect the reverse flux of embryos from $(g^* + 1)$ to g^*, which tends to retard nucleation.

(a) *Evaluation of the condensation rate* ($\vec{\omega}$). The condensation rate ($\vec{\omega}$) refers to the rate at which unbonded molecules collide with and are incorporated into the embryo. Not all colliding molecules will be incorporated into the embryo; the fraction that is accommodated (α_c) has been shown experimentally to be close to 0.04.

Consider the collisions from molecules residing within a cylinder of base ds and length r, inclined θ degrees from the vertical (Fig. 3.6). The total number of molecules within this cylinder is

$$N_1\ r \cos\theta\ ds$$

where N_1 is the concentration of water molecules in the vapor phase.

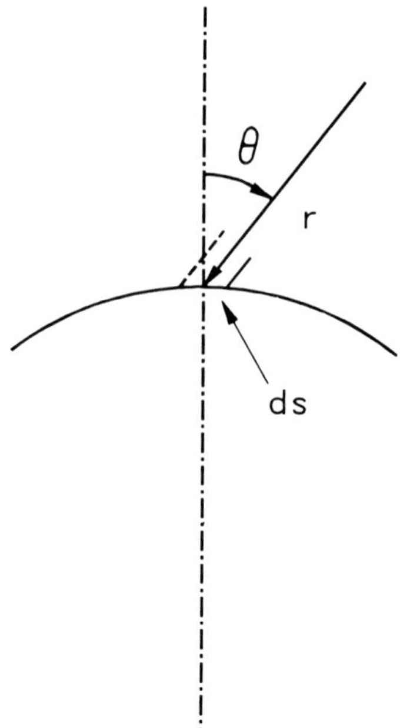

Fig. 3.6. Evaluation of the collision rate per unit area of the drop surface. Molecules within a distance r, moving from direction θ will strike the area ds in a time of $r/\bar{c}$.

Consider the fraction of molecules within this volume moving from directions between θ and θ + $d\theta$ and ϕ and ϕ + $d\phi$ so that they will strike the surface within area ds. This fraction is $d\omega/4\pi$ where $d\omega$ is the differential solid angle and 4π represents the total solid angle. Now, the total number of molecules striking from this direction per unit time is

$$\left.\frac{dN}{dt}\right|_{\theta,\phi} = \frac{N_1\bar{c}\cos\theta}{4\pi}\,d\omega\,ds\,,$$

where $\bar{c}$ is the mean velocity of the molecules. Expressing this in terms of unit surface area,

$$\left.\frac{dN}{dt}\right|_{\theta,\phi} = \frac{N_1\bar{c}\cos\theta}{4\pi}\,d\omega\,. \tag{3.31}$$

In spherical coordinates, $d\omega = \sin\theta\, d\,\theta\, d\phi$. Integrating (3.31) for θ from 0 to $\pi/2$ and ϕ from 0 to 2π,

$$\frac{dN}{dt} = \int_0^{\pi/2}\int_0^{2\pi}\left[\frac{N_1\bar{c}}{4\pi}\right]\cos\theta\,\sin\theta\,d\phi\,d\theta = \frac{N_1\bar{c}}{4}\,.$$

From kinetic theory, the mean molecular velocity is

$$\bar{c} = \left[\frac{8kT}{\pi m} \right]^{1/2} ,$$

where m is the mass of a single water molecule. With this and the ideal gas relation, the condensation rate per unit area follows as

$$\vec{\omega}_{g*} = \frac{\alpha_c \; e}{(2\pi mkT)^{1/2}} . \tag{3.32}$$

(b) *Evaluation of the concentration of critical embryos.* The expected concentration of embryos of the critical size may be deduced from the Boltzmann distribution (3.16),

$$N_{g*} = N_1 \left[\frac{-(\Delta Y)_{g*}}{kT} \right] . \tag{3.33}$$

In order to evaluate N_{g*}, the activation energy $(\Delta Y)_{g*}$ must be determined.

The activation energy may be obtained by evaluating (3.26) at $g = g^*$ and noting that this equation now refers to the equilibrium conditions described by the Kelvin equation (3.15), i.e.,

$$\begin{aligned} (\Delta Y)_{g*} &= -\frac{4}{3}\pi r^{*3} N_L kT \ln \left(\frac{e}{e_{s,w}} \right) + 4\pi r^{*2} \sigma \\ &= -\frac{8}{3}\pi r^{*2} \sigma + 4\pi r^{*2} \sigma \\ &= \frac{4}{3}\pi r^{*2} \sigma . \end{aligned}$$

Substituting once again from the Kelvin equation and using the notation S for the saturation ratio $(e/e_{s,w})$,

$$(\Delta Y)_{g*} = \frac{16\sigma^3\pi}{3 \; (N_L kT \; \ln \; S)^2} . \tag{3.34}$$

The concentration of critical embryos follows directly from (3.33) as

$$N_{g*} = N_1 \exp \left[\frac{-16\sigma^3\pi}{3(kT)^3 \; (N_L \; \ln \; S)^2} \right] . \tag{3.35}$$

(c) *Expression of the homogeneous condensation rate.* Now that the condensation rate (3.32) and the nontransient concentration of embryos of the critical size (3.35) have been evaluated, the homogeneous condensation rate follows directly from (3.29) as

$$J = \frac{4}{N_L}\left(\frac{\sigma}{kT}\right)^{1/2} \frac{\alpha_c\, eN_1}{(2\pi mkT)^{1/2}} \exp\left[\frac{-16\pi\sigma^3}{3(kT)^3\ (N_L\ \ln\ S)^2}\right].$$

Substituting for N_1 from the ideal gas relation and rewriting in terms of the saturation ratio S, the homogeneous condensation rate becomes

$$J = \frac{4\alpha_c}{N_L}\left(\frac{\sigma}{2\pi m}\right)^{1/2}\left(\frac{e}{kT}\right)^2 S^2 \exp\left[\frac{-16\pi\sigma^3}{3(kT)^3\ (N_L\ \ln\ S)^2}\right]. \qquad (3.36)$$

Note that the units of nucleation rate J are events $\text{cm}^{-3}\ \text{s}^{-1}$.

(iii) Comparison between observations and theory. Early observations of homogeneous condensation were made by a rapid expansion of a given volume of saturated air from which all solid or liquid nuclei and ions had been removed. The expansion, and therefore the maximum saturation ratio attained (assuming the expansion is adiabatic), were increased in small increments in successive trials until a cloud was observed. The saturation ratio corresponding to the first appearance of cloud is then called the critical saturation ratio S^*.

Determinations of the critical saturation ratio using a thermal diffusion chamber are described by Heist and Reiss (1973). The diffusion chamber consists of a cooled upper plate and a heated lower plate. The supersaturation in a diffusion chamber may be controlled by varying the temperature difference between the plates.

In practice, many problems arise, such as contamination from the inside walls of the expansion chamber, ion production within the chamber, and the limitation on the maximum saturation ratio as a consequence of drop growth. In addition, the correspondence between nucleation rate and the appearance of a cloud is not at all clear. The visual appearance of a cloud has been associated with nucleation rates from 1 to 10^6 events $\text{cm}^{-3}\ \text{s}^{-1}$. Mason (1971) discusses this problem in his comparison of theoretical and observed homogeneous condensation rates.

Figure 3.7 shows the saturation ratio required to produce homogeneous condensation rates of 1, 10^3, and 10^6 events $\text{cm}^{-3}\ \text{s}^{-1}$ as a function of temperature based on (3.36). Data from several sources are shown for comparison. The observations generally agree with the theoretical calculations although the uncertainties in the observations limit verification of the theory to within five or six orders of magnitude. Thus, it is the exponential term in the nucleation equation that is tested by these observations and not the pre-exponential factor.

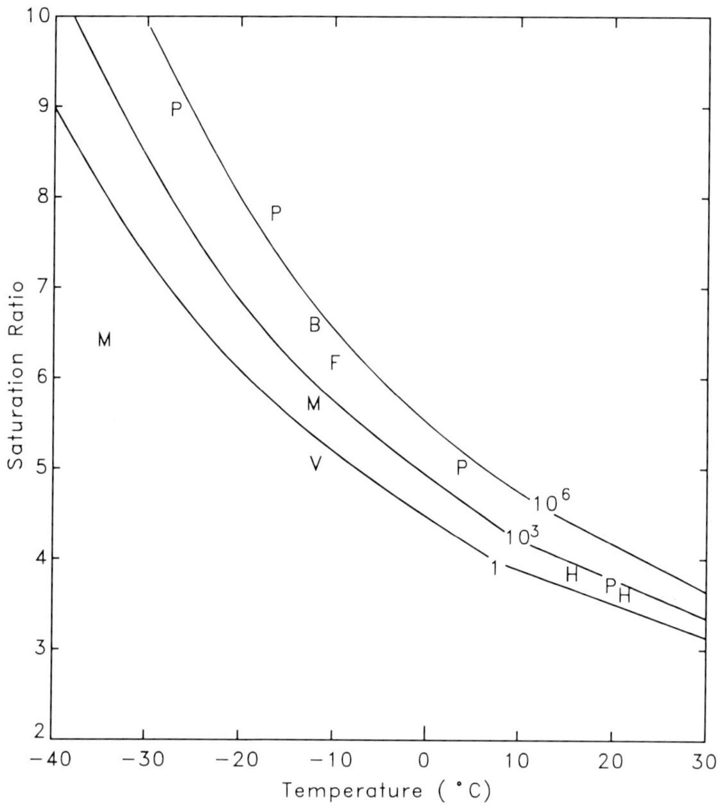

FIG. 3.7. A comparison of theoretical and experimental homogeneous nucleation rates. Calculated values are for 1, 10^3, and 10^6 events cm^{-3} s^{-1}. Experimental data were taken from Mason (1971) for Barnard (B), Frey (F), Madonna et al. (M), Powell (P), and Volmer and Flood (V), and supplemented with data from Heist and Reiss (H) (1973).

When bulk thermodynamic properties are used to describe the behavior of small clusters of molecules such as in this description of homogeneous condensation, one may question the validity of using concepts such as bulk properties and an infinitesimal surface layer. At 0°C, there are 63 molecules in a critical cluster (for $J = 10^3$ events cm^{-3} s^{-1}). Most of the molecules in such a cluster will be surface molecules and the remainder must be within one or two molecules of the surface. Thus, it is remarkable that this treatment shows any agreement with the observations.

The assumption of bulk thermodynamics to describe such systems appears to fail for clusters of fewer than 50 molecules. The observed value of $S^* = 6.4$ at −35°C lies significantly below the theoretical predictions. This observation is supported by additional observations at lower temperatures (not shown). An increase in the interfacial energy as the supercooled water takes on a more ice-like structure would be expected to have the opposite effect. However, since the critical supersaturation increases with decreasing temperature, cluster sizes decrease and the theory may be expected to lose its validity.

(iv) A molecular model of cluster formation. An alternate approach to relying on bulk thermodynamics is that presented by Plummer and Hale (1974). They describe a molecular model which is used to calculate the thermodynamic potential $(\delta Y)_g$ of different sized clusters based on molecular properties. They assume that the clusters have well-defined structures and last long enough to characterize their vibrational spectra. They also assume that the internal properties of the molecules are only slightly affected by the cluster formation and that the (most stable) clusters are roughly spherical.

Clusters consisting of five-membered rings forming closed or partially closed clathrates, as well as clusters arranged in a Ice Ih structure (six-membered rings) are considered. The results agree quite closely with the results presented in Fig. 3.7.

3.3. Condensation on insoluble nuclei

Condensation on a surface is generally more favored than homogeneous condensation since less work is required to form the water–vapor interface for a given volume of condensed water. However, the geometrical treatment becomes more complicated and the interfacial energies between the substrate and water and the substrate and air need to be considered, in addition to the interfacial energy between water and air. These interfacial energies are related through the contact angle, which represents a balance of the interfacial tensions.

The simplest case is that of condensation on a plane surface. In the atmosphere however, nuclei are generally small enough so that the curvature of the nucleus may be important. The concepts of nucleation on insoluble surfaces will be introduced by treating the simplest case first, then this will be extended to the case of spherical, insoluble nuclei.

3.3.1. *Condensation on an insoluble plane surface*

The treatment of condensation on a plane substrate assumes that the water embryo has the shape of a spherical cap, as shown in Fig. 3.8. The symbols *V*, *L*,

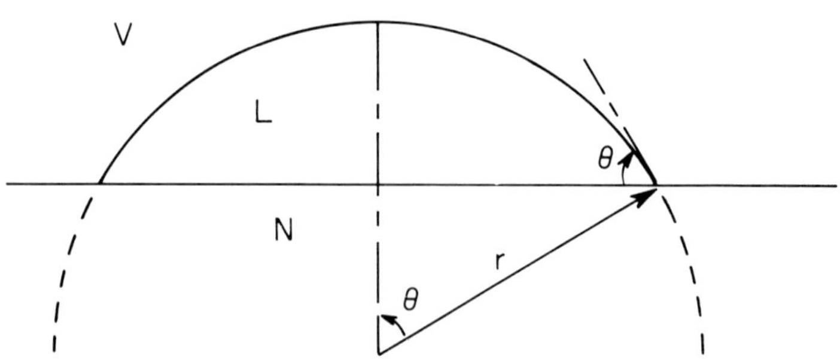

FIG. 3.8. The geometry for a spherical cap embryo on a plane substrate with a contact angle of θ. *V*, *L*, and *N* refer to the vapor, liquid, and nucleating substrate, respectively. *r* is the radius of curvature.

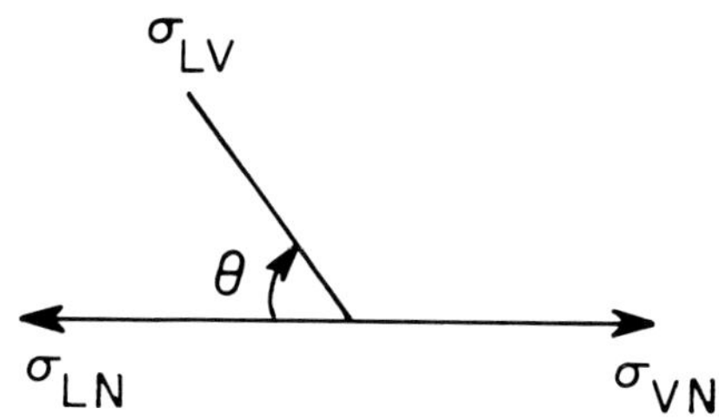

FIG. 3.9. The balance of interfacial tensions between liquid, vapor, and substrate. The interfacial tensions σ_{LV}, σ_{LN}, and σ_{VN} are for the liquid–vapor, liquid–substrate, and vapor–substrate interfaces, respectively. These three interfacial tensions are in balance for a contact angle of θ.

and N refer to the vapor, liquid, and nucleating substrate, respectively. The angle θ that the liquid–vapor interface makes with the substrate is termed the 'contact' angle. The contact angle is determined by the balance of interfacial tensions, as shown in Fig. 3.9. This balance of forces may be expressed as

$$\sigma_{LV} \cos \theta + \sigma_{LN} = \sigma_{VN} . \tag{3.37}$$

The cosine of the contact angle, denoted by $\hat{m}$,

$$\hat{m} = \cos \theta = \frac{\sigma_{VN} - \sigma_{LN}}{\sigma_{LV}},$$

relates the three surface tensions.

(i) Expression of the thermodynamic potential. The change in thermodynamic potential between the initial state (substrate plus vapor) and the final state (spherical water cap on the substrate plus vapor) may be expressed as in (3.26)

$$(\Delta Y)_g = -N_L V_L kT \ln\left(\frac{e}{e_{s,w}}\right) + \sigma_{LV} A_{LV} + (\sigma_{LN} - \sigma_{VN}) A_{LN} , \tag{3.38}$$

where A_{LV} and A_{LN} refer to the areas of the liquid–vapor and liquid–substrate interfaces, respectively, and V_L is the volume of the spherical cap.
The determination of A_{LV}, A_{LS}, and V_L is an exercise in geometry (see Problem 3.5). These may be expressed as

$$A_{LV} = 2\pi r^2 (1 - \hat{m}) ,$$

$$A_{LN} = \pi r^2 (1 - \hat{m}^2) ,$$

$$V_L = \frac{\pi r^3}{3} (2 + \hat{m}) (1 - \hat{m})^2 .$$

Using these, the thermodynamic potential (3.38) may be expressed as

$$(\Delta Y)_g = \frac{-N_L\ kT\ \pi r^3}{3} (2 + \hat{m})(1 - \hat{m})^2 \ln\left(\frac{e}{e_{s,w}}\right) + 2\pi r^2\ \sigma_{LV}\ (1 - \hat{m}) + (\sigma_{LN} - \sigma_{VN})\ \pi r^2\ (1 - \hat{m}^2)\ .$$

Recognizing that $\sigma_{LN} - \sigma_{VN} = -\hat{m}\sigma_{LV}$, this reduces to

$$(\Delta Y)_g = \frac{(\hat{m}^3 - 3\hat{m} + 2)}{4}\left[\frac{-4N_L\ kT\pi r^3}{3} \ln\left(\frac{e}{e_{s,w}}\right) + 4\pi r^2\ \sigma_{LV}\right]. \quad (3.39)$$

Note that (3.39) differs from (3.26) only through the geometrical factor. This factor ranges from 0 to 1, representing no barrier to nucleation for a fully wettable surface, to the homogeneous nucleation barrier for a non-wettable surface. The geometrical factor will be abbreviated as

$$f(\hat{m}) = \frac{(\hat{m}^3 - 3\hat{m} + 2)}{4}\ . \quad (3.40)$$

(ii) The equilibrium vapor pressure. The equilibrium vapor pressure for a spherical cap embryo may be evaluated by determining the extrema of the thermodynamic potential (3.39). Differentiating this with respect to the embryo radius r, equating the result to zero, and solving for the equilibrium saturation ratio gives the Kelvin equation (3.15) as the result.

That the equilibrium vapor pressure over a spherical cap embryo is given by the Kelvin equation is reasonable, since the vapor pressure over any portion of a curved water surface should depend on the curvature at that point. It is also clear that a spherical shape (uniform curvature) is the equilibrium shape since any protuberance (region of higher curvature) would have a higher vapor pressure and would tend to evaporate, while any depression would have a lower vapor pressure and would tend to condense water vapor.

It follows from (3.39) that the activation energy $(\Delta Y)_{g*}$ is the homogeneous activation energy reduced by the geometrical factor, i.e.,

$$(\Delta Y)_{g*} = f(\hat{m})\left[\frac{16\ \sigma_{LV}^3\ \pi}{3(N_L\ kT\ \ln\ S)^2}\right]. \quad (3.41)$$

(iii) The nucleation rate. One may envisage a population of embryos residing on the nucleating substrate. The substrate is assumed to have uniform characteristics, so that these embryos approximate spherical caps with a single contact angle θ, as shown in Fig. 3.8. If the saturation ratio is high enough to give an appreciable nucleation rate, there will be a transient size distribution of embryos on the nucleating substrate similar to that described for homogeneous condensation.

Embryos residing on the surface may gain or lose molecules in two ways. Molecules may be directly exchanged with the vapor as described for homogeneous condensation. In addition, single water molecules may be adsorbed on the substrate and migrate to the edge of the embryo, where they may be incorporated into the embryo. In either event, the nucleation rate, based on (3.21) and (3.22), may be written as

$$J = 2N_{g*}\, A_{g*}\, \vec{\omega}_{g*} \left(\frac{\chi}{\pi}\right)^{1/2}. \tag{3.42}$$

Employing the procedures used in § 3.2.2, part (ii), χ and N_{g*} may be evaluated. From (3.22),

$$\chi = \frac{g^{*-4/3}\, f(\hat{m})\, \sigma_{LV}}{9kT} \left[\frac{36\pi}{N_L^2}\right]^{1/3}.$$

From the Boltzmann distribution and the definition of thermodynamic potential for this system (3.41), the surface concentration of critical embryos becomes

$$N_{g*} = N_1 \exp\left[\frac{-16\, \sigma_{LV}^3\, \pi\, f(\hat{m})}{3\, (kT)^3\, (N_L\, \ln S)^2}\right],$$

where N_1 refers to the surface concentration of adsorbed, single molecules. Incorporating these expressions into (3.42), the nucleation rate becomes

$$J = \frac{N_1\, A_{g*}\, \vec{\omega}_{g*}}{\pi^{1/6}\, N_L^{1/3}} \left(\frac{4}{3g^*}\right)^{2/3} \left[\frac{f(\hat{m})\, \sigma_{\mathrm{LV}}}{\mathrm{kT}}\right]^{1/2} \exp\left[\frac{-16\, \sigma_{LV}^3\, \pi\, f(\hat{m})}{3\, (kT)^3\, (N_{\mathrm{L}}\, \ln S)^2}\right]. \tag{3.43}$$

Evaluation of the nucleation rate as given by (3.43) rests on the relative importance of surface diffusion versus vapor transfer as the principal mechanism for adding molecules to the critical embryos. Pound et al. (1954) conclude that nucleation rates based on surface diffusion exceed those based on vapor transfer by a factor of 100 to 1000. Assuming that surface diffusion is the predominant means by which embryos grow, the pertinent 'area' of the critical embryo is the circumference of the embryo–substrate interface and is proportional to $g^{*1/3}$. Evaluation of the concentration of adsorbed single molecules N_1 and the rate of addition of molecules to the critical embryo $\vec{\omega}_g$ is treated by Pound et al. Fletcher (1962) estimates the pre-exponential factor to be between 10^{24} and $10^{27}\ \mathrm{cm}^{-2}\,\mathrm{s}^{-1}$. Since the nucleation rate is dominated by the exponential term, a more precise evaluation is not required for most purposes.

(iv) Comparison between observations and theory. The nucleation rate for a planar substrate given by (3.43) differs from that for homogeneous nucleation given by (3.36) largely through the geometrical term in the exponential. Although $f(\hat{m})$ only varies between 0 and 1, this has an overwhelming influence on the nucleation rate (see Problem 3.2). Thus, the pre-exponential factor may be considered to be relatively constant (assumed here to be $10^{26}\,\mathrm{cm}^{-2}\,\mathrm{s}^{-1}$) in evaluating nucleation rates.

Figure 3.10 gives the saturation ratio required to produce a nucleation rate of 10^3 events $\mathrm{cm}^{-2}\,\mathrm{s}^{-1}$ as a function of the contact angle. It is clear that nucleation on insoluble nuclei of any size in the atmosphere must be limited to those substances with contact angles less than ~10°. When the contact angle exceeds 150°, the presence of the surface has little affect on nucleation.

It has been argued that the presence of adsorbed water molecules on the surface makes that surface equivalent to a water surface as regards condensation, i.e., there should be no barrier to nucleation on such a surface. Twomey (1959) presents experimental evidence in contradiction to this hypothesis and in support of the classical theory presented here. His data are shown in Fig. 3.10 for comparison. The observed data points tend to lie below the calculated curve since the data were taken at laboratory temperatures and the calculations were made for 0°C.

3.3.2. *Condensation on insoluble nuclei*

Differences between condensation on a plane surface and condensation on a curved surface or spherical particle reside in the different geometries for the two

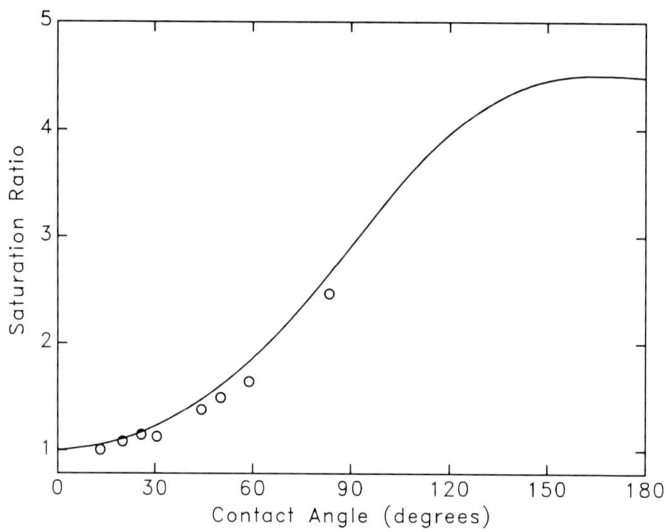

Fig. 3.10. The saturation ratio required to produce a nucleation rate of 10^3 events $\mathrm{cm}^{-3}\,\mathrm{s}^{-1}$ on a plane surface as a function of the contact angle at 0°C. The experimental data are from Twomey (1959).

cases. In the latter case, the geometrical derivations are much more complex and will not be presented here; the reader is referred to Fletcher (1958).[2] The geometrical factor must consider both the contact angle and the particle radius (or radius of curvature). From Fletcher, this factor is

$$f(\hat{m}, x) = \frac{1}{2}\left\{1 + \left(\frac{1 - \hat{m}x}{y}\right)^3 + x^3\left[2 - 3\left(\frac{x - \hat{m}}{y}\right) + \left(\frac{x - \hat{m}}{y}\right)^3\right] - 3\hat{m}x^2\left[1 - \left(\frac{x - \hat{m}}{y}\right)\right]\right\}. \tag{3.44}$$

Here, x is the ratio of the particle radius r_n to the radius of the critical embryo r^* or,

$$x = \frac{r_n}{r^*} = \frac{r_n\ N_L\ kT\ \ln\ S}{2\sigma_{LV}}$$

$$y = (1 + x^2 - 2\hat{m}x)^{1/2}.$$

It is expected that this geometrical factor should reduce to unity as r_n approaches zero (homogeneous condensation) and should reduce to (3.40) as r_n approaches infinity. The former case is straightforward; the latter case does follow although the reduction is lengthy (see Problem 3.9).

The nucleation rate for condensation on a spherical particle is simply that given by (3.43) using the geometrical factor (3.44), and recognizing that this needs to be multiplied by the surface area of the nucleus. Thus, the nucleation rate for condensation on a spherical particle may be expressed as

$$J = \frac{4\pi r_n^2 N_1\ A_{g^*}\ \vec{\omega}_{g^*}}{N_L^{1/3}\ \pi^{1/6}}\left(\frac{4}{3g^*}\right)^{2/3}\left[\frac{f(\hat{m},x)\sigma_{LV}}{kT}\right]^{1/2}\exp\left[\frac{-16\sigma_{LV}^3\ \pi f(\hat{m},x)}{3(kT)^3(N_L\ln S)^2}\right]. \tag{3.45}$$

As the pre-exponential factor in (3.43) was estimated as $\sim 10^{26}\,\mathrm{cm}^{-2}\ \mathrm{s}^{-1}$, this factor will be approximated as $4\pi r_n^2 \times 10^{26}\ \mathrm{s}^{-1}$. Figure 3.11 shows the saturation ratio required to produce a nucleation rate of one event per second per nucleus for several contact angles as a function of the nucleus radius.

For nucleation on insoluble nuclei to be likely in the atmosphere, the contact angle should be less than 10° and the nucleus radius should be larger than 0.05 μm. However, few nuclei consist solely of insoluble material; most have appreciable fractions of soluble materials. The presence of soluble materials in

[2] Note that a factor of 1/2 is missing from Fletcher's equation 11.

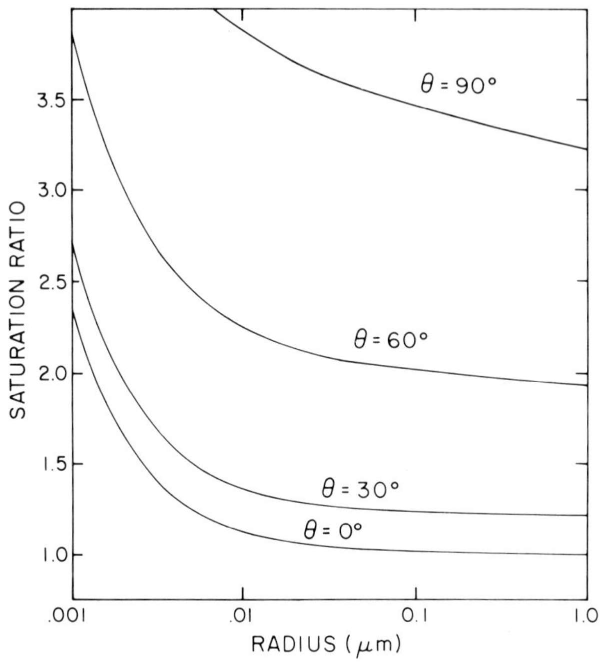

Fig. 3.11. The saturation ratio required to produce a nucleation rate of one event per second per nucleus is plotted as a function of nucleus radius for four different contact angles.

the nucleus will be shown to have an overwhelming effect on its role as a condensation nucleus. The treatment of nucleation on insoluble particles will prove useful in heterogeneous ice nucleation in Chapter 4.

3.4. Condensation on soluble nuclei

It is fairly well established that most condensation in the atmosphere is initiated on hygroscopic nuclei consisting of soluble material. The hygroscopic nature of such nuclei allows water to be condensed on the nucleus, even below water saturation. This is a direct consequence of the lowering of the equilibrium vapor pressure over a solution as compared to the pure solvent (water), in accordance with Raoult's law. This effect tends to counterbalance the curvature effect (Kelvin equation) which inhibits nucleation.

The solution and curvature effects exhibit different dependencies on the drop size. According to the Kelvin equation, the elevation of the vapor pressure due to the drop's curvature is inversely proportional to its radius. The depression of the vapor pressure due to solution effects is inversely proportional to the mass of water, *i.e.*, roughly proportional to the inverse cube of the drop radius. These two effects may be expected to produce stable embryos of rather small size, with an activation barrier which must be surmounted before the embryos can grow without limit.

Since the embryos that will be considered are fairly sizable, the kinetic treatment employed for homogeneous condensation will not be required. Here, a definition of the equilibrium curve serves to define the nucleating characteristics of the nucleus. These curves are traditionally referred to as Köhler curves after Hilding Köhler, who was responsible for the first quantitative descriptions of the behavior of soluble nuclei (see Köhler, 1949).

3.4.1. *Expression of the thermodynamic potential*

The initial and final states for condensation on a soluble particle are shown in Fig. 3.12. The initial state consists of a dry salt particle of mass m_s surrounded by water vapor of mass $m_v + m$. In the final state, the solution drop has mass $m + m_s$ and is surrounded by water vapor of mass m_v. In the initial state, there is a surface energy associated with the dry salt particle ($\sigma_s A_s$); in the final state, the surface energy associated with the solution drop is $4\pi r^2 \sigma'$.

Two contributions to the change in bulk free energy need to be considered, namely those due to water and those due to the salt. The difference in free energy between water vapor and water in bulk solution is

$$-\frac{N_A\ kTm}{M_V} \ln\left(\frac{e}{e_{soln}}\right),$$

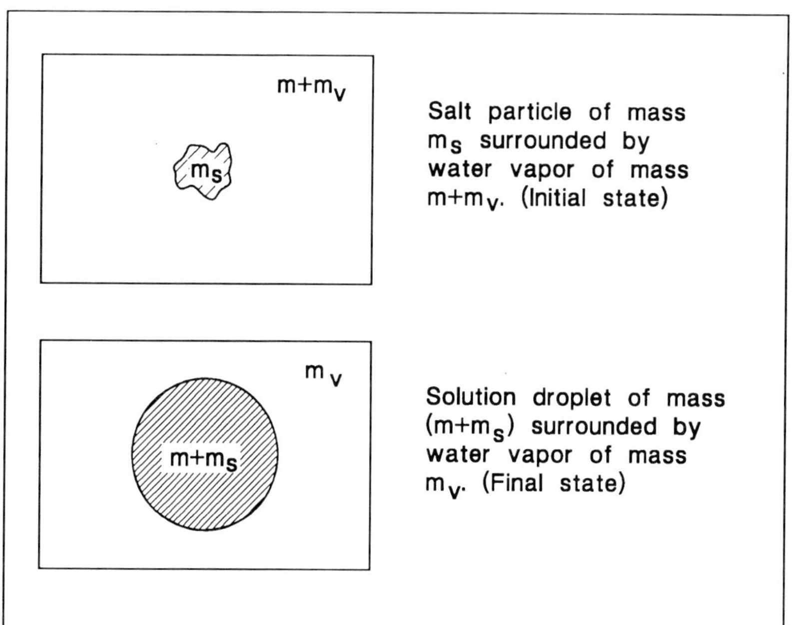

FIG. 3.12. The initial and final states for condensation on a soluble nucleus. From Young and Warren, (1992). *Journal of Atmospheric Sciences*, **49**, 13. American Meteorological Society. Reproduced with permission.

where N_A is Avogadro's number and e_{soln} is the vapor pressure over a plane surface of the solution. This is equivalent to the first term in (3.26), where $kT \ln(e/e_{soln})$ refers to the change in the thermodynamic potential per molecule and mN_A/M_v is the number of molecules. The change in thermodynamic potential is expressed in terms of the condensed water mass rather than the radius in order to simplify the derivation.

The difference in thermodynamic potential between the bulk dry salt and the salt in bulk solution is based only on those ions that are effectively in solution. The effective number of moles of ions in solution per mole of salt is termed the van't Hoff factor (i) and is generally a function of the molality of the solution. For dilute solutions, i approaches the number of ions, e.g., for sodium chloride, NaCl, i approaches 2; whereas for ammonium sulfate, $(NH_4)_2SO_4$, i approaches 3.

The change in thermodynamic potential for the ions that are considered to be in solution is

$$-\frac{N_A\ kT}{M_s} im_s \ln\left(\frac{P_{so}}{P_{ss}}\right).$$

Here, the vapor pressure of the dry salt particle is P_{so} and that of the salt in solution is P_{ss}. The molecular weight of the salt is M_s. The change in thermodynamic potential for the system depicted in Fig. 3.12 follows as

$$(\Delta Y)_m = -\frac{N_A\ kTm}{M_v} \ln\left(\frac{e}{e_{soln}}\right) + \frac{N_A\ kTim_s}{M_s} \ln\left(\frac{P_{ss}}{P_{so}}\right) + 4\pi\sigma'_{LV}\, r^2 - A_s\, \sigma_s\,. \tag{3.46}$$

A_s is the surface area of the dry salt nucleus and σ_s is its surface tension with respect to the vapor phase. The change in thermodynamic potential $(\Delta Y)_m$ is understood to be for a solution drop containing m grams of water.

3.4.2. *Raoult's law*

The change in vapor pressure for a solvent such as water due to the presence of dissolved material such as salt is given by Raoult's law. Simply stated, the vapor pressure is reduced in proportion to the mole fraction of the solute (salt). For the water vapor pressure,

$$\frac{e_{soln}}{e_{s,w}} = \frac{n_w}{n_w + n_s},$$

and for the salt vapor pressure,

$$\frac{P_{ss}}{P_{so}} = \frac{n_s}{n_w + n_s}.$$

Raoult's law may be expressed in terms of the masses of salt and water as follows. The number of moles of water n_w is simply m/M_v; the effective number of moles of salt n_s is im_s/M_s. Thus,

$$\frac{e_{\text{soln}}}{e_{s,w}} = \frac{m}{m + im_s \dfrac{M_v}{M_s}}$$

and

$$\frac{P_{ss}}{P_{so}} = \frac{im_s}{im_s + \mathrm{m} \dfrac{M_s}{M_v}} \tag{3.47}$$

The ratio e/e_{soln} in (3.46) may be expressed in terms of the saturation ratio $(e/e_{s,w})$ and the lowering of the vapor pressure due to the presence of salt (3.47) as

$$\frac{e}{e_{\text{soln}}} = \left(\frac{e}{e_{s,w}}\right)\left(\frac{e_{s,w}}{e_{\text{soln}}}\right) = \frac{e}{e_{s,w}}\left[\frac{m}{m + im_s \dfrac{M_v}{M_s}}\right]^{-1}.$$

The change in thermodynamic potential (3.46) may then be written

$$(\Delta Y)_m = \frac{N_A\ kT}{M_v}\left[-m\ln S + m\ln\left(\frac{m}{m + am_s}\right) + am_s \ln\left(\frac{m}{m + am_s}\right) + 4\pi\sigma'_{LV}\, r^2 - A_s\sigma_s\right], \tag{3.48}$$

where $a = iM_v/M_s$.

3.4.3. *The equilibrium vapor pressure for solution drops*

The equilibrium vapor pressure for solution drops may be determined by differentiating (3.48) with respect to the condensed water mass m. First consider the bulk terms within the brackets of (3.48) as

$$\frac{\partial\ (\text{bulk terms})}{\partial m} = -\ln S + \ln\left(\frac{m}{m + am_s}\right) + m_s \frac{M_v}{M_s}\frac{\partial i}{\partial m} \ln\left(\frac{am_s}{m + am_s}\right). \tag{3.49}$$

Note that the van't Hoff factor i is a function of the condensed water mass.

The surface term in (3.48) is dependent on both the surface tension and the bulk density (which is necessary to solve for r^2 in terms of m). Both are functions of the concentration of the solution, i.e., both are functions of the condensed water mass. The differential of the surface term with respect to m is

$$\frac{\partial(4\pi\sigma'_{LV}\, r^2)}{\partial m} = 4\pi\left[\sigma'_{LV}\frac{\partial r^2}{\partial m} + r^2\frac{\partial\sigma'_{LV}}{\partial m}\right]. \tag{3.50}$$

The solution of this differential equation requires expressions for the surface tension and bulk density as functions of the condensed water mass.

Following the treatment by Young and Warren (1992), the surface tension of the solution may be expressed as

$$\sigma'_{LV} = \sigma_{LV} + \frac{c}{m} \tag{3.51}$$

where σ_{LV} is the surface tension for pure water and $c = c_{\mathrm{o}}m_{\mathrm{s}}/M_{\mathrm{s}}$. Tabulated values of surface tension give $c_{\mathrm{o}} = 1640\,\mathrm{g}^2\,\mathrm{mol}^{-1}\,\mathrm{s}^{-2}$ for sodium chloride and about $2200\,\mathrm{g}^2\,\mathrm{mol}^{-1}\,\mathrm{s}^{-2}$ for ammonium sulfate.

The bulk density may be expressed as

$$\rho'_{\mathrm{L}} = \rho_{\mathrm{L}} + \frac{b}{(m + m_{\mathrm{s}})}, \tag{3.52}$$

where ρ_{L} is the bulk density for pure water and $b = b_{\mathrm{o}}m_{\mathrm{s}}/M_{\mathrm{s}}$. The parameter $b_{\mathrm{o}} = 43\,\mathrm{g}^2\,\mathrm{cm}^{-3}\,\mathrm{mol}^{-1}$ for sodium chloride and $76\,\mathrm{g}^2\,\mathrm{cm}^{-3}\,\mathrm{mol}^{-1}$ for ammonium sulfate.

Now the solution of (3.50) follows from (3.51) and (3.52), and recognizing that

$$r^2 = \left[\frac{3(m + m_{\mathrm{s}})}{4\pi\rho'_L}\right]^{2/3}.$$

The differential of the surface term is

$$\frac{\partial(4\pi\sigma'_{\mathrm{LV}}\, r^2)}{\partial m} = \left[\frac{4\pi}{m + m_{\mathrm{s}}}\right]^{1/3}\left(\frac{3}{\rho'_{\mathrm{L}}}\right)^{2/3}\left\{\frac{2\sigma'_{\mathrm{LV}}}{3}\left[1 + \frac{b}{\rho'_{\mathrm{L}}\,(m + m_{\mathrm{s}})}\right] - \frac{(m + m_{\mathrm{s}})\,c}{m^2}\right\}.$$

Combining this with the solution for the bulk terms from (3.49) and solving for the saturation ratio,

$$\ln S = \ln\left[\frac{m}{m + am_s}\right] + m_s \frac{M_v}{M_s}\frac{\partial i}{\partial m}\, ln\left[\frac{am_s}{m + am_s}\right] + \left[\frac{M_v}{N_A\ kT}\right] \times$$

$$\left[\frac{36\pi}{\rho_L'^2\ (m + m_s)}\right]^{1/3} \left\{\frac{2\sigma_{LV}'}{3}\left[1 + \frac{b}{\rho_L\,(m + m_s)}\right] - \frac{(m + m_s)\ c}{m^2}\right\}. \tag{3.53}$$

Young and Warren (1992) give the empirical expression

$$i = 1.9242 - 0.1844\ \ln M - 0.007931\ (\ln M)^2 \qquad \text{for } M \leq 1.0 \tag{3.54}$$

for the van't Hoff factor for ammonium sulfate, which can be used to determine $\partial i/\partial m$ from the molality (M).

If one assumes that the surface tension and bulk density are not functions of the salt concentration ($b = c = 0$), $m \gg m_s$, and that the van't Hoff factor i is likewise independent of the salt concentration ($\partial i/\partial m = 0$), then (3.53) reduces to the conventional form of the Köhler equation

$$\begin{aligned} \ln S &= \ln\left(\frac{m}{m + am_s}\right) + \frac{2\sigma_{LV}\ M_v}{N_A\ kTr\rho_L} \\ &= \ln\left(\frac{m}{m + am_s}\right) + \frac{2\sigma_{LV}}{N_L\ kTr}. \end{aligned} \tag{3.55}$$

This can be further simplified using a few algebraic approximations to

$$S - 1 = \frac{\hat{a}}{r} - \frac{\hat{b}}{r^3} \tag{3.56}$$

where

$$\hat{a} = \frac{2\sigma_{LV}}{\rho_L\ R_V\ T}, \quad \hat{b} = \frac{3\,im_s M_v}{4\pi\rho_L\ M_s}.$$

3.4.4. *Examination of Köhler curves*

A Köhler curve represents the equilibrium conditions possible for a given salt nucleus. Figure 3.13 presents isopleths of the thermodynamic potential calculated from (3.48) for a 10^{-16} g ammonium sulfate nucleus. In this diagram, values of thermodynamic potential decrease upward; the exact numerical values are unimportant. The Köhler curve represents the locus of points for which the slope of the isopleths is zero. From this, it is clear that the Köhler curve represents a stable equilibrium for sizes smaller than the critical radius of 1.5×10^{-5} cm and an unstable equilibrium for sizes larger than the critical radius. Here, the critical supersaturation is defined as the maximum supersaturation attained along the Köhler curve, and the critical radius is defined as the radius at which the critical supersaturation occurs.

Figure 3.13 may also be used to illustrate the behavior of embryos that may exist in a nonequilibrium environment. For example, an embryo of 2×10^{-5} cm radius in an environment at 0.3% supersaturation ($S = 1.003$) would evaporate until it reached the stable equilibrium radius of 1.0×10^{-5} cm. An embryo of 0.8 $\times$ 10^{-5} cm radius and also at 0.3% supersaturation would grow to the same stable equilibrium radius. Any embryo in an environment with a supersaturation greater than the critical supersaturation would grow without limit. Similarly, any embryo finding itself above the Köhler curve and larger than the critical radius would grow without limit.

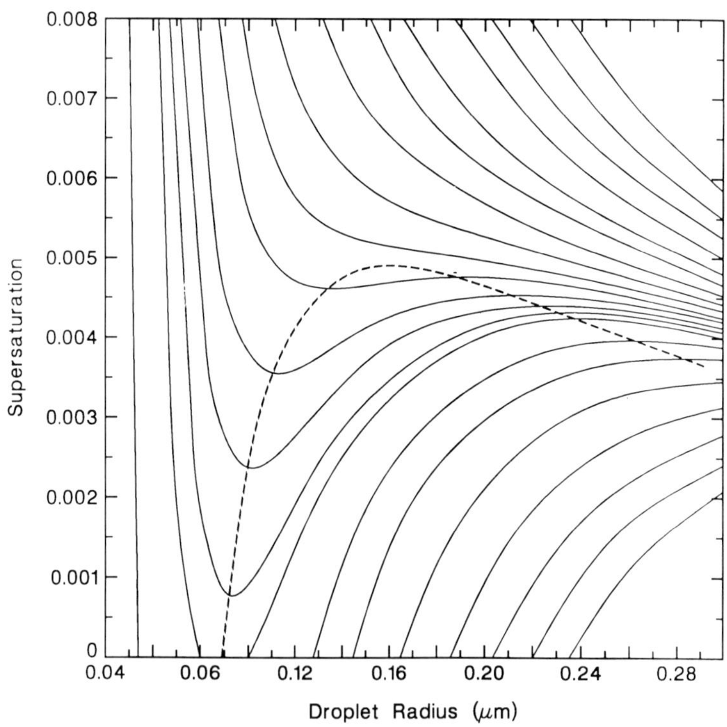

Fig. 3.13. The equilibrium saturation ratio curve for a particle containing 10^{-16} g of ammonium sulfate as deduced from the thermodynamic potential. The thermodynamic potential is calculated from (3.48) for 0°C. Isopleths are plotted at irregular intervals with values decreasing toward the upper portion of the diagram. The broken curve represents the locus of equilibrium points. From Young and Warren (1992). *Journal of Atmospheric Sciences*, **49**, 13. American Meteorological Society. Reproduced with permission.

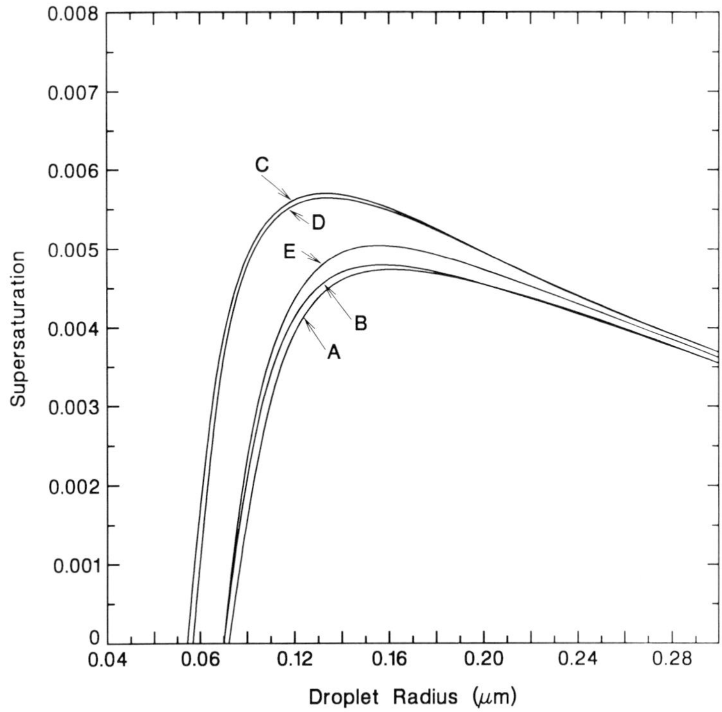

Fig. 3.14. Equilibrium saturation ratio curves for a particle showing the effects of the various assumptions made in the derivation of the Köhler curve. Curve A is based on (3.53); curve B is also based on (3.53) but assuming that the surface tension and density are constant. Curve C is a further simplification of curve B, assuming $\partial i/\partial m = 0$. Curve D is based on (3.56) with a variable van't Hoff factor; curve E is also based on (3.56) but assuming $i = 3$. From Young and Warren (1992). *Journal of Atmospheric Sciences*, **49**, 13. American Meteorological Society. Reproduced with permission.

Figure 3.14 compares the effects on the Köhler curve for a 1×10^{-16} g ammonium sulfate particle, introduced by the various simplifications noted above. Curve A is based on (3.53). Curve B, resulting from considering the surface tension and density as constant, is very close to curve A. Curve C is the same as curve B with $\partial i/\partial m$ is set equal to zero (the van't Hoff factor i is still calculated from (3.54); this is essentially (3.55)). Curve C is significantly shifted to higher supersaturations. Curves D and E are based on (3.56). Curve D treats i as a variable; curve E assumes $i = 3$. Young and Warren (1992) note that this error is greater for larger nuclei, since they have higher molalities at the critical radius than smaller nuclei.

Neglecting the $\partial i/\partial m$ term results in underestimating the stable embryo mass by roughly a factor of two. Such an error would underpredict the depletion of water vapor by the subcritical embryos, leading to an overprediction of the maximum supersaturation attained within the activation zone. This is compensated by overpredicting the critical supersaturation for the nuclei, leading to roughly the same numbers of nuclei being activated.

It should be noted that Köhler curves are temperature dependent. The surface tension increases with decreasing temperature, displacing the Köhler curve toward higher supersaturations. The effects may be significant when comparing cloud condensation nuclei (CCN) spectra obtained under normal laboratory conditions with the activation of those same CCN in clouds at temperatures below 0°C.

Larger nuclei may be expected to exhibit lower critical supersaturations and larger critical radii than smaller nuclei. Figure 3.15 gives the Köhler curves for a range of ammonium sulfate particles that might be expected in the lower troposphere.

Nuclei may or may not be in equilibrium as defined by their Köhler curves, depending on how rapidly the ambient supersaturation is changing. Large nuclei in strong updrafts may be expected to lag considerably behind their equilibrium size. Very large nuclei may never become activated in the ordinary sense and could participate in the collection process before reaching their critical radius. This lag occurs because the rate of growth of an embryo cannot exceed the rate at which water vapor can be transported to the embryo by diffusion. Thus, when considering the initial stages of cloud formation, it may not be adequate to assume that embryos follow their Köhler curves until they reach their critical supersaturation.

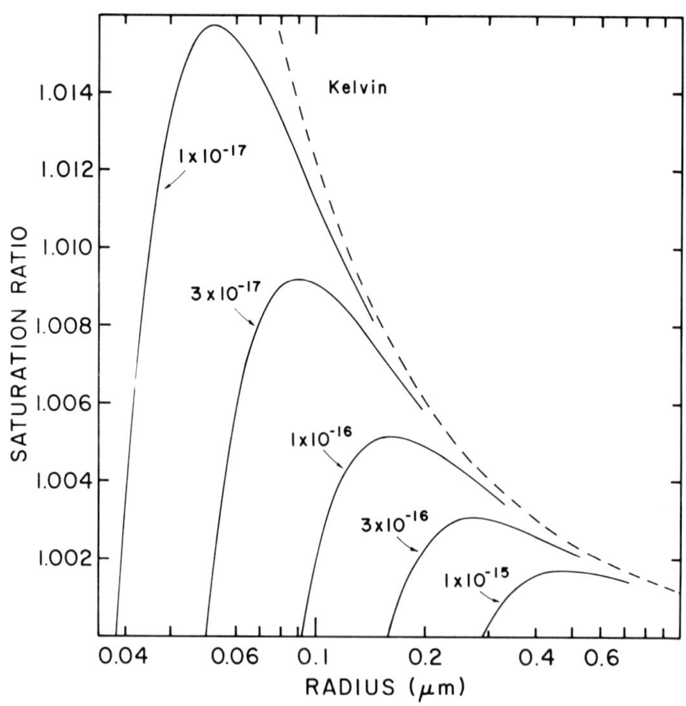

Fig. 3.15. Equilibrium saturation ratio curves for ammonium sulfate CCN with masses ranging from 10^{-15} g to 10^{-17} g. The curves are calculated from (3.53) for $T = 0$°C. The Kelvin curve for pure water is shown for comparison.

3.4.5. *The nature and origin of CCN*

CCN in the atmosphere cannot be expected to consist of a single chemical species. Coagulation among the non-water aerosols present, scavenging by water drops followed by their evaporation, as well as chemical reactions with atmospheric gases, all tend to produce nuclei of mixed composition. The behavior of such mixed nuclei may be expected to be controlled by the behavior of the soluble components of the nucleus.

The predominant chemical species appears to be sulfate in the form of ammonium sulfate or sulfuric acid. Over the oceans, sodium chloride may be expected to represent a significant fraction of the nucleus mass, especially in the larger nuclei. Since the Köhler curves for ammonium sulfate and sodium chloride nuclei of the same masses tend to be quite similar, it is sometimes convenient to treat these as one chemical species with regard to cloud droplet formation.

(i) Production of CCN by bursting of bubbles. The production of CCN consisting primarily of ocean salts is considered to be due to the bursting of bubbles at the ocean surface. Figure 3.16 depicts the sequence of events leading to the formation of CCN by this mechanism. As the bubble reaches the surface, the thin film of water separating the bubble from the atmosphere ruptures, allowing the bubble to collapse. The rupture of this thin film may produce a few hundred small droplets, which, on evaporation, produce nuclei with salt masses on the order of 10^{-14} to 10^{-15} g. Day (1964) shows that the production of film droplets is dependent on the bubble size, with very little production for bubbles of radius <0.02 cm but increasing to 300 to 400 droplets for bubbles of radius 0.2 cm.

The collapse of the bubble produces a rising 'jet' in the center of the depression (Fig. 3.16b,c). This jet separates to form from one to five relatively massive droplets, containing on the order of 10^{-11} g of salt each. Many of the jet drops and a significant fraction of the film droplets fall back to the surface and do not contribute to CCN production. On the basis of a laboratory model, Cipriano et al. (1983) conclude that this mechanism is sufficient to account for most maritime aerosol spectra.

(ii) Production of CCN by gas-to-particle conversion. The production of CCN by gas-to-particle reactions include the oxidation of sulfur dioxide (SO_2), forming sulfate, and the oxidation of nitric oxide (NO), forming nitrate, among others. Here, only the formation of ammonium sulfate will be examined. A more extensive survey is given by Pruppacher and Klett (1978). The primary source of ammonia (NH_3) in the atmosphere is considered to be the decay of organic material. Sulfur dioxide is present from the combustion of sulfur-containing fuels and as a result of volcanic activity.

The sulfur atom has two oxidation states of interest in the formation of sulfate-containing aerosol. Oxidation state IV is noted for sulfur dioxide (SO_2), aqueous sulfur dioxide ($SO_2 \cdot H_2O$), bisulfate (HSO_3^-), and sulfite ($SO_3^=$). Oxidation state VI is noted for sulfuric acid (H_2SO_4), bisulfate (HSO_4^-), and sulfate ($SO_2^=$). The gas phase oxidation of SO_2 to sulfuric acid by hydroxyl radicals is a significant source of sulfate in the atmosphere.

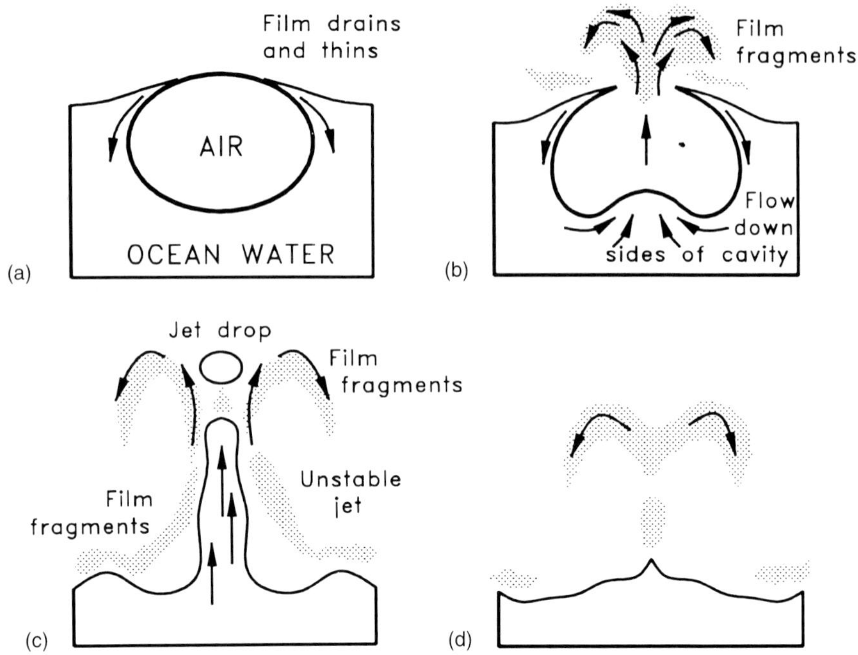

FIG. 3.16. The sequence of the bursting of an air bubble at a water surface. (a) The bubble protrudes from the surface and begins to thin. (b) The film ruptures, forming fragments as the bubble walls begin to collapse. (c) The collapse of the bubble walls creates a jet, which breaks into a small number of larger fragments. (d) Some fragments fall back into the water, others evaporate, forming tiny salt particles. From Day (1965). Adapted from *Natural History*.

The oxidation of sulfur(IV) to sulfur(VI) has been found to be markedly enhanced in clouds. Radke and Hobbs (1969) were among the first to note marked increases in CCN downwind from evaporating clouds and speculated that oxidation of sulfur(IV) to sulfur(VI) in cloud droplets was responsible. Earlier, Scott and Hobbs (1967) suggested that the presence of ammonia enhances the oxidation of sulfur(IV) to sulfur(VI). Sulfur dioxide dissolved in water ($SO_2{\cdot}H_2O$) dissociates, forming sulfite and bisulfite ions,

$$SO_2{\cdot}H_2O \rightleftharpoons H^+ + HSO_3^- \qquad K_{a1}$$

$$HSO_3^- \rightleftharpoons H^+ + SO_3^{2-} \qquad K_{a2}$$

A variety of chemical species have been shown to catalyze the oxidation of sulfite to sulfate, including hydrogen peroxide, ozone, and oxygen in the presence of heavy metal ion catalysts. Many of these catalysts are analyzed by Graedel and Weschler (1981). Seinfeld (1986) compares the rates of sulfate formation by various oxidation paths for typical concentrations of the catalysts as a function of pH (Fig. 3.17). Hydrogen peroxide is the dominant catalyst for sulfate formation for pH values below 5.

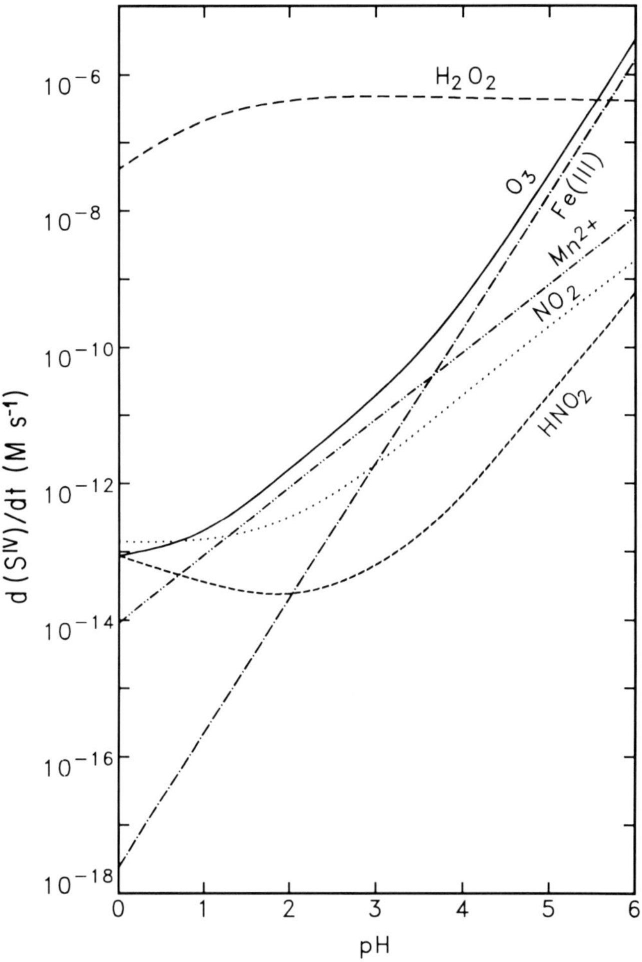

Fig. 3.17. A comparison of the rate of sulfate production by various SO_2 oxidation paths as a function of pH. Assumed concentrations of the constituents are: $[SO_2(g)] = 5$ ppb; $[HNO_2(g)] = 2$ ppb; $[H_2O_2(g)] = 1$ ppb; $[NO_2(g)] = 1$ ppb; $[O_3(g)] = 50$ ppb; $[Fe^{3+}(aq)] = 3 \times 10^{-7}$ M; $[Mn^{2+}(aq)] = 3 \times 10^{-8}$ M. From Seinfeld (1986). From *Atmospheric Chemistry and Physics of Air Pollution*, Seinfeld, J.H. (1968). © 1986. Reprinted by permission of John Wiley & Sons, Inc.

Waltrop et al. (1991) express the rate of sulfate production in droplets as

$$\frac{d[S^{VI}]}{dt} = \frac{7.45 \times 10^{-7}\,[H^+]\,[H_2O_2]\,[S^{IV}]}{(1 + 13\,[H^+])\left(\dfrac{[H^+]}{K_1} + 1\right)}, \tag{3.57}$$

where

$$\log K_1 = \frac{853}{T} - 4.74\,.$$

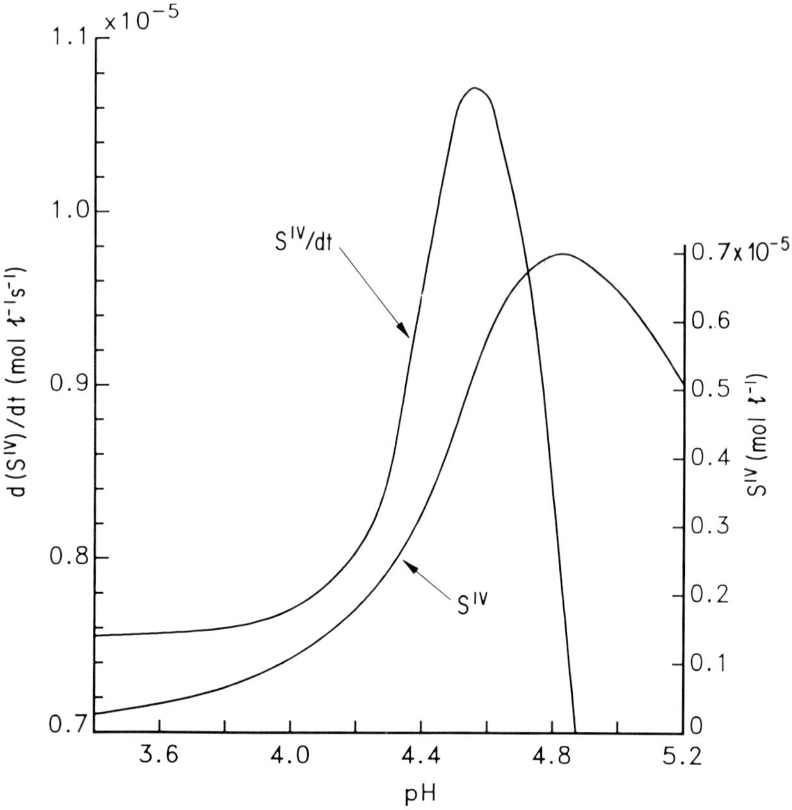

FIG. 3.18. Calculated rate of sulfate production catalyzed by hydrogen peroxide and sulfur(IV) concentration in a drop of 300 μm radius with initial pH = 7. The assumed concentrations are $[SO_2(g)]$ = 100 ppb and $[H_2O_2(aq)] = 10^{-3}$ M. The abscissa is pH decreasing to the left; this may be interpreted as time increasing to the left since the acidity of the droplet increases with time. From Waltrop et al. (1991). Reprinted by permission of Kluwer Academic Publishers.

Figure 3.18 shows the change in sulfur(IV) and rate of production of sulfur(VI) as a drop absorbs SO_2 from the atmosphere. At first, the concentration of sulfur(IV) increases as SO_2 is absorbed. The dissociation as SO_2 forms sulfite increases the hydrogen ion concentration $[H^+]$, decreasing the pH. As the partial pressure of SO_2 at the drop surface rises, the rate of absorption of SO_2 decreases. The oxidation of sulfite to sulfate increases as pH decreases and as concentration of sulfur(IV) increases, according to (3.57). The rate of sulfate production peaks as the concentration of sulfite decreases, leveling off to a fairly constant production rate.

The conversion of sulfur(IV) to sulfur(VI) is most efficient in droplets smaller than 20 μm radius, becoming much less efficient in drops larger than 100 μm radius, a result of changing the ratio of drop surface area to volume. The acidic droplets absorb ammonia; as the droplets evaporate, ammonium sulfate

$\{(NH_4)_2SO_4\}$ aerosol is left behind. Large drops in which the conversion to sulfate is less efficient are more likely to fall out of the cloud as precipitation. The small droplets, in which the conversion is much more efficient, are also more likely to evaporate.

(iii) Supersaturation spectra for CCN. The numbers of CCN that may be activated increase as the supersaturation is increased. This may be interpreted as the progressive activation of smaller nuclei as the supersaturation is increased, in reference to Fig. 3.15. However, natural CCN populations are not limited to a single chemical species and the relation between critical supersaturation and nuclei salt mass is not simple.

A summary of CCN measurements from various locations around the world is presented by Twomey and Wojciechowski (1969). The median of all observations taken over continents and over oceans is shown in Fig. 3.19. The cumulative number of CCN activated as a function of supersaturation may be approximated by

$$N_{\mathrm{CCN}} = c(S-1)^k .$$

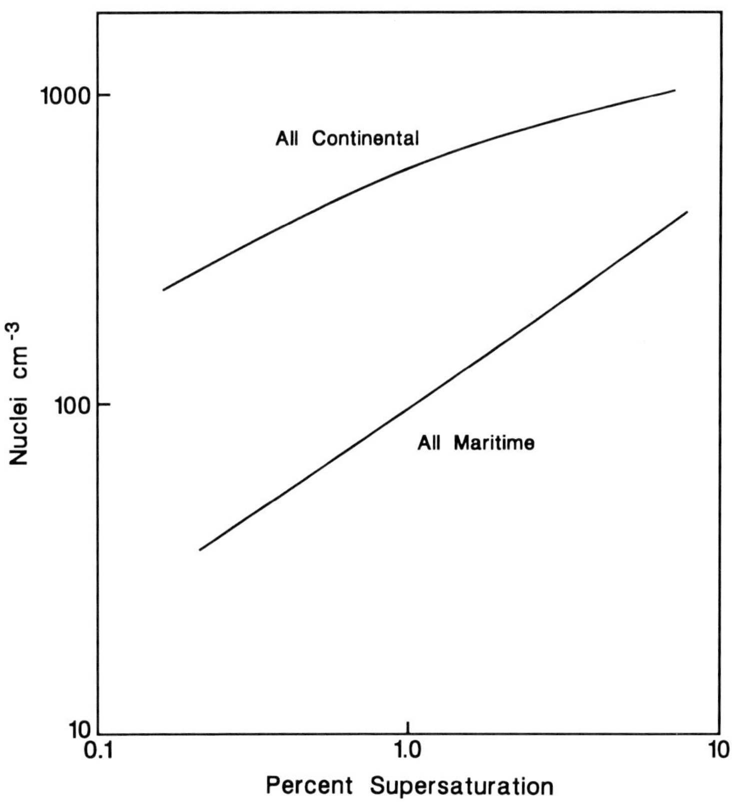

FIG. **3.19.** Median CCN spectra, comparing continental and maritime spectra. From Twomey and Wojciechowski (1969). *Journal of Atmospheric Sciences*, **26**, 4. American Meteorological Society. Reproduced with permission.

Maritime CCN spectra ($c \approx 100\,\mathrm{cm}^{-3}$, $k \approx 0.7$) tend to have lower concentrations of nuclei active at a given supersaturation than do continental spectra ($c \approx 600\,\mathrm{cm}^{-3}$, $k \approx 0.5$) although the exponent tends to be larger for maritime spectra.

CCN spectra are observed to vary markedly in both space and time. Maritime air masses which move inland will exhibit an increase in the numbers of nuclei due to the increased production of sulfate over the continents. Urban regions tend to have exceptionally high concentrations of CCN. Still, the fraction of the total aerosol population that may serve as CCN in clouds (at supersaturations less than 0.5%) is relatively low.

Problems

3.1* Justify the approximation used in deriving the Kelvin equation (3.15) from (3.14).

3.2* Determine the nucleation rate on a unit surface area of a plane substrate for $S = 3.0$. Do for contact angles of 10°, 30°, . . . 170°. Assume $T =$ 0°C. The pre-exponential factor may be assumed to be equal to 10^{26}.

3.3* What is the equilibrium saturation ratio over a saturated solution of ammonium sulfate for:
(a) a plane surface?
(b) a droplet containing 10^{-15} g ammonium sulfate?

3.4** Calculate the number of water molecules in a drop whose vapor pressure corresponds to a saturation ratio of 3.0. Repeat for $S = 4.0$, $S = 5.0$, and $S = 6.0$. Assume $T = 0$°C.

3.5** Derive expressions for the volume (V_L), the water–vapor interface area (A_{LV}), and the water–substrate interface area (A_{LN}) for a spherical cap embryo on a plane substrate.

3.6** Use the Boltzmann distribution (3.16) to estimate the fraction of water molecules which remain unassociated in the vapor for $S = 4.0$ at $T =$ 0°C. Assume that the number of supercritical embryos is zero. Discuss the validity of the approach used in the derivation of the nucleation rate as it pertains to the 'reservoir of unassociated molecules.'

3.7** How many water molecules does a critical cluster contain when $J = 10^3$ events $\mathrm{cm}^{-3}\,\mathrm{s}^{-1}$ at $T = 20$°C. Estimate the fraction of these molecules that would be expected to occupy surface positions. Assume spherical molecules and spherical embryos. Repeat for $T = -20$°C.

3.8*** Show that the work required to open and close the orifice in forming a drop of radius r as shown in Fig. 3.3 is $V(P_2 - P_1)/2$.

3.9*** Show that (3.44) reduces to (3.40) as r_n approaches infinity.

Computer problem

3.10** Determine the critical radius and critical supersaturation using the complete form of the Köhler equation (3.53). You may wish to evaluate the second derivative of (3.48) or you may wish to use an iterative procedure based on (3.53). Assume the CCN are pure ammonium sulfate. Evaluate for the following conditions:

(a) $m_s = 2 \times 10^{-15}$ g, $T = 10°C$
(b) $m_s = 2 \times 10^{-15}$ g, $T = -10°C$
(c) $m_s = 5 \times 10^{-16}$ g, $T = 10°C$
(d) $m_s = 5 \times 10^{-16}$ g, $T = -10°C$

Hint: The mass of the critical embryo is $m + m_s$. Its density must be calculated.

REFERENCES

Cipriano, R.J., Blanchard, D.C., Hogan, A.W., and Lala, G.G., 1983, 'On the production of Aitken nuclei from breaking waves and their role in the atmosphere, *J. Atmos. Sci.* **40**, 469.

Day, J.A., 1964, 'Production of droplets and salt nuclei by the bursting of air-bubble films,' *Quart. J. R. Meteorol. Soc.* **90**, 72.

Day, J.A., 1965, 'The building blocks of clouds: Without dirt, man could not survive,' *Natural History* **74**, 36.

Farley, F.J.M., 1952, 'The theory of the condensation of supersaturated, ion-free vapor,' *Proc. R. Soc. London Ser. A* **212**, 530.

Fletcher, N.H., 1958, 'Size effect in heterogeneous nucleation,' *J. Chem. Phys.* **29**, 572.

Fletcher, N.H., 1962, *The physics of rainclouds*. Cambridge: Cambridge University Press.

Graedel, T.E., and Weschler, C.J., 1981, 'Chemistry with aqueous atmospheric aerosols and raindrops,' *Rev. Geophys. Space Phys.* **19**, 505.

Heist, R.H., and Reiss, H., 1973, 'Investigation of the homogeneous nucleation of water vapor using a diffusion cloud chamber,' *J. Chem. Physics,* **59**, 665.

Köhler, H., 1949, 'On the problem of condensation in the atmosphere,' *Nova Acta R. Soc. Sci. Upsal. Ser. IV* **14**.

Long, A.B., 1969, 'A thermodynamic study of homogeneous nucleation,' *Rev. Geophys.* **7**, 595.

Mason, B.J., 1971, *The physics of clouds* (2nd edition). Clarendon Press: Oxford.

Plummer, B.N., and Plummer, P.L., 1974, 'On nucleation phenomena. I: A molecular model,' *J. Atmos. Sci.* **31**, 1615.

Pound, G.M., Simnad, M.T., and Yang, L., 1954, 'Heterogeneous nucleation of crystals from vapor,' *J. Chem. Phys.* **22**, 1215.

Pruppacher, H.R., and Klett, J.D., 1978, *Microphysics of clouds and precipitation.* Dordrecht: Reidel.

Radke, L.F., and Hobbs, P.V., 1969, 'Measurement of cloud condensation nuclei, light scattering coefficient, sodium-containing particle, and Aitken nuclei in the Olympic Mountains of Washington,' *J. Atmos. Sci.* **26**, 281.

Reiss, H., 1965, *Methods of thermodynamics.* New York: Blaisdell.

Scott, W.D., and Hobbs, P.V., 1967, 'The formation of sulfate in water droplets,' *J. Atmos. Sci.* **24**, 54.

Seinfeld, J.H., 1986, *Atmospheric chemistry and physics of air pollution.* Chichester: John Wiley and Sons.

Twomey, S., 1959, 'Experimental test of the Vollmer theory of heterogeneous nucleation,' *J. Chem. Phys.* **30**, 941.

Twomey, S., and Wojciechowski, T.A., 1969, 'Observations of the geographical variation of cloud nuclei,' *J. Atmos. Sci.* **26**, 684.

Waltrop, A., Mitra, S.K., Flossmann, A.I., and Pruppacher, H.R., 1991, 'On the scavenging of SO_2 by cloud and rain drops: IV. A wind tunnel and theoretical study of the absorption of SO_2 in the ppb(v) range by water drops in the presence of H_2O_2,' *J. Atmos. Chem.* **12**, 1.

Young, K.C., and Warren, A.J., 1992, 'A reexamination of the derivation of the equilibrium supersaturation curve for soluble particles,' *J. Atmos. Sci.* **49**, 1138.

4

NUCLEATION OF THE ICE PHASE

4.1. Introduction

Treatment of the ice phase in the atmosphere is considerably more complex than the preceding treatment of the liquid phase. Ice may originate from supersaturated vapor or from supercooled water. Thus, one may distinguish between *deposition* (vapor to ice) and *freezing* (water to ice) nucleation.

The simple distinction between deposition and freezing nucleation becomes less distinct as it is examined more closely. Below water saturation and without the presence of a macroscopic quantity of liquid water, one may feel fairly confident that a process of adsorption on the nucleus surface is followed by molecular migration across the surface until the molecule becomes incorporated into an ice embryo. For a nucleus present in an environment above water saturation, the rate of condensation nucleation may exceed the rate at which ice embryos can become supercritical. The nucleus may then simply become a water drop or the transient occurrence of a liquid phase may be followed by freezing as the water covers the ice embryos present. This latter process is termed *condensation freezing*.

Freezing of bulk water may occur in two ways. First, a nucleus resident within the drop for a long period of time may nucleate the ice phase. In this case, the drop is generally being cooled, making the likelihood of formation of a supercritical embryo progressively greater. This type of nucleation is termed *immersion freezing* nucleation. This is distinguished from the freezing induced in a drop when an ice nucleus makes contact with a supercooled drop and very quickly initiates freezing. This type of nucleation is termed *contact freezing* nucleation.

Figure 4.1 provides a conceptual picture of the conditions under which each of these modes of ice nucleation may be expected to occur. Below a saturation ratio which is close to water saturation, relatively little liquid water is present and nucleation occurs by deposition nucleation. Both contact freezing nucleation and condensation freezing nucleation are transient in the sense that the time allowed for the embryos to come into equilibrium with their surroundings is very short. If the nucleus does not provide a supercritical embryo within this time, the nucleation opportunity is lost. The difference between these two is that contact freezing involves a macroscopic drop whereas condensation freezing involves a microscopic quantity of water on the nucleus. Immersion freezing takes place

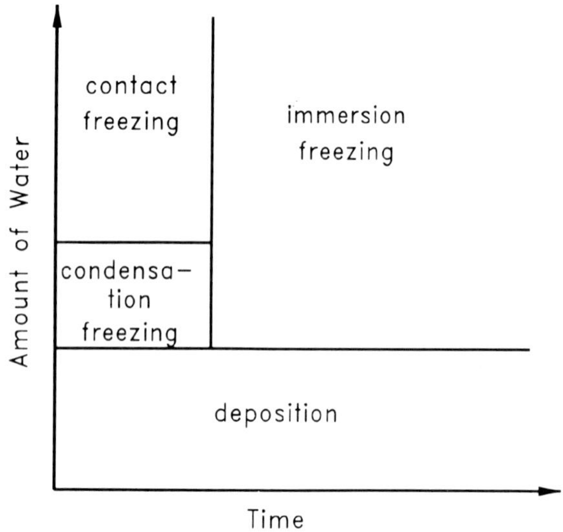

Fig. 4.1. The relationship between the four modes of ice phase nucleation. Contact and condensation freezing nucleation must act quickly or the opportunity to nucleate the ice phase is lost. Immersion freezing requires the nucleus to be immersed in the drop long enough to come to equilibrium with its surroundings. Deposition nucleation normally occurs under conditions with very little liquid water present.

when the time is sufficiently long for the embryos to come into equilibrium with their surroundings.

This chapter will examine each of the four modes of heterogeneous ice phase nucleation: deposition, immersion freezing, contact freezing, and condensation freezing. Homogeneous freezing nucleation will be examined as an introduction to heterogeneous freezing nucleation and to provide a foundation for the treatment of ice particle nucleation in cirrus clouds in Chapter 13. The role of Brownian motion and phoretic forces in promoting (or retarding) contact freezing nucleation will be examined. The relationships between the four modes of nucleation will be discussed in terms of theory and observations. Secondary ice formation, proposed to explain anomalous high ice crystal concentrations in clouds with respect to ice nucleus concentrations, will be analyzed in terms of in-cloud observations, laboratory experiments, and theoretical and modeling studies.

It should be pointed out that much controversy exists in regard to ice phase nucleation. There is disagreement as to the dominant modes of ice nucleation in the atmosphere, and even some argument as to the modes of nucleation which are possible. The methods used to characterize the ice nucleating ability of aerosol are rarely able to examine more than one mode of nucleation for any given aerosol sample. Intercomparisons of the number of ice nuclei active at a given temperature using different instruments on the same aerosol source often vary by several orders of magnitude. The role of secondary ice production in clouds is quite controversial. Here, the attempt will be to synthesize a fairly cohesive, although as yet incomplete, picture of the underlying physics involved and the relationships between the different ice nucleation modes. Both theoretical treatments and observations will be relied on.

4.2. Deposition nucleation

Deposition nucleation is the formation of ice directly from vapor. This implies that there is a population of ice embryos that exchanges water molecules with the vapor in a manner analogous to the condensation nucleation considered in Chapter 3. As noted for condensation, deposition nucleation may be homogeneous or heterogeneous. The case of homogeneous deposition nucleation will not be examined here because it is unlikely to occur under normal atmospheric conditions. For heterogeneous nucleation, only the case analogous to condensation on insoluble particles need be considered, since salts would not be expected to enhance nucleation of the ice phase.[1] This implies that the opportunities for nucleating ice in the atmosphere are more restricted than the opportunities for condensation, since the dominant mode of condensation nucleation relies on the lowering of the vapor pressure of the embryo due to the presence of dissolved salts.

Nucleation of the ice phase introduces two complexities absent in nucleation of the liquid phase. The stable form of ice under atmospheric conditions is not the only possible ice structure. Water molecules could conceivably form ice-like clusters in which the structure is such that the bulk free energy is greater than it is in the parent phase. Such clusters could never become 'nucleated' in the sense of growing without limit beyond their critical size.

Alternatively, ice-like structures may exist for which the bulk free energy is less than in the parent state. Two situations may be envisaged. The molecules may cluster to form a closed structure which is stable but which cannot add molecules without greatly increasing the free energy; or the molecules may cluster to form an open structure which may continue to grow without limit.

These three possibilities are illustrated in Fig. 4.2. Here, it will be assumed that the presence of embryos corresponding to the first two possibilities do not affect our treatment of the nucleation of Ice Ih.

Very small ice particles may be well approximated as spheres. Their small size tends to insure that they maintain an equilibrium shape which is roughly isometric (equal a- and c-axes). In the following treatments, ice embryos will be assumed to be spherical.

The lattice structure of the ice embryo restricts its freedom to bond to the underlying lattice of the nucleating substrate. In the case of condensation on an insoluble substrate, the water molecules were relatively free to orient themselves in response to the surface presented by the substrate. The ice embryo however, must adjust to the substrate lattice by deforming the ice lattice[2] and/or allowing

[1] Enhancement of condensation nucleation results from the solubility of the salt in water. Salts are generally excluded from the ice lattice when water freezes.

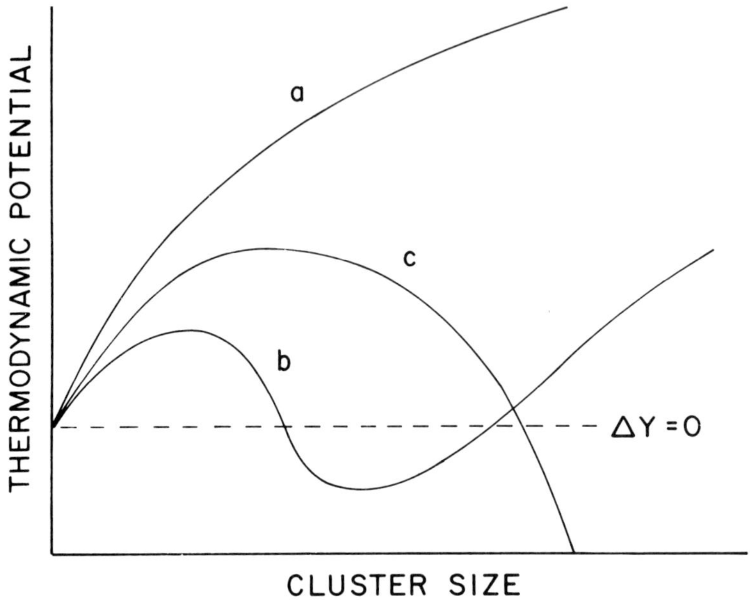

Fig. 4.2. The effect of different ice structures on the growth of embryos of the solid phase. (a) An unfavored lattice structure, i.e., the bulk free energy is greater than in the vapor. (b) A stable, closed (clathrate) structure. (c) The stable, open structure associated with Ice Ih that will be assumed in this chapter.

dislocations at the ice–substrate interface. These two extremes are depicted in Fig. 4.3.

Deformation of the ice lattice results in an increase in the bulk free energy; dislocations at the ice–substrate interface serve to raise the interfacial energy. If the lattice misfit between the substrate and ice is small, few dislocations may be expected, with most of the misfit being accommodated by elastic strain within the bulk ice. If the lattice misfit is large, dislocations may be expected to be fairly numerous. Since both of these effects serve to lower the nucleating efficiency, substrates with lattice parameters similar to those for ice might be expected to be good nucleants. This condition is referred to as epitaxy.

An additional requirement for good ice-nucleating ability of a substrate is related to the electronegativity of the ions and the relative numbers and spacing of the positive and negative ions exposed at the surface of the substrate. Fletcher (1959) suggests that substrates which expose equal numbers of positive and negative ions should be preferred to those which expose different ratios of positive to negative ions. Since the ice lattice projects equal numbers of hydrogens and lone pair orbitals (see Chapter 2), a substrate which could accommodate the alternating hydrogens and lone pair orbitals should have a lower interfacial energy than one which could not.

[2] It is conceivable that the substrate lattice could also adjust to accommodate more readily the ice lattice.

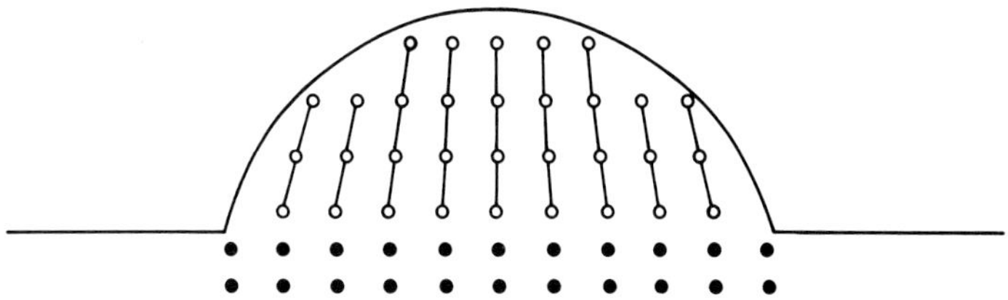
(a)

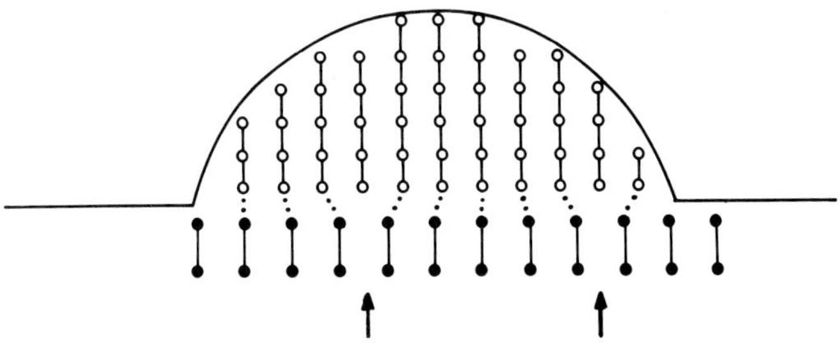
(b)

Fig. 4.3. How the ice lattice adapts itself to the nucleus. (a) Adjustment by straining the ice lattice. Note that each water molecule is lined up with a nucleus lattice position. (b) Adjustment by permitting dislocations. Arrows mark the sites of dislocations.

4.2.1. *The thermodynamic potential*

The expression of the thermodynamic potential for an ice embryo on a substrate follows from (3.38) with the addition of a term to account for the elastic strain in the bulk ice,

$$(\Delta Y)_g = V_\mathrm{I}\left(-N_\mathrm{I}\, kT \ln\left(\frac{e}{e_{\mathrm{s,i}}}\right) + C\epsilon^2\right) + \sigma_{\mathrm{IV}} A_{\mathrm{IV}} + (\sigma_{\mathrm{IN}} - \sigma_{\mathrm{VN}}) A_{\mathrm{IN}}\,. \quad (4.1)$$

Turnbull and Vonnegut (1952) estimate the strain coefficient C to have a value of 1.7×10^{11} dynes cm^{-2} at 0°C.

The elastic strain ϵ may be estimated in the following manner. The relative misfit between the ice and substrate lattices is given by

$$\delta = \frac{(a_n - a)}{a}$$

where a refers to the length a of the ice lattice unit cell (see Fig. 2.3) and a_n refers to the corresponding lattice vector in the substrate. This assumes that ice prefers to nucleate with the basal plane in contact with the substrate.[3] If the misfit is small (the condition of epitaxy), the number of dislocations may be assumed to be small and the elastic strain ϵ equals the misfit δ. If the misfit is large, it may be assumed that the ice lattice does not attempt to accommodate itself to the underlying surface and the elastic strain ϵ is zero.

If one assumes that the elastic strain per molecule is not a function of the embryo size, the treatment in Chapter 3 for condensation on insoluble nuclei may be applied here as well. The thermodynamic potential follows as

$$(\Delta Y)_g = f(\hat{m},x)\left[\frac{4\pi r^3}{3}(-N_{\mathrm{I}}\,kT\ln S_{\mathrm{i}} + C\epsilon^2) + 4\pi r^2\,\sigma_{\mathrm{IV}}\right], \tag{4.2}$$

where S_{i} is the saturation ratio for ice $(e/e_{s,i})$.

The equilibrium vapor pressure may be obtained by finding the extrema for the thermodynamic potential from (4.2) as

$$\ln S_{\mathrm{i}} = \frac{(2/r)\,\sigma_{\mathrm{IV}} + C\epsilon^2}{N_{\mathrm{I}}\,kT}. \tag{4.3}$$

Note that the strain within the ice lattice serves to increase the equilibrium vapor pressure, whereas dislocations only affect the contact angle, which does not affect the vapor pressure.

4.2.2. *The nucleation rate*

The nucleation rate may be determined in the manner in which (3.43) was derived. The concentration of critical embryos follows from the Boltzmann distribution, evaluating the thermodynamic potential (4.2) at r^* using the equilibrium relation (4.3). The activation energy then becomes

$$(\Delta Y)_g = \frac{f(\hat{m},\,x)\;16\pi\sigma_{\mathrm{IV}}^3}{3[N_{\mathrm{I}}\,kT\,\ln S_{\mathrm{i}} - C\epsilon^2]^2}.$$

The addition of water molecules to the ice embryo is similar to the surface migration and incorporation discussed for condensation on insoluble nuclei. The pre-exponential factor is likewise similar to that for condensation, and may be estimated as $4\pi r_o^2 \times 10^{26}$. The nucleation rate for the formation of supercritical ice embryos on a spherical nucleus then follows as

$$J = 4\pi r_o^2 \times 10^{26}\exp\left[\frac{-f(\hat{m},\,x)\;(16/3)\pi\sigma_{\mathrm{IV}}^3}{kT\;(N_{\mathrm{I}}\,kT\;\ln S_{\mathrm{i}} - C\epsilon^2)^2}\right]. \tag{4.4}$$

[3] This is the common observation. If the prism face were in contact with the substrate, one would use the ice lattice vector *c*.

4.2.3. *Deposition nucleation behavior*

Deposition nucleation is dependent on both the temperature and ice supersaturation. However, the ice supersaturation dominates in the same manner as the water supersaturation for condensation. If one were to plot the ice supersaturation required to produce a nucleation rate of one event per nucleus per second, one would obtain curves similar to those given by Fig. 3.10. Such curves are only slightly dependent on temperature (see Problem 4.1).

In general, the initial formation of ice in the atmosphere occurs in supercooled water clouds. Thus, water saturation is a natural reference point for describing ice nucleation behavior. The ice supersaturation at water saturation increases nearly linearly with decreasing temperature (see Problem 4.2). Thus, Fig. 4.4, which shows the temperature required to produce a nucleation rate of one event per nucleus per second, is equivalent to Fig. 3.10. Clearly, deposition nucleation is unlikely to occur on particles much smaller than 0.1 μm radius and with contact parameters much less than 1.

Describing deposition behavior in terms of a temperature dependency at water saturation has led to the erroneous belief that deposition nucleation is primarily a temperature dependent phenomenon, and thus contrasts with condensation, which is supersaturation dependent. If one conducts an experiment in a chamber in which the ice saturation ratio can be controlled independent of temperature, an ice supersaturation dependence will be observed. If one observes deposition nucleation in a chamber maintained at water saturation (in a water cloud), a *temperature dependency* will be observed.

The nucleating behavior shown in Fig. 4.4 excludes the effects of elastic strain within the ice lattice. Lattice strain can be expected to increase the activation energy and decrease the nucleation rate. This effect can be interpreted as a depression of the nucleation threshold, with an elastic strain ϵ of 1% shifting the nucleation curves in Fig. 4.4 by about 1.5°C toward colder temperatures. An elastic strain of 2% would result in a 6°C shift (Turnbull and Vonnegut, 1952). The ice lattice is not likely to deform more than 2% since the contact parameter $\hat{m}$ would tend to decrease due to dislocations at the ice–substrate interface.

4.3. Homogeneous freezing nucleation

Freezing nucleation differs fundamentally from deposition or condensation nucleation in that the embryo is closely surrounded by molecules of the parent phase. Nucleation from the vapor assumes that the embryo is relatively isolated, gaining molecules by random collisions from the vapor. The ice embryo in supercooled water finds itself in close proximity to large numbers of water molecules, which do not have to 'collide' with the embryo in order to become incorporated into the embryonic ice lattice. Instead, molecules of the parent phase must orient themselves in the proper direction for bonding to occur and then must approach to the proper bonding distance. Since a considerable degree of order is present in the liquid, this generally involves the breaking of hydrogen bonds.

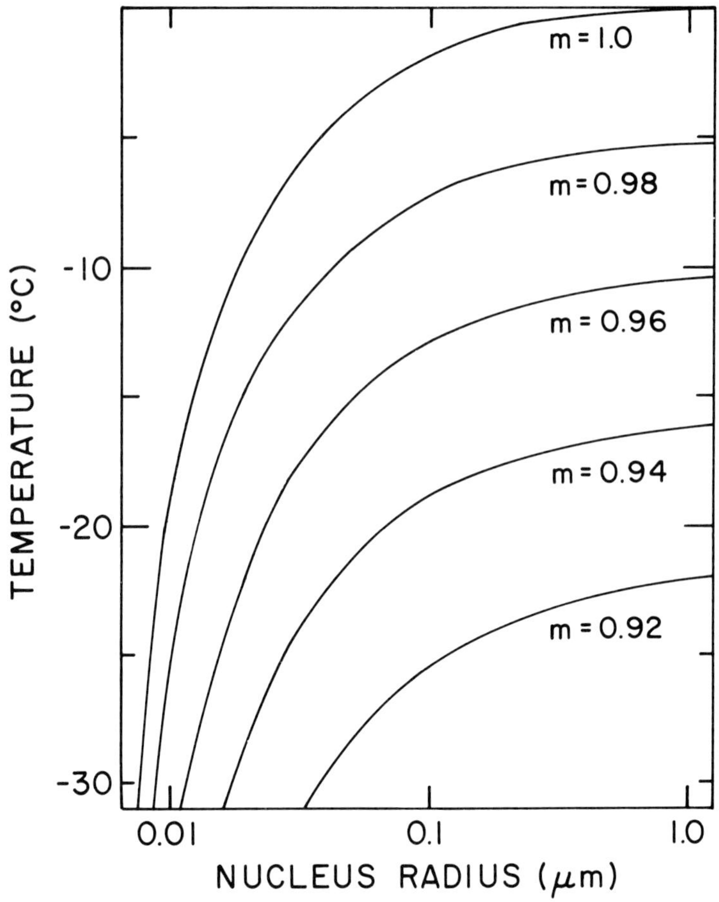

FIG. 4.4. The size dependency of deposition nucleation for various values of the contact parameter ($\widehat{m}_D$), based on (4.4). The ordinate is the temperature at water saturation required to give a nucleation rate of one event per second per nucleus.

The requirements that a molecule must meet in order to become incorporated into the ice embryo impose an energy barrier for the addition of a molecule to the embryonic ice lattice. This implies that only a fraction of the parent molecules in contact with the embryo have sufficient energy to form a bond with the embryo. This fraction is given by

$$\exp\left[\frac{-(\Delta g)}{kT}\right], \tag{4.5}$$

where Δg is the activation energy (of self-diffusion) which must be overcome in order to add a molecule to the embryo. This activation energy is on the order of the energy required to break one hydrogen bond. Based on the data of Hallett (1963) and Hagen et al. (1981), Δg can be approximated as $(3.60 - 0.073\,T) \times 10^{-13}$ erg where T is in °C.

Freezing nucleation also differs from condensation or deposition nucleation in that the change in bulk free energy during freezing is dependent only on temperature. The change in bulk free energy during condensation (or deposition) is determined by the ratio of the given vapor pressure to that at saturation over water (or ice). The change in bulk free energy during freezing is determined by the ratio of the saturation vapor pressure over water to that over ice. Since the saturation vapor pressures over water and ice are uniquely defined functions of temperature,[4] freezing nucleation is solely temperature dependent (see Problem 4.2).

4.3.1. *The nucleation rate*

The change in thermodynamic potential for an ice embryo forming within supercooled water is essentially that for an ice embryo forming in the vapor as given by (4.1) except that the vapor pressure e is replaced by the saturation vapor pressure over water ($e_{s,w}$). Thus,

$$(\Delta Y)_g = -\frac{4}{3}\pi r^3 N_I\, kT \ln\left(\frac{e_{s,w}}{e_{s,i}}\right) + 4\pi r^2 \sigma_{IL}\,. \tag{4.6}$$

The size of the critical embryo may be derived by finding the extrema of the thermodynamic potential from (4.6). The radius of the critical embryo follows as

$$r^* = \frac{2\sigma_{IL}}{N_I\, kT \ln\left(\dfrac{e_{s,w}}{e_{s,i}}\right)}\,. \tag{4.7}$$

In Chapter 3, the general form of the nucleation rate (3.42) was shown to be

$$J = 2N_{g*}\, A_{g*}\, \vec{\omega}_{g*} \left(\frac{\chi}{\pi}\right)^{1/2}.$$

Evaluation of the homogeneous freezing rate requires that each of these factors be evaluated.

The nontransient concentration of critical clusters N_{g*} is given by the Boltzmann distribution (3.16). The number concentration of unassociated molecules is taken to be the number density of water molecules N_L. Unlike the

[4] This ignores a very slight dependence of the saturation vapor pressure on the total atmospheric pressure.

vapor, this assumption for the liquid phase is probably rather poor since the average cluster at 0°C contains 90 molecules and only 22% of the molecules are unassociated (see discussion in Chapter 2). However, this is part of the pre-exponential factor and it may be anticipated that such errors are not consequential.

The free energy of formation of a critical cluster $(\delta Y)_{g*}$ is obtained by substituting $r*$ from (4.7) into the general thermodynamic potential given by (4.6). This yields

$$N_{g*} = N_L \exp\left\{\frac{-16\sigma_{IL}^3}{3\,(kT)^3\left[N_I \ln\left(\frac{e_{s,w}}{e_{s,i}}\right)\right]^2}\right\}. \tag{4.8}$$

The area of a critical cluster A_{g*} follows in a similar manner, substituting $r*$ from (4.7) into the surface area of a spherical embryo as

$$A_{g*} = \frac{16\pi\sigma_{IL}^2}{\left[N_I\, kT \ln\left(\frac{e_{s,w}}{e_{s,i}}\right)\right]^2}. \tag{4.9}$$

The factor χ follows from its definition (3.22) as

$$\chi = \left(\frac{4\pi}{3}\right)^{1/3} \frac{\sigma_{IL}}{3kTN_I^{2/3}}\, g^{*4/3}. \tag{4.10}$$

The rate at which molecules are added to the critical embryo per unit surface area $(\vec{\omega}_{g*})$ is determined by the number of molecules in contact with the embryo which possess sufficient energy to bond with the embryonic ice lattice, and the frequency with which they may do so. The latter is given by kT/h from kinetic theory, where h is Planck's constant. The former follows from the fraction of molecules which possess sufficient energy to bond (4.8) multiplied by the total number of molecules in contact with the embryo.

The number of molecules in contact with a unit surface area of the ice embryo may be estimated as follows. Assume that the molecules in contact with the embryo fill a concentric shell about the embryo with a thickness equal to the diameter of the molecules, as shown in Fig. 4.5. The volume of this concentric shell is

$$\Delta V = \frac{4\pi}{3}\left[(r + 2r_o)^3 - r^3\right],$$

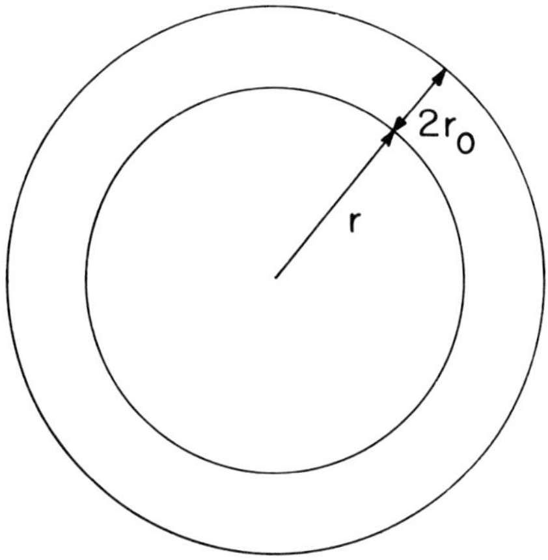

FIG. 4.5. The number of molecules of radius r_o that may surround a sphere of radius r is approximated by dividing the volume of the shell between r and $r + 2r_o$ by the volume occupied by one molecule.

and the total number of molecules contained within this shell is $N_L \Delta V$. To obtain the number of molecules in contact with a unit area of the embryo ξ, simply divide $N_L \Delta V$ by the area of the embryo,

$$\xi = \frac{N_L \ \Delta V}{4\pi r^2} = 2N_L \ r_o \left[1 + \frac{2r_o}{r} + \frac{4r_o^2}{3r^2} \right].$$

Since the embryo is generally much larger than a molecule, this reduces to $\xi = 2N_L r_o$, which, for water, is roughly 1.3×10^{15} molecules cm^{-2}.

The rate at which water molecules incorporate themselves into the ice embryo follows as

$$\vec{\omega}_{g*} = \left(\frac{\xi kT}{h} \right) \exp \left(\frac{-\Delta g}{kT} \right). \tag{4.11}$$

The homogeneous freezing nucleation rate follows from (3.42), substituting from (4.8)–(4.11). The evaluation of $g*^{-4/3}$ requires definition of g as a function of radius and then evaluating $r*$ using (4.7). The algebra is tedious but straightforward. The nucleation rate is

$$J = \frac{N_L}{N_I} \left(\frac{4\xi}{h} \right) (kT\sigma_{IL})^{1/2} \exp \left(\frac{-\Delta g}{kT} \right) \exp \left\{ \frac{-16\pi\sigma_{IL}^3}{3\ (kT)^3 \left[N_I \ln \left(\frac{e_{s,w}}{e_{s,i}} \right) \right]^2} \right\}. \tag{4.12}$$

Substituting appropriate values for the various terms in the pre-exponential factor, the homogeneous freezing rate may be approximated as

$$J = 1 \times 10^{28} \exp\left\{\frac{-16\pi\sigma_{\mathrm{IL}}^3}{3\,(kT)^3\left[N_{\mathrm{I}} \ln\left(\dfrac{e_{\mathrm{s,w}}}{e_{\mathrm{s,i}}}\right)\right]^2}\right\}. \qquad (4.13)$$

4.3.2. *Comparison between theory and observation*

Evaluation of the homogeneous freezing rate is strongly dependent on the value chosen for the water–ice interfacial energy σ_{IL}. This parameter is difficult to measure directly, and estimates of the proper value for the interfacial energy range from 10 to 30 erg cm^{-2}. Ketcham and Hobbs (1969) measured σ_{IL} at 0°C and report a value of 33 ± 3 erg cm^{-2}. Since σ_{IL} enters as the cube in the exponential term, the nucleation rate is highly sensitive to the particular value selected. In practice, the appropriate value is often deduced from homogeneous freezing experiments which render any direct comparisons between theory and observation unconvincing.

Observations of homogeneous freezing require that the water be highly purified and that extreme care be taken to remove any affects the container or support may have on the freezing behavior of the sample. The suspension of small droplets in air is relatively straightforward; the suspension of larger samples of water may be achieved by resting the drops on metal or glass plates coated with a hydrophobic film or by suspending the drops between two immiscible liquids which are both immiscible with water. Since contamination is always possible and would tend to induce freezing at temperatures warmer than the homogeneous freezing temperature, experimental results representing the coldest temperatures reported were selected from Fletcher (1970a) as representing homogeneous freezing. These results are presented in Fig. 4.6.

The probability that a given drop will freeze is directly related to its volume. Given a drop of volume V, the probability the drop will freeze in t seconds is

$$\Pr\,(\text{freezing event}) = 1 - \exp\,(-JVt),$$

where J is the rate of formation of supercritical embryos per cm^3 per second. Thus, the probability of a larger volume of water freezing is greater than for a smaller drop at the same temperature. Pruppacher and Klett (1978) extend this treatment for a constant cooling rate and show that the temperature threshold for homogeneous freezing increases as the cooling rate is decreased.

One may use (4.12) to estimate the ice–water interfacial energy σ_{IL} by evaluating the homogeneous freezing rate for various values of σ_{IL}. Using a value of 10^{28} for the pre-exponential factor, the value of 23.5 erg cm^{-2} seems to fit best the data presented in Fig. 4.6. This 'theoretical' threshold for homogeneous freezing is also shown in Fig. 4.6, evaluated for that temperature at which the probability of freezing (the product of J and V) is equal to one event

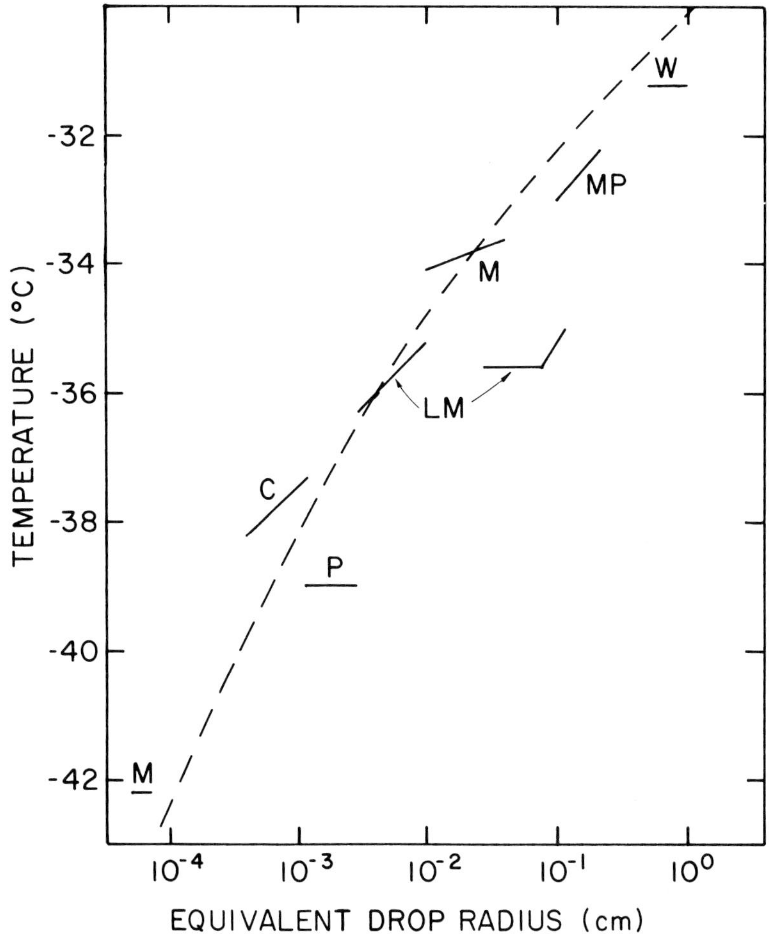

FIG. 4.6. Comparison between the theoretical and observed homogeneous freezing nucleation rates. The broken curve represents the solution of (4.12) with the ice–water interfacial energy assumed to be 23.5 erg cm^{-2}. The observed values are taken from Fletcher (1970a), representing Carte (C), Langham and Mason (LM), Mossop (M), Meyer and Pfaff (MP), Pound (P), and Wylie (W).

per second. The degree of similarity between the theory and observations in Fig. 4.6 should not be construed as verification of the theory.

The sizes of the critical embryos involved in homogeneous freezing are relatively small. From (4.7), a critical embryo at −40°C contains about 250 water molecules. The model of liquid water discussed in Chapter 2 suggests an average cluster size on the order of 100 molecules near 0°C. Since the appearance of a supercritical cluster is necessarily a rare event, even at −40°C, it seems plausible that the majority of such clusters are not ice-like. This conclusion is supported by Stillinger and Rahman (1972), whose molecular dynamics model for liquid water failed to find Ice Ih structures for temperatures as low as −8°C.

4.4. Immersion freezing nucleation

Water generally freezes at temperatures much warmer than those required for homogeneous freezing. In fact, it is generally difficult to cool large water drops below −20°C unless precautions are taken to remove particles from the water and to isolate the drop from any effects its support may have on its freezing behavior. The effect of dissolved salts on the freezing behavior is generally to depress the freezing threshold by the same amount that the melting point is depressed for bulk solutions (Hoffer, 1961). However, the presence of solid particles can have a marked effect on freezing nucleation.

The formation of ice embryos on an insoluble spherical substrate within bulk supercooled water involves the concepts of lattice strain introduced for deposition nucleation, and the orientation and bonding requirements for incorporation of molecules into the ice embryo introduced for homogeneous freezing. Unfortunately, the required contact parameters for potential freezing nuclei are not known, and a direct comparison between theory and observation is not possible. The theoretical treatment may still provide some insights into the requisites for effective immersion freezing nuclei.

4.4.1. *The nucleation rate*

The derivation of the nucleation rate for immersion freezing closely follows the treatments for homogeneous freezing and deposition on insoluble nuclei. The thermodynamic potential for a spherical cap ice embryo with radius of curvature r forming on a spherical nucleus of insoluble material with contact parameter $\hat{m}$ and radius r_n is quite similar to that for heterogeneous deposition given by (4.2),

$$(\Delta Y)_g = f(\hat{m},x)\left\{\left(\frac{4\pi r^3}{3}\right)\left[-N_{\mathrm{I}}\,kT\ln\left(\frac{e_{\mathrm{s,w}}}{e_{\mathrm{s,i}}}\right)+C\epsilon^2\right]+4\pi r^2\,\sigma_{\mathrm{IL}}\right\}. \tag{4.14}$$

Again, $e_{\mathrm{s,w}}$ represents the equilibrium vapor pressure of the parent phase, noting that immersion freezing behavior is solely temperature dependent. The contact parameter $\hat{m}$ now refers to the liquid–ice–substrate system describing freezing. The radius of curvature for a critical embryo follows by determining the extrema for (4.14),

$$r^* = \frac{2\sigma_{\mathrm{IL}}}{N_{\mathrm{I}}\,kT\ln\left(\dfrac{e_{\mathrm{s,w}}}{e_{\mathrm{s,i}}}\right)-C\epsilon^2}. \tag{4.15}$$

The evaluation of the general expression for the nucleation rate (3.42) proceeds in much the same fashion as for homogeneous freezing. The rate at which molecules are incorporated into the critical embryo $\vec{\omega}_g$ is given by (4.11); the activation energy $(\Delta Y)_{g*}$ and χ must be multiplied by the geometrical factor given by (3.44). The area of the critical embryo is now the area of the ice–water interface. Fletcher's (1958) treatment allows this to be expressed as

$$A_{g*} = 4\pi r^2 \left\{ \frac{1}{2} \left[1 + \frac{(1 - \hat{m}x)}{y} \right] \right\}$$

where x and y have the same definitions as in (3.44).

The equilibrium concentration of critical embryos on the surface of the nucleus is given by the Boltzmann distribution (3.16). The surface density of unassociated molecules may be approximated as $\xi = 2N_L r_o$ according to the treatment in the previous section. Thus, the nontransient surface concentration of critical embryos is given by

$$N_{g*} = \xi \exp \left\{ \frac{-16\pi\sigma_{IL}^3 f(\hat{m},x)/3}{kT \left[N_I\, kT \ln \left(\dfrac{e_{s,w}}{e_{s,i}} \right) - C\epsilon^2 \right]^2} \right\}. \tag{4.16}$$

Substitution of these into the general nucleation expression yields a result very similar to (4.12),

$$J = \frac{4\xi^2}{hN_I} (kT\sigma_{IL})^{1/2} \exp\left(\frac{-\Delta g}{kT} \right) \frac{1 + (1 - \hat{m}x)/y}{2[f(\hat{m},x)]^{1/6}} \times$$

$$\exp \left\{ \frac{-16\pi\sigma_{IL}^3 f(\hat{m},x)/3}{kT \left[N_I\, kT \ln \left(\dfrac{e_{s,w}}{e_{s,i}} \right) - C\epsilon^2 \right]^2} \right\}. \tag{4.17}$$

where the nucleation rate is given per unit surface area on the nucleus. Considering the geometrical factors as being close to unity and choosing $\Delta g/kT$ for a temperature close to $-10°C$, this reduces to

$$J = 2 \times 10^{24}\, r_n^2 \exp \left[\frac{-\Delta Y)_{g*}}{kT} \right],$$

for a nucleus of radius r_n. For more exacting purposes, one should consider the temperature variation of $\delta g/kT$ which ranges over three orders of magnitude between 0°C and −40°C, and the geometrical factors in the pre-exponential term, which may become quite small for $\hat{m}$ close to 1.

4.4.2. *Immersion freezing nucleation behavior*

Examination of (4.17) for the nucleation rate allows one to deduce that a wider variety of substances should provide effective immersion freezing nuclei than those serving as effective deposition nuclei. Since freezing is dependent on the water–ice interfacial energy, which has a value near 23 erg cm^{-2}, the barrier to nucleation is considerably reduced. This reduces the supersaturation (supercooling) required for any given contact parameter.

Figure 4.7 presents the temperatures required to give a nucleation rate of one event per nucleus per second for a range of nuclei sizes and contact

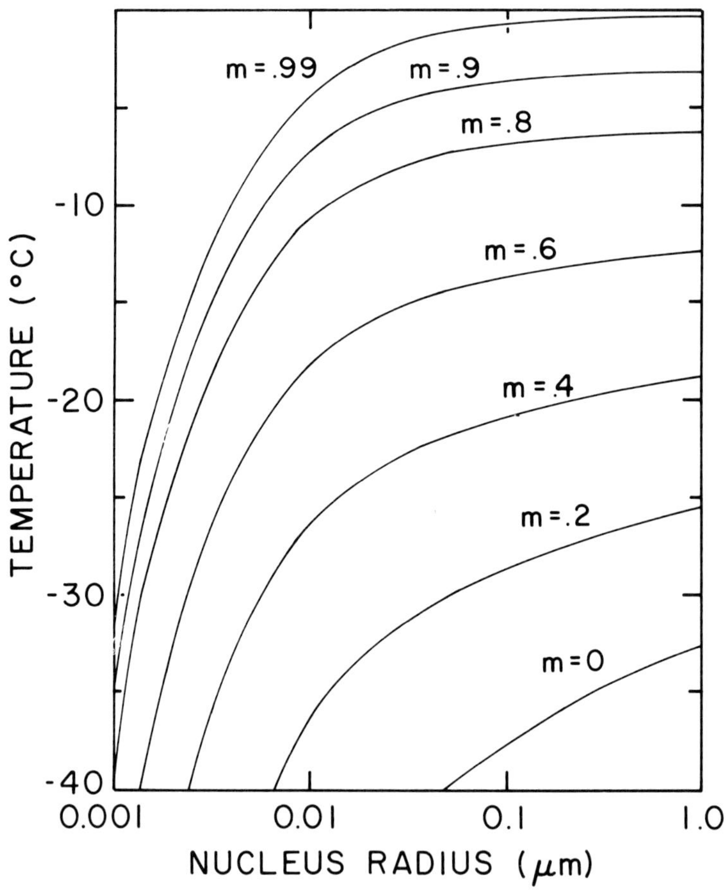

Fig. 4.7. The size dependency of immersion freezing nucleation for various values of the contact parameter ($\hat{m}_F$), based on (4.17). The ordinate represents the drop temperature required to give a nucleation rate of one event per second per nucleus.

parameters. Comparing Figs. 4.4 and 4.7, a contact parameter of 0.94 is required to nucleate ice on a nucleus of radius 0.1 μm at −20°C by deposition, whereas a contact parameter of 0.4 is required for the same size nucleus by immersion freezing. Although a given substance cannot be expected to have the same contact parameter for the two processes, it seems likely that the relatively restricted range of the contact parameter for efficient deposition nuclei will be met by relatively fewer substances.

A further consequence of the lower interfacial energy for immersion freezing nucleation is that the sizes of the critical embryos are much smaller than for deposition, according to (4.15). Since the geometrical factor scales as r_n/r^*, this implies that effective immersion freezing nuclei may be smaller than deposition nuclei for the same values of the contact parameter. In other words, a nucleus that is too small to be an efficient deposition nucleus may serve as an effective immersion freezing nucleus. This relation also plays an important role in contact freezing nucleation.

4.5. Contact freezing nucleation

Gokhale and Goold (1968) demonstrate that freezing of droplets upon contact with a nucleus could occur at temperatures considerably higher than those at which the same nuclei could initiate ice particles by deposition or immersion freezing nucleation. Whereas the mechanism for contact nucleation remains somewhat controversial, the mechanism advanced by Cooper (1974)[5] is based on classic theory and appears to be supported by laboratory observations. In addition, Cooper's treatment lays the foundation for developing relationships between the four heterogeneous modes of ice nucleation; a necessary step in furthering our understanding of ice nucleation in the atmosphere.

Contact freezing nucleation can occur if the population of embryos on the nucleus in equilibrium with the vapor is such that no embryos of the critical size for deposition nucleation exist, but embryos equal to or larger than the critical size for immersion freezing nucleation are present. Thus, when the nucleus is flooded by water at the moment of contact, the newly supercritical embryos cause the drop to freeze. This is possible since the critical size for deposition nucleation is greater than that for immersion freezing nucleation.

The population of embryos on the surface of the nucleus at the moment of contact is assumed to be in equilibrium with the vapor at water saturation corresponding to the drop temperature. Cooper (1974) estimates that the time required to form a critical ice embryo is on the order of 10^{-4} seconds and suggests that this is sufficient to insure the above condition. This assumption has been questioned by Pruppacher and Klett (1978). Fukuta (1975a), on the other hand, argues that the ice embryos in equilibrium with the vapor will adjust too rapidly to permit freezing as they are flooded on contact.

[5] This mechanism was first suggested by Guenadiev (1972).

Several alternate mechanisms for contact freezing nucleation have been advanced. Fletcher (1970b) suggested that the etching of active sites when the nucleus is immersed in water could account for the observed difference in activity. Alternately, he suggested that when the nucleus is present as an aerosol in an environment near water saturation, any dissolved salts might lower the freezing point by as much as 10°C. When the nucleus is flooded, the salt concentration is rapidly reduced, permitting freezing at a higher temperature. Fukuta (1975b) suggests that a high energy zone is present due to the moving water–nucleus interface boundary. He postulates that this increases the available free energy, thereby promoting freezing, although the relation between the energy available from the difference in interfacial surface energies and the free energy involved in forming a supercritical ice embryo is not clear.

4.5.1. *The nucleation probability*

The population of ice embryos on the nucleus is given by the Boltzmann distribution, using the thermodynamic potential defined for deposition nucleation (4.2). The number of ice embryos containing g molecules (and having radius r) may be expressed as

$$N_g = \xi \exp \left\{ f(\hat{m}_{\mathrm{D}},x) \left[\frac{4}{3} \pi r^3 N_{\mathrm{I}} \ln \left(\frac{e_{\mathrm{s,w}}}{e_{\mathrm{s,i}}} \right) - \frac{4\pi r^2 \sigma_{\mathrm{IV}}}{kT} \right] \right\}, \qquad (4.18)$$

ignoring the effects of elastic strain in the ice lattice. It has been assumed that a monolayer of water molecules has been adsorbed onto the nucleus, and that virtually none of these adsorbed molecules are included within the ice embryos. Thus, the surface density of unincorporated molecules is roughly ξ (~1.3×10^{15} molecules cm^{-2}).

The number of embryos equal to or greater than the critical size for immersion freezing will be approximated as simply the number equal to the critical size. The number of embryos of a given size decreases fairly rapidly with increasing size since the embryos involved are considerably smaller than the critical size for deposition. Since there is a tendency for the embryos to lose molecules immediately following contact with water, some embryos that are slightly supercritical may actually lose water molecules rapidly enough to fail to nucleate the ice phase. In addition, the number of single molecules per unit area on the nucleus surface has been overestimated by assuming that this is given by ξ. These factors tend to balance each other.

The number of embryos exceeding the critical size may then be approximated by evaluating (4.18) at r^* as given by (4.15),

$$N_{g*} = \xi \exp \left\{ \frac{-16\pi\sigma_{\mathrm{IL}}^3 f(\hat{m}_{\mathrm{D}},x)\,(\sigma_{\mathrm{IV}} - 2\sigma_{\mathrm{IL}}/3}{kT \left[N_{\mathrm{I}}\, kT \ln \left(\dfrac{e_{\mathrm{s,w}}}{e_{\mathrm{s,i}}} \right) \right]^2} \right\}. \qquad (4.19)$$

Note that the contact parameter and critical radius for deposition must be used in the evaluation of the geometrical factor $f(\hat{m}_D, x)$. The expected number of supercritical embryos on a given nucleus is simply $4\pi r^2 N_{g*}$.

4.5.2. *Nucleus collection rates*

Contact freezing nucleation is governed by the processes that bring the nucleus into contact with the drop, in addition to the processes that determine whether the nucleus will contain an ice embryo of sufficient size to initiate freezing on contact. In the atmosphere, various processes serve to promote or suppress the collection of aerosol which may include potentially active contact freezing nuclei. In this section, three processes by which nuclei are collected by drops will be examined.

(i) Brownian collection. The random motion of aerosol particles under bombardment by air molecules results in a net transport in the direction of the smallest aerosol concentration. This applies to any class or subset of the aerosol population under consideration, including potentially active contact freezing nuclei.

The collection of aerosol particles of radius r_a and concentration n_a by a water drop of radius r_w is given by Pruppacher and Klett (1978) as

$$\left.\frac{1}{n_a}\frac{dN_a}{dt}\right|_{Br} = 4\pi r_w D_a f_a , \tag{4.20}$$

where D_a refers to the 'diffusivity' of the aerosol particles,[6]

$$D_a = \frac{kT_\infty (1 + N_{Kn})}{6 r_a \eta_\infty} \tag{4.21}$$

with N_{Kn} being the Knudsen number (the ratio of the mean free path for air molecules to the radius of the aerosol), η refering to the dynamic viscosity and the subscripts ∞ refering to conditions at an 'infinite' distance from the drop (see Chapter 5), i.e., ambient conditions. N is the number of aerosol particles collected; n_a is the concentration of aerosol particles (cm^{-3}).

The factor f_a in the Brownian collection rate corrects for the motion of the drop as it falls. These are given by Pruppacher and Klett as

$$f_a = 1 + 0.5\,(N_{Sc} N_{Re})^{1/3} \quad \text{for } N_{Re} \ll 1 ,$$
$$f_a = 1 + 0.3\, N_{Re}^{1/2} N_{Sc}^{1/3} \quad \text{for } N_{Re} \gg 1$$

where the Reynolds number (ratio of inertial to viscous forces) and the Schmidt number (ratio of viscosity to diffusivity) are given in Appendix 3. The effect of the factor f_a is to enhance the collection rate as the drop's motion carries it into

[6] From Young (1974).

regions of higher aerosol concentrations and away from the regions it has depleted.

(ii) Thermophoresis. Thermophoresis refers to the transport of aerosol in a temperature gradient, from warm to cold. This is commonly thought of as being due to a greater transfer of momentum by air molecules colliding from the warm side. However, such momentum transfer is manifest as pressure, and, since the pressures on both sides are equal, there can be no net transfer of momentum in this fashion.

Actually, there are two processes responsible for thermophoresis. First, there is a molecular transfer from cold to warm along the edge of the aerosol within one mean free path of its surface. Second, there is an exchange of momentum as 'cold' molecules strike a warmer surface (and vice versa, see Fig. 4.8a).

The net flux ff within one mean free path can be determined as follows. Consider the flux from cold to warm ff_1 and from warm to cold ff_2. The flux is simply the number of molecules times their velocity,or

$$\mathrm{ff}_1 = n_1 v_1$$

and

$$\mathrm{ff}_2 = n_2 v_2.$$

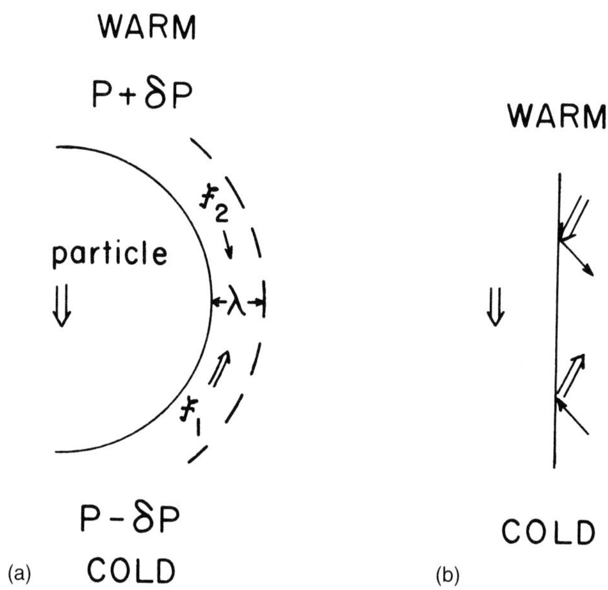

Fig. 4.8. Two mechanisms which result in thermophoresis. (a) The molecular fluxes within one mean free path of the surface. Since $\mathrm{ff}_1 > \mathrm{ff}_2$, there is a net flux from cold to warm, resulting in a pressure change (δP) which produces a hydrodynamic flow carrying the aerosol towards the cold side. (b) The exchange of momentum as particles arriving from the warm side lose momentum and particles arriving from the cold side gain momentum. The net result is to impart momentum to the aerosol, directed toward the cold side.

The kinetic energy per molecule in the direction of interest is $kT/3$ and the velocity follows as

$$v_1^2 = \frac{2kT_1}{3m},$$

where m is the mass of an air molecule. Thus, the fluxes are

$$\mathrm{ff}_1 = n_1 \left(\frac{2kT_1}{3m} \right)^{1/2}$$

and

$$\mathrm{ff}_2 = n_2 \left(\frac{2kT_2}{3m} \right)^{1/2}.$$

Since the total pressure is assumed to be constant,

$$P_1 = P_2 = n_1 kT_1 = n_2 kT_2 .$$

From this, the ratio of fluxes is

$$\frac{\mathrm{ff}_1}{\mathrm{ff}_2} = \left(\frac{T_2}{T_1} \right)^{1/2},$$

or, since $T_2 > T_1$, $\mathrm{ff}_1 > \mathrm{ff}_2$, i.e., there is a net flux from cold to warm.

This net flux from the cold to the warm side of an aerosol particle results in a pressure excess on the warm side of the aerosol. This pressure excess tends to push the aerosol toward colder temperatures.

The second process is shown in Fig. 4.8b. Molecules which arrive from the 'cold' side tend to pick up energy as they collide with the warmer surface and rebound with a greater velocity than they had when they arrived. The conservation of momentum requires that the surface is given an impetus in the opposite direction, i.e., toward colder temperatures. Similarly, molecules arriving from the 'warm' side tend to lose energy to the colder surface and impart momentum to the surface, again directed toward colder temperatures.

The collection rate due to thermophoresis is

$$\left. \frac{1}{n_a} \frac{dN_a}{dt} \right|_{\mathrm{Th}} = 4\pi r_w^2 f_t \frac{Q}{P}, \tag{4.22}$$

where P is the total pressure and Q is the heat flux [erg cm^{-2} s^{-1}] given by

$$Q = 4.185 \times 10^7 \left(\frac{K_{\mathrm{air}}}{r_w} \right) (T_\infty - T_w)\,(1 + 0.3\, N_{\mathrm{Re}}^{1/2}\, N_{\mathrm{Pr}}^{1/3}) .$$

where N_{Pr} is the Prandtl number (ratio of viscosity to conductivity). The thermophoretic factor f_t is given by Waldmann and Schmidt (1966) as

$$f_t = \frac{0.4\left[1 + 1.45N_{Kn} + 0.4N_{Kn}\exp\left(\frac{-1}{N_{Kn}}\right)\right](K_{air} + 2.5K_aN_{Kn})}{(1 + 3N_{Kn})(2K_{air} + 5K_aN_{Kn} + K_a)},$$

where K_a and K_{air} are the thermal conductivities for the aerosol and air, respectively. Waldmann and Schmidt show that this is a good approximation of thermophoretic transfer over most of the range of Knudsen number, although it may overestimate the transfer by as much as 30% for N_{Kn} near unity.

(iii) Diffusiophoresis. Diffusiophoresis refers to the transport of aerosol particles in association with a flux of a condensing vapor. The water vapor field surrounding a growing drop is such that the vapor pressure increases with increasing distance from the drop. Since the total pressure is assumed to be roughly constant, the pressure of the dry air must decrease with increasing distance from the drop (see Fig. 4.9).

Water vapor diffuses toward the drop, where it condenses on the drop. The air molecules also find themselves in a concentration gradient and diffuse away from the drop. Since there is no source of air molecules at the drop surface, the partial pressure of the dry air (and also the total pressure) decreases in the vicinity of the drop and increases further away from the drop. This sets up a pressure gradient, resulting in a hydrodynamic flow called Stefan flow. The Stefan flow is in the same direction as the diffusive flux of condensing water molecules.

The aerosol particle is carried with the Stefan flow, i.e., in the same direction as the diffusive flux of water molecules. It is incorrect to assume that the flux of water molecules is directly responsible for the movement of the

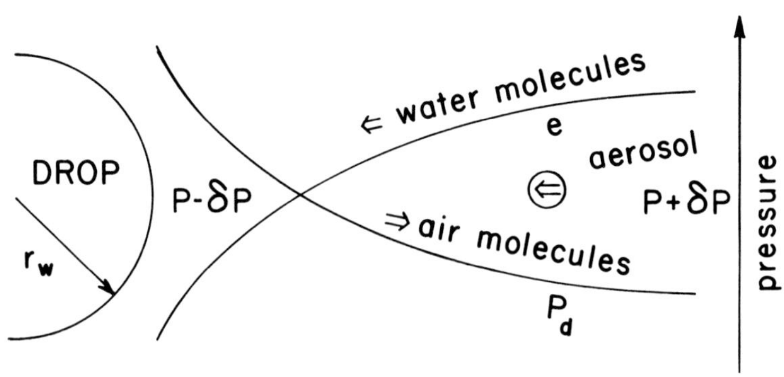

FIG. 4.9. The mechanism which results in diffusiophoresis. Since the concentration of water vapor molecules is lower closer to a growing drop and the total pressure is assumed constant, there is a concentration gradient of air molecules, decreasing away from the drop. This results in a diffusive flux of air molecules away from the drop. Since there is no source of air molecules at the drop surface, a hydrodynamic return flow (Stefan flow) carrying the aerosol particles is established.

aerosol particle. In addition to the Stefan flow, a mechanism similar to that depicted in Fig. 4.8b for thermophoresis, but due to the concentration gradients of molecules having different masses, contributes to diffusiophoresis.

The collection rate due to diffusiophoresis is

$$\frac{1}{n_a}\frac{dN_a}{dt}\bigg|_{Df} = 4\pi r_w^2\, g_d W \left[\frac{m_v^{1/2}}{n_v m_v^{1/2} + n_{air} m_{air}^{1/2}}\right], \tag{4.23}$$

where n and m refer to the number and mass of the vapor (v) and air molecules, respectively. W is the flux of water vapor, given by

$$W = N_L\left(\frac{D_w}{r_w}\right)(\rho_\infty - \rho_{s,w})\,(1 + 0.3 N_{Re}^{1/2} N_{Sc,w}^{1/3}),$$

where D_w is the diffusivity of water vapor in air and ρ refers to the water vapor density at the drop surface (s,w) and ambient (∞). Goldsmith and May (1966) find good experimental verification of (4.23) with $g_d = 1.0$.

(iv) Net collection rates. Of the three mechanisms for the collection of aerosols discussed here, only Brownian motion is strongly dependent on the size of the aerosol particles, the collection rate decreasing rapidly with increasing particle size. Diffusiophoresis is essentially independent of the aerosol size, whereas thermophoresis shows a slight tendency toward decreased rates with increasing aerosol size. Assuming that these rates are additive, it may be expected that Brownian collection dominates for relatively small aerosol particles and the phoretic processes are more important for the larger aerosol particles.

If attention is restricted to the collection of aerosol particles by freely falling water drops, the relationship between thermophoresis and diffusiophoresis is determined. Water drops are generally close to their equilibrium wet-bulb temperature, thereby fixing both the heat and vapor fluxes in (4.22) and (4.23). Under these conditions, the absolute magnitude of thermophoresis is generally greater than that for diffusiophoresis.

Figure 4.10 presents the individual and net collection rates for growing ($S = 1.003$) and evaporating ($S = 0.98$) drops of 10 μm radius. Note that thermophoresis tends to retard collection for growing drops and, for this drop radius, actually suppresses collection of nuclei larger than 0.15 μm radius. The net collection rates for growing drops as a function of the drop radius are shown in Fig. 4.11. This shows that growing drops smaller than 20 μm radius do not collect nuclei larger than radius 0.15 μm.

Leong et al. (1982) present results of laboratory measurements of the collection of aerosol particles by thermophoresis and diffusiophoresis, and find good agreement for larger aerosol sizes but the observed collection rate was greater than predicted for the smaller sizes. They also give a comprehensive summary of thermophoresis and diffusiophoresis, giving a more detailed theoretical treatment with laboratory observations of the various parameters that affect the phoretic forces.

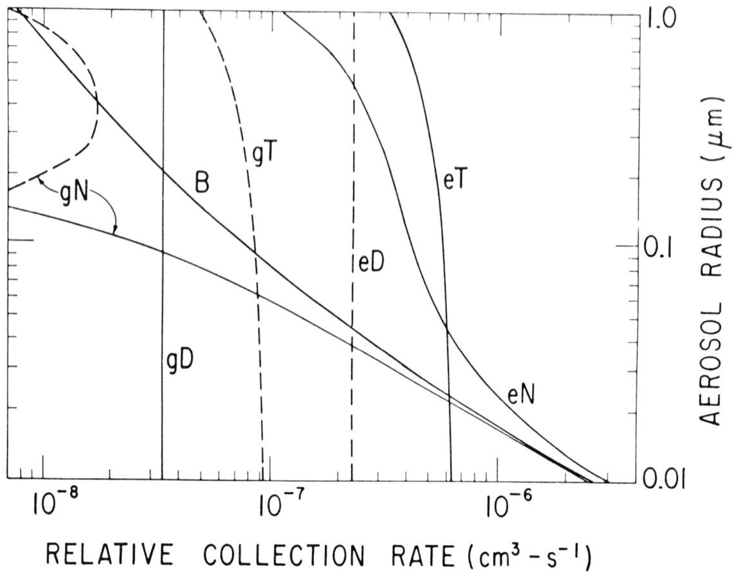

Fig. 4.10. Relative collection rates for Brownian motion (B), thermophoresis (T), diffusiophoresis (D), and the net collection (N) as a function of the aerosol radius. Growing drops (g) are at 0.3% supersaturation; evaporating drops (e) are at 2% subsaturation. The drop radius is 10 μm. Ambient conditions are P = 600 mbar and T = −5°C. From Young (1974). *Journal of Atmospheric Sciences*, **31**, 3. American Meteorological Society. Reproduced with permission.

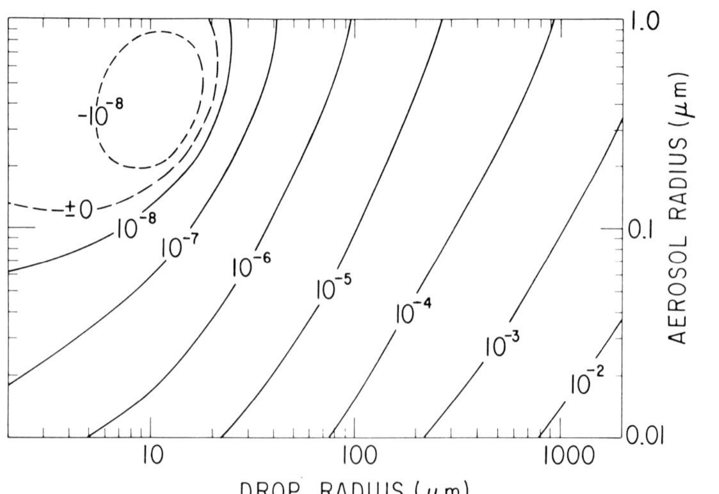

Fig. 4.11. Relative net collection rates as functions of the aerosol and drop radii for growing drops. Broken isopleths indicate negative values, i.e., no collection. The growing drops are at 0.3% supersaturation. The ambient conditions are P = 600 mbar and T = −5°C. From Young (1974). *Journal of Atmospheric Sciences*, **31**, 3. American Meteorological Society. Reproduced with permission.

4.5.3. *Evidence for contact freezing nucleation in clouds*

Calculations such as those presented in Figs. 4.10 and 4.11 indicate that contact freezing nucleation should be reduced or suppressed in updrafts (since updrafts generally contain small droplets which are growing under supersaturated conditions), and enhanced in downdrafts and at cloud boundaries, where drops are evaporating, and enhanced when large drops are present. Young (1974) summarizes a number of observations which tend to support this hypothesis, although it should not be construed that this is the only hypothesis which could explain the cited observations.

Koenig (1963) summarizes his and other observations, and concludes that ice in cumulus clouds is most likely found in regions of downdrafts and at cloud edges. He notes that ice is more likely to form when large drops are present and that the size spectrum of the ice particles tends to resemble the size spectrum of the drops. He also notes that large drops are more likely to be frozen than small drops.

More recently, Cooper and Vali (1981) suggested that contact freezing nucleation was responsible for their observations of ice formation at the leading edge of mountain cap clouds, ruling out deposition nucleation and secondary ice production. Hobbs and Rangno (1985) note the observation of high ice crystal concentrations in cumulus clouds shortly after the clouds had reached their maximum vertical extent. They conclude that the mixing of ambient air into the cloud top results in subsaturated conditions, promoting contact freezing nucleation. Rangno and Hobbs (1991) describe two stages of ice enhancement, as shown in Fig. 4.12. They postulate that the first stage is due to contact freezing nucleation.

4.6. Condensation freezing nucleation

The idea that condensation followed almost immediately by freezing is a distinct mode of ice nucleation is fairly recent. As such, the theory is not well developed. Two similar mechanisms for condensation freezing will be examined: condensation followed by immersion freezing, and condensation followed by contact freezing.

4.6.1. *Condensation followed by immersion freezing*

Fukuta and Schaller (1982) propose that heterogeneous condensation on aerosol particles immerses the nucleus, forming a droplet. Provided that the temperatures are sufficiently cold, ice embryos then form on the nucleus and may initiate freezing via the immersion freezing mechanism. Thus, the probability of a nucleus initiating an ice particle by condensation freezing may be expressed as the product of the probability of condensation (Pr_c) times the conditional probability of immersion freezing given condensation on the nucleus (Pr'_{if}).

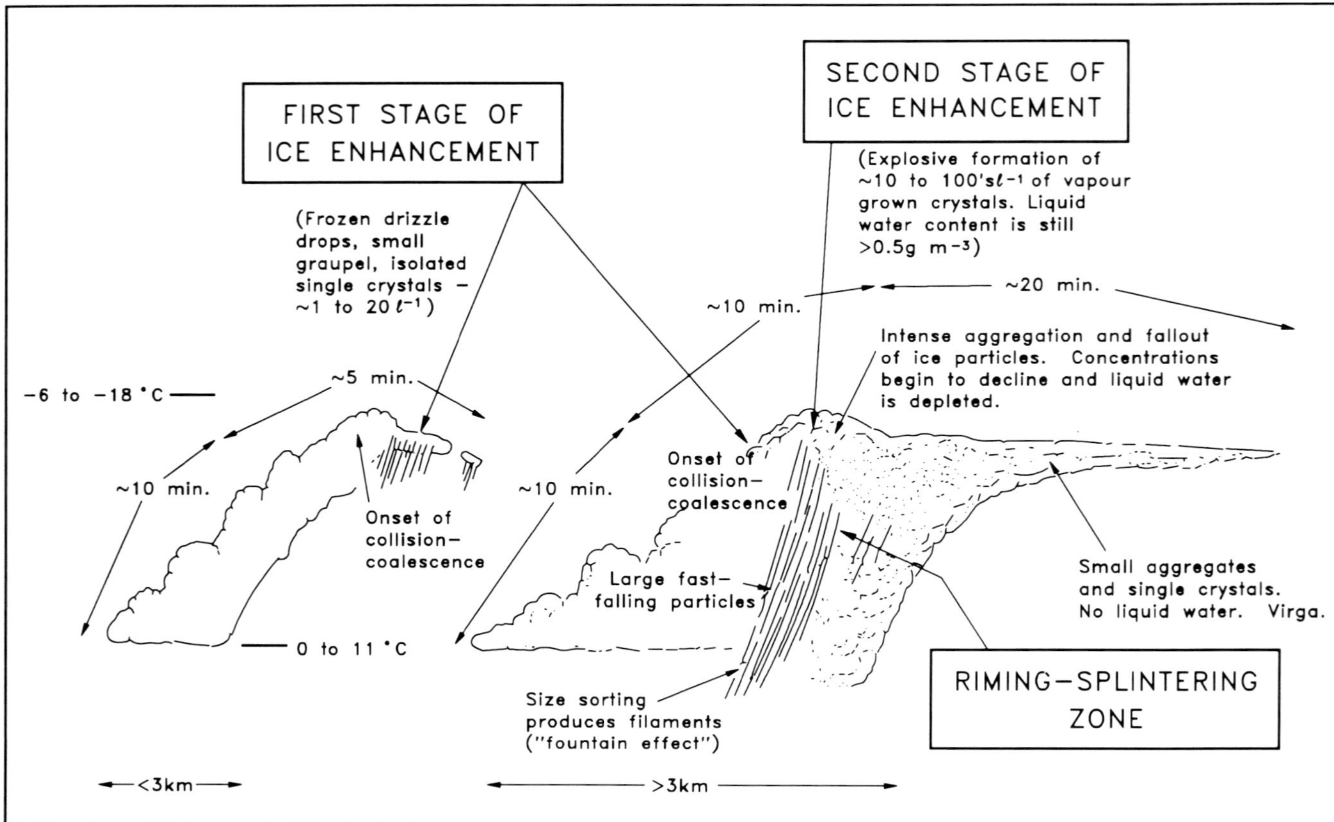

Fig. 4.12. Schematic diagram of observations and speculations regarding the formation of ice in small polar maritime cumuliform clouds. The first stage of ice enhancement is hypothesized to result from contact freezing nucleation; the second stage from condensation freezing nucleation. From Rangno and Hobbs (1991).

The probability of condensation may be determined from the condensation nucleation rate (J_c from (3.45)) as follows. Given the number of nuclei (N_o), the rate at which condensation occurs is

$$dN_c = (N_o - N_c)\, J_c dt\,.$$

Integrating this, the number of nuclei on which condensation has occurred is

$$N_c = N_o\,[1 - \exp\,(-J_c t)\,.$$

Similarly, if N_c is constant, the number of nuclei on which immersion freezing would occur is

$$N_{if} = N_c\,[1 - \exp\,(-J_{if} t)]\,.$$

The probability of condensation freezing (Pr_{cf}) then follows as

$$\mathrm{Pr}_{cf} = \frac{N'_{if}}{N_o} = \frac{N_c}{N_o}\frac{N'_{if}}{N_c} = \mathrm{Pr}_c \mathrm{Pr}'_{if}\,,$$

where N'_{if} incorporates the time dependence of N_c. The solution for N'_{if} and the probability of condensation freezing are given by Fukuta and Schaller (1982) as follows:

$$\mathrm{Pr}_{cf} = \frac{N'_{if}}{N_o} = 1 + \frac{1}{J_c - J_{if}}\,[J_{if} \exp\,(-J_c t) - J_c \exp\,(-J_{if} t)]\,.$$

For sufficiently small times (t) or small $J_c t$ and $J_{if} t$, then

$$\mathrm{Pr}_{cf} = \frac{1}{2}\, J_c J_{if} t^2\,. \tag{4.24}$$

4.6.2. *Condensation followed by contact freezing*

An alternate hypothesis is that populations of water and ice embryos coexist on the surface of the nucleus. Condensation freezing occurs when a water embryo becomes supercritical, flooding the nucleus with water. Should any of the ice embryos be larger than the critical size for immersion freezing, ice nucleation would occur in a manner quite similar to that proposed for contact freezing nucleation. This suggests that condensation freezing is a *self-contact* freezing nucleation.

The probability of condensation freezing is the product of the probability of condensation (Pr_c) and the probability that a given nucleus will contain a supercritical embryo for immersion freezing. The expected number of ice embryos exceeding the critical size for immersion freezing is given by (4.19). If

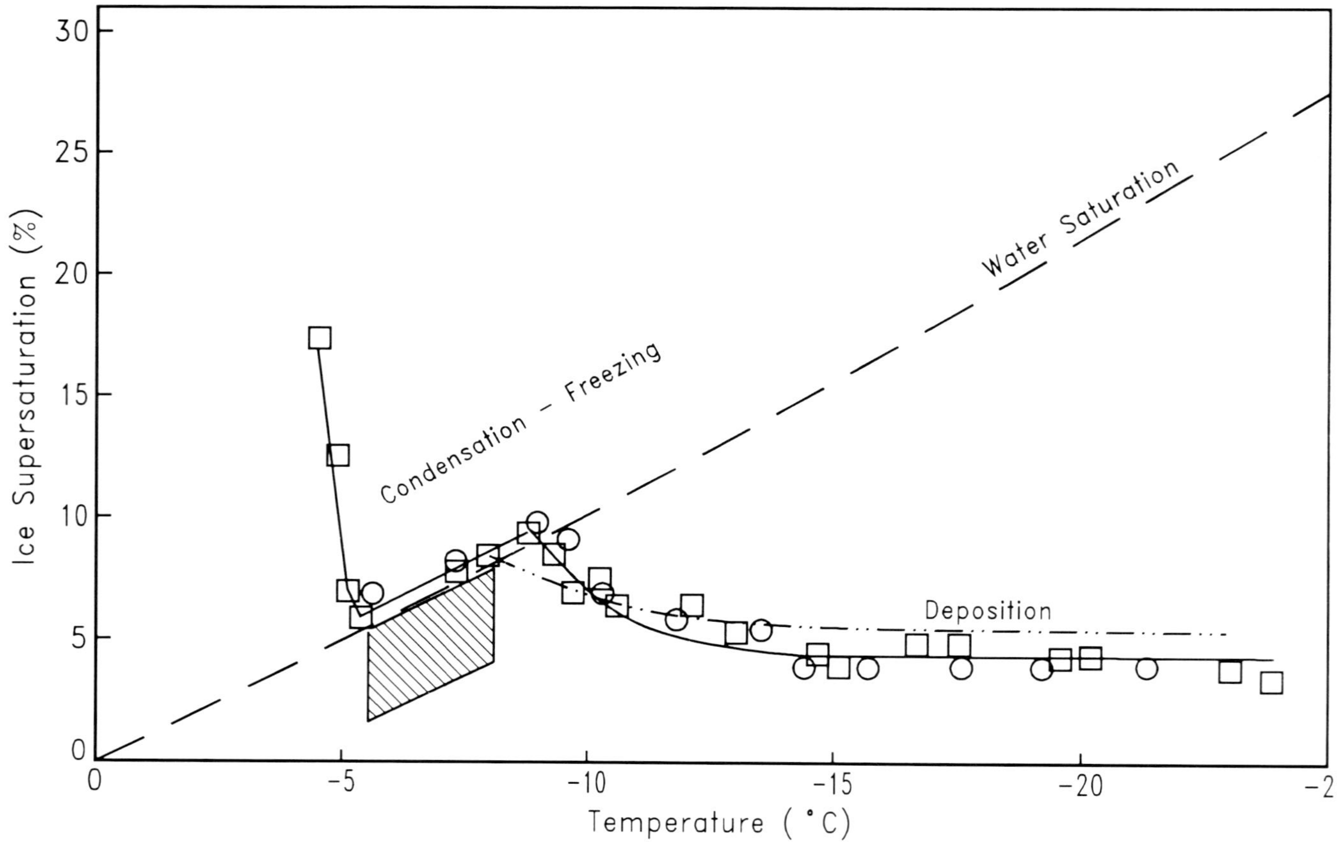

Fig. 4.13. Threshold curves for ice nucleation by silver iodide. The hatched region represents observations of contact freezing nucleation. The threshold for condensation freezing nucleation near water saturation is noted to be virtually identical to the contact freezing nucleation threshold. (□), Prepared in CCN-free air; (○), prepared in room air. From Schaller and Fukuta (1979). *Journal of Atmospheric Sciences*, **36**, 9. American Meteorological Society. Reproduced

the expected number of supercritical embryos per nucleus (N_{crit}) is much smaller than 1, then

$$Pr_{cf} = Pr_c N_{crit} .$$

If N_{crit} is not much smaller than 1, then

$$Pr_{cf} = Pr_c \, [1 - \exp(-N_{crit})] .$$

4.6.3. *Condensation freezing nucleation behavior*

These two proposed mechanisms for condensation freezing nucleation may be evaluated in light of the available experimental evidence given by Schaller and Fukuta (1979) and shown in Fig. 4.13. They used a wedge-shaped ice thermal diffusion chamber through which a prepared aerosol sample may be passed. As the sample passes through the chamber, it encounters an increasing ice–water supersaturation. Observation of the location where ice crystals are first produced yields both the temperature and supersaturation.

For the silver iodide (AgI) aerosol analyzed in Fig. 4.13, three modes of ice nucleation are evident. At subsaturation with respect to water, deposition nucleation is noted for the colder temperatures. The hatched region indicates contact freezing nucleation, which was initiated on haze droplets formed from an ammonium chloride smoke introduced into the chamber. The ice nucleation at water supersaturations is considered to be condensation freezing since it was coincident with or following appearance of a water cloud, and the discussion in the previous section on aerosol collection rates would suggest suppression of contact freezing nucleation.

Schaller and Fukuta note that a small but distinct supersaturation with respect to water is required for condensation freezing nucleation. This is shown in Fig. 4.14, comparing deposition and condensation freezing behavior as a function of water saturation. Since the condensation freezing mode requires water supersaturation, the condensation freezing rate must be zero at water saturation, rising rapidly with water supersaturation. For temperatures in Fig. 4.13 where deposition nucleation is observed, one may expect a discontinuity in the nucleation rate just above water saturation as condensation freezing becomes dominant over deposition.

The classic theory would predict that the deposition activity thresholds depicted in Fig. 4.13 should be at a constant ice supersaturation, intersecting the water saturation curve close to −5°C. However, as the temperature is increased and the water saturation curve is approached, the deposition nucleation rate at a given ice supersaturation decreases. This behavior remains unexplained although it probably reflects the inability of the classical approach to handle high concentrations of water molecules adsorbed on the nucleus surface.

The condensation freezing threshold shown in Fig. 4.13 suggests that at higher water supersaturations, condensation freezing nucleation can occur at higher temperatures. Fukuta and Schaller (1982) explain this in terms of (4.24),

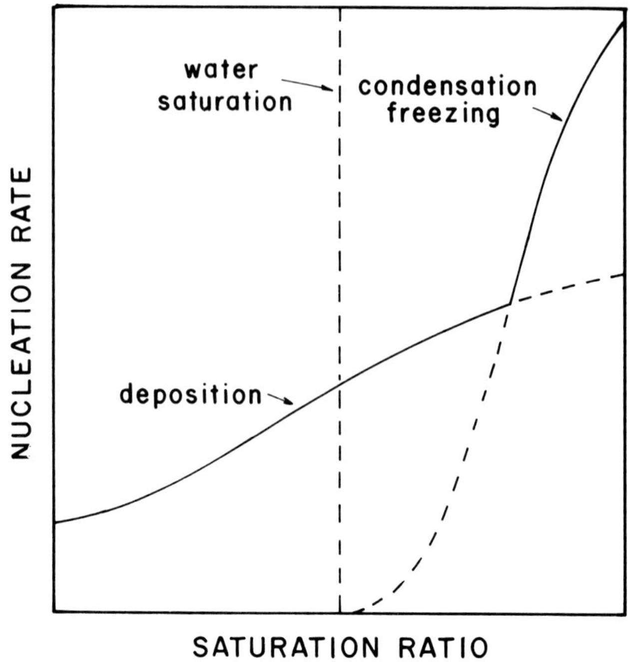

Fig. 4.14. A schematic representation of the relation between deposition and condensation freezing nucleation. The deposition nucleation rate exceeds that for condensation freezing nucleation below some value of water supersaturation, above which condensation freezing is dominant. From Fukuta and Schaller (1976).

which suggests that an inverse relationship between J_c and J_{if} exists along the nucleation threshold curve. At warmer temperatures, J_{if} decreases as the immersion freezing activity decreases. This must be compensated by an increase in J_c, which requires higher water supersaturations. Note that this behavior does not distinguish between the two proposed mechanisms for condensation freezing nucleation.

One key piece of information which does serve to distinguish the two proposed mechanisms for condensation freezing nucleation is the temperature threshold for condensation freezing. The temperature threshold for condensation freezing may be expected to parallel the temperature threshold for the ice nucleation step of the proposed mechanism. Table 4.1 presents temperature thresholds for several substances for the four ice nucleation modes. The close correspondence between contact freezing and condensation freezing nucleation thresholds, and the lack of correspondence between immersion freezing and condensation freezing nucleation thresholds argues strongly for the 'self-contact freezing' nucleation picture.

Rangno and Hobbs (1991) present evidence for ice production in natural clouds resulting from condensation freezing nucleation. They note a second stage

Table 4.1 Comparison of the ice nucleation thresholds for the different modes. Thresholds represent the warmest temperatures (°C) at which the given mode of nucleation has been observed. Care has been taken to insure that the experimental procedures were sensitive to only one particular mode of nucleation. Data are taken from Cooper (1974), Young (1974), and Schaller and Fukuta (1979).

Substance	Contact freezing	Condensation freezing	Deposition	Immersion freezing
Silver iodide	−3 to −5	−4	−8	−13 to −16
Cupric sulfide	−6	n/a	−13	−16
Lead iodide	−6	−7	−15	n/a
Cadmium iodide	−12	n/a	−21	n/a
Metaldehyde	−3	−2	−10	n/a
1,5-Dihydroxynaphlene	−6	−6	−12	n/a
Phloroglucinol	n/a	−5	−9	n/a
Kaolinite	−5 to −12	−10	−19	−32

of ice enhancement (shown in Fig. 4.12) which rapidly follows the growth of large particles by collision and coalescence. As will be discussed in Chapter 7, the conversion from high concentrations of small droplets to low concentrations of large drops is associated with a substantial increase in the ambient water supersaturation. Thus, conditions become favorable for condensation freezing nucleation, provided the ambient temperatures are below the condensation freezing nucleation threshold. Ice crystal concentrations were observed to increase from near zero to 350 l^{-1} within 9 minutes, with cloud top temperatures of −8°C.

4.7. Relationships between modes of ice nucleation

A common difficulty in comparing theoretical predictions with observations of ice nucleation is that the appropriate contact parameters cannot be defined independently with sufficient accuracy. However, following Cooper's (1974) treatment, both deposition and contact freezing nucleation may be assumed to be governed by the same contact parameter. Thus, it is possible to predict the relative spread between the threshold nucleation activity for these two modes.

The deposition and contact freezing nucleation temperature thresholds for a nucleus of radius 0.1 μm are shown as a function of the contact parameter for deposition nucleation ($\hat{m}_D$) in Fig. 4.15. Note that the supercooling[7] required for deposition activity is roughly 2.3 times that required for contact freezing activity. An examination of Table 4.1 suggests this relationship holds quite well.

As noted in the previous section, condensation freezing nucleation may be described as a self-contact process, with a nucleation threshold virtually identical

[7] Supercooling is defined as the depression of the temperature below 0°C, i.e., a temperature threshold of −10°C represent a supercooling of 10°C.

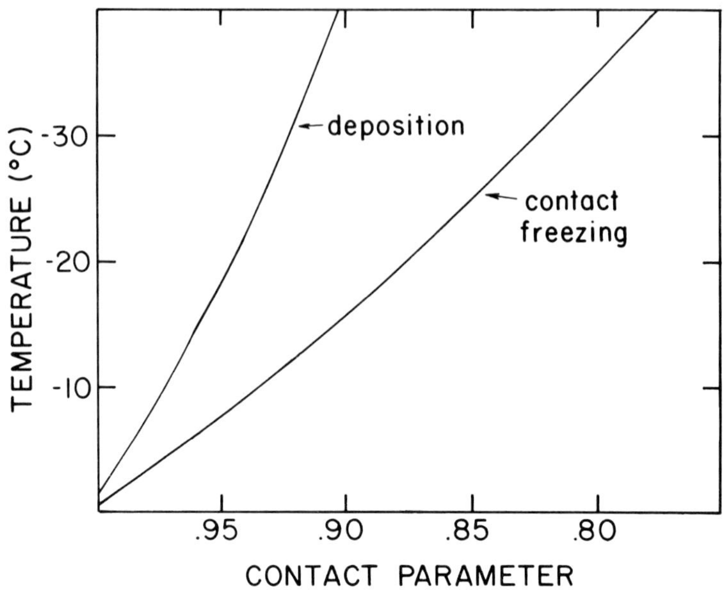

FIG. 4.15. Comparison of the nucleation thresholds for contact freezing and deposition nucleation at water saturation for a particle of radius 0.1 μm as a function of the contact parameter ($\hat{m}_D$).

to that for contact freezing nucleation. Again, the data presented in Table 4.1 support this contention.

A theoretical relationship between the contact parameters for deposition ($\hat{m}_D$) and immersion freezing ($\hat{m}_F$) nucleation has not been established. Most substances seem to yield immersion freezing thresholds at lower temperatures than their deposition thresholds. This implies that $\hat{m}_F$ must be much smaller than $\hat{m}_D$. Hoffer (1961) reports a deposition threshold at water saturation of -9°C for silver iodide and an immersion freezing threshold of -16°C. This suggests that $\hat{m}_D = 0.97$ and $\hat{m}_F = 0.4$.

Substances which are good deposition nuclei also appear to be good immersion freezing nuclei. The limited amount of experimental data presented in Table 4.1 tends to support this contention. The ratio of contact parameters for deposition and immersion freezing nucleation may be expressed as

$$\frac{\hat{m}_D}{\hat{m}_F} = \left(\frac{\sigma_{VN} - \sigma_{IN}}{\sigma_{LN} - \sigma_{IN}} \right) \frac{\sigma_{IL}}{\sigma_{IV}}.$$

If only substances exhibiting epitaxy with ice are considered, then σ_{IN} may be neglected. The ratio of contact parameters then reduces to

$$\frac{\hat{m}_D}{\hat{m}_F} \approx \frac{\sigma_{VN}}{\sigma_{LN}} \frac{\sigma_{IL}}{\sigma_{IV}}.$$

The ratio of the ice–water and ice–vapor interfacial energies is fixed with a value near 1/4. The observation that $\hat{m}_D$ tends to be much larger than $\hat{m}_F$ implies that the interfacial energies between the nucleus substrate and liquid water (σ_{LN}) must be very small. Since only epitaxial substances are being considered, it is not surprising that substances which exhibit a low interfacial energy for ice also exhibit a low interfacial energy for water. Thus, it may be concluded that substances which are relatively good ice nuclei will exhibit deposition activity at temperatures significantly warmer than those at which they exhibit immersion freezing behavior. Relatively poor ice nuclei are much less likely to exhibit such behavior.

4.8. Secondary ice production

The observed concentration of ice crystals within clouds at relatively warm temperatures has been found on numerous occasions to exceed greatly the measured concentration of ice nuclei active at the coldest temperatures within the cloud. Mossop (1970) summarizes a number of his observations in maritime cumuli off Tasmania; Koenig (1963) presents comparisons for summertime cumuli in Missouri. A number of other observations are cited by Young (1974). Hobbs (1969) points out that the observed discrepancy is greatest for the warmest temperatures, and counts of ice nuclei and ice crystals are in good agreement for temperatures colder than −20°C. Most of the observations which indicate large discrepancies were obtained in cumulus clouds and in the presence of large drops.

4.8.1. *The Hallett–Mossop mechanism*

A wide variety of possible 'ice multiplication' mechanisms has been advanced; Mossop (1985) presents a comprehensive summary of these mechanisms. The most likely mechanism for secondary ice production appears to be that ice splinters formed as droplets larger than 12 μm in radius are collected by a large graupel particle at temperatures between −3°C and −8°C, as advanced by Hallett and Mossop (1974). They show that an average of 60 splinters are produced for every milligram of rime ice formed under these conditions. This corresponds roughly to one splinter for every 200 accreted droplets.

The droplet size requirement and the narrow range of temperatures for which the Hallett–Mossop mechanism is active place rather stringent bounds on explanations for secondary ice production by this mechanism. The explanation advanced by Dong and Hallett (1989) appears to account for these limits. Thermal stress due to temperature differences between the rimer (the large graupel particle) and the surface of the newly accreted droplet after it has frozen may result in fracturing and splinter formation. The thermal stress is proportional to the base radius of the accreted droplet and the temperature difference between the base and top of the accreted droplet. The base may be assumed to be at the rimer temperature; the top may be assumed to be close to 0°C.

At warm temperatures, the temperature difference across the accreted droplet is small, resulting in little stress and no splinter formation. If the accreted droplet is too small, the base radius is too small and little stress results. As the temperature of the rimer particle decreases or the size of the accreting droplet increases, the thermal stress increases, resulting in splinter formation. This explains the warm temperature threshold near $-3°C$.

The cold temperature threshold is explained on the basis of the increased solubility of air in water with decreasing temperature. The amount of dissolved air in the accreting droplets is determined largely by the ambient temperature. As the droplet freezes, the air is forced out of solution, forming air bubbles which tend to weaken the ice structure. Dong and Hallett postulate that this renders the ice structure more plastic at temperatures colder than $-10°C$, reducing the formation of splinters. Note that the rimer temperature generally will be warmer than the ambient temperature, hence the observed cold temperature threshold near $-9°C$.

4.8.2. *Field evidence for secondary ice production*

Several attempts have been made to estimate secondary ice production for clouds in which microphysical observations have been conducted. Beheng (1987) finds good agreement between the results of a simplified microphysical model and observations of ice particle concentrations, yielding a multiplication factor of 10^4 within 50 minutes. However, this model included only collection processes and did not consider alternate hypotheses to explain the high ice crystal concentrations. Harris-Hobbs and Cooper (1987) present similar studies for clouds in three geographically dissimilar locations and find the observed ice particle production rate roughly three times that estimated assuming a Hallett–Mossop mechanism was active. The uncertainties involved in their estimations could not rule out secondary ice production by this mechanism.

It seems likely that there is no single explanation for 'anomalous high' ice particle concentrations in natural clouds. The previously cited observations of Hobbs and Rangno (1985) strongly suggest that contact freezing nucleation was responsible for the high ice crystal concentrations they observed. The lack of large particles and the apparent *in situ* formation of the ice, together with the cloud top mixing, rule out condensation freezing or secondary ice production by the Hallett–Mossop mechanism.

The observations of Rangno and Hobbs (1991) suggest that conditions favorable for condensation freezing nucleation existed in the polar maritime cumuli they observed. The rapid increase in ice crystal concentrations is much too rapid to be attributed to a Hallett–Mossop mechanism or to contact freezing nucleation. The observations by Cooper and Vali (1981) of ice particle formation at the leading edge of a mountain wave cloud are less clear, but may reflect condensation freezing, since the droplet sizes and updraft conditions were probably not favorable for contact freezing nucleation.

There are also observations cited above that may be explained by secondary ice production resulting from the Hallett–Mossop mechanism. However, it is difficult to rule out primary ice production by means of condensation freezing or contact freezing nucleation. The presence of graupel particles may be expected to have a similar effect on the water supersaturation as would the presence of large water drops. Thus, to distinguish a Hallett–Mossop mechanism from condensation freezing, one would need to observe 'ice multiplication' in the absence of updrafts. In addition to the difficulty of supporting the graupel particle in the absence of updrafts, one would also have to rule out contact freezing nucleation. Thus, it may not be possible to identify situations in which the Hallett–Mossop mechanism may be active while ruling out the possibility of condensation freezing nucleation.

4.9. The behavior of ice nuclei in nature

The observed behavior of ice nuclei is not always consistent with theoretical expectations. The preceding section illustrates some of the uncertainties associated with characterizing the concentration of ice crystals based on observations of the 'active ice nuclei concentrations.' This section will consider sources of natural ice nuclei, the active sites concept and pre-activation, some other sources of ice crystals in clouds, and will then summarize some of the methods used to determine ice nuclei activity.

4.9.1. *Sources of natural ice nuclei*

Ice nuclei may consist of organic or inorganic substances. Early studies suggested that ice nuclei consisted almost exclusively of inorganic substances which were largely derived from soil particles. Later studies showed certain organic substances were also very active as artificial ice nuclei. More recent studies have also shown that naturally occurring organic substances may exhibit very warm ice nucleation thresholds. These latter nuclei are termed *biogenic* nuclei.

The ice nucleating ability of inorganic nuclei is reasonably well understood in terms of the theoretical treatments presented earlier in this chapter. The reasons why certain organic compounds and biogenic nuclei exhibit excellent ice nucleation characteristics is not well understood and, in particular, mechanisms for explaining the activity of biogenic nuclei seem to be quite lacking.

(i) Inorganic ice nuclei. Perhaps the most extensive study of the nuclei upon which ice crystals have grown was conducted by Kumai (1961). Kumai collected ice crystals as they fell through a hole in the roof of an igloo, which served to maintain an ice saturated environment. Each single crystal was placed on an electron microscope slide and carefully sublimed while being photographed periodically to ascertain the crystallographic center of the crystal. Kumai collected a total of 271 ice crystals at a site in northern Michigan (Kumai, 1961), and later collected 356 ice crystals on the Greenland ice cap (Kumai and Francis, 1962).

These studies showed that natural ice nuclei consist predominantly of clay minerals (85%), principally of the kaolinite (51%) and montmorillonite (15%) groups. Roughly 10% of the samples contained nuclei whose composition could not be determined; less than 5% did not appear to contain a nucleus.

The sizes of the nuclei reported are also informative. The nuclei observed in Michigan were generally much larger than those observed in Greenland. This may be attributed to Greenland being a greater distance from the sources of such nuclei (soils). Needle crystals tended to contain larger nuclei than plate or dendritic crystals. Since needle crystals tend to form near −5°C and plate and dendritic crystals generally grow between −10°C and −20°C, this supports the concept that larger nuclei tend to be active at warmer temperatures. The nuclei sizes ranged from 0.1 to 6 μm in radius, although the most common sizes were smaller than one micrometer.

Certain anthropogenic sources of ice nuclei have been identified. For example, Langer and Rosinski (1967) report high concentrations of ice nuclei emanating from a steel mill near Pueblo, Colorado. Hobbs and Locatelli (1969) report high concentrations of ice nuclei from a natural forest fire, suggesting that certain types of combustion processes may serve as sources of ice nuclei.

(ii) Biogenic and organic ice nuclei. Three sources of biogenic nuclei have been described by Schnell (1976). These are leaf-derived nuclei (LDN), ocean-derived nuclei (ODN), and bacteria-derived nuclei (BDN). Vali et al. (1976) suggest that the activity of the LDN derives from a species of bacterium, and that the bacterial cells themselves may act as the nuclei. Later, Schnell (1977) relates the ODN to bacterial activity as well. This suggests that all the biogenic ice nuclei are due to bacterial cells, either living or dead.

Fukuta (1966) tested a wide variety of organic substances for their ice nucleating ability. From these studies, he deduced that organic substances which are good ice nucleants possess hydrogen bonding capability and exhibit rotational symmetry, thus insuring the hydrogen bonds will be available at the nucleus surface. He also noted the positions of the hydrogen bonding sites and the direction of the bonds needed to match those found in ice.

Thus, it is likely that a wide variety of organic substances may serve as effective ice nuclei. Power and Power (1962) find that certain amino acids, including *l*-leucine and *l*-tryptophan, are effective ice nucleants. Since cell coatings are often made up of amino acids, it would seem the biogenic nuclei derive their ice nucleating ability in a manner similar to that found for artificial organic ice nuclei.

4.9.2. *The active sites concept and pre-activation*

The idea that nuclei have smooth, homogeneous surfaces without small irregularities would seem to have little relation to what one would expect from naturally occurring nuclei. Considerable experimental evidence exists supporting the view that ostensibly identical nuclei possess a range of activation thresholds, and that the threshold exhibited by a given nucleus is reproducible, implying that some characteristic of that particular nucleus is responsible for its nucleation behavior. Anderson and Hallett (1976) conclude that ice crystals form at specific sites on freshly-cleaved surfaces of AgI and CuS crystals.

A morphological inhomogeneity in the surface may serve as a preferred site for ice formation. Fletcher (1969) presents a theoretical treatment of a size distribution of active sites ($\hat{m} = 1$) on the nucleus. These active sites are assumed to arise from the reduction of the equilibrium vapor pressure over a concave surface of water (or ice) held in a surface inhomogeneity or hole. This may be expected to favor the formation of ice, resulting in a small area on the surface of the nucleus which is ice, i.e., has a contact parameter of unity.

A given cavity containing ice will nucleate the ice phase if the external surface area presented by the ice equals or exceeds the area of contact between a critical ice embryo and the nucleus. If the cavity presents an area smaller than the required critical embryo area, the shape of the embryo and its nucleating behavior is dictated by the contact parameter of the nucleus surface as shown in Fig. 4.16c. If the cavity presents an area larger than the required critical embryo area, as shown in Fig. 4.16b, the site nucleates a macroscopic ice phase. As the ice supersaturation (or supercooling) is increased, the critical embryo size decreases, allowing smaller cavities to serve as active sites and initiate the ice phase.

The reduction of the equilibrium vapor pressure in concavities on the nucleus surface suggests that previously formed ice crystals could evaporate, leaving microscopic amounts of ice in such concavities. These may be expected to result in enhanced ice nucleation activity, provided the nucleus was not subjected to temperatures above 0°C or to prolonged conditions of low humidity. Higuchi and Fukuta (1966) report such 'pre-activated' behavior when nuclei were cooled to temperatures below −35°C, comparing their ice nucleating ability before and after treatment.

Studies by Edwards and Evans (1971) suggest these observations are better explained by considering regions on the nucleus surface in which the adsorbed water molecules are ordered (ice-like). They postulate that the adsorbed water molecules tend to form disordered clusters at warmer temperatures, transforming to an ordered state below some critical temperature (specific to each nucleating

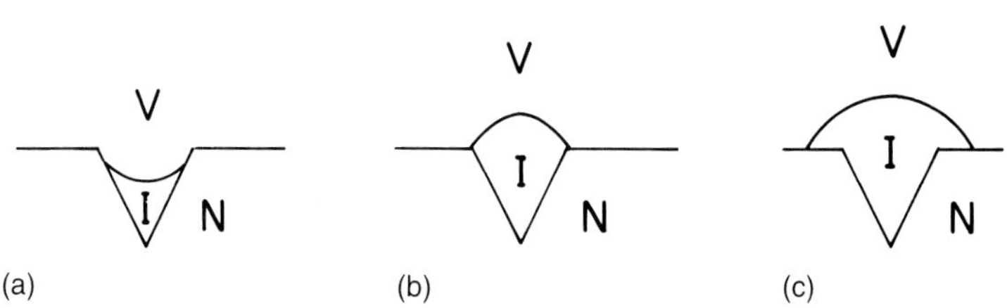

FIG. 4.16. Representation of active sites due to a morphological inhomogeneity. (a) The survival of ice in a pit due to negative curvature in a subsaturated environment. (b) A supercritical embryo contained within a pit. (c) A critical embryo which is too large for the pit. Its activity is determined by the contact parameter of the surface rather than by the pit.

substrate). They observe that the ice nucleating ability is enhanced by cooling to −30°C at humidities below ice saturation, followed by drying the sample for 2 minutes at 80% humidity at the test temperature and then exposing the sample to humidities above water saturation.

4.9.3. *Other sources of ice crystals in clouds*

In addition to the various ice multiplication mechanisms that have been advanced to explain the 'anomalous high' ice crystal concentrations in natural clouds (see Mossop, 1985), the production of ice crystal by shock waves and by aircraft has been suggested. The evidence is ambiguous but suggestive. The theoretical basis for these mechanisms is not understood.

(i) Aircraft-produced ice particles. Repeated aircraft penetrations of the same volume of supercooled cloud have, on occasion, shown marked increases in the concentration of ice crystals. In an effort to elucidate the conditions under which aircraft-produced ice particles (APIPs) may occur, Kelly and Vali (1991) penetrated a number of clouds under a wide variety of conditions. They were able to document only one case of APIPs, which occurred under conditions of heavy airframe icing. They postulate a splinter mechanism resulting from the airframe icing since similar cloud conditions without icing did not result in APIPs.

Sassen (1991) reported APIPs in a supercooled altocumulus cloud at temperatures between -28°C and -31°C without airframe icing. These observations are attributed to the cooling of air in the vortices associated with the wing tips and propellers, as suggested by Vonnegut (1986). The importance of APIPs lies in the potential for contamination of the cloud microstructure that one wishes to observe.

(ii) Shock-induced freezing. A variety of studies have suggested that subjecting water drops to mechanical shock could result in freezing the drops. Czys (1989) summarizes several of these studies and presents additional evidence that mechanical shock can induce freezing in large drops. Czys postulates that collisions between millimeter sized drops may provide sufficient shock to induce freezing and thereby account for the appearance of high concentrations of ice particles in relatively warm clouds when large drops are present.

Czys relates the shock-induced freezing to the occurrence of cavitation within the drops. The cavitation results as the external pressure drops suddenly and the resulting forces are great enough to exceed the tensile strength of the liquid. He notes that the occurrence of cavitation is highly sensitive to the purity of the water, with a pressure drop of 500 mbar required for homogeneous cavitation but as small as 5 mbar for aerated tap water. Thus, the presence of dissolved gases and other impurities has a marked effect on cavitation.

The mechanism by which cavitation produces freezing is unclear. Evaporation into the bubble may produce some cooling; the rapid expansion as the bubble forms may also cool the water vapor sufficiently to form ice.

The importance of such a process in the atmosphere is also unclear. If millimeter sized drops are required, then the role of shock-induced freezing would be to increase the concentrations of ice particles and promote glaciation of the cloud. However, as noted in § 4.8.2, the presence of large drops in ascending air may be expected to produce high concentrations of ice crystals by condensation freezing. Likewise, proposed efforts to clear supercooled fog using ultrasonic generators are likely to be ineffective since the fog droplets may be too small to induce cavitation.

4.9.4. *Laboratory characterization of ice nuclei*

There are four basic methods for determining the ice nucleation activity for a given sample of nuclei. The most common method is to expose the nuclei as aerosol to conditions close to water saturation at a predetermined temperature. A cloud is formed, either by expansion of moist air (Warner, 1957) or by introduction of water vapor into a cold chamber (Bigg, 1957). The aerosol is already present and as the ice crystals are formed, they fall onto a collecting surface, which may be a tray of supercooled sugar solution which is nucleated at the sites where ice crystals fall into the solution. Provided the collection time is not excessive, a count of the nucleation sites provides a count of the ice crystals formed.

A more convenient method for determining ice nucleation behavior involves collecting the nuclei on a surface, typically a filter or a metal foil (Bigg et al., 1963). Air is drawn through the filter, and nuclei larger than the pore size are trapped on the filter surface. Deposition on a metal foil is usually accomplished by a centrifuge but this may also be effected by a thermal precipitator. These have the advantage of concentrating the aerosol sample and allowing the sample to be stored for analysis elsewhere.

Once the nuclei have been deposited on the substrate, they are exposed to supersaturated conditions, and the ice crystals formed are allowed to grow to visible size and then counted. Several methods by which supersaturated conditions may be maintained are in use. The most common method is to place the sample in a thermal gradient diffusion chamber. The temperatures of the warmer, upper plate which is the water vapor source, and the lower, colder plate on which the sample rests are both controlled so that both the sample temperature and ice supersaturation are known and maintained while the crystals grow.

A third technique for evaluating ice nuclei activity is the drop freezing technique. Drops, usually of rainwater, are placed on a polished metal plate and the plate is cooled at a constant rate. The temperature at which each drop freezes is noted. This gives a freezing spectrum of the nuclei contained within the rain water (e.g., Vali, 1971).

A fourth technique is to use a chamber in which the temperature and saturation ratio can be varied under controlled conditions as a function of time. DeMott (1990) describes the use of the dynamic cloud chamber at Colorado State University (CSU) (Ft. Collins) to characterize the ice nucleating ability of silver iodide aerosol. In this chamber, the temperature can be varied and, knowing the initial saturation ratio and aerosol spectrum, the saturation ratio can be modeled as a function of time.

Attempts to 'count' the number of active ice nuclei as a function of temperature have produced widely disparate results. Vali (1975) discusses the need to identify the dominant mode of ice nucleation in a particular instrument before assuming that the measurements are comparable. Even for a given instrument, the ice nucleation mode may change, depending on the operating conditions. A chamber between ice and water saturation may be expected to form ice crystals *via* deposition nucleation but if the chamber is above water saturation, condensation freezing and possibly contact freezing nucleation may be expected.

The CSU dynamic cloud chamber was used by DeMott (1990) to describe the ice nucleating characteristics of a silver iodide aerosol under each of the four modes of ice nucleation. Conditions within the chamber could be controlled to isolate deposition, contact freezing, and immersion freezing nucleation modes.

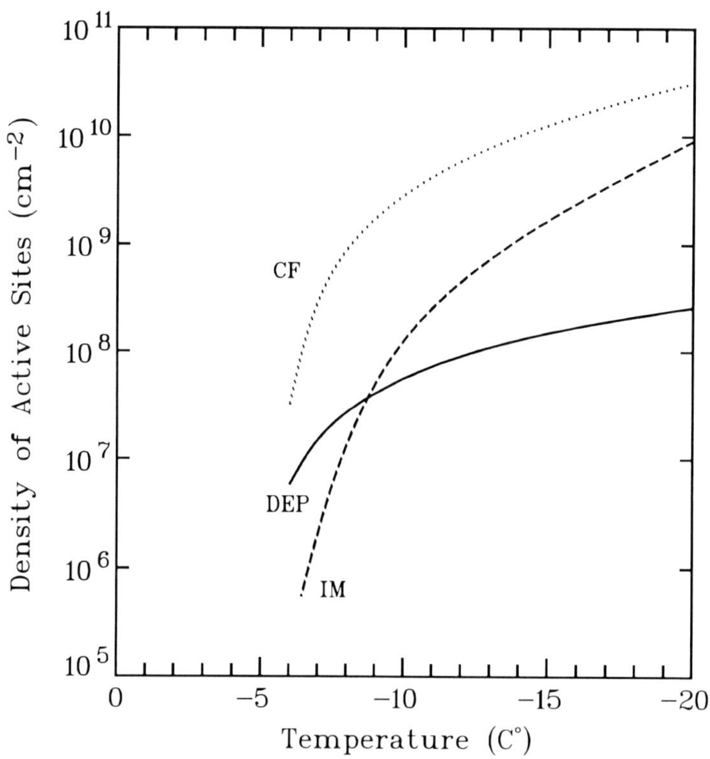

Fig. 4.17. Comparison of active site densities by deposition (DEP), contact freezing (CTF), and immersion freezing (IMF) nucleation for AgI–AgCl–4NaCl aerosols. The number of nucleation sites for a given size nucleus can be computed from the areal density of active sites. Water saturation is assumed. From DeMott (1990). Reproduced with permission.

Condensation freezing nucleation activity was examined by estimating the number of ice crystals produced by deposition and contact freezing nucleation based on previous experiments, and subtracting this number from the total number of ice crystals produced.

Figure 4.17 illustrates the relative activity of a silver iodide aerosol due to deposition, immersion freezing, and contact freezing nucleation in terms of the areal density of nucleation sites as a function of temperature, assuming water saturation. The greatest density of nucleation sites is noted for contact freezing; the lowest density for deposition. DeMott (1990) notes that the curves for immersion freezing and deposition nucleation should be considered minimum values. The relation between contact freezing and deposition nucleation for a nucleation site density of 10^8 cm^{-2} is in rough agreement with Cooper's (1974) theory. Observed deposition and condensation freezing activity as functions of the ambient temperature and ice supersaturation are shown in Fig. 4.18. These curves are similar to those shown in Fig. 4.13 although a range of nucleation site densities are provided.

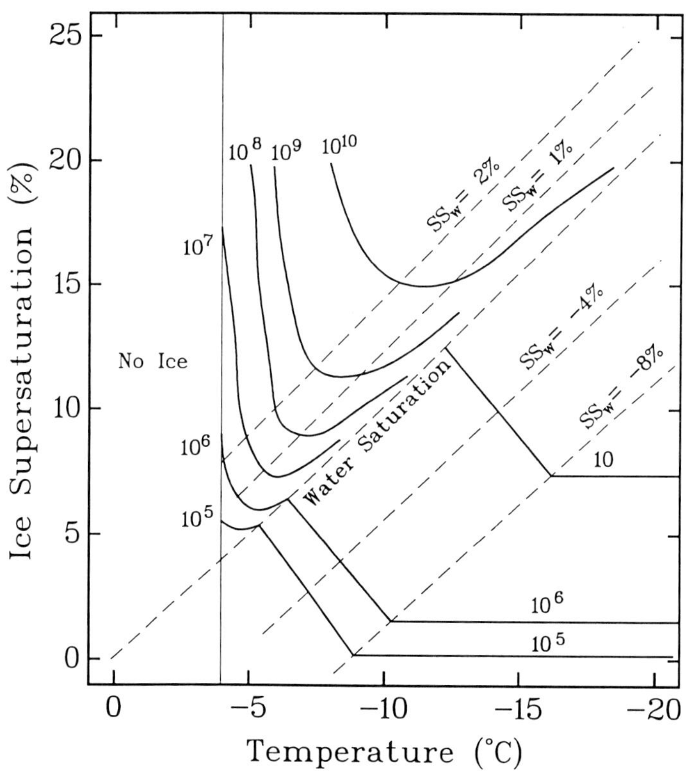

FIG. 4.18. Summary of combined deposition and condensation freezing nucleation site densities for AgI–AgCl aerosols as a function of the ambient temperature and ice supersaturation. Isopleths of water supersaturation are provided for comparison. From DeMott (1990). Reproduced with permission.

Applying laboratory observations of the ice nucleating ability of a given aerosol sample to the development of ice within the cloud from (or near) which the aerosol sample was taken is difficult at best. A series of measurements of ice nucleating ability via a single mode of nucleation cannot be said to describe adequately the ice nucleation behavior of that aerosol. There is no guarantee that the ice nucleation mode observed in the instrument will be the dominant mode in the cloud. One may find that the dominant mode of ice nucleation in the cloud is contact freezing nucleation, whereas the measurements relate to immersion freezing nucleation. Since the parameters governing the nucleation rates by these two modes are quite different, one cannot apply the laboratory measurements directly to the cloud and expect to achieve a correspondence between the ice nucleus and ice crystal concentrations.

As Figure 12 from Rangno and Hobbs (1991) suggests, the dominant mode of ice nucleation may depend on location within the cloud and the stage of development of the cloud. Thus, to understand fully the formation of ice in the cloud, it is necessary to characterize the aerosol sample for all four modes of ice phase nucleation, as described by DeMott (1990). The theoretical relationships developed in this chapter allow one to describe the contact freezing activity based on observed deposition nucleation activity, for example. Still, one would need to determine the size distribution of the nuclei in order to predict contact freezing nucleation behavior in the cloud.

The complexities associated with the activation of ice nuclei in clouds strongly suggest that a numerical modeling approach should be incorporated into the interpretation of observations of ice production in clouds. For example, it is not possible to measure water supersaturations within clouds to anything close to the precision required to predict condensation freezing nucleation behavior, i.e., the water supersaturation must be estimated from the observed particle spectra and updraft velocity. The contact freezing nucleation behavior is also dependent on water supersaturation, since the phoretic forces which play a major role in determining contact rates are quite sensitive to the supersaturation.

In summary, the problem of comparing ice nuclei with ice crystals in clouds will require an approach which joins field and laboratory observations with sophisticated numerical modeling. Until this is done, it seems unlikely that the relative importance of the four modes of ice nucleation and the role of secondary ice production can be assessed.

Problems

4.1* Calculate deposition nucleation rates for a nucleus of radius 0.1 μm with $\hat{m} = 0.96$. Do this for ice saturation ratios of 1.12, 1.14, and 1.16 at temperatures of -5°C and -15°C. Is deposition nucleation predominately dependent on temperature or on ice supersaturation?

4.2* Plot the ice supersaturation ($S_i - 1$) at water saturation as a function of temperature between 0°C and -40°C.

4.3** Calculate the ice supersaturation over a spherical ice cap with a radius of curvature of 0.1 μm, neglecting the effects of elastic strain. Compare this to an embryo with 2% strain ($\epsilon = 0.02$). Assume $T = -10°C$.

4.4** What is the height of a critical ice embryo for deposition nucleation at water saturation and −10°C on a plane surface with a contact parameter of 0.95? How many molecular layers does this correspond to? You may assume that the ice nucleates with its basal face in contact with the substrate.

4.5** Considering the nucleation rate given by (4.42) to represent the product of the number of critical embryos and the rate at which they add molecules to become supercritical, what is the ratio between the actual number of critical embryos and that which would be expected in the steady state distribution described by the Boltzmann distribution (4.16)? Consider a nucleus with radius 1.0 μm and $m = 0.8$, immersed in supercooled water at −6°C.

4.6*** Consider a water drop of radius 0.1 cm. Initially the drop is at 0°C. The drop is frozen, becoming an ice particle at 0°C, and then melted, in the following manner:

(a) The drop cools reversibly from 0°C to −10°C with no ice being formed.
(b) Ice is nucleated within the drop. The latent heat released warms the drop to 0°C, freezing a fraction of the water without exchange of heat between the drop and its environment. Assume that all of the ice is formed at 0°C.
(c) The remainder of the water freezes reversibly at 0°C.
(d) The ice particle is melted reversibly at 0°C.

Note that this represents a cyclic process, i.e., the initial and final states of the drop are the same. Calculate the change in entropy for the drop for each of the four steps.

Calculate the total change in entropy for the drop.

Calculate the total change in entropy for the environment.

Explain your results in terms of the second law of thermodynamics. Make sure the signs for the entropy changes are correct.

REFERENCES

Anderson, B.J., and Hallett, J., 1976, 'Supersaturation and time dependence of ice nucleation from the vapor on single crystal substrates,' *J. Atmos. Sci.* **33**, 822.

Beheng, K.D., 1987, 'Microphysical properties of glaciating cumulus clouds: Comparison of measurements with a numerical simulation,' *Quart. J. R. Meteorol. Soc.* **113**, 1377.

Bigg, E.K., 1957, 'A new technique for counting ice-forming nuclei in aerosols,' *Tellus* **9**, 394.

Bigg, E.K., Mossop, S.C., Meade, R.T., and Thorndike, N.S.C., 1963, 'The measurement of ice nucleus concentrations by means of millipore filters,' *J. Appl. Meteorol.* **2**, 266.

Cooper, W.A., 1974, 'A possible mechanism for contact nucleation,' *J. Atmos. Sci.* **31**, 1832.

Cooper, W.A., and Vali, G., 1981, 'The origin of ice in mountain cap clouds,' *J. Atmos. Sci.* **38**, 1244.

Czys, R.R., 1989, 'Ice initiation by collision-freezing in warm-based cumuli,' *J. Appl. Meteorol.* **28**, 1098.

DeMott, P.J., 1990, *Quantifying ice nucleation ny silver iodide aerosols.* Ph.D. Dissertation, Colorado State University, Paper #466.

Dong, Y.Y., and Hallett, J., 1989, 'Droplet accretion during rime growth and the formation of secondary ice crystals,' *Quart. J. R. Meteorol. Soc.* **115**, 127.

Edwards, G.R., and Evans, L.F., 1971, 'The mechanism of activation of ice nuclei,' *J. Atmos. Sci.* **28**, 1443.

Fletcher, N.H., 1958, 'Size effect in heterogeneous nucleation,' *J. Chem. Phys.* **29**, 572.

Fletcher, N.H., 1959, 'Entropy effect in ice crystal nucleation,' *J. Chem. Phys.* **30**, 1476.

Fletcher, N.H., 1969, 'Active sites and ice crystal nucleation,' *J. Atmos. Sci.* **26**, 1266.

Fletcher, N.H., 1970a, *The chemical physics of ice*. Cambridge: Cambridge University Press.

Fletcher, N.H., 1970b, 'On contact nuclei,' *J. Atmos. Sci.* **27**, 1098.

Fukuta, N., 1966, 'Experimental studies of organic ice nuclei,' *J. Atmos. Sci.* **23**, 191.

Fukuta, N., 1975a, 'Comments on 'A possible mechanism for contact nucleation',' *J. Atmos. Sci.* **32**, 2371.

Fukuta, N., 1975b, 'A study of the mechanism of contact nucleation,' *J. Atmos. Sci.* **32**, 1597.

Fukuta, N., and Schaller, R.C., 1976, 'Ice nucleation by aerosol particles,' *Proc. Int. Conf. Cloud Physics 1976*. American Meteorological Society, p. 47.

Fukuta, N., and Schaller, R.C., 1982, 'Ice nucleation by aerosol particles: Theory of condensation-freezing nucleation,' *J. Atmos. Sci.* **39**, 648.

Gokhale, N.R., and Goold, J., Jr., 1968, 'Droplet freezing by surface nucleation,' *J. Appl. Meteorol.* **7**, 870.

Goldsmith, P., and May, F.G. 1966, 'Diffusio- and thermophoresis in water vapour systems,' *Aerosol Science*. New York: Academic Press, p. 163.

Guenadiev, N., 1972, 'Congélation des grosses gouttes d'eau au contact des particules d'iodure d'argent,' *Pure Appl. Geophys.* **99**, 251.

Hagen, D.E., Anderson, R.J., and Kassner, J.L., 1981, 'Homogeneous condensation-freezing nucleation measurements for small water droplets in an expansion cloud chamber,' *J. Atmos. Sci.* **38**, 1236.

Hallett, J., 1963, 'The temperature dependence of the viscosity of supercooled water,' *Proc. Phys. Soc.* **82**, 1046.

Hallett, J., and Mossop, S.C., 1974, 'Production of secondary ice particles during the riming process,' *Nature* **249**, 26.

Harris-Hobbs, R.L., and Cooper, W.A., 1987, 'Field evidence supporting quantitative predictions of secondary ice production rates,' *J. Atmos. Sci.* **44**, 1071.

Higuchi, K., and Fukuta, N., 1966, 'Ice in the capillaries of solid particles and its effect on their nucleating ability,' *J. Atmos. Sci.* **23**, 187.

Hobbs, P.V., 1969, 'Ice multiplication in clouds,' *J. Atmos. Sci.* **26**, 315.

Hobbs, P.V., and Locatelli, J.D., 1969, 'Ice nuclei from a natural forest fire,' *J. Appl. Meteorol.* **8**, 833.

Hobbs, P.V., and Rangno, A.L., 1985, 'Ice particle concentrations in clouds,' *J. Atmos. Sci.* **42**, 2523.

Hoffer, T.E., 1961, 'A laboratory investigation of droplet freezing,' *J. Meteorol.* **18**, 766.

Kelly, R.D., and Vali, G., 1991, 'An experimental study of the production of ice crystals by a twin-turboprop aircraft,' *J. Appl. Meteorol.* **30**, 217.

Ketcham, W.M., and Hobbs, P.V., 1969, 'An experimental determination of the surface energies of ice,' *Phil. Mag.* **19**, 1161.

Koenig, L.R., 1963, 'The glaciating behavior of small cumulonimbus clouds,' *J. Atmos. Sci.* **20**, 29.

Kumai, M., 1961, 'Snow crystals and the identification of the nuclei in the northern United States of America,' *J. Meteorol.* **18**, 139.

Kumai, M., and Francis, K.E., 1962, 'Nuclei in snow and ice crystals on the Greenland ice cap under naturally and artificially stimulated conditions,' *J. Atmos. Sci.* **19**, 474.

Langer, G., and Rosinski, J., 1967, 'A continuous ice nucleus counter and its application to tracking in the atmosphere,' *J. Appl. Meteorol.* **6**, 114.

Leong, K.H., Beard, K.V., and Ochs, H.T., III, 1982, 'Laboratory measurements of particle capture by evaporating drops,' *J. Atmos. Sci.* **39**, 1130.

Mossop, S.C., 1970, 'Concentrations of ice crystals in clouds,' *Bull. Am. Meteorol. Soc.* **51**, 474.

Mossop, S.C., 1985, 'The origin and concentration of ice crystals in clouds,' *Bull. Am. Meteorol. Soc.* **66**, 264.

Power, B.A., and Power, R.F., 1962, 'Some amino acids as ice nucleators,' *Nature* **194**, 1170.

Pruppacher, H.R., and Klett, J.D., 1978, *Microphysics of clouds and precipitation.* Dordrecht: Reidel.

Rangno, A.L. and Hobbs, P.V., 1991, 'Ice particle concentrations and precipitation development in small polar maritime cumuliform clouds,' *Quart. J. R. Meteorol. Soc.* **117**, 207.

Sassen, K., 1991, 'Aircraft-produced ice particles in a highly supercooled altocumulus cloud,' *J. Appl. Meteorol.* **30**, 765.

Schaller, R.C., and Fukuta, N., 1979, 'Ice nucleation by aerosol particles: Experimental studies using a wedge-shaped ice thermal diffusion chamber,' *J. Atmos. Sci.* **36**, 1788.

Schnell, R.C., 1976, 'Bacteria acting as natural ice nucleants at temperatures approaching −1°C,' *Bull. Am. Meteorol. Soc.* **57**, 1356.

Schnell, R.C., 1977, 'Ice nuclei in seawater, fog water and marine air off the coast of Nova Scotia: Summer 1975,' *J. Atmos. Sci.* **34**, 1299.

Stillinger, F.H., and Rahman, A., 1972, 'Molecular dynamics study of temperature effects on water structure and kinetics,' *J. Chem. Phys.* **57**, 1281.

Turnbull, D., and Vonnegut, B., 1952, 'Nucleation catalysis,' *Industr. Engng. Chem. (Industr.)* **44**, 1292.

Vali, G., 1971, 'Quantitative evaluation of experimental results on the heterogeneous freezing nucleation of supercooled liquids,' *J. Atmos. Sci.* **28**, 402.

Vali, G., 1975, 'Workshop summary: Ice nucleation workshop 1975,' *Bull. Am. Meteorol. Soc.* **56**, 1180.

Vali, G., Christensen, M., Fresh, R.W., Galyan, E.L., Maki, L.R., and Schnell, R.C., 1976, Biogenic ice nuclei. Part II: Bacterial sources,' *J. Atmos. Sci.* **33**, 1565.

Vonnegut, B., 1986, 'Nucleation of ice crystals in supercooled clouds caused by passage of an airplane,' *J. Clim. Appl. Meteorol.* **25**, 98.

Waldmann, L., and Schmidt, K.H., 1966, 'Thermophoresis and diffusiophoresis of aerosols,' *Aerosol Science*. New York: Academic Press, p. 137.

Warner, J., 1957, 'An instrument for the measurement of freezing nucleus concentration,' *Bull. Obs. Puy de Dôme*, No. 1, 33.

Young, K.C., 1974, 'The role of contact nucleation in ice phase initiation in clouds,' *J. Atmos. Sci.* **31**, 768.

5

DIFFUSION GROWTH OF WATER DROPS

5.1. Introduction

The growth of a drop by diffusion involves two processes. First, the water vapor is transferred to the drop surface by molecular diffusion in the vapor density gradient established around the drop. The condensation of water vapor onto the drop releases latent heat, thereby warming the drop. Since the rate of mass transfer is directly related to the difference in vapor density between the drop surface and its environment, the release of latent heat, which warms the drop, reduces its growth rate.

Clearly, heat must be removed from the drop in order to permit the drop to continue to grow. This excess heat can be removed by conduction as the drop becomes warmer than its environment. Thus, both the mass flux of water molecules and the conduction of heat between the drop and its environment must be considered.

The diffusional growth of droplets will be introduced by considering the simplest case, that of a stationary water drop growing in isolation. The effects of competition with other drops, kinetic effects for small drops, and the enhancement of mass and heat transfer as the drop falls will be considered as they affect the growth (or evaporation) of the drop.

5.2. The equations governing drop growth by diffusion

The molecular flux of water molecules to the growing drop and the conduction of heat away from the drop are readily solved assuming spherical symmetry. The two equations must be solved in tandem. One common method of solution is to compute the drop temperature[1] by balancing the sensible and latent heat fluxes. The mass and sensible heat flux solutions then follow directly. An alternate method will be presented, which relates the mass and heat transfer equations through the Clausius–Clapeyron equation, which, through certain simplifying assumptions, leads to an analytical solution for the mass growth rate.

[1] The drop temperature is often referred to as the 'particle wet-bulb' temperature.

5.2.1. *The mass flux equation*

Consider the simple case of a single drop, growing in isolation in a vapor field of infinite extent, and stationary with respect to its environment. Ignoring the convective transfer of vapor,[2] the diffusion equation for water vapor is

$$\frac{\partial \rho}{\partial t} = \nabla^2 D\rho \,. \tag{5.1}$$

The mass flux of water vapor through any concentric shell of radius r surrounding the drop is then proportional to the radial gradient of the vapor density (ρ), to the diffusion coefficient for water vapor in air (D), and to the area of the shell. This may be expressed as

$$\frac{dm}{dt} = 4\pi r^2 D \frac{d\rho}{dr} \,.$$

If it is assumed that there are no sources or sinks of water vapor except at the drop surface and at some infinite distance from the drop ($r = \infty$), (5.1) may be readily integrated from the drop surface to infinity, recognizing that the mass flux through any concentric shell is constant,

$$\frac{dm}{dt} \int_{r_w}^{\infty} \frac{dr}{r^2} = 4\pi D \int_{\rho_{s,x}}^{\rho_\infty} d\rho$$

or

$$\frac{dm}{dt} = 4\pi r_w D(\rho_\infty - \rho_{s,w}) \,. \tag{5.2}$$

Note that this assumes the diffusivity of water vapor in air (D) is not a function of radial distance from the drop. The simultaneous existence of a temperature gradient requires that the diffusion coefficient be a function of distance.[3] The total air density (rather than total air pressure) is also assumed constant; otherwise it would be necessary to express the transport in terms of the mixing ratio. The drop growth is assumed to be quasi-static, i.e., the drop radius does not change with time. These effects are generally assumed to be negligible.

This treatment also assumes that the value of the vapor density at the drop surface is equal to the saturation value for water at the drop temperature. The

[2] See, for example, Pruppacher and Klett (1978).

[3] Hall and Pruppacher (1976) provide an empirical fit for the diffusivity of water vapor in air as

$$D = 0.211 \frac{P_o}{P} \left(\frac{T}{T_o}\right)^{1.94} ,$$

where $T_o = 273.15$ K and $P_o = 1013.25$ mbar.

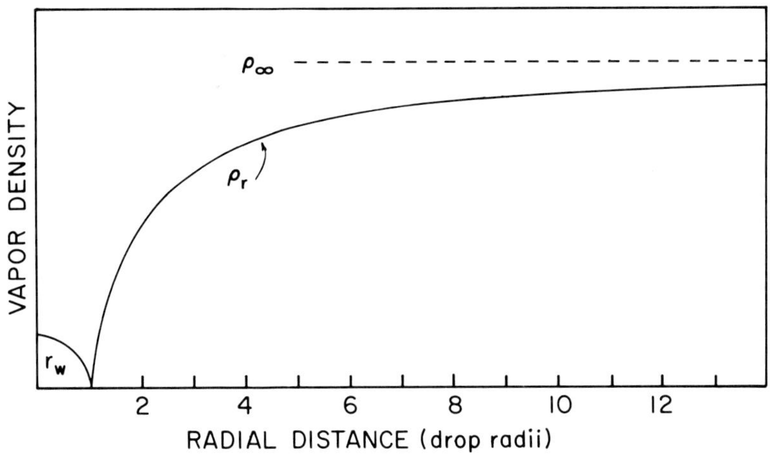

FIG. 5.1. The vapor density field surrounding a growing water drop shows an inverse radius dependence, increasing from the saturation value at the drop surface and approaching some limiting value at an 'infinite' distance from the drop (ρ_∞).

effects of curvature and soluble material on the saturation vapor pressure will be treated later.

The integration of (5.1) leads naturally to the concept of an ambient vapor density (ρ_∞) or some limiting value of vapor density at a great distance from the drop. If the vapor density is plotted as a function of radial distance, as shown in Fig. 5.1, the discrepancy between the vapor density at r and that at infinity decreases quite rapidly with distance. It can be shown (Problem 5.5) that the vapor density approaches the ambient vapor density as $1/r$ as given by

$$\frac{\rho_\infty - \rho_r}{\rho_\infty - \rho_{s,w}} = \frac{r_w}{r} \tag{5.3}$$

where ρ_r is the vapor density at radial distance r from the center of the drop and r_w is the drop radius.

5.2.2. *The sensible heat flux equation*

The conduction of heat away from a growing drop may be treated in a manner analogous to the transfer of mass to the drop. Again, ignoring the convective transfer of heat, the diffusion equation for heat is

$$\frac{\partial T}{\partial t} = \nabla^2 KT \tag{5.4}$$

where K is the thermal conductivity for air. Assuming that the thermal conductivity is constant, the heat flux through any concentric shell of radius r is

$$\frac{dQ}{dt} = 4\pi r^2 K \frac{dT}{dt}.$$

Again, assuming no sources or sinks of heat except at the drop surface and some infinite distance, the heat flux equation for a drop of radius r_w is

$$\frac{dQ}{dt} = 4\pi r_w K(T_\infty - T_x), \tag{5.5}$$

defining heat transfer to the drop as positive.[4]

This treatment neglects the transfer of heat by the flux of water vapor molecules (see Staley, 1972) and assumes that the thermal conductivity is not dependent on the temperature or on the composition of the conducting air. Again, these effects are generally considered to have only a small effect on the heat flux equation.

The integration of (5.4) leads to the concept of an ambient temperature (T_∞) in the same fashion that integration of (5.1) led to the concept of the ambient vapor density (ρ_∞). The radial temperature gradient has the same radius dependence as the vapor density, and obeys the inverse radius dependence described by (5.3).

5.2.3. *The particle wet-bulb*

The release of latent heat at the surface of a growing drop and the conduction of heat away from the drop serve to determine the drop's temperature. The drop temperature, in turn, fixes the vapor density at the drop surface which controls the mass flux and, thereby, the release of latent heat (through (5.2)). Thus, solution of either the mass flux equation (5.2) or the heat flux equation (5.5) requires determination of the drop temperature. Since the solutions are of the same form for both water drops and ice particles,[5] the more general approach is used in this and the following sections.

Under the assumption that changes in the particle temperature are small, the release of latent heat at the particle surface may be equated with the heat transfer away from the particle. This equality may be expressed as

$$L_x \frac{dm}{dt} = -\frac{dQ}{dt} = 4\pi r_x D L_x (\rho_\infty - \rho_{s,x}) = 4\pi r_x K(T_x - T_\infty) \tag{5.6}$$

[4] Note that inward fluxes are defined as positive for both mass and heat transfer.
[5] For non-spherical ice particles, the shape factor C is used in place of the particle radius r_x.

where the subscripts x refer to the particle, which may be either water or ice. The particle wet-bulb relation is

$$\frac{\rho_\infty - \rho_{s,x}}{T_x - T_\infty} = \frac{K}{DL_x} . \tag{5.7}$$

In principle, the Clausius–Clapeyron relation (A.1.17) may be used to define the relation between the particle temperature (T_x) and the saturation vapor density at that temperature ($\rho_{s,x}$). In practice, the algebra becomes tedious and an iterative solution is usually more rapid and more accurate.[6]

Young (1974) presents a two-step iterative solution based on a linear approximation of the saturation vapor density,

$$\rho_{s,x} = sT_x + b ,$$

where the slope s and intercept b are determined directly from saturation vapor density tables. The solution for the particle temperature follows as

$$T_x = \frac{DL_x (\rho_\infty - b) + KT_\infty}{DL_x s + K} , \tag{5.8}$$

where L_x represents the latent heat of vaporization for water or the latent heat of sublimation for ice. The initial estimates for s and b are based on a 1°C range centered on the ambient temperature T_∞; the final estimates for s and b are based on a 0.2°C range centered on the initial solution for the particle temperature.[7]

5.2.4. *The combined growth equation*

Mason (1953) presents a formulation of the growth equation in a form which does not require a separate solution for the particle temperature. This is accomplished through use of the Clausius–Clapeyron relation as follows. The mass flux equation (5.2) needs to be rewritten in terms of the vapor pressure,

$$\frac{dm}{dt} = \frac{4\pi r_x D}{R_v}\left[\frac{e_\infty}{T_\infty} - \frac{e_{s,x}}{T_x}\right] = \frac{4\pi r_x D}{R_v T_x}\left[e_\infty\left(\frac{T_x}{T_\infty}\right) - e_{s,x}\right] .$$

The saturation ratio at the ambient temperature ($e_\infty/e_{s,\infty}$) is obtained by dividing through by $e_{s,\infty}$,

$$\frac{e_\infty}{e_{s,\infty}}\left(\frac{T_x}{T_\infty}\right) = \frac{e_{s,x}}{e_{s,\infty}} + \left(\frac{R_v T_x}{4\pi r_x D e_{s,\infty}}\right)\frac{dm}{dt} . \tag{5.9}$$

[6] The Clausius–Clapeyron is an idealized and somewhat simplified equation. The Goff–Gratch formulation is the basis of the tables presented in the Smithsonian Meteorological Tables. Pruppacher and Klett (1978) present a more convenient approximation by Lowe and Ficke. For numerical modeling purposes, the saturation vapor density table is generated once and values are obtained by a table lookup.

[7] The full equation given by Young)1974) includes terms for the change in particle temperature with time, the effects of accretion of water drops on an ice particle, and the effects of radiative transfer of heat from the particle.

The ratio $e_{s,x}/e_{s,\infty}$ must be evaluated, after which the solution for dm/dt follows.

Integrating the Clausius–Clapeyron equation (A.1.17) from $(T_\infty, e_{s,\infty})$ to $(T_x, e_{s,x})$,

$$\ln\left(\frac{e_{s,x}}{e_{s,\infty}}\right) = -\frac{L_x}{R_v}\left(\frac{1}{T_x} - \frac{1}{T_\infty}\right) = \frac{L_x}{R_v T_x T_\infty}(T_x - T_\infty)\,.$$

From the derivation of the wet-bulb relation,

$$(T_x - T_\infty) = \frac{L_x}{4\pi r_x K}\frac{dm}{dt}\,,$$

reducing (5.9) to

$$\frac{e_\infty}{e_{s,w}}\left(\frac{T_x}{T_\infty}\right) = \exp\left[\frac{L_x^2\; dm/dt}{4\pi r_x K R_v T_x T_\infty}\right] + \left[\frac{R_v T_x}{4\pi r_x D e_{s,\infty}}\right]\frac{dm}{dt}\,. \tag{5.10}$$

Since the value of the exponent in (5.10) is much less than 1 for typical growth rates, the exponential may be approximated as $e^x \approx 1 + x$, allowing an explicit solution for the mass flux,

$$\frac{dm}{dt} = \frac{4\pi r_x\left[\left(\dfrac{e_\infty}{e_{s,\infty}}\right)\dfrac{T_x}{T_\infty} - 1\right]}{\dfrac{L_x^2}{K R_v T_x T_\infty} + \dfrac{R_v T_x}{D e_{s,\infty}}}\,. \tag{5.11}$$

Since the purpose of this exercise is to avoid calculation of the particle temperature T_x, the further approximation $T_x \approx T_\infty$ is made, and (5.11) reduces to

$$\frac{dm}{dt} = \frac{4\pi r_x\,(S - 1)}{\dfrac{L_x^2}{K R_v T_\infty^2} + \dfrac{R_v T_\infty}{D e_{s,\infty}}}\,. \tag{5.12}$$

The growth rate is now expressed entirely in terms of the ambient conditions. Note that the approximation $T_x \approx T_\infty$ does *not* imply that the drop temperature has been assumed to be at the ambient temperature for the overall derivation.

The approximations made in this derivation result in errors which are in direct relation to the wet-bulb elevation or depression. The growth equation (5.12) may be assumed to be within a few percent for growing drops, but may be in error by 10% or more for evaporating drops (see Problem 5.4).

5.3. Competition and the meaning of infinity

The growth of droplets in clouds occurs in a population of droplets and not in isolation, as the treatments of the previous section have assumed. Since the separation of droplets cannot be infinite, drops do compete with each other for the available water vapor. If there is an appreciable overlap between the vapor density (and temperature) fields around two drops, the 'ambient' vapor density the drops 'see' will be reduced, thereby reducing their respective growth rates.

Competition may occur between droplets of similar size or between droplets of quite different sizes. In addition, the vapor density fields of water drops and ice crystals may overlap, although in this case, the 'ambient' vapor density is generally between water and ice saturation such that the water drops are evaporating and the ice crystals are growing. Thus, the presence of water drops within the vapor density 'well' surrounding an ice particle provide sources of water vapor and thereby enhance the vapor density gradient and the growth of the ice particle. This topic will be treated in some detail in Chapter 6.

The concepts of competition will be introduced using a 'back-of-the-envelope' approach, which will demonstrate that the effects of competition are negligible for most purposes. The simplified approach will then be extended to demonstrate that competition for the available water vapor in an updraft may be expected to broaden the droplet spectrum.

5.3.1. *Competition between like particles*

Consider a typical continental cumulus updraft containing 1000 drops cm^{-3} of 5 μm radius each. The liquid water content of such a cloud is ~0.5 g m^{-3}. If these drops were uniformly spaced, they would be 0.1 cm apart, or 200 drop radii. Equation (5.3) predicts that the relative difference between the actual vapor density ρ at this distance and the ambient vapor density ρ_∞ would be 1%. One could argue that this difference is negligible for most purposes.

One conceptual problem with this picture is that nowhere in the volume does the vapor density ever achieve its 'ambient' value. Calculations of the water mass balance assume that the *average* vapor density is the same as the *ambient* vapor density. How valid is this approximation?

If one requires that the relative difference between the actual and ambient vapor densities be <2% before considering that point to be at 'infinity,' the volume within the vapor density fields of the drops in the above example is 6.5%. This implies that 93.5% of the volume is within 2% of the ambient vapor density, which suggests that the approximation of the ambient vapor density as the average value is reasonably good. It may be noted that the fraction of the volume influenced by the vapor fields is proportional to the liquid water content, and is not dependent on the drop sizes.

Since the droplets are randomly distributed in space, a probabilistic approach is required. The probability that any drop in the above example would have a vapor field that overlaps another drop would be 6.5%. Thus, most of the time, a given drop is growing essentially in isolation. This may be expected to have a negligible effect on their growth as an ensemble. However, the droplet spectrum would be expected to broaden, since growth for a few drops may be quite favored (rarely encountering other drops), and others, by chance, may encounter many drops and their growth may be retarded.

5.3.2. *Competition between unlike particles*

Consider the case of an occasional large drop falling through a cloud of much smaller drops. A raindrop of radius 0.1 cm affects the vapor density in a volume of 524 cm^3 (within a 2% relative difference between the actual and ambient vapor densities). If the cloud contains hundreds of droplets per cm^3, there are tens of thousands of droplets within the vapor field of the large drop at all times.

The effect of the small drops on the growth rate of the larger drop may be neglected since the coalescence growth rate in such a case would greatly exceed the growth rate by diffusion. Small corrections in the diffusion growth rate would be negligible by comparison.

Likewise, the effect the large drop has on the growth of small droplets is negligible. The fall velocity of a 0.1 cm radius drop is ~5 m s^{-1}. The diameter taken for its volume of influence is 100 radii or 10 cm. This yields an interaction time of 0.02 seconds for any given cloud droplet. Since the concentration of such raindrops is quite low, their effect on the growth of small droplets is negligible (unless the small drop is collected by the large drop).

5.3.3. *Drop spectra broadening in updrafts*

The growth of cloud droplets generally occurs in updrafts. Assuming that the updraft is steady, water vapor is being made available for condensation on drops uniformly throughout the volume. If the average supersaturation remains constant and there is no transfer of water vapor from one drop's volume of influence into another's, then the growth rate of each droplet is proportional to the volume it controls. Srivastava (1969) shows that under these assumptions the droplet size spectrum is defined by a probability density function.

If the updraft is not steady, but instead fluctuates in space and time, what effect may this have on the drop spectrum? The simple picture is that the supersaturation is increased in regions of higher updraft, thereby increasing the growth rate of the droplets in that volume. However, the time required to reach a given level in the cloud is correspondingly decreased. Thus, the resulting sizes of the drops at any given level might be expected to be independent of the drops' histories.

Srivastava (1989) has developed a model of droplet growth which distinguishes between the 'microscopic' supersaturation that an individual droplet 'sees' and the 'macroscopic' supersaturation as calculated from the overall water mass balance (the average water vapor density discussed above). In this model, fluctuations in the updraft and transfers of water vapor between different drop volumes are permitted. The essential conclusion remains that the droplets with the larger volumes of influence grow bigger.

It has been argued that (turbulent) diffusion would be sufficient to smooth out the effects of the drop's volume on the microscopic supersaturation that it sees. Srivastava argues that if turbulent diffusion were effective in this regard, then it would also smooth out supersaturation fluctuations due to fluctuations in the updraft velocity. Accordingly, the sizes of droplets at a given level then would depend on their histories with the droplets taking the longest time to reach a given level having the largest sizes.

Srivastava's (1989) results also suggest that competition between like droplets can be neglected if

$$\frac{l_*}{r_o}(D\tau)^{1/2} \ll 1$$

where l_* is the fraction of the volume occupied by drops, r_o is the typical drop radius, and τ is the age of the process. This implies that broadening of cloud droplet spectra through competition may be expected to increase with the age of the cloud.

5.4. Factors modifying the basic flux equations

Several factors may affect the solutions for the basic mass and heat flux equations. The presence of dissolved salts and the curvature affect the vapor density over the droplet surface, and directly affect only the mass flux equation. The kinetic and ventilation effects are generally reduced to multiplicative factors that can be applied directly to both the mass and sensible heat flux equations. These are discussed in the following sections. The final section discusses some non-steady state effects that may become significant under certain conditions.

5.4.1. *Solution and curvature effects*

The combined effects of the presence of dissolved salt and the elevation of the saturation vapor pressure over a curved surface can be approximated using the simplified version of the Köhler equation

$$\frac{e_{\mathrm{drop,x}}}{e_{\mathrm{s,x}}} = \frac{\rho_{\mathrm{drop,x}}}{\rho_{\mathrm{s,x}}} = 1 + \frac{\hat{a}}{r_{\mathrm{w}}} - \frac{\hat{b}}{r_{\mathrm{w}}^3} \tag{5.13}$$

where the subscript drop refers to the vapor pressure or vapor density over the drop surface at its (particle) temperature (x).

Applying the correction to the vapor density over the drop surface (5.13) to the diffusion equation for mass flux, (5.2) becomes

$$\frac{dm}{dt} = 4\pi r_{\mathrm{w}} D\left\{\rho_\infty - \rho_{\mathrm{s,x}}\left(1 + \frac{\hat{a}}{r_{\mathrm{w}}} - \frac{\hat{b}}{r_{\mathrm{w}}^3}\right)\right\}. \tag{5.14}$$

Note that the net effect of curvature and solute through (5.14) is to retard growth for droplets larger than their critical radius.[8] Srivastava (1991) points out that since this effect is more pronounced for smaller droplets than for larger droplets, this will lead to a broadening of the droplet spectrum.

The effects of dissolved salts and curvature are readily incorporated into the combined growth (5.12). This exercise is left for the student (Problem 5.8). The resulting combined growth equation is

$$\frac{dm}{dt} = \frac{4\pi r_w \left(S - 1 - \dfrac{\hat{a}}{r_w} + \dfrac{\hat{b}}{r_w^3} \right)}{\left[\dfrac{L_v^2}{K R_v T_\infty^2} + \dfrac{R_v T_\infty}{D e_{s,\infty}} \right]}. \tag{5.15}$$

5.4.2. *Kinetic effects*

The growth process has been assumed to be described by Fickian diffusion, whereas in reality it consists of the migrations of individual molecules. As the drop size approaches the mean free path of the water vapor molecules, the growth rate predicted by Fickian diffusion exceeds the rate at which molecules can be conveyed to the growing drop (or conveyed away from the evaporating drop). A similar difficulty occurs with heat transfer although at somewhat smaller drop sizes than for mass transfer. Thus, both the mass and heat transfers for small droplets are less than would be predicted by simple Fickian diffusion.

The rate at which water molecules condense on a drop per unit surface area is given by (3.32). Thus, the kinetic theory predicts the growth rate for a drop of radius r_w as

$$\frac{dm}{dt} = 4\pi r_w^2\, \rho(r_w + \Delta)\, \alpha_c R_v \left(\frac{m T_\infty}{2\pi k} \right)^{1/2} \tag{5.16}$$

where $\rho(r_w + \Delta)$ refers to the vapor density at one mean free path (for water molecules) distance from the drop surface $(r_w + \Delta)$ and α_c refers to the condensation coefficient.[9]

Note that the growth rate according to the kinetic treatment is proportional to the square of the drop radius whereas for Fickian diffusion (5.2), it is proportional to the drop radius. Thus, it is clear that the requirements of Fickian diffusion exceed the mass transfer possible under kinetic theory for drops smaller than some particular size. This size is on the order of the mean free path of the molecules.

[8] Note that there is no corresponding effect on the sensible heat flux. Thus, the reduced mass flux results in a reduced latent heat flux and a drop temperature closer to the ambient temperature.

[9] The condensation coefficient is the number of molecules that are actually incorporated into the drop divided by the number of molecules that collide with the drop. This is sometimes confused with the accommodation coefficient, which refers to the transfer of heat by collisions with air molecules.

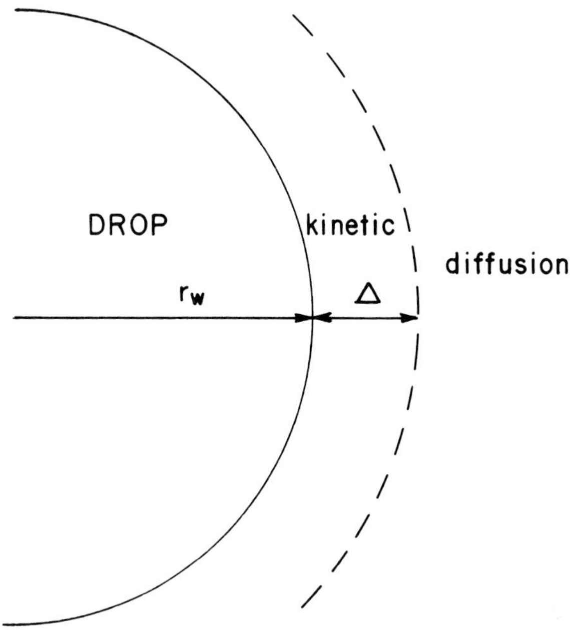

FIG. 5.2. The solution for the mass flux to a small drop is based on the diffusion equation (5.2) from $r = \infty$ to $r = r_w + \Delta$, and is based on the kinetic transfer equation (5.17) from $r = r_w + \Delta$ to $r = r_w$.

Fuchs (1959) considers a boundary layer of thickness Δ where $\Delta \approx \lambda$, through which water molecules are transferred according to the kinetic process and outside of which the transfer is according to Fickian diffusion, as shown in Fig. 5.2.[10] The net rate of transfer of water molecules through the boundary layer Δ is the difference between the influx given by (5.16) and the evaporation rate

$$\frac{dm}{dt} = 4\pi r_w^2 \, \rho(r_w) \, \alpha_e R_v \left(\frac{mT_\infty}{2\pi k}\right)^{1/2},$$

where α_e is the evaporation coefficient. Assuming the condensation and evaporation coefficients to be equal, the net rate of transfer of water vapor molecules is

$$\frac{dm}{dt} = 4\pi r_w^2 \left(\frac{mT_\infty}{2\pi k}\right)^{1/2} \alpha \{\rho(r_w + \Delta) - \rho(r_w)\} . \tag{5.17}$$

Note that the temperatures at r_w and $r_w + \Delta$ have been assumed to be the same.

[10] Robinson and Scott (1981) present a treatment that provides for a smooth transition between kinetic and diffusive fluxes, which reduces to the conventional form as a first approximation. The largest differences are about 5% for drops smaller than 1 μm radius.

The net kinetic transfer from $r_w + \Delta$ to the drop according to (5.17) must equal the diffusive transfer from r_∞ to $r_w + \Delta$ since no sources or sinks of water molecules are permitted at $r_w + \Delta$. Thus,

$$4\pi(r_w + \Delta)\, D\{\rho_\infty - \rho(r_w + \Delta)\} = 4\pi r_w^2\, \alpha R_v \left(\frac{mT_\infty}{2\pi k}\right)^{1/2} \{\rho(r_w + \Delta) - \rho(r_w)\}\,.$$

Solving for the vapor density at $r_w + \Delta$,

$$\rho(r_w + \Delta) = \frac{Ar_w^2 \rho(r_w) + (r_w + \Delta)\, D\rho_\infty}{Ar_w^2 + (r_w + \Delta)D},$$

where

$$A = \alpha R_v \left(\frac{mT_\infty}{2\pi k}\right)^{1/2}.$$

Substituting this into the diffusion equation

$$\frac{dm}{dt} = 4\pi(r_w + \Delta)\, D \left\{ \rho_\infty - \frac{\rho(r_w)Ar_w^2 + (r_w + \Delta)D\rho_\infty}{Ar_w^2 + (r_w + \Delta)D} \right\},$$

which, after some algebraic manipulation, may be written as

$$\frac{dm}{dt} = 4\pi r_w D(\rho_\infty - \rho_{s,x}) \left\{ \frac{Ar_w(r_w + \Delta)}{Ar_w^2 + (r_w + \Delta)D} \right\}. \tag{5.18}$$

The term in braces in (5.18) represents a multiplicative factor (f_1) which may be expressed as

$$f_1 = \left[\left(\frac{r_w}{r_w + \Delta}\right) + \frac{D}{\alpha r_w} \left(\frac{2\pi}{R_v T_\infty}\right)^{1/2} \right]^{-1} \tag{5.19}$$

A similar treatment for heat transfer yields an analogous factor

$$f_1^* = \left[\left(\frac{r_w}{r_w + \Delta_H}\right) + \frac{\nu_c}{\alpha_H r_w} \left(\frac{2\pi}{R_d T_\infty}\right)^{1/2} \right]^{-1} \tag{5.20}$$

where ν_c is the thermometric conductivity and α_H is the accommodation coefficient.

Experimental values for the condensation coefficient generally range from 0.01 to 0.07. Pruppacher and Klett (1978) list a variety of experimental

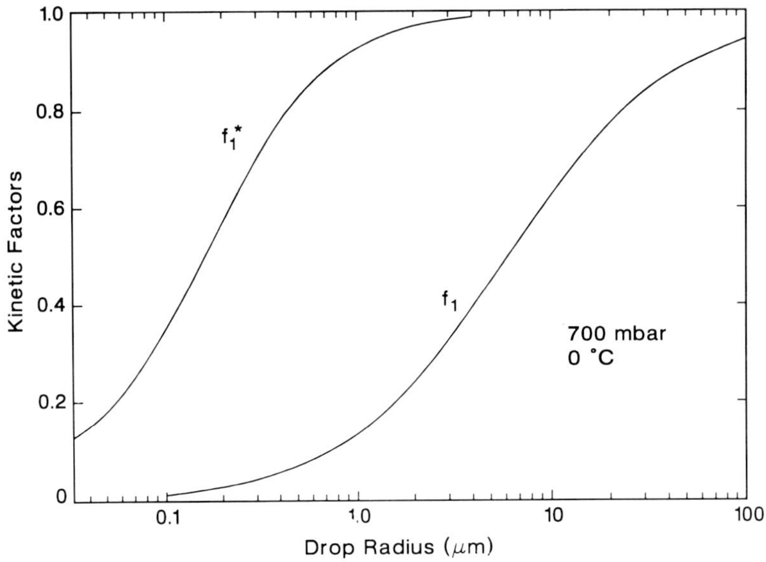

FIG. 5.3. The kinetic factors for mass (f_1) and sensible heat (f_1^*) transfer as a function of drop radius. Values are for P = 700 mbar and T_∞ = 0°C.

determinations of the condensation coefficient; a value of 0.036 is generally used. Fukuta and Walter (1970) suggest that the condensation coefficient is temperature dependent, although the data they present suggest that this effect is not significant over the range of temperatures normally encountered in the atmosphere. The accommodation coefficient α_H is usually assumed to be close to unity. The thickness of the boundary layer Δ is roughly 1.0×10^{-5} cm; that for heat transfer Δ_H is roughly 2.0×10^{-5} cm.

These kinetic effects significantly affect the growth rates for water drops smaller than 10 μm radius (see Fig. 5.3). Since these coefficients are an integral part of the mass and heat flux equations, they must also appear in the particle wet-bulb equation (5.7). Because $f_1^* > f_1$, the wet-bulb elevation for growing drops (or the wet-bulb depression for evaporating drops) will be reduced for smaller drops compared to larger drops.

5.4.3. *Ventilation effects*

A falling drop tends to fall into the vapor density and temperature fields surrounding it, as shown in Fig. 5.4. Since these fields cannot adjust instantaneously to the drop's changing position, the gradients tend to be enhanced below the drop and spread out behind the drop. If this effect is symmetrical, i.e., a doubling of the gradient below the drop is matched by a halving of the gradient above the drop, it is clear that this effect will serve to enhance the net growth (or evaporation) of the drop.

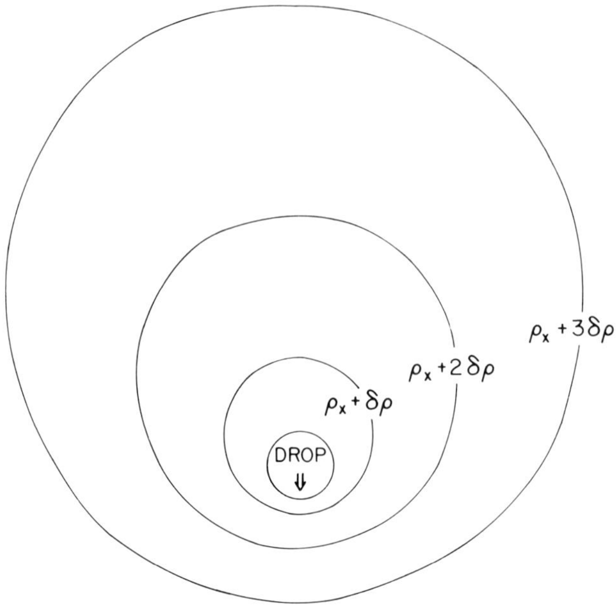

FIG. 5.4. A falling drop tends to pack the water vapor isopynics (and isotherms) beneath it and spread them out above it. The net effect is to increase the growth (evaporation) rate.

An engineering approach to the ventilation is usually relied on. The ventilation factors for mass and sensible heat transfer are expressed in terms of the Sherwood (N_{Sh}) and Nusselt (N_{Nu}) numbers, respectively, as

$$f_2 = \frac{1}{2} N_{Sh}$$

and

$$f_2^* = \frac{1}{2} N_{Nu} .$$

Woo and Hamielec (1971) present a numerical solution for the ventilation factors for freely falling drops. Their results are in good agreement with experimentally determined values.

A solution for the ventilation factors for suspended drops is given by Dye and Hobbs (1968). The presence of temperature and water vapor gradients in the vicinity of a drop result in convection due to radial differences in air density. In the near vicinity of an evaporating drop, the reduced temperature increases the air density, whereas the enhanced water vapor content decreases the air density. Since the effect of temperature is greater, air in the near vicinity of a suspended drop which is evaporating, is sinking. This enhances the mass and sensible heat transfers, although to a much smaller degree that for freely falling drops.

The ventilation factors for mass and sensible heat transfer are usually assumed to be equal. For drops smaller than about 60 μm radius, Beard and Pruppacher (1971) provide the empirical expression

$$f_2 \approx f_2^* \approx 1.00 + 0.108\, X^2 \qquad \text{for } X < 1.4\,. \tag{5.21}$$

For drops larger than about 60 μm radius, Pruppacher and Rasmussen (1979) provide the empirical expression

$$f_2 \approx f_2^* \approx 0.78 + 0.308\, X \qquad \text{for } X > 1.4\,. \tag{5.22}$$

The parameter X in both (5.21) and (5.22) is

$$X = N_{\mathrm{Sc}}^{1/3}\, N_{\mathrm{Re}}^{1/2}\,,$$

where the Schmidt number (N_{Sc}) and the Reynolds number (N_{Re}) are given in Appendix 2.

The ventilation factor is shown in Fig. 5.5 as a function of drop radius for a pressure of 700 mbar and an ambient temperature of 0°C. For drops smaller than about 30 μm radius the ventilation factor can be assumed to be unity. For drops between 100 μm and 500 μm radius, the ventilation factor is very nearly a linear function of drop size.

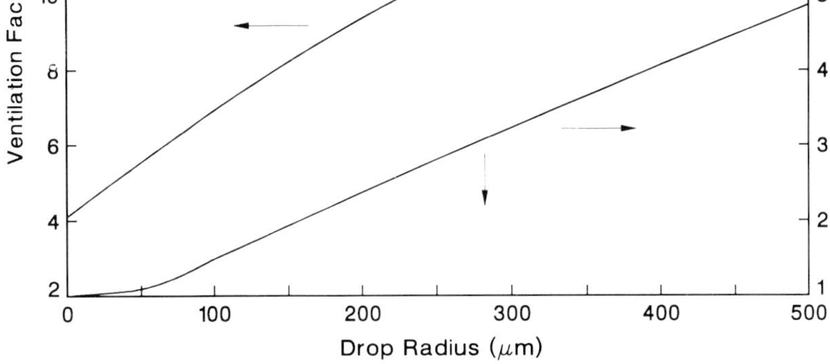

FIG. 5.5. The ventilation factor is calculated from (5.21) and (5.22) as a function of equivalent drop radius for P = 700 mbar and T_∞ = 0°C. The fall velocities are based on Chapter 8. The upper curve is referenced by the top and left scales; the lower curve by the bottom and right scales.

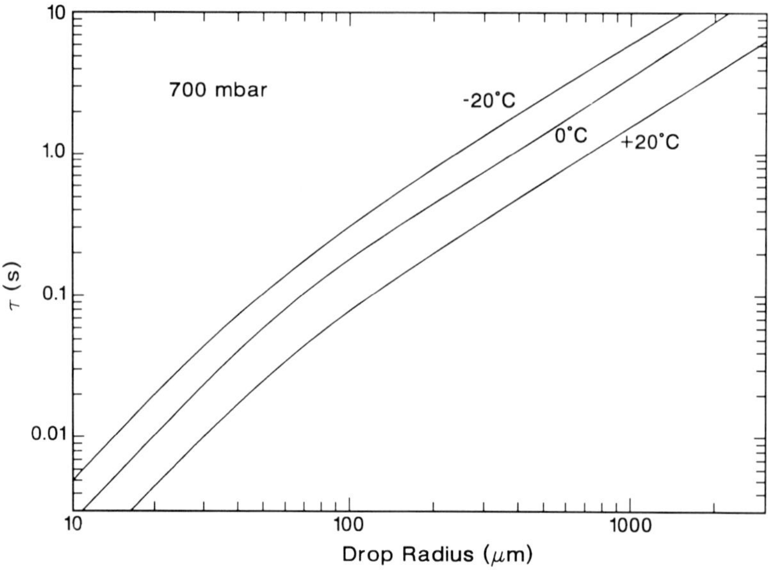

Fig. 5.6. The relaxation time for water drops calculated from (5.23) for $P = 700$ mbar and different ambient temperatures. Drops adjust more rapidly at warmer temperatures since both the water vapor and thermal diffusivities are greater.

5.4.4. *Non-steady state effects*

The presentations in this chapter have been based on the assumption that the temperature and vapor density fields are steady state and do not change with time. Clearly, as drops fall through temperature or humidity gradients in the atmosphere, the temperature and vapor density fields surrounding the drops must change in response to these effects.

Kinzer and Gunn (1951) show that the relaxation time[11] for freely falling water drops is

$$\tau = \frac{r_w^2 \rho_w c_w}{3 f_2 \left[K + L_v D \dfrac{\partial \rho_{s,w}}{\partial T} \right]}, \tag{5.23}$$

where ρ_w is the density of water and c_w is the specific heat for water. Figure 5.6 suggests that the relaxation times are generally less than 0.1 seconds for cloud droplets ($r_w < 50\ \mu$m), and even for millimeter-sized drops, the relaxation times are on the order of a few seconds. Thus, for most conditions encountered in the atmosphere, the assumption that the drop is in steady state with its environment is probably a good one.

[11] The relaxation time is the time required for the difference between the steady state drop temperature and its actual temperature to decrease by a factor of $1/e$.

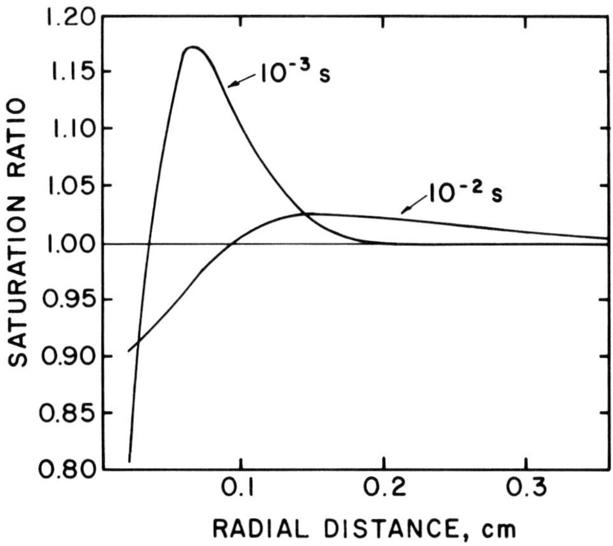

FIG. 5.7. As a drop freezes, it warms quickly to 0°C. As the water vapor density and temperature fields respond to the new conditions, transient water supersaturations arise. The drop in this case was initially at -15°C (r_w = 100 μm). From Rosinski and Kerrigan (1972). *Zeit. ang. Math. Phys.*, **23**. Birkhäuser Verlag AG. Reproduced with permission.

In the case of the freezing of a supercooled water drop, rapid changes in the vapor density and temperature at the particle surface occur as it warms to near 0°C. The water vapor density and temperature of the near environment must increase for these fields to approach steady state. The rate at which the vapor density and temperature fields approach steady state is determined by the water vapor (D) and thermal (κ) diffusivities respectively. Since $D > \kappa$, the temperature at a given distance from the particle will lag behind the water vapor density, creating a transient supersaturated condition.

Calculations by Rosinski and Kerrigan (1972) show that such transient supersaturations may reach values greater than 10% over water in the vicinity of the freezing drop. Figure 5.7 indicates the rapidity with which the fields adjust to their new conditions. The occurrence of similar transient supersaturations within thermal gradient diffusion chambers under certain conditions is detailed by Fitzgerald (1970).

Problems

5.1* Calculate the particle wet-bulb depression for water drops of 1, 10, 100, and 1000 μm radius. Assume P = 700 mbar, T_∞ = 0°C, and S = 0.8.

5.2* What is the maximum wet-bulb depression possible for a drop at an ambient temperature of 10°C? (Neglect kinetic and ventilation effects.)

5.3* Derive dr_w/dt for condensation growth. Discuss the implications this has on the width of the droplet size spectrum.

5.4* Compare the solutions for dm/dt using (5.2) versus (5.12). Choose values for the ambient pressure and temperature, and evaluate the relative error over a range of saturation ratio.

5.5** Derive (5.3) for the radial dependence of vapor density around a drop.

5.6** Given a 1 mm radius drop at $T_\infty = -15°\text{C}$ and $P = 700\,\text{mbar}$, how long would such a drop take to freeze completely after the initiation of nucleation? What fraction of its mass would it lose during the freezing process? Assume the environment is at water saturation.

5.7** What is the latent heat of vaporization for a drop of radius 0.1 μm at 0°C?

5.8** The drop growth equation as derived by Mason (1953) can be adjusted for small drops by subtracting a curvature term and adding a solute term to the supersaturation, as shown in (5.15). Rederive Mason's equation using

$$\frac{e_{\text{drop},x}}{e_{s,\infty}}$$

rather than

$$\frac{e_{s,x}}{e_{s,\infty}}.$$

5.9*** Plot the vapor density as a function of radial distance from a drop of radius 1 μm with $S = 1.0$ and $T_\infty = 15°\text{C}$. Consider the kinetic effects.

5.10*** How long would it take to melt an ice sphere of radius 0.1 cm in an environment at 700 mbar, 10°C, and 20% relative humidity? What would the final radius of the melted particle be? It may be assumed that the particle is stationary and that ventilation effects can be ignored.

Computer problem

5.11** How far would a water drop of radius 0.1 cm fall in a completely dry (and isothermal) atmosphere at 0°C? How far would a drop of radius 10 μm fall under the same conditions? Start at 600 mbar and assume the surface is at 1000 mbar.

REFERENCES

Beard, K.V., and Pruppacher, H.R. 1971, 'A wind tunnel investigation of the rate of evaporation of small water drops falling at terminal velocity in air,' *J. Atmos. Sci.* **28**, 1455.

Dye, J.E., and Hobbs, P.V.,1968, 'The influence of environmental parameters on the freezing and fragmentation of suspended water drops,' *J. Atmos. Sci.* **25**, 82.

Fitzgerald, J.W., 1970, 'Non-steady state supersaturations in thermal diffusion chambers,' *J. Atmos. Sci.* **27**, 70.

Fuchs, N.A., 1959, *Evaporation and droplet growth in gaseous media.* New York: Pergamon Press.

Fukuta, N., and Walter, L.A., 1970, 'Kinetics of hydrometeor growth from a vapor-spherical model,' *J. Atmos. Sci.* **27**, 1160.

Hall, W.D., and Pruppacher, H.R., 1976, 'The survival of ice particles falling from cirrus clouds in subsaturated air,' *J. Atmos. Sci.* **33**, 1995.

Kinzer, G.D., and Gunn, R., 1951, 'The evaporation, temperature and thermal relaxation-time of freely falling waterdrops,' *J. Meteorol.* **8**, 71.

Mason, B.J., 1953, 'The growth of ice crystals in a supercooled water cloud,' *Quart. J. R. Meteorol. Soc.* **79**, 104.

Pruppacher, H.R., and Klett, J.D., 1978, *Microphysics of clouds and precipitation.* Dordrecht: Reidel.

Pruppacher, H.R., and Rasmussen, R., 1979, 'A wind tunnel investigation of the rate of evaporation of large water drops falling at terminal velocity in air,' *J. Atmos. Sci.* **36**, 1255.

Robinson, N.F., and Scott, W.T., 1981, 'Two-stream Maxwellian kinetic theory of cloud droplet growth by condensation,' *J. Atmos. Sci.* **38**, 1015.

Rosinski, J., and Kerrigan, T.C., 1972, 'Mechanism of ice formation in seeded convective storms,' *Zeit. Ang. Math. Phys.* **23**, 277.

Srivastava, R.C., 1969, 'Note on the theory of growth of cloud drops by condensation,' *J. Atmos. Sci.* **26**, 776.

Srivastava, R.C., 1989, 'Growth of cloud drops by condensation: A criticism of currently accepted theory and a new approach,' *J. Atmos. Sci.* **46**, 869.

Srivastava, R.C., 1991, 'Growth of cloud drops by condensation: Effect of surface tension on the dispersion of drop sizes,' *J. Atmos. Sci.* **48**, 1596.

Staley, D.O., 1972, 'Heat conduction in the vicinity of a growing drop,' *J. Atmos. Sci.* **29**, 206.

Woo, S., and Hamielec, A.E., 1971, 'A numerical method of determining the rate of evaporation of small water drops falling at terminal velocity in air,' *J. Atmos. Sci.* **28**, 1448.

Young, K.C., 1974, 'A numerical simulation of wintertime, orographic precipitation. Part I. Description of model microphysics and numerical techniques,' *J. Atmos. Sci.* **31**, 1735.

6

DIFFUSION GROWTH OF ICE CRYSTALS

6.1. Introduction

The growth of ice crystals within a cloud is the mechanism by which much of the precipitation in mid-latitude frontal systems forms, as well as being important in the development of precipitation in many convective clouds. Ice crystals may grow to relatively large sizes and form aggregates, or they may grow through accretion to form rimed crystals, graupel, and hail. Small ice crystals may be collected by water drops and the resulting frozen drop (droxtal) may grow further by accretion.

The complexity of ice crystal habits makes it difficult to describe accurately ice crystal growth under many natural conditions. The habit (shape) of an ice crystal is determined primarily by its temperature, the ambient vapor density excess over ice saturation, and the crystal's prior history. Even under fixed ambient conditions, the crystal's aspect ratio[1] changes as the crystal grows.

The way in which an ice crystal grows is referred to as its growth mode. The integration of the growth equation for ice crystals must account for the crystal habit as well as the growth mode. It will be shown that the growth mode has a dominant influence on the growth of ice crystals.

Ice crystal growth is regulated to a large extent by the surface properties of the crystal lattice. Under many conditions, water molecules cannot be incorporated into the crystal lattice at their arrival position. Instead, molecules must be incorporated into steps or corners in the ice lattice, and must migrate across the crystal surface until they either reach such a site or return to the vapor. The necessity of incorporating incoming water molecules into the growing ice lattice may substantially reduce the growth rate of the ice crystal.

6.2. Ice crystal habits

In Chapter 2, it was pointed out that ice crystals possess a hexagonal symmetry, growing in the shape of a right hexagonal prism. Ice crystals observed in nature generally conform to the basic hexagonal symmetry, but exhibit a wide range of aspect ratios and are frequently found to have incomplete or skeletal structures.

[1] The aspect ratio is defined as the c-axis length divided by the a-axis length, i.e., the distance between the basal faces divided by the distance between opposite prism faces.

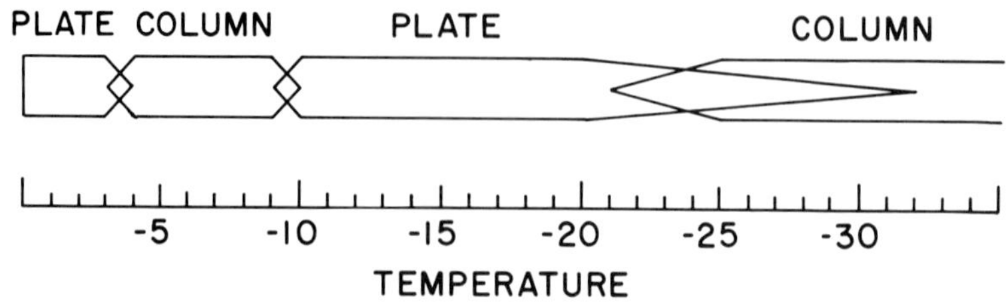

FIG. 6.1. The principal ice crystal habits as a function of temperature. Overlapping regions represent transition zones. From Ono (1970). *Journal of Atmospheric Sciences*, **27**, 4. American Meteorological Society. Reproduced with permission.

Photographs of natural ice crystals usually represent the most photogenic examples collected at the ground. Although such photographs demonstrate the variety that ice crystals may exhibit, such examples yield relatively little information regarding the conditions under which the crystals formed. Observations of ice crystal growth under carefully controlled conditions in the laboratory are more useful, although limited to fairly small crystals or to suspended crystals. Suspended crystals lack the ventilation effects experienced by freely falling crystals. Thus, collection of ice crystals from the clouds within which they are growing is a requisite for understanding ice crystal growth in clouds.

Crystal habits may be classified as simple or complex in the following manner. The habit attained by a single ice crystal grown under constant environmental conditions may be considered to be a simple habit. Complex habits arise when a single ice crystal is transferred from one growth regime to another, e.g., column to plate, or when the ice crystal forms on a frozen drop consisting of two or more different crystallographic orientations.

Crystal habits may also be classified as growth habits or sublimation habits. Growth habits for ice crystals are generally quite dissimilar to the equilibrium habits in which the aspect ratio is close to unity. Sublimation habits tend to be more similar to the equilibrium habits.

6.2.1. *The growth of simple ice crystals*

The two primary influences on the habit assumed by a growing ice crystal are the temperature and the vapor density excess over ice saturation. The temperature determines whether a crystal grows as a plate (a-axis > c-axis) or as a column (c-axis > a-axis). There are three transitions between plate and column growth, occurring at temperatures of $-4°C$, $-10°C$, and near $-23°C$, as shown in Fig. 6.1. The transitions at $-4°C$ and $-10°C$ are well-defined and crystals growing near the transition zones tend to be isometric.[2] The transition zone between $-21°C$ and $-32°C$ is rather gradual, and crystals of either type may occur.

The vapor density excess largely determines the aspect ratio and the degree of skeletal growth. The dependence of the crystal habit on temperature and vapor

[2] Isometric crystals have an aspect ratio of unity.

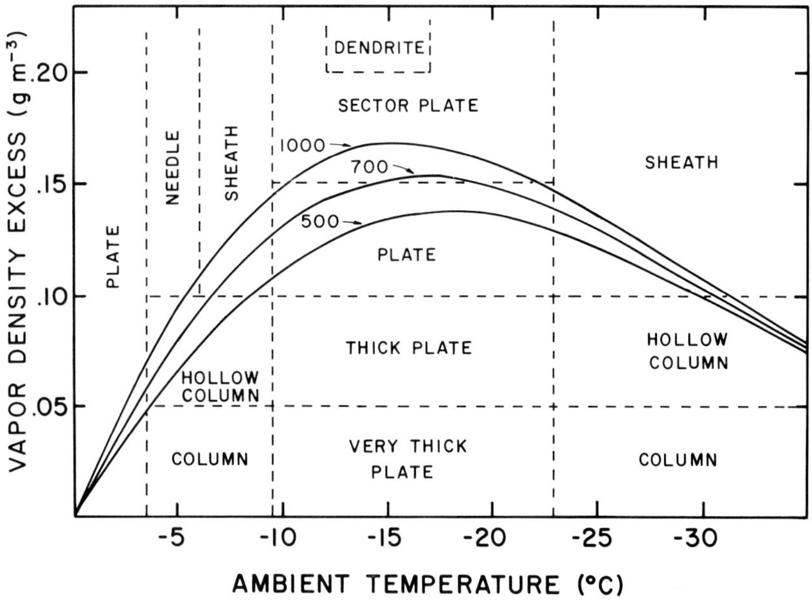

FIG. 6.2. Simple crystal habits as a function of temperature and the vapor density excess between the environment and the crystal surface. These represent the habits attained by large ice crystals. The vapor density excesses corresponding to water saturation at 1000 mbar, 700 mbar, and 500 mbar are shown to illustrate how ice crystal habits may change with altitude.

density excess is shown in Fig. 6.2. In the c-axis growth region, solid columns grow at low vapor density excesses. At higher vapor density excesses, the basal faces tend to be incomplete, forming hollow columns or sheaths. Needles represent sheaths whose prism faces are also incomplete. Examples of c-axis growth ice crystals are illustrated in Fig. 6.3.

In the a-axis growth region, thick plates grow at low vapor density excesses. Thin plates grow just below water saturation, and sector plates and dendrites grow at relatively high vapor density excesses. Examples of a-axis growth ice crystals are illustrated in Fig. 6.4.

Ice crystals usually encounter three growth regimes as they grow from small to large sizes. When the crystals are very small ($r < 5$ μm), they tend to grow nearly isometrically. Once this size is exceeded, crystals tend to grow in a habit characteristic of the temperature, i.e., into either plates or columns. While in this growth regime, the crystal tends to extend both the c-axis and the a-axes with a constant aspect ratio.

Columnar ice crystals will grow along both the c-axis and the a-axes until the a-axes reach a limiting size. Once this limiting dimension is reached, growth is limited to the c-axis, extending the length of the crystal but not its diameter. Plate crystals encounter a similarly limiting dimension for the c-axis. Once plate crystals reach a certain thickness, they grow by extending their diameter at

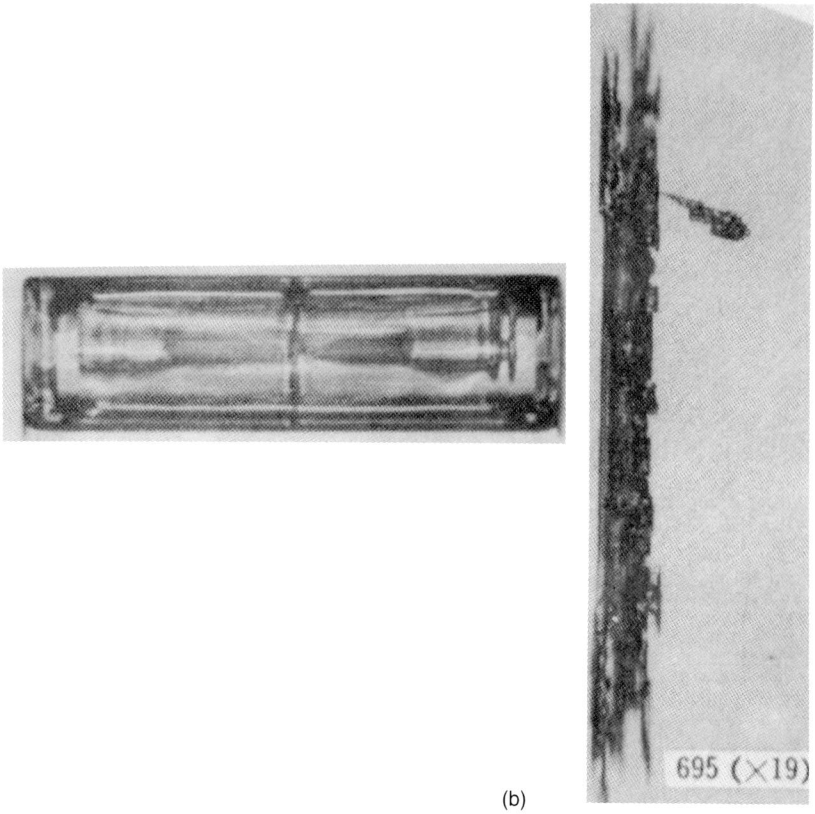

(a) (b)

Fig. 6.3. Examples of c-axis growth regime ice crystals. (a) Banded column (C1e); (b) needle (N1b). Reprinted by permission of the publishers from *Snow Crystals: Natural & Artificial*, Ukichio Nakaya, Cambridge, Mass.: Harvard University Press, Copyright © 1954 by the President and Fellows of Harvard College.

constant thickness. This behavior is shown in Fig. 6.5, based on the observations of natural ice crystals collected in clouds by Ono (1970).

From Ono's data, the following limiting dimensions may be deduced. In the needle growth region between −4°C and −6°C, the crystal a-axes (radius) are limited to a value between 25 μm and 60 μm. In the sheath growth region between −6°C and −8°C, the a-axes are limited to a value between 40 μm and 80 μm. Columns grown between −8°C and −10°C are in the transition zone between column and plate growth, and exhibit limiting a-axes between 60 μm and 100 μm. Plate crystals near −16°C are limited to a c-axis of about 15 μm, with corresponding values at −11°C and −18°C of 30 μm and 23 μm, respectively.[3]

[3] The convention in this text is to refer to the lengths of the axes as the perpendicular distances from the center of the crystal to the center of the relevant crystal fae. This corresponds to using the radius to describe the size of a spherical drop.

Small ice crystals tend to grow as complete prisms with densities close to that for bulk ice. As an ice crystal grows, its density may tend to decrease even prior to reaching its limiting dimension. Columnar ice crystals grown by Fukuta (1969) under laboratory conditions exhibit skeletal growth clearly prior to reaching their limiting dimension. As the ice crystal grows to large sizes, its bulk

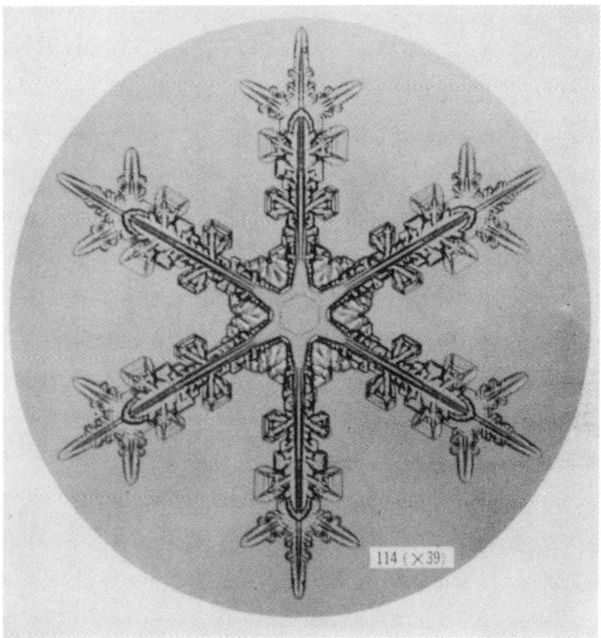

(a)

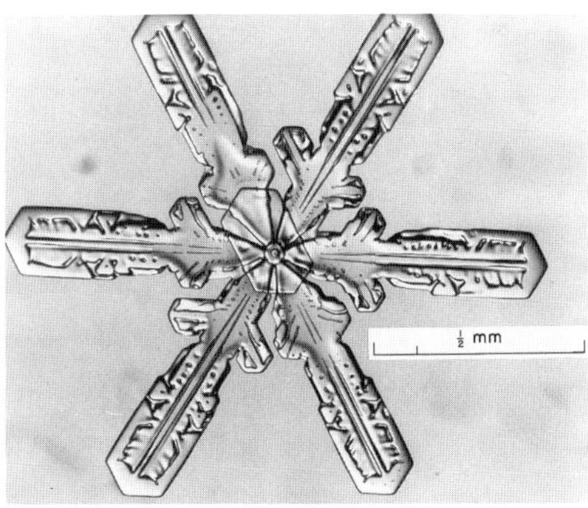

(b)

Fig. 6.4. Examples of a-axis growth regime ice crystals. (a) Dendrite (P1c), (b) sector plate (P1b) (courtesy of C.A. Knight). Note the characteristic pointed ends of the branches for dendrites, compared to the flatter ends of sector plates. (a) Reprinted by permission of the publishers from *Snow Crystals: Natural & Artificial*, Ukichio Nakaya, Cambridge, Mass.: Harvard University Press, Copyright © 1954 by the President and Fellows of Harvard College.

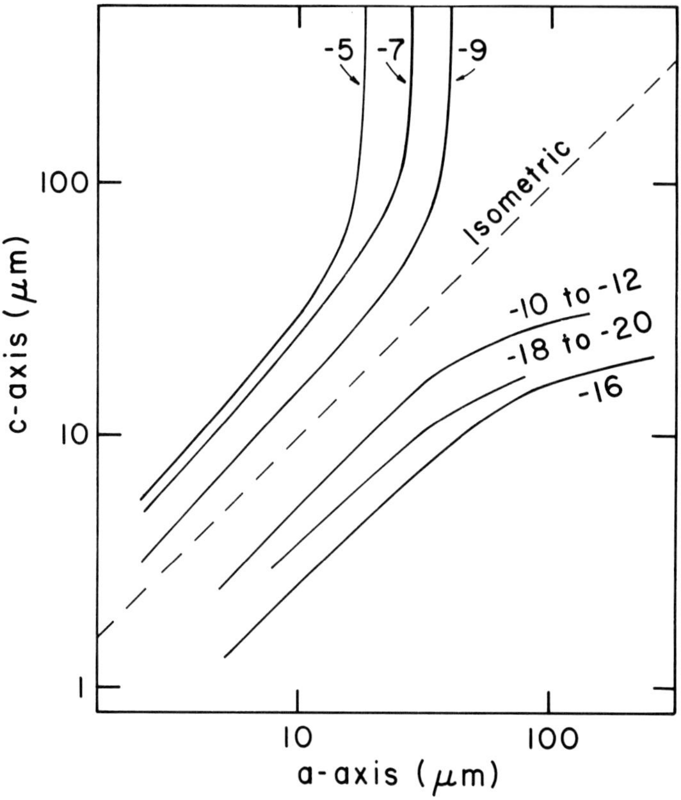

FIG. 6.5. Ice crystal dimensions for natural ice crystals for various temperatures. Note that in all cases, the ice crystals tend to grow with a constant aspect ratio until they reach a limiting dimension. Once the limiting dimension is reached, growth along that axis ceases. After Ono (1970). *Journal of Atmospheric Sciences*, **27**, 4. American Meteorological Society. Reproduced with permission.

density[4] approaches a limiting value, as noted by Ono (1970). Values for the bulk densities are given in Table 6.1.

6.2.2. *The growth of complex ice crystals*

If a columnar ice crystal is transferred from its original growth region into a region where the predominant growth is along the a-axis, it will attempt to grow as a plate. Since the c-axis length of columns is generally much greater than the limiting value for plates, the crystal will grow plates at each end of the crystal, forming a *capped column*. Similarly, a plate crystal transferred into a c-axis growth region tends to sprout needles along its edges.

Crystals that grow from frozen drops which consist of more than one crystallographic orientation tend to develop several branches. The number of

[4] Bulk density is defined as the crystal mass divided by the volume of the right hexagonal prism defined by the *a*-axes and *c*-axis dimensions.

Table 6.1 Bulk densities for large ice crystals of various habits.

Crystal habit	Bulk density ($g\ cm^{-3}$)
Solid columns	0.9
Hollow columns	0.7
Sheaths	0.6
Needles	0.3
Plates	0.9
Sector plates	0.5
Dendrites	0.16

branches and their separation is probably determined by competition between them for the available water vapor. If two branches attempt to grow too close together, they compete for the water vapor and only one is likely to grow. The branches are generally distributed in three dimensions, forming spatial dendrites in the plate growth regime and clusters of needles or bullet rosettes in the column growth regime. Some complex ice crystal habits are illustrated in Fig. 6.6.

The unusual shapes of bullet and pyramid ice crystals which arise from bullet rosettes are also apparently due to competition for the available water vapor while the individual bullets are still small. This serves to limit their a-axes, and as the bullet extends further away from the parent frozen drop, competition decreases and the a-axes increase, until reaching their limiting dimension for columns. If the limiting dimension is not reached, the result is a pyramid. As bullet rosettes sublimate, the bullets and pyramids tend to separate as the necks connecting them to the parent frozen drop weaken. Thus, one can find individual bullet and pyramid ice crystals as well as the complete rosettes.

The existence of these types of ice crystals depends on the number of individual crystallographic orientations that arise from the freezing of a supercooled water drop. Hallett (1964) found that drops which freeze at temperatures warmer than −5°C will freeze monocrystalline. Drops of radius 1 mm freezing near −10°C tend to produce between two and five crystals, increasing to hundreds of crystals for drops freezing below −30°C. Bullet rosettes are fairly common in cold clouds (−30°C and colder) but are virtually non-existent in clouds warmer than −10°C.

Frozen drops may also grow crystals of the same orientation from opposite portions of the frozen drop. In the a-axis growth regime, these tend to grow into *tsuzumi* crystals, so named for their drum-like appearance of two closely separated plates.[5] In the c-axis growth regime, the frozen drop forms a dark band parallel to the a-axes, dividing the crystal into roughly equal parts. These crystals are termed *banded columns*.

[5] Tsuzumi is the Japanese word for drum.

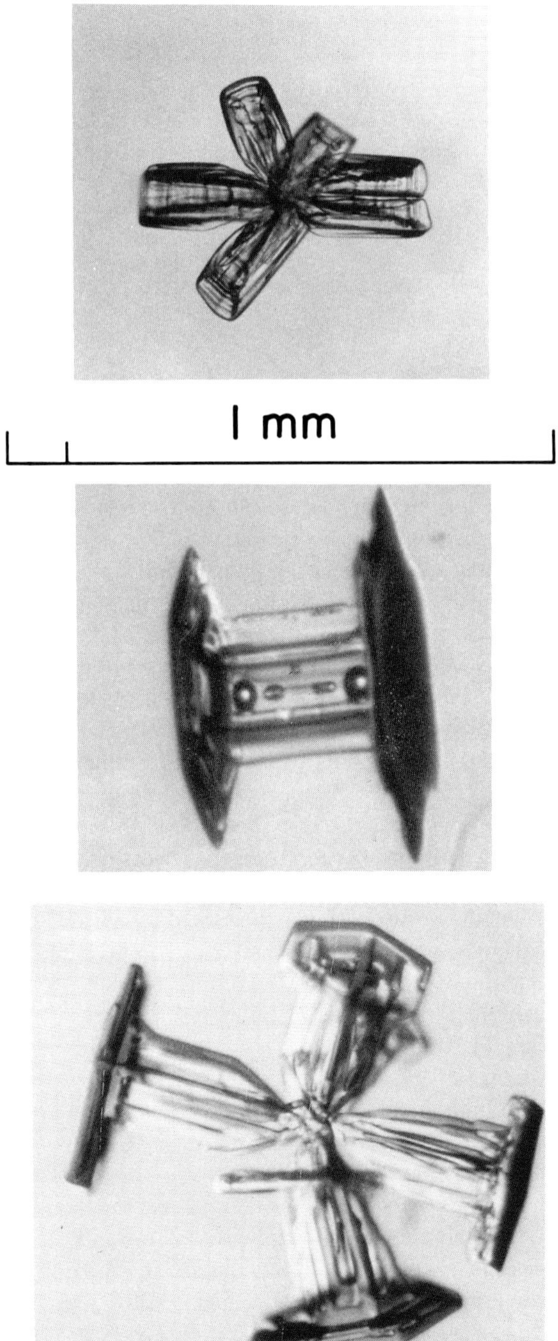

FIG. 6.6. Examples of complex ice crystal habits. (a) Bullet rosette (C2a); (b) capped column (CP1a); (c) bullet rosette with plates (CP2a) (courtesy of C.A. Knight).

6.2.3. *Classification of ice crystals*

A general classification system for ice particles which has gained widespread acceptance is given by Magono and Lee (1966). The classification system is summarized in Table 6.2 and Fig. 6.7. Their paper includes 159 photographs, giving examples of virtually all of their 80 different crystal types.

The classification system considers three types of simple ice crystals (needles, columns, and plates), two types of complex ice crystals, rimed particles, irregular particles, and small 'germs' of ice crystals. The basic type is given a capital letter (two for combinations of plates and columns). Each basic type is divided into several categories (number) and most of these are further subdivided (lower case letter). For example, a hollow column is C1f, indicating a columnar crystal (C), which consists of a single crystal (1) and is of the hollow column subcategory (f).

6.3. The equations governing ice crystal growth

The wide variety of shapes for ice crystals and the fact that ice crystals tend to have relatively pointed edges and corners, renders a general analytic solution to the growth equation (5.1) an extremely difficult task. The common solution relies on the analogous use of capacitance in dealing with irregularly shaped objects in electrostatics. In the following, the *electrostatic analogy* will be introduced, and shape factors which replace r_w in (5.2) and (5.5) will be presented. The growth equation will then be integrated for different growth modes. A comparison of the numerical integration of the growth equation with observations of ice crystal growth will be presented.

6.3.1. *The electrostatic analogy*

The current (i) which flows to an object with capacitance C through a medium with conductivity Λ, given a potential difference between the object (at V_o) and 'infinity' (at V_∞) is given by

$$i = 4\pi C\Lambda\,(V_\infty - V_o)\,.$$

It is clear that this is of the same form as (5.2) for mass transfer and (5.5) for sensible heat transfer, since the capacitance of a sphere is equal to its radius. The electrical conductivity is noted to be the analog of the diffusivity of water vapor in air (D) and the thermal conductivity of air (K).

The derivation of capacitance in electrostatics assumes that the surface of the object is at a constant potential. This is equivalent to the surface of an ice crystal being at a constant temperature or there being a constant vapor density at all points at the surface. Thus, in order for the analogy to hold, the surface of the ice crystal must be at a constant temperature.

Table 6.2 Classification of ice crystals.

N	Needle crystal	1. Simple needle	a. Elementary needle b. Bundle of elementary needles c. Elementary sheath d. Bundle of elementary sheaths e. Long solid column
		2. Combination of needle crystals	a. Combination of needles b. Combination of sheaths c. Combination of long solid columns
C	Columnar crystal	1. Simple column	a. Pyramid b. Cup c. Solid bullet d. Hollow bullet e. Solid column f. Hollow column g. Solid thick plate h. Thick plate of skeleton form i. Scroll
		2. Combination of columns	a. Combination of bullets b. Combination of columns
P	Plane crystal	1. Regular crystal developed in one plane	a. Hexagonal plate b. Crystal with sectorlike branches c. Crystal with broad branches d. Stellar crystal e. Ordinary dendritic crystal f. Fernlike crystal
		2. Plane crystal with extensions of different form	a. Stellar crystal with plates at ends b. Stellar crystal with sectorlike ends c. Dendritic crystal with plates at ends d. Dendritic crystal with sectorlike ends e. Plate with simple extensions f. Plate with sectorlike extensions g. Plate with dendritic extensions
		3. Crystal with irregular number of branches	a. Two-branched crystal b. Three-branches crystal c. Four-branches crystal
		4. Crystal with twelve branches	a. Broad branch crystal with twelve branches b. Dendritic crystal with twelve branches
		5. Malformed crystal	
		6. Spatial assemblage of plane branches	a. Plate with spatial plates b. Plate with spatial dendrites c. Stellar crystal with spatial plates d. Stellar crystal with spatial dendrites
		7. Radiating assemblage of plane branches	a. Radiating assemblage of plates b. Radiating assemblage of dendrites

Table 6.2 (*continued*)

CP	Combination of column and plate crystals	1. Column with plane crystals at both ends	a. Column with plates b. Column with dendrites c. Multiple capped column
		2. Bullet with plane crystals	a. Bullet with plates b. Bullet with dendrites
		3. Plane crystal with spatial extensions at ends	a. Stellar crystal with needles b. Stellar crystal with columns c. Stellar crystal with scrolls at ends d. Plate with scrolls at ends
S	Columnar crystal with extended side planes	1. Side planes	
		2. Scalelike side planes	
		3. Combination of side planes, bullets and columns	
R	Rimed crystal	1. Rimed crystal	a. Rimed needle crystal b. Rimed columnar crystal c. Rimed plate or sector d. Rimed stellar crystal
		2. Densely rimed crystal	a. Densely rimed plate or sector b. Densely rimed stellar crystal c. Stellar crystal with rimed branches
		3. Graupellike snow	a. Graupellike snow of hexagonal type b. Graupellike snow of lump type c. Graupellike snow with unrimed extensions
		4. Graupel	a. Hexagonal graupel extensions b. Lump graupel c. Conical graupel
I	Irregular snow crystal	1. Ice particle	
		2. Rimed particle	
		3. Broken piece from crystal	a. Broken branch b. Rimed broken branch
		4. Miscellaneous	
G	Germ of ice crystal	1. Minute column	
		2. Germ of skeleton form	
		3. Minute hexagonal plate	
		4. Minute stellar crystal	
		5. Minute assemblage of plates	
		6. Irregular germ	

Source: After Magono and Lee (1966).

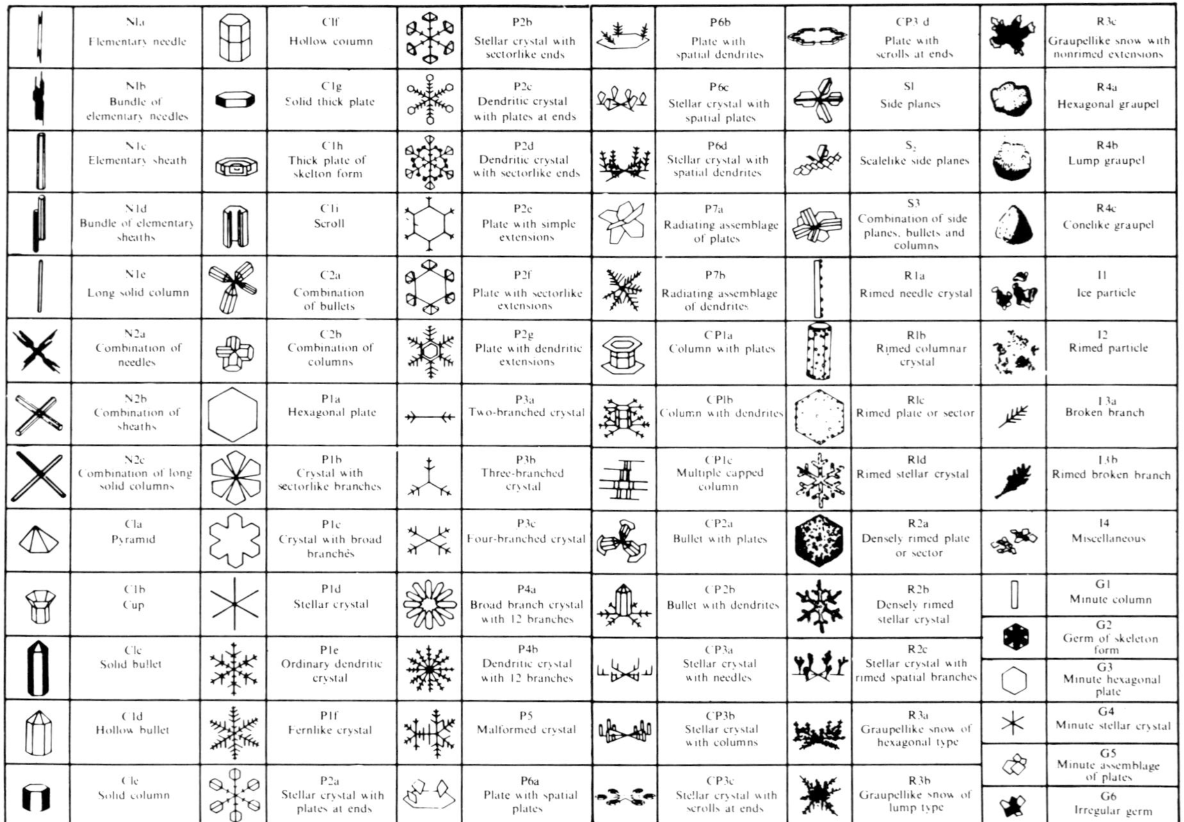

Fig. 6.7. Classification of ice crystals. After Magono and Lee (1966).

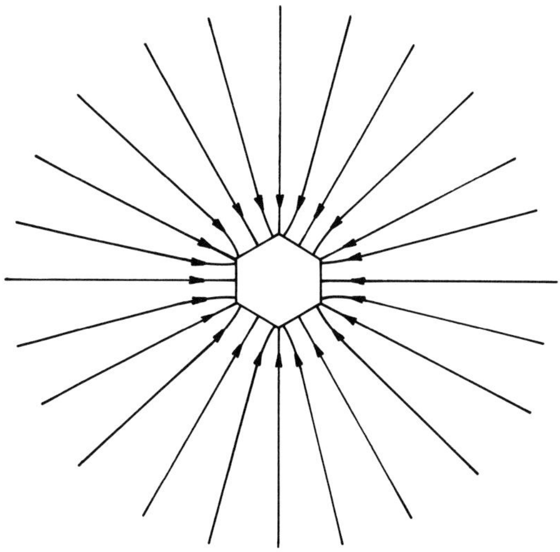

Fig. 6.8. Flux lines around a plate crystal in two dimensions (based on electrolytic tank experiments).

Figure 6.8 illustrates the flux lines surrounding a hexagonal plate in two dimensions. It is clear that the corners receive more current and, by analogy, more water molecules than do the faces. Thus, one might expect that the corners would be warmer due to a greater release of latent heat. However, the conduction of heat away from the corners is enhanced to the same degree, since the same flux lines apply to the sensible heat transfer. Thus, each section of the ice crystal is at the ice-bulb temperature and the crystal has a uniform temperature everywhere on its surface.

The angular shape of ice crystals leads to variations in the curvature on the crystal surface, particularly at the crystal edges and corners. Since the equilibrium vapor pressure is related to the curvature through the Kelvin equation (3.15), this effect would lead to variations in vapor density over the surface given a uniform temperature (or variations in temperature given a uniform vapor density). These effects are usually assumed to be small and the shape factor is taken to be the capacitance.

Assumption of the electrostatic analogy allows the mass transfer equation (5.2) and heat transfer equation (5.5) to be expressed as

$$\frac{dm}{dt} = 4\pi CD(\rho_\infty - \rho_{s,x}) \tag{6.1}$$

and

$$\frac{dQ}{dt} = 4\pi CK(T_\infty - T_x)\,, \tag{6.2}$$

where C is the shape factor, approximated as the electrostatic capacitance.

6.3.2. *Shape factors for ice crystals*

Analytic evaluations of the capacitance are available for most shapes which approximate common ice crystal habits. Plate crystals may be approximated as oblate spheroids. The capacitance for an oblate spheroid is given by

$$C = \frac{ae}{\sin^{-1} e},$$

where the eccentricity e is

$$e = \left(1 - \frac{c^2}{a^2}\right)^{1/2}.$$

For very thin plates, a is much larger than c and the capacitance reduces to

$$C = \frac{2a}{\pi}.$$

Column crystals may be approximated as prolate spheroids. In this case, the capacitance is given by

$$C = \frac{ce}{\ln\left[\dfrac{(1 + e)c}{a}\right]},$$

where the eccentricity is now defined as

$$e = \left(1 - \frac{a^2}{c^2}\right)^{1/2}.$$

For large aspect ratios ($c \gg a$) such as for needles or sheath crystals, the capacitance can be approximated as

$$C = \frac{c}{\ln\left[\dfrac{2c}{a}\right]}.$$

These theoretical expressions for capacitance were tested by McDonald (1963), who used a walk-in Faraday cage and brass models of various crystal shapes. McDonald found that hexagonal plates had very nearly the same capacitance as thin discs, with their thickness being relatively unimportant. Deeply insected plates (simulating dendrites) also had capacitances very close to discs of the same radius. He noted that needle-like cylinders had capacitances about 70% to 80% of those calculated.

6.3.3. *Integration of the growth equation*

Integration of the growth equation (5.1) for ice crystals requires that the growth mode for that particular ice crystal be specified. Small crystals initially grow nearly isometrically, and subsequently grow at a constant aspect ratio. Thus, the crystal extends all its axes in three-dimensional growth. A plate crystal undergoing dimension-limited growth extends only its a-axes, thereby growing in two dimensions. A column crystal undergoing dimension-limited growth, extends only its c-axis, thereby growing in one dimension.

Three-dimensional growth of an ice crystal with a constant aspect ratio may be integrated as follows. The mass of an ice crystal with a bulk density of ρ_x is

$$m = 4\sqrt{3}\, a^2 c \rho_x \,.$$

Given a constant aspect ratio of Γ, the differential of this equation is

$$dm = 12\sqrt{3}\, \rho_x \Gamma a^2 da \,.$$

Assuming the crystal is a thin plate (and using the capacitance for a thin plate), the growth equation (6.1) in this case is

$$dm = 8aD(\rho_\infty - \rho_{s,x})dt \,.$$

Equating these two expressions, integrating from some initial size a_o to some final size a and solving for a,

$$a = \left\{ a_o + \frac{4D(\rho_\infty - \rho_{s,x})}{3\sqrt{3}\, \Gamma \rho_x} t \right\} .$$

If the initial crystal was very small, the mass of the crystal as a function of time becomes

$$m = 4.6794 \left\{ \frac{[D(\rho_\infty - \rho_{s,x})]^3}{\Gamma \rho_x} \right\}^{1/2} t^{3/2} \,. \qquad (6.3)$$

Note that the smaller the aspect ratio (more extreme) and the lower the bulk density, the greater the mass of the crystal after a given growth time.

Now consider the case of a plate crystal growing with a constant thickness ($2c$), i.e., two-dimensional growth. In this case,

$$dm = 8\sqrt{3}\, ac\rho_x da = 8aD(\rho_\infty - \rho_{s,x})dt \,.$$

Once again, integrating from some initial size a_o,

$$a = a_o + \frac{D(\rho_\infty - \rho_{s,x})}{\sqrt{3}\, c\rho_x} t .$$

For very small initial crystal sizes, the mass of the crystal becomes

$$m = \frac{4D^2(\rho_\infty - \rho_{s,x})^2}{\sqrt{3}\, c_* \rho_x} t^2 . \tag{6.4}$$

Clearly, a crystal growing at constant thickness will eventually dominate over a crystal growing at constant aspect ratio. The thinner and less dense the plate, the more massive it will eventually become. Thus, dendritic crystals become more massive than solid plates, both because their bulk density is lower and their limiting thickness is smaller.

Consider the case of a sheath or needle crystal growing with a constant diameter. Using the shape factor for very long prolate spheroids,

$$dm = 4\sqrt{3}\, a^2 \rho_x dc = \frac{4\pi c D(\rho_\infty - \rho_{s,x})}{\ln\,(2c/a)} dt .$$

Integrating as before from c_o to c,

$$c = \frac{a}{2} \exp\left\{ \ln^2\left(\frac{2c_o}{a}\right) + \frac{2\pi D(\rho_\infty - \rho_{s,x})}{\sqrt{3}\, \rho_x a^2} t \right\}^{1/2} .$$

For very small initial crystal sizes, the mass of the crystal becomes

$$m = 2\sqrt{3}\, a^3 \rho_x \exp\left\{ \frac{2\pi D(\rho_\infty - \rho_{s,x})}{\sqrt{3}\, \rho_x a^2} t \right\}^{1/2} . \tag{6.5}$$

Thus, a crystal growing at constant diameter increases its mass exponentially with time and must eventually dominate crystals growing in two or three dimensions.

6.3.4. *Comparison between theory and observations*

Water saturation provides a convenient reference point for ice crystal growth since much of the growth of natural ice crystals takes place in water clouds. The vapor density difference between water saturation at a given temperature and the ice saturation value at the ice-bulb temperature reaches a maximum near $-15°$C at 1000 mbar (see Fig. 6.2) and the maximum is shifted toward colder temperatures as the pressure decreases (see Problem 6.1). This suggests that for

Table 6.3 Comparisons of calculated versus observed ice crystal mass.

Data source	Temperature (°C)	Experimental condition	Growth time (s)	Observed mass (10^{-8} g)	Calculated mass (10^{-8} g)
Mason	−2.5	Cold room	120	1.5–4.8	1.4
Isono et al.	−4.4	Field	300	60–180	23
Mason	−5	Cold room	160	10–11	9
Ono	−7	Cold chamber[a]	1400	280–400	820
Fukuta	−10	Cold chamber	45–50	1.2	0.9–1.1
Shaw	−13	Cold room	60	1.8–11	4
Ono	−15	Field	200–500	220–1400	80–940
Reynolds	−17	Cold chamber	130	26–65	21

[a] Ice crystals were suspended while growing and were not freely falling.

Source: After Miller and Young (1979).

short growth times, crystals grown near −15°C will be the largest, and that the relative mass of the crystals will tend to follow the water saturation curves in Fig. 6.2.

Crystals that reach the limited growth stage earliest may be expected to grow more rapidly and attain the largest sizes. This suggests that maxima in the plot of crystal mass as a function of temperature will appear near −5°C (needles) and −15°C (dendrites). These correspond to the most favored one-dimensional and two-dimensional growth regimes respectively.

Miller and Young (1979) present a numerical model for the diffusional growth of ice crystals based on the concept of constant aspect ratio growth and limited growth regimes, and compare their results with eight observations of crystal growth. They adjusted the limiting dimension and the sublimation density[6] within the observed values given by Ono (1970) until the calculated crystal masses were close to the observed crystal masses in each case. The transitions within each growth regime were required to be smoothly varying.

Table 6.3 details the various sources of observed crystal growth and conditions that were used by Miller and Young. For the most part, there is good agreement between the observed and calculated crystal masses, as would be expected. The data obtained by Ryan et al. (1976) are compared to the calculated ice crystal masses in Fig. 6.9 between −5°C and −20° as a function of time. Agreement between the calculated crystal masses and those observed by Takahashi and Fukuta (1988) and Takahashi et al. (1991) is good for smaller ice crystals, but in error by as much as a factor of two after 10 minutes for crystals growing near −5°C and near −15°C.

[6] The sublimation density is defined here as the density of the mass that is added to the existing crystal from which the change in crystal dimensions can be calculated. In the limit as the crystal grows large, the bulk density of the crystal equals the sublimation density.

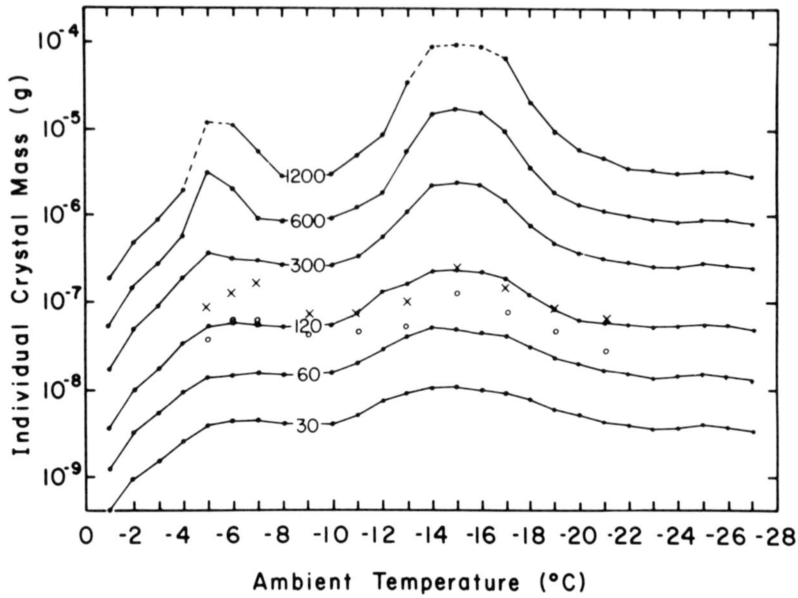

FIG. 6.9. Predicted crystal mass after 30, 60, 120, 300, and 1200 seconds of growth at water saturation and constant ambient temperature. Points indicated by circles and crosses are observations by Ryan et al. (1976) at 100 and 150 seconds, respectively. From Miller and Young (1979). *Journal of Atmospheric Sciences*, **36**, 3. American Meteorological Society. Reproduced with permission.

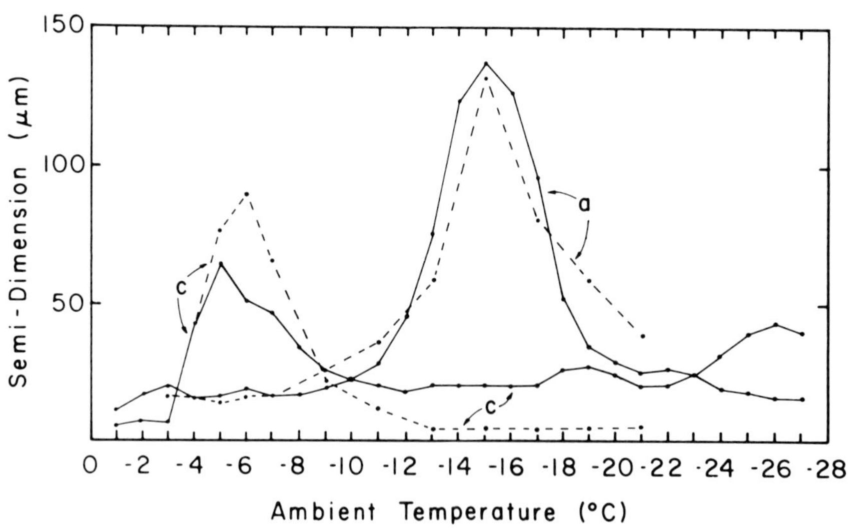

FIG. 6.10. Predicted crystal semi-dimensions after 120 seconds of growth at water saturation are shown by the solid lines. The broken lines represent the observations by Ryan et al. (1976). From Miller and Young (1979). *Journal of Atmospheric Sciences*, **36**, 3. American Meteorological Society. Reproduced with permission.

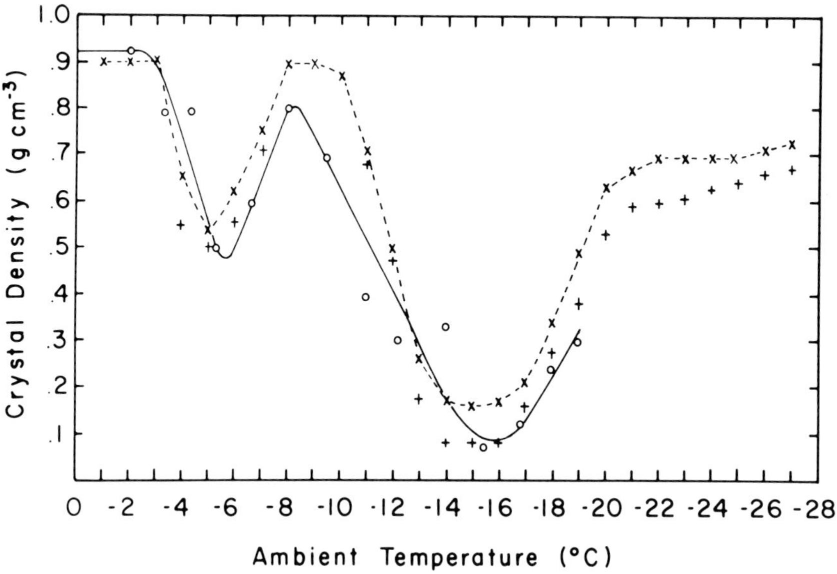

Fig. 6.11. Predicted crystal densities after 60 seconds of growth at water saturation are shown by the crosses and the broken line. Observations by Fukuta (1969) at 45 to 50 seconds are shown by the circles and the solid line. Points marked with plus symbols are predicted crystal densities after 300 seconds. From Miller and Young (1979). *Journal of Atmospheric Sciences*, **36**, 3. American Meteorological Society. Reproduced with permission.

The data from Ryan et al. also permit comparison between the calculated and observed a-axes and c-axis dimensions, as shown in Fig. 6.10. There is good agreement except for the plate thickness for temperatures colder than −12°C. The experiments of Fukuta (1969) provide a check of the calculated bulk densities. These comparisons are shown in Fig. 6.11.

From the foregoing, it is clear that specification of the growth mode and the definition of the limiting dimensions are crucial in determining the mass of ice crystals as a function of time. These specifications must rely on observations of the growth of ice crystals since our theoretical understanding is inadequate at providing useful guidance.

6.3.5. *Ventilation effects*

In a manner similar to that described in Chapter 5 for drops, (6.1) and (6.2) should be modified for the effects of ventilation. Unfortunately, relatively few studies of ventilation factors have been made for ice crystals. Thorpe and Mason's (1966) work remains the only experimental study; the few theoretical studies have used spheroidal approximations.

The results of Hall and Pruppacher (1976) provide reasonable approximations to the ventilation factors for both mass and sensible heat transfer. Hall and Pruppacher note that the Sherwood number for mass transfer correlates well with the Reynolds number when the latter is based on a characteristic length, defined as the surface area divided by the perimeter normal to the fall axis. The characteristic lengths (L^*) for plates and columns are

$$L^* = a + 2c \tag{6.6}$$

and

$$L^* = \frac{a + 2c}{\frac{2}{3} + \frac{\Gamma}{3^{1/2}}}, \tag{6.7}$$

where Γ is the aspect ratio, c/a. These characteristic lengths may be substituted for the characteristic dimension of a sphere $2r$ in the definition of the Reynolds number (see Appendix 2).

Hall and Pruppacher give the ventilation factor for mass transfer as

$$\begin{aligned} f_2 &= 1.0 + 0.14X^2 && \text{for } X < 1.0 \\ &= 0.86 + 0.28X && \text{for } X \geq 1.0\,, \end{aligned}$$

where

$$X = N_{\text{Sc}}^{1/3}\, N_{\text{Re}}^{1/2}\,.$$

The ventilation factor for sensible heat transfer is

$$\begin{aligned} f_2^* &= 1.0 + 0.14\, X_{\text{Q}}^2 && \text{for } X_{\text{Q}} < 1.0 \\ &= 0.86 + 0.28\, X_{\text{Q}} && \text{for } X_{\text{Q}} \geq 1.0\,, \end{aligned}$$

where

$$X_{\text{Q}} = N_{\text{Pr}}^{1/3}\, N_{\text{Re}}^{1/2}\,.$$

The Schmidt and Prandtl numbers are given in Appendix 2 and have values of roughly 0.63 and 0.72, respectively. This suggests that the ventilation factor for sensible heat transfer is slightly larger than that for mass transfer.

6.3.6. *Ice crystal growth in a water cloud*

The presence of water drops in the vapor density and temperature fields surrounding a (growing) ice particle serve to enhance the mass and sensible heat transfers. This may be seen by examination of Fig. 6.12, in which the drop (wet-bulb) temperatures remain at the ambient temperature, serving as sources of water vapor and sinks for sensible heat. The drops within the field of influence of the ice particle will experience subsaturated conditions and evaporate. Under the assumption that the ambient conditions are at water saturation, the depression

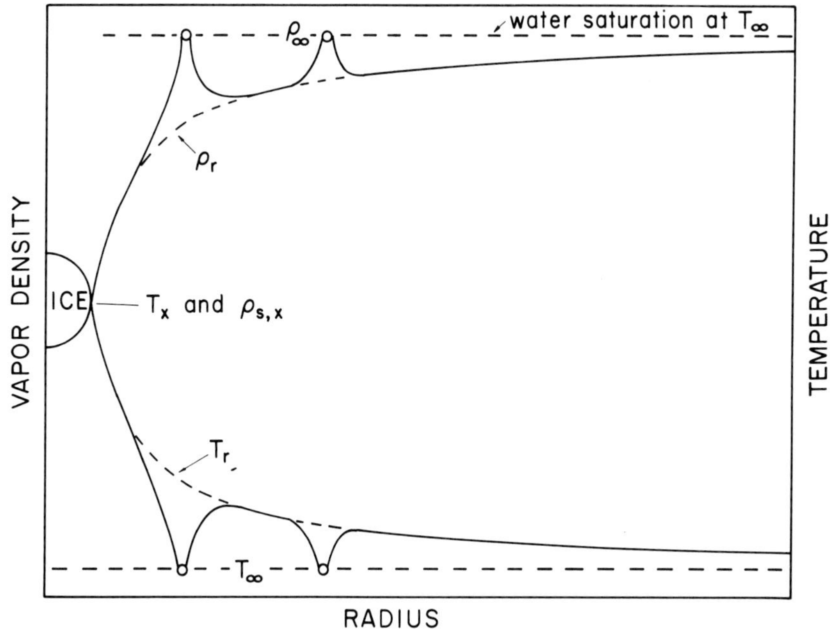

Fig. 6.12. The vapor density and temperature fields surrounding a spherical ice particle within a water cloud. Two small water drops modify the vapor density and temperature fields as shown. Note that the drops are at the ambient temperature regardless of their position within the fields (assuming the environment is at water saturation). The drops serve to enhance the vapor density and temperature gradients 'seen' by the ice particle, thereby enhancing its growth.

of the drop temperature due to evaporation just balances the elevation in temperature above the ambient temperature due to conduction of sensible heat away from the growing ice particle.

Marshall and Langleben (1954) present a derivation for an isolated ice particle growing in a cloud of small water drops. They find that the growth rate of the ice particle is enhanced by a factor of

$$f_3 = 1 + r_x(4\pi\Sigma r_w)^{1/2} \tag{6.8}$$

where Σr_w represents the sum of the drop radii and r_x is the radius of the ice particle, which has been assumed to be spherical for this derivation.

Since the water drop temperatures are depressed below that of their immediate environment, the drops also act as sensible heat sinks in the same manner as they act as water vapor sources. Thus, a 'sink' factor (f_3^*), defined as in (6.8), should be applied to the sensible heat transfer equation (6.2) in the same manner that the source factor should be applied to the mass transfer equation (6.1).

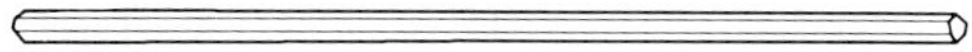

FIG. 6.13. Example of a column with an extreme aspect ratio (N1e) collected at −20°C near the base of thick cirrostratus by Heymsfield. The crystal is 0.1 cm in length. This extreme aspect ratio is apparently produced by sublimation.

The effect of these terms on the diffusional growth rate of an ice particle may be as large as 5%. However, should an ice particle find itself in such a water cloud, its growth rate due to the collection of water drops would probably exceed its diffusional growth rate by a large margin. Thus, for most purposes, these factors may be expected to have a relatively small effect on the overall growth of ice particles growing in water clouds.

6.3.7. *Kinetic effects*

For small ice crystals, kinetic effects may be expected to reduce the growth rates in a manner similar to that for water drops. Since small ice crystals tend to be nearly isometric, (5.18) and (5.19) may be applied. Isono (1968) gives a value of 0.06 for the deposition coefficient (analogous to the condensation coefficient in (5.18)).

For ice crystals growing at temperatures colder than −20°C and relatively low supersaturations with respect to ice, the necessity of nucleating steps at which water molecules may be incorporated into the crystal lattice reduces the growth rate of macroscopic crystals. Yokoyama and Kuroda (1988) note that under these conditions the effects of the surface kinetic processes on the growth become more pronounced for larger crystals and for smaller ice supersaturations. These effects are discussed further in § 6.4.3.

6.3.8. *Subliming ice crystals*

Ice crystals undergoing sublimation are commonly believed to tend toward the equilibrium habit for ice, *i.e.*, that of a nearly isometric column. The rationale is that this habit will tend to minimize the surface energy of the ice crystal. In contrast, rapidly growing ice crystals exhibit aspect ratios that are often far from equilibrium. Relatively few observations of subliming ice crystals are available to substantiate this hypothesis.

A columnar ice crystal collected near the base of a thick layer of cirrostratus at a temperature of −20°C is shown in Fig. 6.13. This crystal is approximately 0.1 cm in length but its aspect ratio is extreme, even for growing ice crystals. This extreme aspect ratio has apparently resulted from sublimation. Since columnar crystals tend to fall with their long axis horizontal, the effects of ventilation may have been to accentuate the sublimation from the prism faces.

6.4. Growth of the ice crystal lattice

The classical view of the incorporation of water molecules arriving from the vapor into the ice crystal lattice has two steps. First, a water molecule must diffuse to a position roughly one mean free path from the crystal surface, from whence it may be adsorbed onto the surface of the crystal. Second, the molecule must then migrate across the crystal face to a step boundary, along which it migrates until it can find a kink in the step where it can be incorporated into the lattice. It is usually assumed that once the molecule migrates to a step, it will not desorb.

Molecules incorporated into a step on the crystal surface cause the step to propagate across the surface. Once the step reaches the edge of the crystal, growth ceases unless new steps are formed. Thus, propagation of the face in the direction normal to its surface requires some means of creating new steps. The difference in the face propagation velocity between the basal and prism faces determines the crystal habit.

6.4.1. *Migration of molecules on the crystal surface*

A water molecule adsorbed onto an ice crystal surface may be pictured as being hydrogen-bonded to water molecules firmly embedded in the crystal lattice. If the molecule is held to the surface by two hydrogen bonds, it is not free to change its position. If it can gain enough energy to break one of the hydrogen bonds, it is then free to rotate about the remaining hydrogen bond and reform the second hydrogen bond in a new location. By successively breaking and reforming one hydrogen bond at a time, the molecule can wander around on the crystal surface. If the molecule gains enough energy to break both hydrogen bonds, it will desorb and return to the vapor state.

Einstein's equation relates the surface diffusion length ℓ_s to the surface diffusion coefficient D_s and the surface adsorption lifetime τ_s as

$$\ell_s = (D_s \tau_s)^{1/2} . \tag{6.9}$$

The surface diffusion coefficient is given by

$$D_s = j^2 \nu \exp\left(\frac{-H_1}{kT}\right) ,$$

where j is the jump distance for molecules adsorbed on the surface, ν is the frequency with which jumps can be made,[7] and H_l is the activation energy for surface diffusion (roughly the energy required to break one hydrogen bond). The exponential factor gives the fraction of potential jumps for which the molecule will possess sufficient energy to break at least one of the hydrogen bonds.

The surface adsorption lifetime is the product of the jump frequency and the probability that the molecule will have enough energy to break both hydrogen bonds, given by

$$\frac{1}{\tau_s} = \nu \exp\left(\frac{-H_2}{kT}\right) .$$

[7] ν is on the order of the atomic vibration frequency.

The surface diffusion length is then given by

$$\ell_s = j \exp\left[\frac{H_2 - H_1}{2kT}\right] . \tag{6.10}$$

The surface diffusion length is inversely related to the temperature, increasing with decreasing temperature as may be expected. Values for the surface diffusion length are typically a few micrometers. It should be noted that this treatment assumes that the concentration of adsorbed molecules is very small.

6.4.2. *Propagation of steps on a crystal face*

The simple picture of step growth assumes that all the adsorbed molecules within a certain distance of the step (or on the face of the step) are incorporated into the step and any molecules which arrive at the crystal outside of this area are not involved in the propagation of the step. The distance within which the adsorbed molecules are collected is termed the *collection distance* x_c, as shown in Fig. 6.14. The step propagates with a velocity given by

$$u = \frac{\mathrm{ff}_n}{h\rho_i}(2x_c + h) , \tag{6.11}$$

where ff_n is the net flux of water molecules to the crystal surface (per unit surface area) and h is the step height.

A more general treatment is given by Hobbs and Scott (1965). They assume that a fraction of the surface is covered by adsorbed molecules and that incoming molecules are adsorbed if they strike an open position and reflected otherwise. Their treatment yields a step propagation velocity of

$$u = \frac{2\mathrm{ff}_n \ell_s}{h\rho_i}\left[1 + \frac{\tau_s \mathrm{ff}_n}{C_s}\right]^{-1/2} . \tag{6.12}$$

For negligible adsorption,[8] (6.12) reduces to a form essentially the same as (6.11). Note that the term in brackets represents the correction resulting from the partial occupation of adsorption sites.

The temperature dependence of the step propagation velocity u may be predicted qualitatively in the following manner. Both the surface diffusion length ℓ_s and the surface adsorption lifetime τ_s increase with decreasing temperature. The surface diffusion length increases with decreasing temperature since $H_2 > H_1$ in (6.10). Assuming a constant net flux of molecules to the crystal, the increasing surface adsorption lifetime with decreasing temperatures implies a corresponding increase in the fraction of occupied adsorption sites.[9]

[8] Negligible adsorption occurs for a net flux of water molecules (ff_n) much less than the surface concentration of adsorbed molecules (C_s).

[9] Note that a constant flux of molecules implies a relative constant vapor density difference, which implies an increase in ice supersaturation.

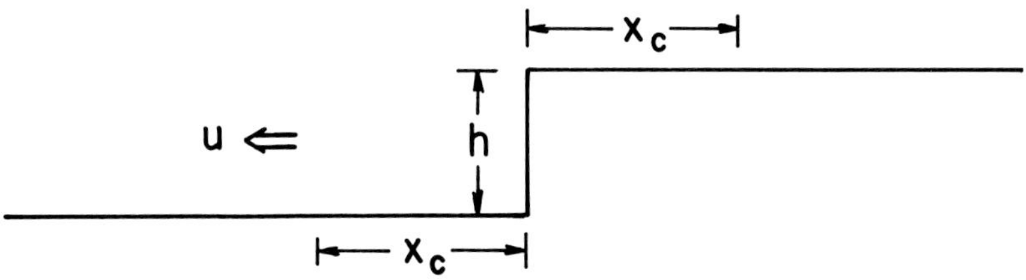

Fig. 6.14. Single step on a crystal face. The step of height h collects molecules within a collection distance x_c and propagates to the left with velocity u.

At warmer temperatures, the step propagation velocity follows (6.11) since the fraction of occupied sites is low. This predicts an increasing step propagation velocity with decreasing temperature. At colder temperatures, the step propagation velocity follows (6.12) as the term $\tau_s \mathrm{ff}_n/C_s$ begins to dominate. From (6.9), the step propagation velocity now follows the surface diffusion coefficient D_s and decreases with decreasing temperature.

The propagation velocity of steps on the basal faces of ice has been observed by changes in the interference patterns produced by very thin crystals grown on substrates such as covellite (CuS), as described by Bryant et al. (1959) and Hallett (1961). The step propagation velocity on the basal face is observed to decrease with decreasing temperature to −4°C, to increase from −4°C to −6°C, and to decrease at colder temperatures. Equation (6.11) is used to estimate the collection distance as a function of temperature, as shown in Fig. 6.15. This behavior at temperatures colder than −4°C agrees with the qualitative expectations in the previous paragraph. At warmer temperatures, the model is probably invalid due to the presence of a quasi-liquid layer, as discussed in Chapter 2.

The propagation of steps on the prism faces of ice crystals has not been observed because it has not been possible to grow crystals thin enough in the a-axis dimension to observe the interference patterns. However, since the linear growth rate (dc/dt) on the basal face is in approximate proportion to the observed temperature dependence of the step propagation velocity, the step propagation velocity on the prism faces may be deduced from observations of the linear growth rate (da/dt).

Observations of the linear growth rates of both the basal and prism faces as determined by Lamb and Hobbs (1971) are shown in Fig. 6.16. One interesting feature of this figure is that the linear growth rate curves (and presumably the step propagation velocity curves) intersect at −5.3°C and at −9.5°C These correspond fairly closely to two of the three habit transitions observed for growing ice crystals.

6.4.3. *Propagation of the crystal face*

In the preceding discussion, the ice crystal has been tacitly assumed to be ideal, without dislocations, such that a step could propagate entirely across the crystal

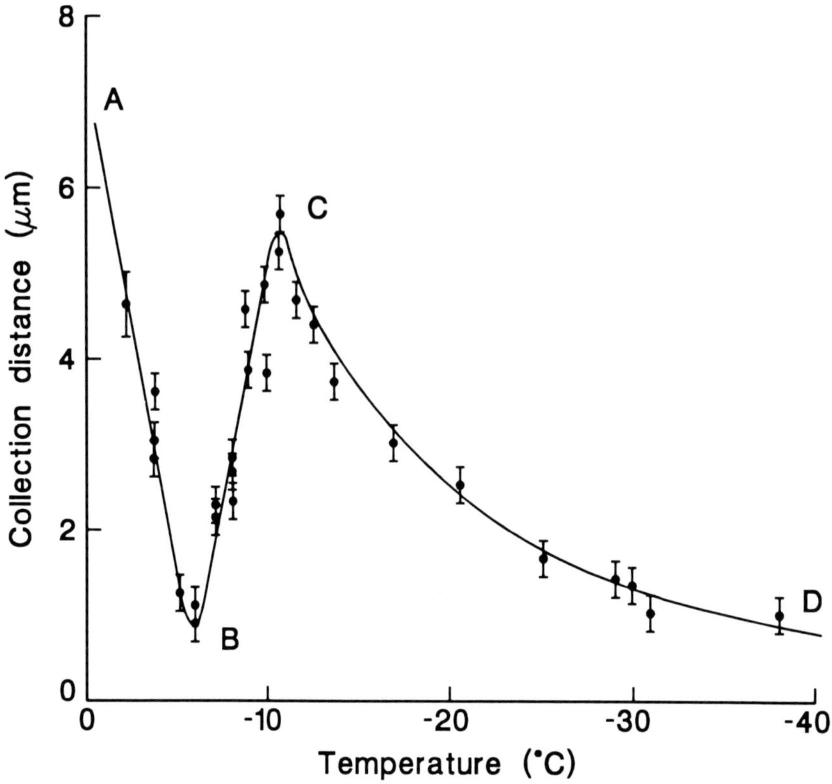

Fig. 6.15. The collection distance of a step on the basal face of ice is shown as a function of temperature, based on experiments by Mason et al. (1963). © Taylor and Francis Ltd. Reproduced with permission.

face and complete the face. Thus, a new step would have to be nucleated in order to allow the face to propagate, and the propagation velocity of the face (linear growth rate) would be directly related to the rate at which new steps could be nucleated. This type of nucleation is referred to as *two-dimensional* nucleation. The propagation of a crystal face through an emergent screw dislocation would obviate the need for repeated nucleation. Either of these mechanisms would be expected to occur when the surface concentration of adsorbed molecules is relatively low.

When the surface concentration of adsorbed molecules is relatively large but not all of the surface lattice sites are occupied, incoming molecules may be readily incorporated into the crystal lattice. Propagation of the crystal face in this manner is termed *rough interface* growth. As the surface concentration of adsorbed molecules exceeds one, a quasi-liquid layer exists on the ice crystal surface and the ice lattice grows as if from bulk water (freezing).

(i) Two-dimensional nucleation. The formation of a two-dimensional 'island' on a crystal face may be approached in the same manner as previous treatments of homogeneous nucleation. Consider an embryo of height h and radius r on the crystal surface, as shown in Fig. 6.17. The change in thermodynamic potential

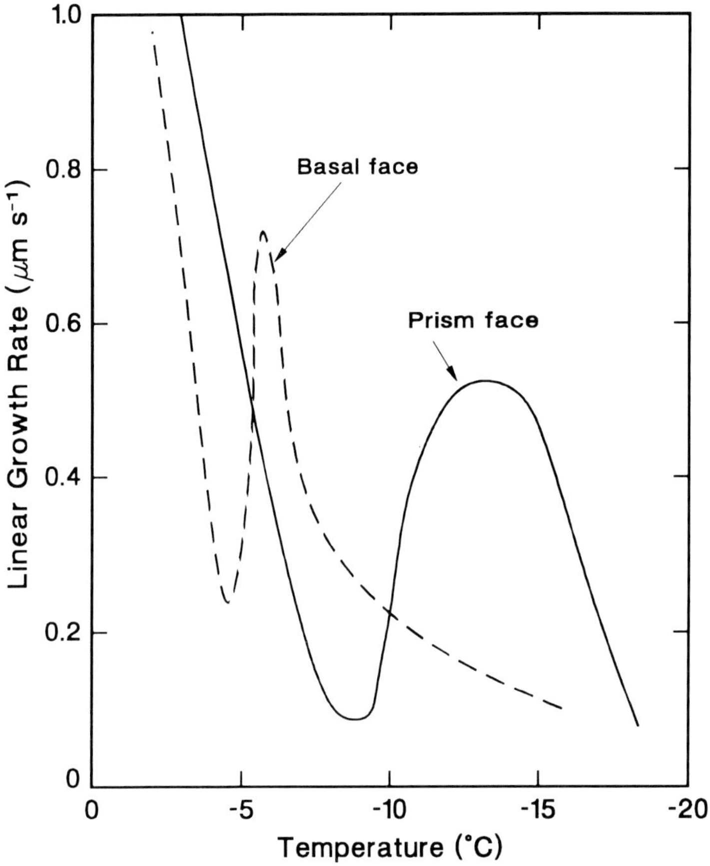

FIG. 6.16. Linear growth rates on the basal and prism faces of ice crystals as a function of temperature. Curves represent the best fit to experimental measurements of linear growth rates at a constant vapor pressure in excess of 0.013 mbar. From Hobbs (1974). Reproduced with permission.

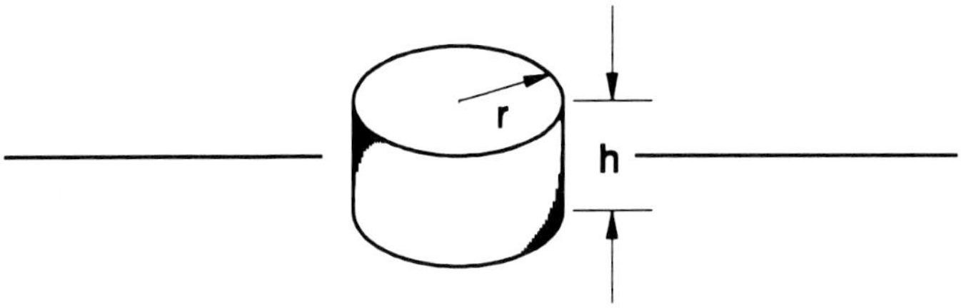

FIG. 6.17. Two-dimensional nucleation requires formation of an embryo on an ideal crystal surface. The embryo is approximated by a cylinder of height h and radius r.

required to form such a cylindrical island is the sum of the bulk and surface terms

$$(\Delta Y)_g = -\pi r^2 h N_s kT \ln\left(\frac{e_\infty}{e_i}\right) + 2\pi r h \sigma_{IV}.$$

The critical embryo radius[10] is obtained by finding the extrema for the thermodynamic potential,

$$r^* = \frac{\sigma_{IV}}{N_s kT \ln\left(\frac{e_\infty}{e_i}\right)}. \tag{6.13}$$

The change in thermodynamic potential for the critical embryo is

$$(\Delta Y)_{g^*} = \frac{\pi h \sigma_{IV}^2}{N_s kT \ln\left(\frac{e_\infty}{e_i}\right)},$$

and the two-dimensional nucleation rate is

$$J = A_g \vec{\omega}_g N_o \exp\left(\frac{-(\Delta Y)_{g^*}}{kT}\right). \tag{6.14}$$

From this, the linear growth rate under the assumption of two-dimensional nucleation would be an exponentially increasing function of the supersaturation.

The two-dimensional nucleation of steps on an ideal crystal implies that there must be some critical supersaturation that must be achieved before a crystal may grow. No such barrier to crystal growth seems to exist at temperatures warmer than $-20°C$, suggesting some other means of propagating the crystal face is taking place.

(ii) Emergent screw dislocations. Growth by means of an emergent screw dislocation has been suggested as one method of eliminating the need for repeated nucleation of new steps in order to propagate a crystal face. This concept is developed more fully by Burton et al. (1951). Bryant and Mason (1960) present experimental evidence for such emergent screw dislocations on ice crystals.

The growth spiral for an emergent screw dislocation is approximately given by $r = 2r^*\theta$ in polar coordinates. The spiral is assumed to originate from a two-dimensional critical embryo of radius r^*. The distance between successive turns of the spiral is simply $4\pi r^*$. This represents the distance that a step must propagate in order to advance the screw by one step height.

[10] The critical embryo radius is independent of height.

The propagation rate of the face is the product of the step height and the frequency with which steps can cover the distance between successive turns of the spiral. This is given by

$$v = \frac{uh}{4\pi r^*} = \frac{2\mathrm{ff}_n x_c}{4\pi r^* \rho_i} \tag{6.15}$$

using the step propagation velocity from (6.11), assuming a negligible concentration of adsorbed molecules. The critical radius is given by (6.13), reducing the face propagation velocity to

$$v = \frac{x_c\, \mathrm{ff}_n N_s kT \ln\left(\dfrac{e_\infty}{e_i}\right)}{2\pi\sigma_{IV}\rho_i}. \tag{6.16}$$

Note that the collection distance x_c is used in (6.15) and (6.16) rather than the surface diffusion length ℓ_s. If the distance between successive turns of the spiral is less than $2\ell_s$, then the collection distance is $2\pi r^2$ and the face propagation velocity is a linear function of supersaturation (ff_n), from (6.15). If the distance between successive turns of the spiral is greater than $2\ell_s$, the face propagates at a velocity which is proportional to the supersaturation squared, from (6.16).

Since the distance between successive turns depends on r^*, which in turn decreases with increasing supersaturation (6.13), it is clear that the linear growth rate for a crystal face is proportional to the square of the supersaturation for small supersaturations and becomes a linear function of the supersaturation at higher supersaturations.

The linear growth rate for the crystal face is proportional to the step propagation velocity. Only at low supersaturations, however, is the step propagation velocity proportional to the surface diffusion length because the steps are wider than $2\ell_s$. At higher supersaturations, the step propagation velocity is determined by the actual collection distance, which is limited by the width of the steps. Thus, the face propagates at a rate proportional to the area over which adsorbed molecules are incorporated into steps. Once the surface diffusion length is large enough, all of the adsorbed molecules are incorporated, regardless of the magnitude of the surface diffusion length.

(iii) Growth of a rough interface. For relatively high surface concentrations of adsorbed molecules, incoming molecules do not need to migrate very far to locate a suitable site for bonding. In this case, the face propagation velocity may be derived from (3.32) as

$$v = \frac{\alpha(e_\infty - e_{s,i})V_m}{(2\pi mkT)^{1/2}}, \tag{6.17}$$

where V_m is the volume of a single molecule of mass m. Kuroda (1982) suggests that the accommodation coefficient α_c may be set equal to 1 and that rough interface growth occurs when more than 2% of the surface lattice sites are occupied by adsorbed molecules.

(iv) Growth in the quasi-liquid layer. The growth of the ice crystal lattice in the quasi-liquid layer at temperatures near the melting point is preceded by condensation of water vapor into the quasi-liquid layer. Since the thickness of the quasi-liquid layer must be relatively constant for a given temperature, the rate of condensation must equal the rate of growth of the ice crystal lattice

The equation describing the condensation rate into the quasi-liquid layer is similar to (6.17), yielding the face propagation velocity

$$v = \frac{\alpha_c [e_\infty - e(\delta)] V_m}{(2\pi mkT)^{1/2}},$$

where $e(\delta)$ represents the equilibrium vapor pressure of a quasi-liquid layer of thickness δ. Kuroda (1982) shows that the equilibrium vapor pressure increases as the thickness of the quasi-liquid layer increases, thereby decreasing the face propagation velocity. At the same time, the growth of ice from the quasi-liquid layer increases as the thickness of the layer increases. Since the two rates must be equal, it is possible to determine the thickness of the quasi-liquid layer for given conditions, and hence the face propagation velocity. Due to the relatively low nucleation barrier to two-dimensional nucleation in the liquid, the quasi-liquid layer presents relatively little resistance to the propagation of the face as the ice crystal grows from the vapor.

6.4.4. *The formation of blocking steps*

The propagation velocity of a step is inversely proportional to the step height. This implies that a short step will tend to catch up with a tall step. When the shorter step catches up with the taller step, they may amalgamate to form an even taller step, which would propagate even more slowly. If enough short steps are amalgamated into a tall step, its propagation velocity may be so slow that the step may be regarded as stationary. Such a step is termed a *blocking* step since it blocks any propagation of short steps past its location.

Mason et al. (1963) suggest a number of ways in which blocking steps may originate. Two steps of unequal height will amalgamate if the shorter step follows the taller step. In addition, they show that steps of equal height will tend to amalgamate if they lie within $0.6\ell_s$ of each other. The amalgamation of equal-sized steps results from competition for the adsorbed molecules with the overlapping diffusion lengths.

Mason et al. also demonstrate that blocking steps cannot form within a distance of roughly $20\ell_s$ of the crystal edge. This implies that skeletal growth resulting from the formation of blocking steps cannot commence until the ice crystal has reached a certain critical size.

6.5. Ice crystal growth modes

The problem of why ice crystals display the wide variety of shapes observed has intrigued scientists for centuries. Although there is no consensus at present regarding the underlying reasons why ice crystals grow as they do, enough is known to provide some insight into the mechanisms. Explanations of the changes in ice crystal habits with temperature must consider the quasi-liquid surface layer on crystals at temperatures close to the melting point as well as the mechanisms for nucleating steps and propagating faces.

On important aspect of ice crystal growth is that there are three basic growth regimes. The transitions between these regimes are a function of the crystal size. According to earlier discussions, very small ice crystals grow nearly isometrically; intermediate-sized ice crystals grow with roughly constant aspect ratios; large ice crystals tend to grow with one axis fixed and to exhibit skeletal growth.

6.5.1. *Isometric growth*

If the crystal dimensions are small enough to allow individual water molecules to migrate over a significant portion of the surface before desorbing, the crystal may be expected to grow in a manner which minimizes the total surface energy. In other words, if the crystal dimensions are on the order of ℓ_s or smaller, then an appreciable fraction of the adsorbed molecules can migrate between the basal and prism faces, becoming incorporated in locations which serve to minimize the total surface energy.

The concept of the equilibrium shape of an ice crystal as being that shape which minimizes the total surface energy is due to Wulff (1901). Wulff's theorem states that the distance of any face from the center of the crystal is proportional to the surface free energy per unit area of that face. Fletcher (1962) estimates that the surface energy of the prism faces exceeds that for the basal faces by 6%. This yields an aspect ratio of 0.94 for ice crystals (see Problem 6.5).

6.5.2. *Constant aspect ratio growth*

Measurements of the linear growth rates for both the basal and prism faces of ice crystals suggest that temperature variations of the surface diffusion length are responsible for the basic changes in habit for temperatures colder than −5°C. The intersection of the linear growth rates on the basal and prism faces at temperatures closely corresponding to the transitions between column and plate growth (see Fig. 6.16) make this a particularly attractive hypothesis.

This leaves unanswered the fundamental question as to why the surface diffusion length is different on the basal and prism faces. The surface diffusion length is dependent on the jump distance for surface molecules (6.10). Since the energies required to break one (H_1) and two (H_2) hydrogen bonds must be quite similar on the two faces, the jump distance must be different. Inspection of the available hydrogen-bonding sites on the basal and prism faces suggests that the average jump distance on the prism face is about 20% less than the jump distance on the basal face.

The enhanced vapor density gradients at the edges and corners of the crystal makes two-dimensional nucleation more favored at these sites. The two-dimensional nucleation rate is given by (6.14). The thermodynamic potential for the formation of a critical embryo on the basal faces is slightly larger than that on the prism faces (Kuroda, 1982). However, the two-dimensional nucleation rate is proportional to the rate at which molecules are added to the critical embryo $\vec{\omega}_g$ which is strongly face dependent. The rate at which molecules are added to the embryo is given by

$$\vec{\omega}_{\mathrm{g}} = D_{\mathrm{s}} \left. \frac{dC_{\mathrm{s}}}{dx} \right|_{\mathrm{x}=0}$$

where C_{s} is the surface concentration of adsorbed molecules. Hobbs and Scott (1965) provide a solution for the gradient of adsorbed molecules,

$$\left. \frac{dC_{\mathrm{s}}}{dx} \right|_{\mathrm{x}=0} = \mathrm{ff}_{\mathrm{n}} \left(\frac{\tau_{\mathrm{s}}}{D_{\mathrm{s}}} \right)^{1/2},$$

and the rate at which molecules are added to the embryo reduces to

$$\vec{\omega}_{\mathrm{g}} = \mathrm{ff}_{\mathrm{n}} \, (D_{\mathrm{s}} \tau_{\mathrm{s}})^{1/2} = \mathrm{ff}_{\mathrm{n}} \, \ell_{\mathrm{s}} \, .$$

Since the linear growth rate is proportional to the two-dimensional nucleation rate, the aspect ratio is approximately given by the ratio of the surface diffusion length on the basal face to that on the prism face.[11] The observations presented in Fig. 6.5 suggest aspect ratios in this growth regime range from 0.25 for plates at $-16°$C to 3.0 for columns at $-5°$C. This is in fair agreement with the ratios of the linear growth rates presented in Fig. 6.16 although the temperatures at which the most extreme aspect ratios occur do not match.

6.5.3. *Limited and skeletal growth*

The formation of blocking steps appears to be responsible for both skeletal growth and for the existence of a limiting dimension. Mason et al. (1963) have shown that blocking steps may form as close as $20\ell_{\mathrm{s}}$ to the edge of the crystal face. This implies that the crystal will be complete if it is less than $40\ell_{\mathrm{s}}$ across, regardless of the supersaturation.

If the crystal face is greater than $40\ell_{\mathrm{s}}$ across, the onset of skeletal growth will depend on the separation between successive steps. For an emergent screw

[11] This ignores the differences in thermodynamic potential for the critical embryo on the basal and prism faces.

dislocation, the separation of the steps decreases as supersaturation increases, according to (6.13). For two-dimensional nucleation, the separation between steps (Δ) is the step propagation velocity (u) divided by the nucleation rate for new steps (J)

$$\Delta = \frac{u}{J} = \frac{2/h\rho_i A_g}{\left(1 + \frac{\tau_s ff_n}{C_s}\right)^{1/2} N_o \exp\left(\frac{-(\Delta Y)_{g^*}}{kT}\right)},$$

Thus, since the separation between steps is inversely related to the supersaturation through the exponential term, blocking steps will tend to form as the supersaturation increases.

As an ice crystal grows, its ventilation factor also increases. This increases the flux of water vapor molecules to the crystal and serves to decrease the separation between steps. Thus, as an ice crystal grows under fixed environmental conditions, the likelihood of skeletal growth increases due to increased ventilation.

Since the step separation is inversely related to the surface diffusion length through τ_s, blocking steps will form first on the face with the largest surface diffusion length. Equal-sized steps will amalgamate when the steps are separated by less than 0.6 ℓ_s. It is this amalgamation which leads to the formation of blocking steps.

Once skeletal growth has commenced on a given face, the vapor flux to the edges and corners becomes still more pronounced. This serves to promote nucleation of new steps, accentuating the growth of the 'limited' face. Apparently, the accentuated growth of the limited face and the concentration of the water vapor flux to the rapidly growing corners and edges markedly reduces the growth of the other face, i.e., the growth of the other face appears to have ceased.

6.5.4. *Growth near the melting point*

Kuroda (1982) points out that the prism faces have a higher surface tension than do the basal faces and thereby obtain a larger reduction in the surface free energy by maintaining a quasi-liquid layer to lower temperatures than do the basal faces. He postulates two transition temperatures for each face as the fraction of surface sites occupied by adsorbed molecules decreases with decreasing temperatures. Ice crystal growth would progress from growth in the quasi-liquid layer at warm temperatures, to growth of a rough interface, and finally to two-dimensional nucleation at relatively cold temperatures. These are shown in Fig. 6.18.

For temperatures warmer than −4°C, both faces are covered with a quasi-liquid layer, which is thicker on the prism faces. Since the thermodynamic potential to form a critical two-dimensional embryo on the prism faces is less than on the basal faces, the growth rate of the prism faces is greater than that of the basal faces, resulting in plate crystals. The basal faces transit to rough

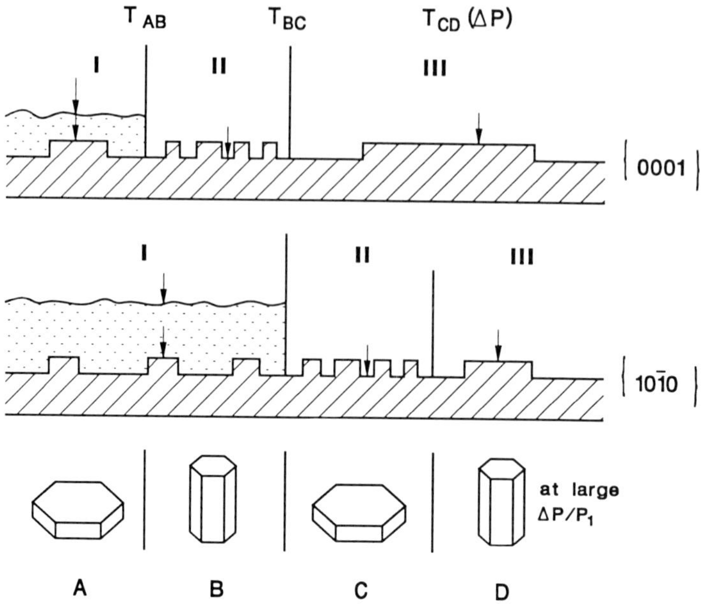

FIG. 6.18. Schematic representation of ice crystal growth regimes on the basal and prism faces. Crystal faces growing in region I propagate by freezing in the quasi-liquid layer; crystal faces growing in region II advance through deposition onto a rough interface; crystal faces in region III grow through two-dimensional nucleation. After Kuroda (1982).

interface growth at temperatures near -4°C while the prism faces continue to grow with a quasi-liquid layer for temperatures warmer than -10°C. In this region, since there is virtually no barrier to the incorporation of incoming water molecules into the crystal lattice with the rough interface, the basal faces propagate faster than the prism faces, resulting in columnar crystals.

At temperatures colder than -10°C, the propagation of the basal and prism faces appears to be governed by the surface diffusion length, as discussed above. This picture of the mechanisms governing the changes in ice crystal habits and growth modes with temperatures seems to provide a suitable explanation for the observed crystal habits.

Problems

6.1* Plot the quantity $(\rho_{s,w} - \rho_{s,x})$ as a function of temperature between 0°C and -25°C for ice crystals growing at water saturation. Assume an ambient pressure of 700 mbar. Note that $\rho_{s,w}$ refers to water saturation at the ambient temperature and $\rho_{s,x}$ refers to ice saturation at the particle temperature.

6.2* Derive the characteristic lengths (L^*) for plates and columns given by (6.6) and (6.7), noting that plates fall with their a-axes horizontal and columns fall with their c-axis horizontal.

6.3* How rapidly will a step 10^{-6} cm in height propagate across the basal face of an ice crystal given a collection distance of 4 μm and a net flux to the crystal of 0.001 g cm^{-2} s^{-1}?

6.4. What is the temperature of a plate crystal growing in a water cloud at −16°C:

(a)* for a plate with a = 50 μm and c = 10 μm?
(b)** for a plate with a = 500 μm and c = 30 μm?

6.5** Wulff's theorem states that, in equilibrium, the distance of any face from the center of the crystal is proportional to the surface free energy per unit area of that face. Fletcher (1962) finds that the surface free energy per unit area on the prism face exceeds that on the basal face by 6%.

(a) Prove Wulff's theorem for Ice Ih.
Hint: Find the aspect ratio which gives a minimum total surface energy for a given volume crystal.

(b) Determine the equilibrium aspect ratio for ice crystals.

Computer problem

6.7** How long would it take a planar ice crystal growing at constant thickness to increase its a-axis length from 0.01 cm to 0.1 cm growing as:

(a) a dendrite at −16°C?
(b) a sector plate at −18°C?
(c) a simple plate at −12°C?

REFERENCES

Bryant, G.W., and Mason, B.J., 1960, 'Etch pits and dislocations in ice crystals,' *Phil. Mag.* **5**, 1221.

Bryant, G.W., Hallett, J., and Mason, B.J., 1959, 'The epitaxial growth of ice on single crystalline substrates,' *J. Phys. Chem. Solids,* **12**, 189.

Burton, W.K., Cabrera, N., and Frank, F.C., 1951, 'The growth of crystals and the equilibrium structure of their surfaces,' *Phil. Trans. R. Soc.* **A243**, 299.

Fletcher, N.H., 1962, *The physics of rainclouds.* Cambridge: Cambridge University Press.

Fukuta, N., 1969, 'Experimental studies on the growth of small ice crystals,' *J. Atmos. Sci.* **26**, 522.

Hall, W.D., and Pruppacher, H.R., 1976, 'The survival of ice particles falling from cirrus clouds in subsaturated air,' *J. Atmos. Sci.* **33**, 1995.

Hallett, J., 1961, 'The growth of ice crystals on freshly cleaved covellite surfaces,' *Phil. Mag.* **6**, 1073.

Hallett, J., 1964, 'Experimental studies of the crystallization of supercooled water,' *J. Atmos. Sci.* **21**, 671.

Hobbs, P.V., 1974, *Ice physics*. Oxford: Clarendon Press.

Hobbs, P.V., and Scott, W.D., 1965, 'Step growth on single crystals of ice,' *Phil. Mag.* **11**, 1083.

Isono, K., 1968, 'Formation and growth of ice crystals at low pressure,' *Proc. Int. Conf. Cloud Physics* 1968. American Meteorological Society, p. 270.

Isonop, K., Komabayashi, M., Yamanaka, Y., and Fujita, H., 1956; 'An experimental investigation of the growth of ice crystals in a supercooled fog,' *J. Meteorol. Soc. Jpn, Ser. 2*, **34**, 34–9.

Kuroda, T., 1982, 'Growth kinetics of ice single crystal from vapour phase and variation of its growth form,' *J. Meteorol. Soc. Jpn,* **60**, 520.

Lamb, D., and Hobbs, P.V., 1971, 'Growth rates and habits of ice crystals grown from the vapor phase,' *J. Atmos. Sci.* **28**, 1509.

Magono, C., and Lee, C.W., 1966, 'Meteorological classification of natural snow crystals,' *J. Fac. Sci. Hokkaido Univ. Ser. VII* **2**, 321.

Marshall, J.S., and Langleben, M.P., 1954, 'A theory of snow-crystal habit and growth,' *J. Meteorol.* **11**, 104.

Mason, B.J., 1953; 'The growth of ice crystals in a supercooled water cloud,' *Quart. J. Roy. Meteorol. Soc.*, **79**, 104–20.

Mason, B.J., Bryant, G.W., and Van den Heuvel, A.P., 1963, 'The growth habits and surface structure of ice crystals,' *Phil. Mag.* **8**, 505.

McDonald, J.E., 1963, 'Use of the electrostatic analogy in studies of ice crystal growth,' *Zeit. Angew. Math. Phys.* **14**, 610.

Miller, T.L., and Young, K.C., 1979, 'A numerical simulation of ice crystal growth from the vapor phase,' *J. Atmos. Sci.* **36**, 458.

Nakaya, U., 1954, *Snow crystals natural and artificial*. Cambridge, MA: Harvard University Press.

Ono, A., 1970, 'Growth mode of ice crystals in natural clouds,' *J. Atmos. Sci.* **27**, 649.

Reynolds, S.E., 1952; 'Ice-crystal growth,' *J. Meteorol.,* **9**, 36–40.

Ryan, B.F., Wishart, E.F., and Shaw, D.E., 1976, 'The growth rates and densities of ice crystals between −3°C and −21°C,' *J. Atmos. Sci.* **33**, 842.

Shaw, D., 1954; *Ice crystal growth from vapour*, PhD. thesis, The University of London.

Takahashi, T., and N. Fukuta, 1988, 'Supercooled cloud tunnel studies on the growth of snow crystals between -4 and -20°C,' *J. Meteorol. Soc. Japan,* **66**, 841.

Takahashi, T., Endoh, T., Wakahama, G., and Fukuta, N., 1991, 'Vapor diffusional growth of free-falling snow crystals,' *J. Meteorol. Soc. Jpn* **69**, 15.

Thorpe, A.D., and Mason, B.J., 1966, 'The evaporation of ice spheres and ice crystals,' *Brit. J. Appl. Phys.* **17**, 541.

Wulff, G., 1901, 'Zue Frage der Geschwindigkeit des Wachtums und der Auflösung der Krystallflächen,' *Zeit. Krystallogr.* **34**, 449.

Yokoyama, E., and Kuroda, T., 1988, 'Quantitative investigation of rate determining process of growth of snow crystals – two-dimensional nucleation growth and spiral growth,' *J. Meteorol. Soc. Jpn* **66**, 927.

7

GROWTH OF DROPS BY COLLECTION

7.1. Introduction

The growth of a population of droplets by diffusion leads to a broadening of the spectrum of fall velocities and increases the likelihood that two drops will collide with each other. These drops may either bounce apart or coalesce into one. Such coalescences serve to broaden further the spectrum of fall velocities, increasing the likelihood of more coalescences, and so on, until raindrops are produced, often within a fairly short span of time.

The rate at which drops collect smaller drops is determined by the volume the larger drop sweeps out with respect to the smaller drop per unit time and the concentration of smaller drops. The flow field relative to the falling drop tends to carry smaller drops around the larger drop. This tends to reduce the number of collections in comparison to what might be expected if the volume swept out was solely determined by the drop sizes (geometric collection). The ratio of the volume actually swept to the volume traversed by the drop is termed the collection efficiency.

The collection equation yields a rate at which a large drop may be expected to collect small drops of a given size. The physical interpretation of this rate is of fundamental importance in determining the growth of a population of drops by collection. The differential probability that a large drop will collect a small drop in time dt defines a stochastic process whose complete solution would not be practical. Continuous and quasi-stochastic models will be introduced and compared to simple versions of the pure stochastic model.

Coalescences between fairly large drops may not always result in continued growth. Depending on the geometry of the collision event, drops may coalesce only temporarily. In the process of breaking apart again, these drops usually produce small fragments (satellites). Thus, large drops may lose as well as gain mass from collisions with other drops. The likelihood that drops will lose rather than gain mass increases with increasing drop size, effectively limiting the largest drops to a radius of roughly 0.25 cm.

7.2. The collection equation

The basic form of the collection equation considers the interaction between two drops. The larger or faster falling drop is termed the collector drop (of radius r_L);

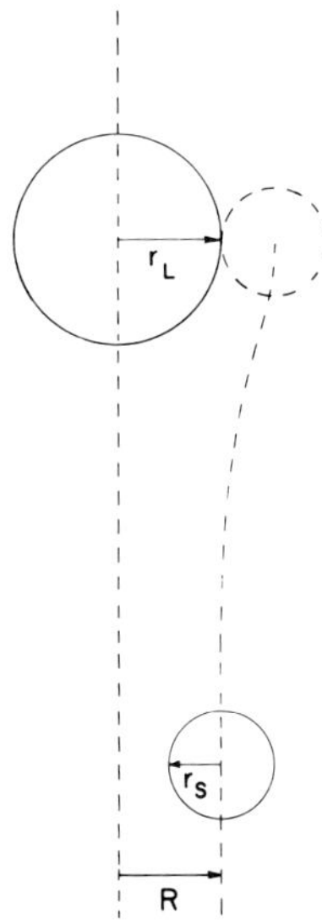

FIG. 7.1. A grazing trajectory for the collected drop (r_S) as the flow field around the collector drop (r_L) tends to move the smaller drop aside. All drops whose centers lie within distance R of the fall axis of the collector drop are assumed to be collected; all drops that lie without are not collected.

the smaller or slower falling drop is termed the collected drop (of radius r_S). These drops are usually assumed to be falling at their terminal fall velocity, which represents a balance between the viscous drag and inertial (gravitational) forces. The volume from which the collector drop removes all the collected drops (per unit time) is termed the swept volume or the collection kernel.

The relation between the drop sizes and the swept volume is shown in Fig. 7.1. The trajectory of a collected drop which just grazes the collector drop is shown. All small drops whose centers lie within a distance R of the axis of fall of the collector drop are assumed to be collected; all drops outside the cylinder defined by radius R are assumed not to be collected. The collection kernel is given by

$$\Gamma = \pi R^2 (v_L - v_S), \tag{7.1}$$

and the collection rate is simply the product of the collection kernel and the collected drop concentration,

$$\frac{dN_s}{dt} = \pi R^2 n_s (v_L - v_S), \tag{7.2}$$

where v_L and v_S are the fall velocities of the collector and collected drop, respectively and n_s is the number concentration of the collected drops.

7.2.1. *Terminal fall velocities for water drops*

A water drop released in the atmosphere will accelerate until it reaches a velocity at which the inertial force due to gravity is balanced by the resistance or drag force of the air. For typical tropospheric conditions, this steady or terminal velocity is reached quite quickly, and water drops may be assumed to be falling at their terminal velocity.

The inertial force is simply the buoyancy, given by

$$F_{\mathrm{G}} = g(\rho_{\mathrm{w}} - \rho_{\mathrm{a}}) \frac{4}{3} \pi r^3 \approx g \rho_{\mathrm{w}} \frac{4}{3} \pi r^3 ,$$

where g is the acceleration due to gravity and the density of water ρ_{w} is much greater than the density of air ρ_{a}. The drag force may be derived from the Navier–Stokes equation (Pruppacher and Klett, 1978) as

$$F_{\mathrm{D}} = \frac{\pi}{2} r_{\mathrm{w}}^2 v^2 \rho_{\mathrm{a}} C_{\mathrm{D}} ,$$

where v is the fall velocity and C_{D} is the drag coefficient. The drag force can be rewritten in terms of the Reynolds number N_{Re} as

$$F_{\mathrm{D}} = 6\pi\eta r_{\mathrm{w}} v \frac{C_{\mathrm{D}} N_{\mathrm{Re}}}{24} .$$

where η is the dynamic viscosity.

The drop falls at constant velocity (terminal fall velocity) when the inertial and drag forces are balanced. The terminal fall velocity is given by

$$v_{\mathrm{t}} = \frac{2\rho_{\mathrm{w}}}{9\eta} g r^2 \frac{24}{C_{\mathrm{D}} N_{\mathrm{Re}}} , \tag{7.3}$$

whose solution is dependent on the flow regime (Beard, 1976). Figure 7.2 illustrates the streamlines around a falling drop for different flow regimes.

For Reynolds numbers less than 0.01 (radii less than 10 μm), the Stokes approximation[1] $C_{\mathrm{D}} N_{\mathrm{Re}} = 24$ is valid and the terminal fall velocity equation (7.3) reduces to

$$v_{\mathrm{t}} = \frac{2\rho_{\mathrm{w}}}{9\eta} g r^2 .$$

Beard (1976) introduces a slip correction factor to account for non-continuum flow since the Knudsen number is not negligible. For drops larger than 10 μm radius, this factor is very close to unity; for drops of radius 0.5 μm, it is 1.17.

[1] See Lamb (1945), p. 598.

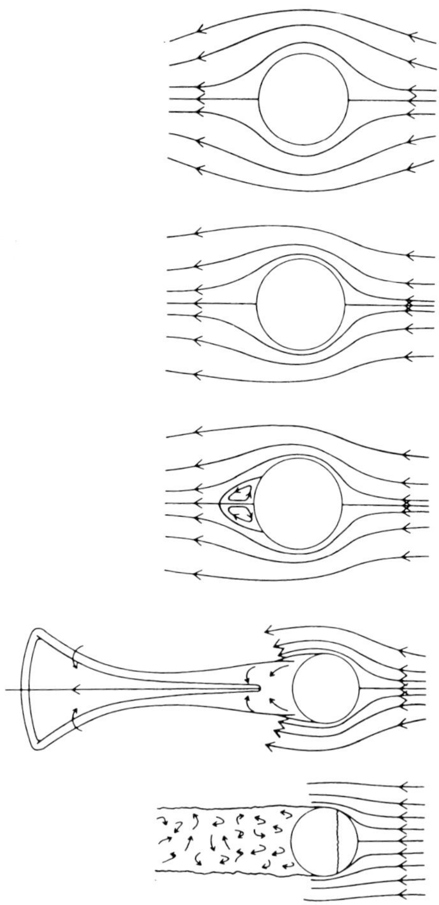

Fig. 7.2. Flow regimes about a rigid sphere. (a) Stokes region exhibits laminar, steady flow with axial and fore–aft symmetry. (b) Low Reynolds numbers ($0 < N_{Re} < 20$) exhibit laminar, steady, axisymmetric flow, but flow no longer exhibits fore–aft symmetry. (c) Intermediate Reynolds numbers ($20 < N_{Re} < 200$) exhibit laminar, steady, axisymmetric flow, separated with an enclosed vortex ring. (d) Moderate Reynolds numbers ($300 < N_{Re} < 450$) exhibit vortex loops formed by roll-up and partial detachment of wake vortex ring. Vortex loops detach periodically from diametrical opposite sides of the wake axis, producing a zig-zag motion. (e) Large Reynolds numbers ($N_{Ne} > 450$) exhibit an increasingly chaotic wake. From Beard (1976). *Journal of Atmospheric Sciences*, **33**, 5. American Meteorological Society. Reproduced with permission.

Drops in the size range from 10 μm to 500 μm radius (N_{Re} from 0.01 to 300) are essentially spherical and have a steady wake. Beard (1976) provides a best fit to a combination of experimental and numerical studies in terms of the Davies number N_{Da}[2]

$$N_{Da} = \frac{32\rho_a(\rho_w - \rho_a)gr^3}{3\eta^2}.$$

[2] The Davies number is also known as the Best number, although it was first described by Davies.

The terminal fall velocity is derived from the Reynolds number given by

$$N_{\mathrm{Re}} = C_{\mathrm{sc}} \exp\left[b_{\mathrm{o}} + \sum_{\mathrm{i}=1}^{6} b_{\mathrm{i}} x^{\mathrm{i}} \right],$$

where x is the natural logarithm of the Davies number. The slip correction factor is

$$C_{\mathrm{sc}} = 1 + 1.255\, \ell / r$$

where

$$\ell = \ell_{\mathrm{o}} \left(\frac{\eta}{\eta_{\mathrm{o}}} \right) \left(\frac{P_{\mathrm{o}}}{P} \right) \left(\frac{T}{T_{\mathrm{o}}} \right)^{1/2}$$

and $\ell = 6.63 \times 10^{-6}$ cm, $P_{\mathrm{o}} = 1013.25$ mbar, $\eta_{\mathrm{o}} = 1.818 \times 10^{-4}\,\mathrm{g\,cm^{-1}\,s^{-1}}$, and $T_{\mathrm{o}} = 293.15$ K. The coefficients are given in Table 7.1.

Raindrops larger than 500 μm radius tend not to be spherical, deforming in response to hydrodynamic forces. Studies have also suggested that the shapes of very large raindrops may fluctuate in response to turbulence, although the effects of these fluctuations on the fall velocity appears to be fairly small. The shapes of raindrops of different sizes falling at their terminal fall velocity are shown in Fig. 7.3, based on the calculations of Beard and Chuang (1987).

Beard (1976) provides a polynomial relation for drops in this regime based on the Bond number N_{Bo} and an unnamed physical parameter number N_{P} obtained by eliminating the radius from the Davies and Bond numbers. The Bond number is given by

$$N_{\mathrm{Bo}} = \frac{16(\rho_{\mathrm{w}} - \rho_{\mathrm{a}})\, g r^2}{3\sigma_{\mathrm{LV}}},$$

Table 7.1. Coefficients for the solutions of the terminal fall velocities for water drops.

	10 μ m < *r* ≤ 500 μm	**500 μm < *r* < 0.35 cm**
b_0	−3.186 57	−5.000 15
b_1	0.992 696	5.237 78
b_2	$-1.531\,93 \times 10^{-3}$	−2.049 14
b_3	$-9.870\,59 \times 10^{-4}$	0.475 294
b_4	$-5.788\,78 \times 10^{-4}$	−0.054 281 9
b_5	$-8.551\,76 \times 10^{-5}$	$2.384\,49 \times 10^{-3}$
b_6	$-3.278\,15 \times 10^{-6}$	

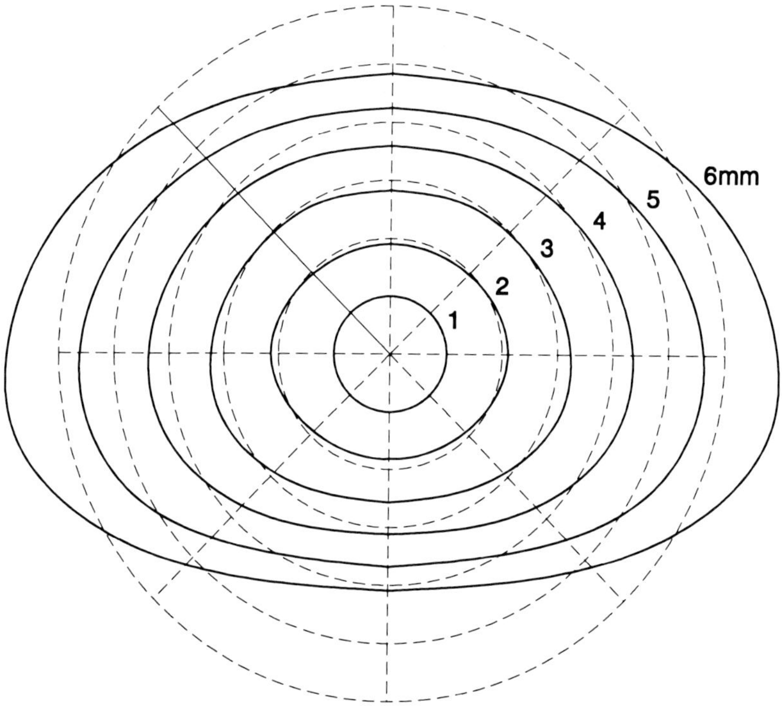

FIG. 7.3. Computed shapes of freely falling water drops at terminal velocity in air for diameters of 1, 2, 3, 4, 5, and 6 mm. From Beard and Chuang (1987). *Journal of Atmospheric Sciences*, **44**, 11. American Meteorological Society. Reproduced with permission.

where σ_{LV} is the surface tension of the drop. The physical parameter number is

$$N_p = \frac{\sigma_{LV}^3 \rho_a^2}{\eta^4(\rho_w - \rho_a)g}.$$

The Reynolds number is given in terms of these two dimensionless numbers as

$$N_{Re} = N_P^{1/6} \exp\left[b_o + \sum_{i=1}^{5} b_i x^i\right],$$

where

$$x = \ln[N_{Bo} N_P^{1/6}]$$

and the coefficients b_i are given in Table 7.1.

The terminal fall velocities for water drops at 700 mbar and 0°C are shown in Fig. 7.4. Note how the flattening of the drops at larger sizes causes the drag force (for the same fall velocity) to increase at nearly the same rate as the inertial force, resulting in a nearly constant terminal fall velocity for millimeter-sized rain drops.

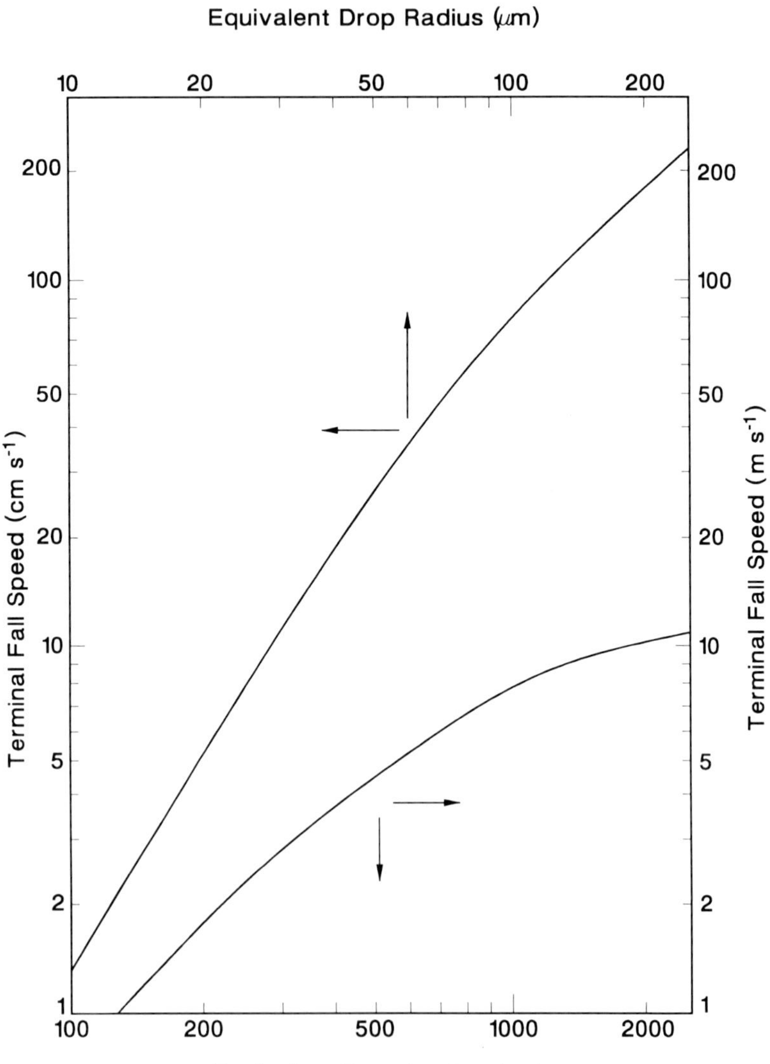

FIG. 7.4. The terminal fall velocity for water drops in air at 0°C and 700 mbar based on Beard's (1976) treatment. The arrows indicate the scales to be read.

7.2.2. *Collection efficiencies*

The concept of collection efficiency is used in solutions of (7.1) and (7.2) rather than a specification of the radius of the swept cylinder. This permits easy visualization of the effects that the hydrodynamic forces between the drops have on their mutual collection. The problem of determining collection efficiencies has been approached theoretically and experimentally. In general, theoretical approaches yield 'collision' efficiencies rather than collection efficiencies, whereas laboratory observations tend to yield collection efficiencies.

(i) Definition of collection efficiency. The collection efficiency is commonly defined in relation to the volume which would be swept out in the absence of any hydrodynamic forces between the two drops, i.e.,

$$E_{\mathrm{LS}} = \frac{R^2}{(r_{\mathrm{L}} + r_{\mathrm{S}})^2} . \tag{7.4}$$

The collection efficiency defined in this manner is often termed the geometric collection efficiency. Values of the geometric collection efficiency less than 1 represent mutual repulsion as the drops approach and, values greater than 1 represent mutual attraction.

For computational purposes, it is more convenient to define the collection efficiency as

$$E_{\mathrm{LS}} = \frac{R^2}{r_{\mathrm{L}}^2} , \tag{7.5}$$

which is simply the ratio of the area swept to the area traversed by the collector drop. This obviates the need to include the collected drop radius in the collection equation (7.2), which may then be written

$$\frac{dN_{\mathrm{s}}}{dt} = \pi r_{\mathrm{L}}^2 E_{\mathrm{LS}} n_{\mathrm{s}} (v_{\mathrm{L}} - v_{\mathrm{S}}) . \tag{7.6}$$

The geometrical collection efficiency may be converted to the collection efficiency as defined by (7.5) by multiplying by

$$\left(\frac{r_{\mathrm{L}} + r_{\mathrm{S}}}{r_{\mathrm{L}}} \right)^2 .$$

A linear collection efficiency is sometimes used and is defined as the square root of the collection efficiency as defined by (7.5).

(ii) Collision versus collection. A collection event is generally pictured as being the result of a collision followed by a coalescence. Thus, the collection efficiency can be considered to consist of the product of a collision efficiency and a

coalescence efficiency (given a collision). Consider the case in which two drops collide and bounce apart. There is a collision but no collection, since coalescence did not occur. However, even in this seemingly straightforward example, conceptual difficulties are encountered in trying to determine exactly what constitutes a collision and how a collision differs from a coalescence.

When two drops approach, they tend to deform and trap air between them. In order for two drops to actually touch, this film of air must be squeezed out from between the two drops, otherwise, the two drops will bounce apart again without ever having their surfaces in contact. To the casual observer, this may appear to be a collision. Is it really a 'collision?' How close do two drops have to approach before it may be called a 'collision?'

The orientation of the surface molecules in each drop also serves to keep the drops apart. As was noted in Chapter 2, the surface layer is ordered in such a way that more hydrogens are oriented outward, i.e., lone-pair orbitals, giving a slightly positive charge at the surface. As the two drops approach, this electrostatic repulsion must be overcome before contact can be made.

Once contact is made, does a true interface between the drops exist? Once this repulsion is overcome, perhaps the drops simply coalesce. If collision is defined as making contact in this fashion, perhaps all 'collisions' result in coalescences. Coalescence has clearly occurred if there is a single air–water interface that encompasses the masses of the two drops. However, it is quite difficult to define what a collision is or is not.

Two drops may collide in such a manner that they remain together temporarily and then pull apart again. In this case, the criterion for coalescence is met (and obviously the criterion for collision as well) but has a collection event occurred? It would not seem so. One may ask how long must two drops remain together before one decides they will not separate? Once again, there are no well-defined answers to these questions.

If two drops can collide and coalesce without a collection event taking place, then the view of collection as being the result of a collision followed by a coalescence is erroneous. For these reasons, it is preferable to consider a two-step process. Consider first that the formation of a single air–water interface (a collision followed by a coalescence) results in a collection event. Second, consider the fraction of these 'collections' which results in separation or breakup. This treatment is preferred since two drops which temporarily coalesce tend to produce satellites (fragments), whereas those which fail to coalesce, but rather bounce apart, generally do not.

(iii) Theoretical estimates of collision efficiencies. As a large drop approaches a small drop, the flow fields around the drops tend to deflect the small drop from the path of the large drop, reducing its likelihood of being collected compared to geometric collection. The deflection should be less for larger drops since the inertia becomes relatively more important than the hydrodynamic forces. Thus, for $r_L \gg r_S$, the collision efficiency should increase as r_L increases for constant r_S. Similarly, drops much smaller than the large drop tend to remain with the flow

and are carried around the large drop. This becomes less important as the small drop size is increased, i.e., the collision efficiency should increase as r_S increases for constant r_L.

The problem of modeling the flow around two spheres falling at different velocities is highly complex, and complete solutions have been accomplished only in the regime where the Stokes approximation is valid. Davis (1972) presents numerical solutions, which include kinetic effects at close separations. For larger drops, collision efficiencies are obtained by a method of superposition in which the forces acting on one drop as a result of the flow around the other drop are modeled but the flow fields do not interact. These methods are described in some detail by Pruppacher and Klett (1978).

(iv) Observations of collection efficiencies. A variety of experimental approaches have been used in efforts to determine collection efficiencies between collector and collected drops of various sizes. Due to inherent limitations in the design of the apparatus, it usually is not possible to observe drop collection over a wide range of fall velocities with a single experimental apparatus. Observed collection efficiencies are thus restricted to relatively limited portions of the field of collection efficiencies.

One of the more useful experiments was conducted by Woods and Mason (1964) for collector drops between 34 μm and 55 μm radius and collected drops between 1 μm and 10 μm radius. This range of drop sizes is important in the initial stages of coalescence growth and coincides with several theoretical treatments, enabling an evaluation of the validity of the theoretical approaches. Beard and Pruppacher (1971) used a vertical wind tunnel to determine collection efficiencies for somewhat larger collector drops between 70 μm and 300 μm radius and collected drops between 4 μm and 6 μm radius.

Several experiments have been designed to examine collection between relatively large drops of similar sizes (> 50 μm radius). Some experimental results suggested that such drop pairs were mutually attracted, resulting in geometric collection efficiencies greater than 1. Some theoretical studies of collision efficiencies also suggested such a possibility. Collection efficiencies greater than unity are not considered realistic at present. Experiments by Woods and Mason (1965) remain perhaps one of the more reliable sources of collection efficiencies for drops in this range.

Discrepancies between the observed collection efficiencies and theoretical collision efficiencies for drops between 50 μm and 100 μm radius led Beard and Ochs (1983) to suggest that the coalescence efficiency is significantly less than 1. Later studies by Ochs et al. (1986) suggest that the coalescence efficiency is a function of the Weber number (N_{We}), defined in Appendix 2. The dependence of the coalescence efficiency as a function of the Weber number is shown in Fig. 7.5.

The field of geometric collection efficiencies based on both observed collection efficiencies and theoretical collision efficiencies is shown in Fig. 7.6. The regions examined by Woods and Mason (1964,1965), Beard and Ochs

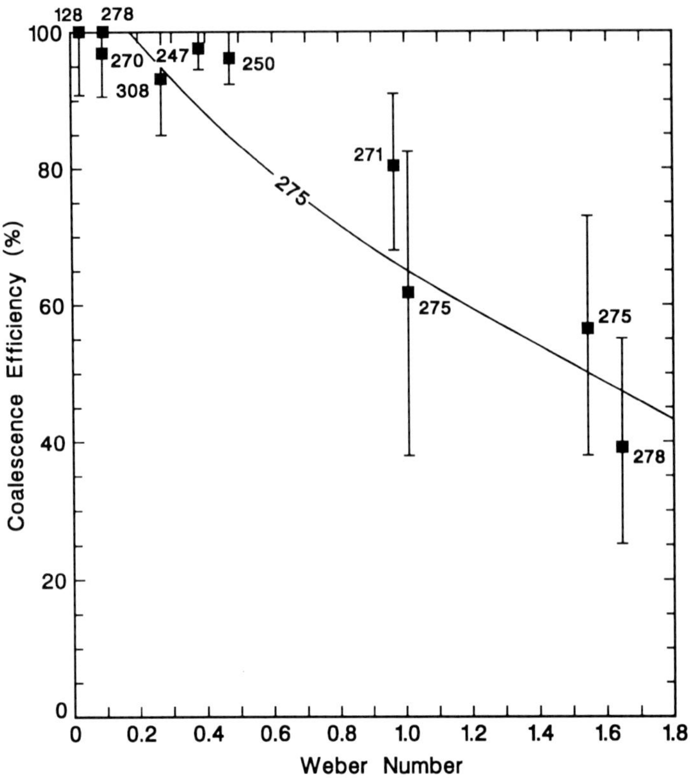

FIG. 7.5. Coalescence efficiencies as a function of the Weber number, as determined experimentally by Ochs et al. (1986). (■), Present data; (—), Beard and Ochs (1983). *Journal of Atmospheric Sciences*, **43**, 3. American Meteorological Society. Reproduced with permission.

(1983), and Beard and Pruppacher (1971) are cross-hatched. Theoretical collision efficiencies based on Schlamp et al. (1976) were used to fill in the isopleths for collector drop radii less than 80 μm. The field of theoretical collision efficiencies has been adjusted to match the observations in the regions of overlap. Theoretical collision efficiencies for drops larger than 100 μm radius were obtained from Mason (1971).

(v) Other effects on collection efficiencies. The relatively low collection efficiencies for cloud droplets has led to suggestions that other effects may enhance the mutual collection of cloud droplets by increasing their collection efficiencies. de Almeida (1977, 1979) presents a model comparing the growth of a population of cloud droplets by coalescence both with and without the effects of turbulence. His approach has been criticized by Pruppacher and Klett (1978).

More recently, Reuter et al. (1988) examine the collection kernel for drops subject to random fluctuations using stochastic differential equations. They find a slight increase in the collection efficiency for collector drops smaller than

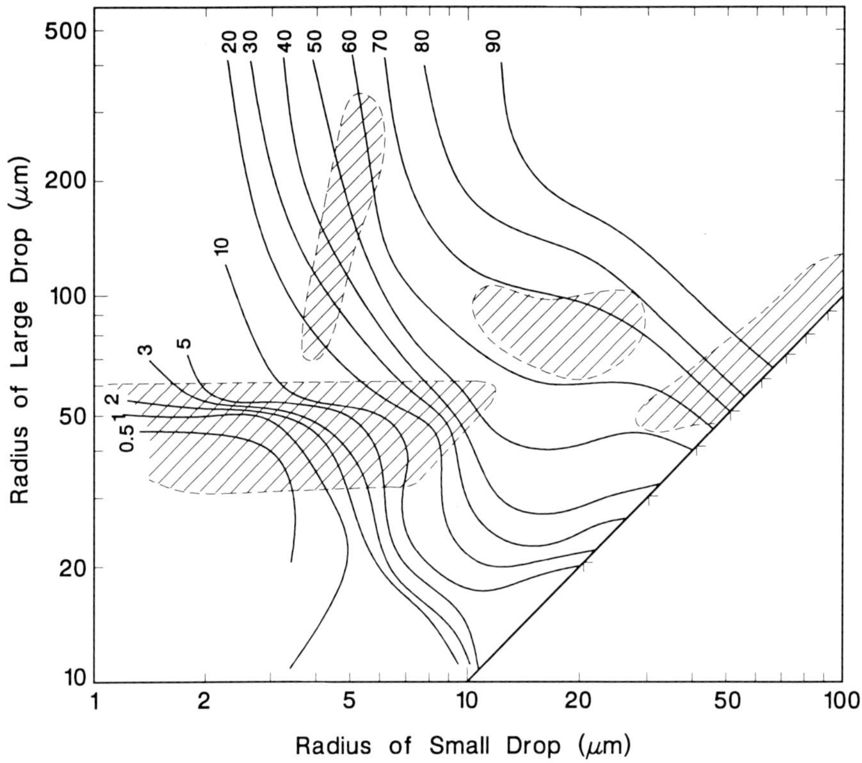

FIG. 7.6. The field of collection efficiencies (per cent) based on experiment and theory, as discussed in the text. Isopleths in the cross-hatched regions are based on experimental values from (clockwise from lower left) Woods and Mason (1964), Beard and Pruppacher (1971), Beard and Ochs (1983), and Woods and Mason (1965).

100 μm radius. Cooper and Baumgardner (1989) suggest that these effects are much smaller than claimed by Reuter et al. since the turbulent diffusion coefficient affecting the drops must decrease as the two drops approach. They suggest turbulence may affect the relative fall velocities of the drops and thereby influence their mutual collection efficiency.

Rogers and Davis (1990) suggest that van der Waals attractions may substantially increase the collection efficiencies between drops smaller than 10 μm radius. Based on a numerical model, they find that the time required to produce a given mass of droplets with radii greater than 25 μm is reduced by between 20% and 25% when these effects are included.

Schlamp et al. (1976) examined the effects of electric charges on the drops and the presence of vertical external electric fields on the collision efficiencies on the basis of numerical simulations. Their results suggest that these effects are negligible for drops larger than 50 μm radius, although such effects may increase the collision efficiencies for drops smaller than 12 μm radius by as much as two orders of magnitude (values exceeding 100%) for electric fields of 3000 V cm^{-1} and for oppositely charged drops carrying ± 2 esu cm^{-2} (6.7×10^{-2}).

Sartor and Abbott (1968) found that coalescence would always occur provided the electric field strength exceeded a critical value. Bradley and Stow (1978b) examine the drainage of the air film separating two drops as they approach for uncharged and charged drops. They conclude that drainage is enhanced as the charge on the drops is increased, allowing coalescence above a critical charge.

Very little work on the effects that changing ambient conditions might have on the collection efficiency has been carried out. de Almeida (1977) presents one computation showing a substantial increase in the collision efficiency with decreasing pressure. The reduced air density may be expected to reduce the tendency of the small drop to be carried out of the collector drop path. This suggests that the collection efficiencies presented in Fig. 7.6 may underestimate those relevant to typical cloud conditions.

7.3. Solutions of the collection equation

There are three basic interpretations as to how to solve the collection equation (7.2). The earliest and simplest solution in essence assumed it was possible to collect fractions of drops in a continuous manner. The times required to produce raindrops by this method were inordinately long, leading to suppositions such as Langmuir's (1948) chain reaction process resulting from drop breakup.

Telford (1955) was the first to recognize the importance of stochastic processes in the growth of drops by collection. In a stochastic process, the collection equation provides the differential *probability* that a given drop will collect a (smaller) droplet. This has often been interpreted as the *fraction* of drops that would undergo a collection event in dt, leading to solutions which were completely deterministic. Gillespie (1975) gives a proper stochastic treatment of the collection equation and demonstrates the relationships between three methods of solution. Gillespie's treatment forms the basis of the following sections.

7.3.1. *Three models for collection processes*

The three interpretations of the collection equation described above are termed 'continuous,' 'quasi-stochastic,' and 'pure-stochastic.' The continuous model interprets dN_s as the number of droplets (which may be and usually is fractional) collected by each and every drop in time dt. Thus each drop grows at the same rate. The physical interpretation of this model is that the mass of the droplets is uniformly distributed (on a molecular scale).

The quasi-stochastic model interprets dN_s as the *fraction* of drops that receive a droplet in time dt. Clearly, a drop cannot collect a fractional droplet, i.e., collections must occur as discrete events. Hence, this model assumes that a definite fraction of the drops will collect one (or more) droplets and the remainder will collect none during the time dt. Thus, different drops may follow different growth histories although the time evolution of the drop spectrum is determined and only one outcome (and history) is possible for a given initial condition.

The quasi-stochastic model can be interpreted in two ways. The droplets may be uniformly distributed in space or they may be randomly distributed. If the droplets are uniformly distributed in space, the drops will either collect one droplet or none.[3] If the droplets are randomly distributed, the number of droplets collected by a given fraction of the collector drops is Poisson-distributed, with an expected value given by the collection equation. Thus, while most drops may not collect any droplets, some may collect one, two, three, or more droplets in a single time step. These interpretations of the quasi-stochastic model are illustrated in Fig. 7.7. Young (1975) shows that the differences between these two treatments are small for time steps of 5 seconds.

The pure-stochastic model interprets dN_s as the *probability* that a given collector drop will experience a collection event in time dt. The randomly changing positions of the droplets lends a probabilistic nature to the collection process. The Poisson interpretation of the quasi-stochastic model led to a single outcome because it assumed a time invariant random positioning of the droplets. In the pure-stochastic model, the positions of the droplets are not fixed but are randomly changing with time. Thus, different histories and different outcomes may be expected for each possible sequence of positionings of the droplets. What may actually occur for any given occasion is randomly determined. The set of all possible histories and outcomes is referred to as a stochastic process.

These three models may be summarized as follows:

1. the continuous model predicts that all the collector drops will have the same size at time t;
2. the quasi-stochastic model predicts that all clouds will have the same drop size distribution at time t;
3. the pure-stochastic model predicts that each cloud will have its own unique distribution of drop sizes at time t.

7.3.2. *The pure-stochastic model*

This treatment of stochastic coalescence makes a number of simplifying assumptions which yield tractable solutions but which make the model rather unrealistic. The collector drop is assumed to have a fixed collection kernel, regardless of its size. In reality, the collection kernel increases so rapidly after the first few collection events that growth approaches the continuous growth model. The value of this exercise lies in the concepts presented and in demonstrating the relation between the quasi-stochastic and pure-stochastic models.

Consider that there are n_s droplets of mass μ each with n_s very large so that depletion may be neglected. Consider also that there are L large drops which are initially of mass m_o and which grow by collecting the small droplets. In this

[3] If the average number of droplets collected is greater than 1, say 3.1, either 3 or 4 droplets are collected.

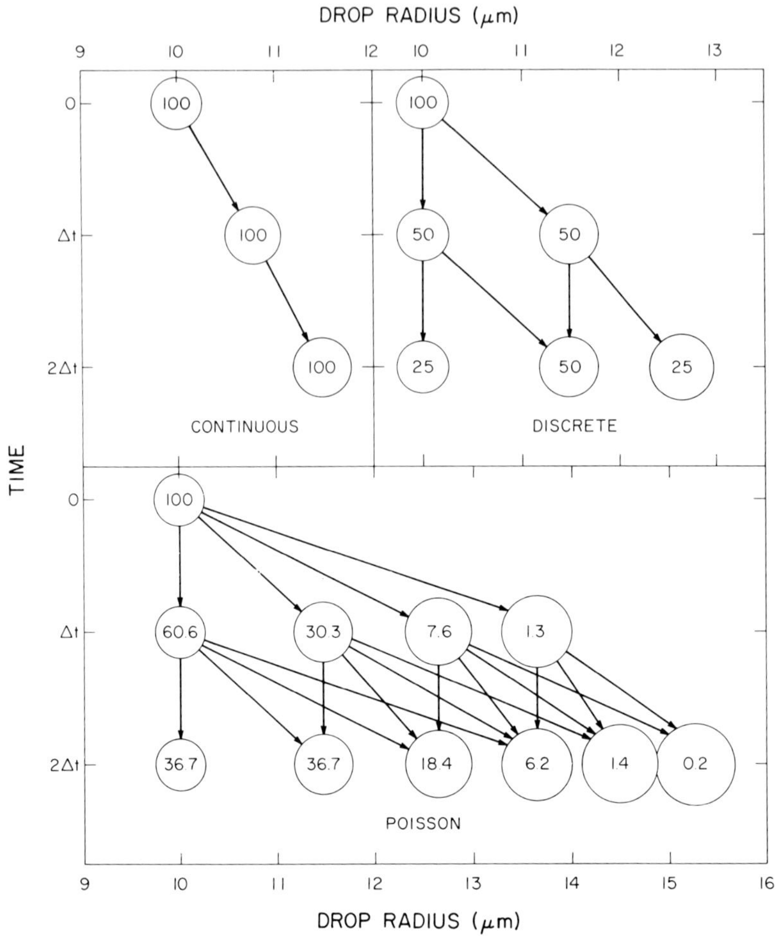

Fig. 7.7. The growth of 10 μm radius drops collecting 8 μm radius drops for the continuous, discrete, and Poisson collection models. The expected number of collection events within a given time step (Δt) is 0.5. Numbers within the circles reflect the percentage of drops of that size; arrows show growth paths. From Young (1975). *Journal of Atmospheric Sciences*, **32**, 5. American Meteorological Society. Reproduced with permission.

model, L and μ are fixed with n_s and the collection kernel Γ assumed constant, and m varies in discrete units of μ.

The probability that any drop will collect a droplet in time dt is $n_s \Gamma dt$. The expression $\Pr(n,m,t)$ will be used to denote the probability that exactly n drops have mass m at time t. Here, n may take on values from 0 to L and m must be m_o plus an integer multiple of μ.

First, consider the probability that any given drop will collect k droplets in time t, denoted by $\Pr(k,t)$. For $k = 0$,

$$\Pr(0, t + dt) = \Pr(0,t)\,(1 - n_s \Gamma dt)\,,$$

which states that the probability of not collecting any droplets in time $t + dt$ is the product of not collecting any droplets in time t and the probability of not collecting any in time dt. This may be expressed as

$$\frac{\Pr(0,t + dt) - \Pr(0,t)}{\Pr(0,t)} = -n_s \Gamma dt = \frac{d[\Pr(0,t)]}{\Pr(0,t)} .$$

Integrating this expression, the probability of not collecting any droplets in time t is[4]

$$\Pr(0,t) = \exp(-n_s \Gamma t) . \tag{7.7}$$

For $k \geqslant 1$, a recursive technique of the form

$$\Pr(k,t) = \int_0^t \Pr(k - 1,t') \Pr(0,t - t') n_s \Gamma dt' \tag{7.8}$$

may be employed. This may be read as the probability of collecting k droplets in time t is the integral over time of the probability of collecting the kth droplet in the interval dt' after having collected $k - 1$ droplets up to time t' and then not collecting any more droplets from t' to t. For $k = 1$, (7.8) may be written as

$$\Pr(1,t) = \int_0^t \Pr(0,t') \Pr(0,t - t') n_s \Gamma dt' .$$

Substituting for the probabilities within the integral from (7.7),

$$\Pr(1,t) = \int_0^t \exp[-n_s \Gamma t'] \exp[-n_s \Gamma (t - t') n_s \Gamma dt' \tag{7.9}$$

$$= n_s \Gamma t \exp(-n_s \Gamma t) .$$

The expression for $k = 2$ then follows from application of (7.7) and (7.8). It quickly becomes apparent that the general solution is

$$\Pr(k,t) = \frac{(n_s \Gamma t)^k \exp(-n_s \Gamma t)}{k!} . \tag{7.10}$$

The probability that a particular arrangement of n drops will collect exactly k droplets in time t is

$$[\Pr(k,t)]^n [1 - \Pr(k,t)]^{L-n} .$$

[4] Note that Pr (0,0) = 1.

However, particular combinations of drops are not of interest but rather the probability of having any n drops having exactly k collections. The total number of arrangements of L drops taken n at a time is

$$\frac{L!}{n!\,(L-n)!}$$

and the probability of having exactly n drops having mass $m = m_o + k\mu$ at time t is

$$\Pr(n,m,t) = \frac{L!}{n!\,(L-n)!}\,[\Pr(k,t)]^n\,[1-\Pr(k,t)]^{L-n}\,. \tag{7.11}$$

Equation (7.11) represents a complete and exact solution to the pure-stochastic model, given our initial assumptions. Note that at any time t, there exists a finite probability that any number of drops, from 0 to L, may have any specified mass from m_o and greater in multiples of μ. This is in contrast to the quasi-stochastic model, which states that a definite number of drops will have mass m in time t.

A somewhat more general solution is possible for the pure stochastic model by allowing the collection kernel to vary with the number of collection events. It is still necessary to assume that drops only collect droplets and that the concentration of droplets is very large. If Γ_i is the collection kernel for the ith collection event, then the recursive technique can be employed as before. The general form is complex and uninformative. The result for $k = 1$ (see Problem 7.7) is

$$\Pr(1,t) = \frac{\Gamma_1}{\Gamma_2 - \Gamma_1}\,[\exp(-n_s\Gamma_1 t) - \exp(-n_s\Gamma_2 t)]\,. \tag{7.12}$$

7.3.3. *Quasi-stochastic versus pure-stochastic models*

Gillespie (1975) shows that the expected result of the pure-stochastic model (the average over all possible results) is the same as the result for the quasi-stochastic model. The expected number of drops of mass m at time t may be expressed as

$$E(n) = \sum_{n=0}^{L} n\Pr(n,m,t)$$

$$= \sum_{n=0}^{L} \frac{nL!}{n!(L-n)!}\,\Pr^n\,(1-\Pr)^{L-n}\,.$$

By factoring out $L\Pr$, the summation is simply the summation of a binomial distribution over its range, which is equal to unity. The expected number $E(n)$

then becomes LPr. Gillespie (1975) shows this to be the result for the quasi-stochastic model as well.

The variance of n can also be derived in a similar manner (see Problem 7.6). The result is

$$V(n) = E(n)\left[1 - \frac{E(n)}{L}\right]. \tag{7.13}$$

For relatively long times, the expected number of drops which have exactly mass m becomes very small, and the variance is equal to the expected number of drops. This implies that the quasi-stochastic model can be used to give not only the expected outcome but also a measure of the spread in the outcomes due to the stochastic nature of the collection process.

As an illustrative example, consider the coalescence process taking place within a cloud with a volume of 1 km^3. Assume that the entire volume of the cloud can be sampled and there are 10^{12} drops of radii between 48 μm and 50 μm. This represents an average drop concentration of 1000 drops per m^3. What is the expected range in the concentration of drops of this size due to the stochastic nature of the process?

The variance of the number is simply the expected number from (7.13). The standard deviation is the square root of this or 10^6 and the average drop concentration is

$$n = \frac{10^{12} \pm 10^6}{10^9} = 1000 \pm 10^{-3} \text{ drops m}^{-3}.$$

It may be concluded that the differences due to the stochastic nature of the collection processes are negligible and that the cloud behavior may be adequately described by the quasi-stochastic treatment.

7.4. Drop breakup

It can be shown (see Problem 7.3) that the radius of a drop growing by collection will increase at a rate roughly proportional to its fall velocity. Since the fall velocity generally increases with increasing size, this implies that the larger the drop, the faster it increases its radius. Since drops larger than 0.25 cm are rarely observed in rainfall, some mechanism which limits the growth of drops must be operative.

Early ideas held that large drops were inherently unstable and would break up when they reached some limiting size. Although this concept has merit, the fact that drops as large as 0.45 cm radius have been suspended in a vertical airstream for considerable periods of time argues that some other mechanism must serve to limit the sizes of raindrops. Virtually all of the growth of raindrops is *via* the collection process and such collisions are quite likely to introduce unstable oscillations in a large drop. Thus, drop breakup due to collisions with smaller drops may be expected to be responsible for limiting the growth of raindrops.

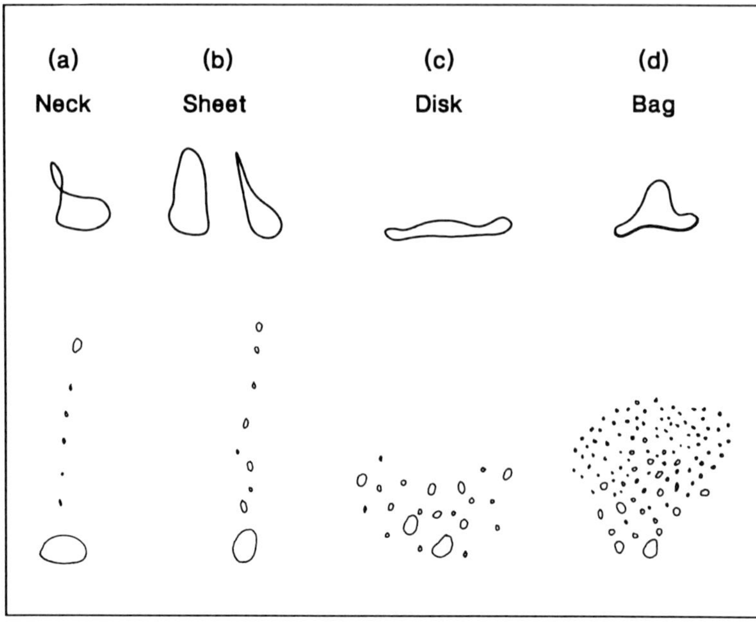

FIG. 7.8. The four observed breakup modes as described in the text. (a) Filament breakup; (b) sheet breakup; (c) disk breakup; (d) bag breakup. From McTaggart-Cowan and List (1975). *Journal of Atmospheric Sciences*, **32**, 7. American Meteorological Society. Reproduced with permission.

7.4.1. *Modes of drop breakup*

One of the more extensive series of observations conducted on the breakup of large drops and the fragments produced by the breakup process is described by McTaggart-Cowan and List (1975), and Low and List (1982a,b). Both the 1975 and 1982 experiments produced over 700 cases of breakup each, with roughly 2% to 5% of the attempts resulting in collision. McTaggart-Cowan and List classified their observations into four distinct modes, as shown in Fig. 7.8.

(i) Filament breakup. Filament breakup occurs as a result of a glancing collision. The small drop makes contact with the large drop, forming a neck (filament), which becomes elongated and eventually breaks up as the two drops separate. As the drops separate, the small drop becomes elongated, although the large drop appears to remain unaffected. After separation, the large and small drops retain much of their identity, losing a small portion of their mass, typically forming 2 to 10 fragments.[5] The number of fragments formed was noted to increase as the collisional energy increased. Filament breakup was observed over the range of drop pairs tested.

[5] The fragments include the parent drops.

(ii) Sheet breakup. Sheet breakup occurs when the small drop strikes the side of the large drop in such a manner that a sheet of water is torn off the large drop. The onset of sheet breakup was noted at slightly larger small drop sizes than was required to produce filament breakup. After breakup, the small parent drop generally is indistinguishable from the other small fragments. The large drop is noticeably distorted and is smaller after separation. The number of fragments formed typically ranged from 2 to 15, increasing with increasing collisional energy.

(iii) Disk breakup. Disk breakup occurs when the small drop strikes the large drop near its center. As the small drop begins to coalesce with the large drop, a small disk begins to spread out from the point of impact. The large drop becomes deformed into a disk extending two to three times its original dimensions. This results in an increased drag force and the disk curves upward, forming a bowl. As the disk begins to contract, it frequently sheds drops and the entire disk tends to break up. If the collisional energy is low, relatively few fragments are formed and the disk contracts into a drop. If the collisional energy is high, many fragments (up to 50) may be observed.

(iv) Bag breakup. Although bag breakup is rarely observed, it tends to produce the most fragments. Bag breakup is similar to disk breakup with the exception that a toroid is formed with a thin sheet of water across its center. This thin sheet forms a bag due to the pressure differential between the stagnation points. Breakup of the toroid forms a few large fragments; breakup of the bag produces a large number of small fragments.

7.4.2. *Drop breakup probabilities*

The probability that two drops, having collided and coalesced, will separate is determined by the energy of the collision tending to pull them apart and the surface tension tending to hold them together. Brazier-Smith et al. (1972) postulate that separation will occur when the rotational energy is greater than the available surface energy. Their treatment is presented and compared to the experimental coalescence efficiencies determined by Low and List (1982a). Bradley and Stow (1978a,b,c) extend this treatment to consider deformation of the drops during the collision and (1984) present a useful parameterization of the breakup probability.

(i) A theoretical treatment of drop breakup. Considering a given drop pair, the surface energies prior to and following coalescence are fixed. The requirement that the rotational energy exceed the surface energy for breakup to occur implies that the collision must be off-center by some critical amount. This is shown in Fig. 7.9, where x_c represents the critical separation between the fall axes of the two drops. Drops whose centers lie within the distance x_c will be collected since their rotational energy is too small to result in separation; those whose centers lie outside x_c but still collide with the large drop will result in breakup.

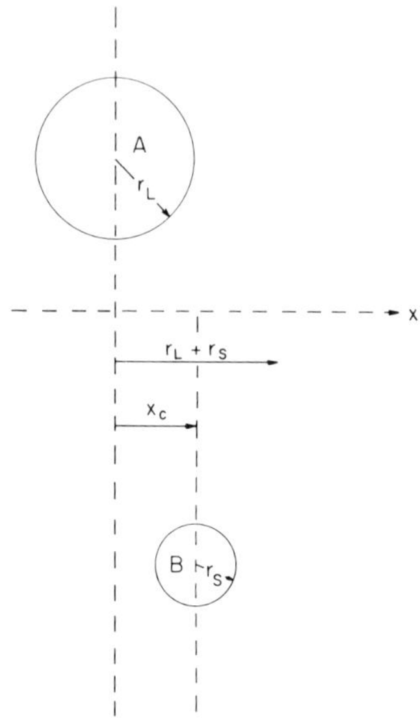

FIG. 7.9. Critical value of the impact parameter for drop breakup. Drops A and B are assumed to collide and coalesce. If the fall axes for the drops lie closer than x_c, the rotational energy is less than the excess surface energy and the drops will remain together. If they lie without x_c but within $r_L + r_S$, they will temporarily coalesce before separating.

The surface energy of the parent drops prior to collision and the combined drop after collision are approximated as proportional to the surface area of the equivalent spheres, even though the combined drop may depart noticeably from sphericity. The excess surface energy (ESE) is

$$\mathrm{ESE} = 4\pi\sigma\,(r_L^2 + r_S^2 - r_C^2)$$

where the subscripts L, S, and C refer to the large, small, and combined drops, respectively. The combined radius is simply $(r_L^3 + r_S^3)^{1/3}$. Defining $r_L/r_S = \gamma$, the ESE reduces to

$$\mathrm{ESE} = 4\pi r_S^2\sigma\,[1 + \gamma^2 - (1 + \gamma^3)^{2/3}]\,. \tag{7.14}$$

The definition of rotational energy (RE) is

$$\mathrm{RE} = \frac{I_c\omega^2}{2} = \frac{\Omega^2}{2I_c}, \tag{7.15}$$

where I_C refers to the moment of inertia of the combined system, ω is its angular velocity and Ω is its angular momentum. The angular momentum for the 'dumbbell' formed by the two drops is, by conservation of angular momentum, the same as for the combined drop. The moment of inertia for the combined drop is that for a sphere of radius r_C,

$$I_c = \frac{2mr_c^2}{5}.$$

The angular momentum of the two-drop dumbbell is $I\omega$, where the angular velocity ω is simply the difference in fall velocities divided by their separation ($\Delta v/x_C$). The moment of inertia for this system is

$$I = m_L x_g^2 + m_S (x_c - x_g)^2,$$

where x_g refers to the distance between the fall axis of the large drop and the center of gravity of the two-drop dumbbell. The location of the center of gravity is simply the mass weighted position

$$x_g = \frac{m_S x_c}{m_L + m_S}.$$

The moment of inertia for the two-drop dumbbell then reduces to

$$I = \frac{x_c^2 m_S m_L}{m_L + m_S}.$$

The angular momentum may then be written as

$$\Omega = \frac{4\pi\rho_w r_S^3 r_L^3 x_c \,\Delta v}{3(r_S^3 + r_L^3)}.$$

The rotational energy then follows by substituting the angular momentum in the definition of rotational energy given by (7.15),

$$\mathrm{RE} = \frac{5\pi\rho_w r_L^6 (\Delta v)^2 x_c^2}{3r_S^5 (1 + \gamma^3)^{11/3}}. \tag{7.16}$$

Equating the RE (7.16) and the ESE (7.14), the solution for the critical separation distance is

$$x_c^2 = \frac{12\sigma r_S^7 [1 + \gamma^2 - (1 + \gamma^3)^{2/3}](1 + \gamma^3)^{11/3}}{5(\Delta v)^2 \rho_w r_L^6}.$$

The coalescence efficiency may be defined as the fraction of collisions in which the drops remain together, assuming a geometric sweepout (not unreasonable for drops of this size),

$$E_c = \left[\frac{x_c}{r_L + r_S} \right]^2$$

or

$$E_c = \frac{2.4\sigma(1 + \gamma^3)^{11/3}[1 + \gamma^2 - (1 + \gamma^3)^{2/3}]}{(1 + \gamma)^2 \gamma^6 r_S \rho_w (\Delta v)^2}. \tag{7.17}$$

The probability of breakup given a collision is simply $1 - E_c$.

(ii) Comparison with observations. Experimental breakup probabilities determined by Low and List (1982a) and theoretical breakup probabilities determined from (7.17) are compared in Table 7.2. Agreement between the observed and theoretical breakup probabilities is best when the ratio of the large drop to small drop radius is less than 3.0. This is noted by Bradley and Stow (1978c), who derive the condition of stability for coalescing drops without assuming spherical drops.

When the radius ratio is large, the difference between the surface energy of the original drops and that of the combined drops is very small. In such situations, relatively small deviations from sphericity may be expected to produce large changes in the calculated breakup probability. The last column of Table 7.2 shows the increase in surface area for the combined drop required to

Table 7.2. Comparison between observed and theoretical breakup probabilities. The factor represents the increase in surface area required to bring the theoretical values into agreement with the observed values.

Drop pairs		**Probability of breakup**			
r_L **(cm)**	r_S **(cm)**	**Observed**	**Fitted**	**Theory**	**Factor**
0.09	0.0198	0.412	0.528	−1.341	1.031
0.20	0.0198	0.523	0.418	−11.979	1.009
0.22	0.0198	0.348	0.396	−15.927	1.007
0.09	0.0358	0.707	0.727	0.650	1.018
0.09	0.05	0.787	0.681	0.788	1.000
0.15	0.05	0.9133	0.878	0.723	1.058
0.18	0.05	1.00	0.860	0.599	1.062
0.23	0.05	1.00	0.783	0.228	1.040
0.18	0.09	1.00	0.918	0.890	1.155
0.23	0.09	1.00	0.947	0.830	1.109

Source: Observed values are from Low and List (1982a); the fitted values are from Bradley and Stow (1984); theoretical breakup probabilities are based on Brazier-Smith et al. (1972).

bring the theoretical treatment into agreement with the experimental values. In general, the required increases in the surface area are less than 5%, a requirement that can be met if the two-drop system is a prolate spheroid with a ratio of major to minor axes of 1.7.

Low and List (1982a) define a collision kinetic energy (CKE) as

$$\mathrm{CKE} = \frac{2\pi\rho_w r_L^3 r_S^3}{3(r_L^3 + r_S^3)} (\Delta v)^2 ,$$

which they use to define the energy of the two-drop system. Bradley and Stow (1984) suggest that the probability of breakup may be described by

$$\Pr\{\text{breakup}\} = 0.67 \left[\frac{E_T}{S_C}\right]^{0.36} , \tag{7.18}$$

where the total excess energy (E_T) is

$$E_T = \mathrm{CKE} \left[1 - \frac{2.3\, r_L^3 r_S^3 (r_L + r_S)^2}{(r_L^3 + r_S^3)^{8/3}} X^2 \right] + \mathrm{ESE} \tag{7.19}$$

and the surface energy of the combined drop (S_C) is

$$S_C = 4\pi\sigma_{LV}(r_L^3 + r_S^3)^{2/3} .$$

The parameter X is termed the impact parameter and is defined in reference to Fig. 7.9 as the distance between the fall axes of the two drops divided by the sum of their radii. In order to solve (7.18) for breakup probabilities, (7.19) must be integrated from $X = 0$ to $X = 1.0$, remembering that the probability of occurrence for a given value of X is proportional to X^2.

The breakup probabilities calculated from (7.18) are shown in Fig. 7.10. The drop pairs for which Low and List (1982a) determined breakup probabilities are indicated by dots. The extension of the experimentally determined breakup probabilities using (7.18) is preferable to a strictly empirical approach since it incorporates the rotational and surface energies in a rational fashion.

7.4.3. *Fragments resulting from drop breakup*

Specification of the fragments resulting from the breakup of drops is crucial to modeling the evolution of the drop size spectrum. Subsequent to drop breakup, the fragments are rarely close to spherical in shape, making it difficult to ascertain their mass with any certainty. Bradley and Stow (1978a) and Low and List (1982a,b) present experimental observations on the fragments resulting from drop breakup.

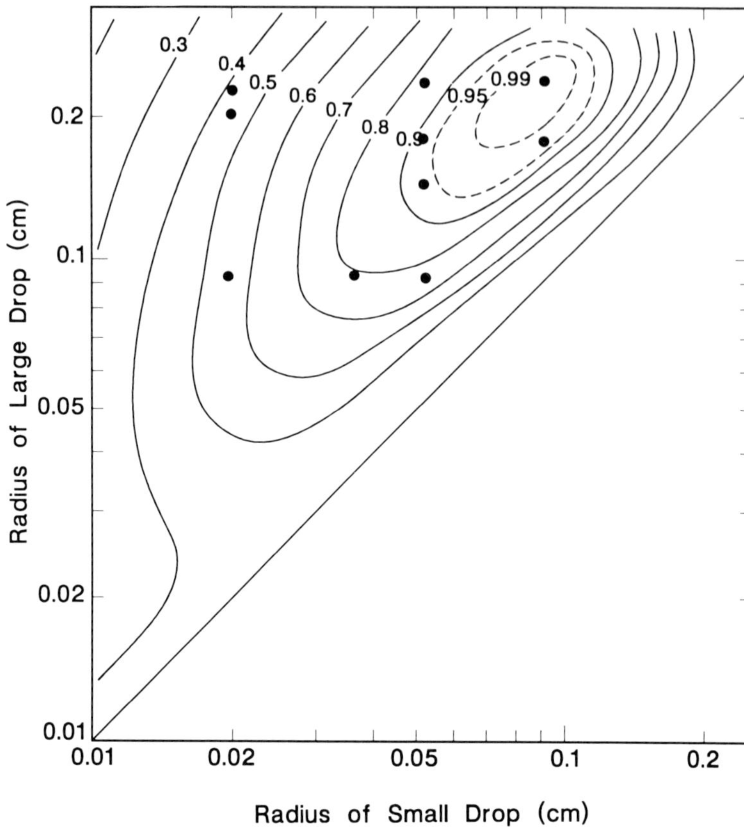

FIG. 7.10. The field of drop breakup probabilities based on (7.18). Dots represent drop pairs for which observed breakup probabilities from Low and List (1982a) are available.

Three cases are possible for two coalescing drops. The drops may remain together (one fragment), they may separate without producing satellites (two fragments), or they may separate producing satellites (more than two fragments). If the drops separate without producing satellites, does each drop retain its original mass or is there a transfer of mass between the drops? Bradley and Stow suggest that the small drop may lose mass to the larger drop for more direct collisions (small values of X) but could not provide a quantitative measure.

For the case in which drop separation produces satellites, one needs to know the number of satellites and their sizes as well as how much mass is lost by each parent drop. Low and List (1982b) consider the fragments produced by filament, sheet, and disk breakup separately. The average numbers of fragments for each breakup mode as a function of the total excess energy (the integration of (7.19)) are shown in Fig. 7.11. Each of these can be represented by a semilogarithmic fit as shown by the broken lines.

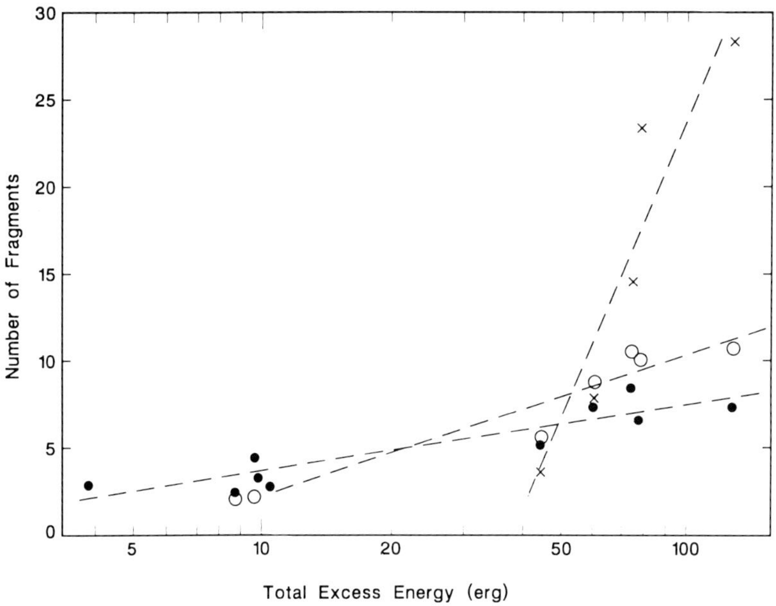

Fig. 7.11. The number of fragments resulting from filament, sheet, and disk breakup are shown as a function of the total excess energy based on experimental data from Low and List (1982a). The broken lines represent least squares fits. For filament breakup, $n = 1.62 \ln E_T - 0.05$; for sheet breakup, $n = 3.52 \ln E_T - 5.90$; and for disk breakup, $n = 24.2 \ln E_T - 88.3$. Only values of $n \geqslant 2$ are permitted. (●), Filament; (○), sheet; (×), disk.

Clearly, the number of average fragments depends on the impact parameter, X, as noted by Bradley and Stow (1978a). Based on the descriptions of the breakup modes and the fractions of breakups of each type given by McTaggart-Cowan and List (1975), values of $X > 0.85$ result in filament breakup, $0.4 < X \leq 0.85$ result in sheet breakup, and $X \leq 0.4$ result in disk or bag breakup. Thus, for a given breakup, the probability of each breakup mode and the resulting average number of fragments due to that mode can be specified.

Low and List (1982a) provide information on the distribution of fragment numbers for each drop pair interaction they studied. From this it is clear that some of the breakups observed at relatively low values of the total excess energy did not result in the formation of satellites. From the published data, it does not appear possible to ascertain the total mass of the satellite drops formed which should also be a function of the total excess energy and the impact parameter. The parameterizations provided by Low and List (1982b) have been noted to fail to conserve mass by Valdez and Young (1985), and in some cases of sheet breakup, the mass requirements for the satellite drops exceeded the total mass of the parent drops. Valdez and Young provide several alternate approaches based on the Low–List parameterizations which provide mass conservation and maintain physical consistency.

Problems

7.1* Langmuir's chain reaction hypothesis assumes that the collection kernel of the fragments of a large drop is significantly larger than the collection kernel of the original drop, i.e., even though the collection kernels of the fragments are individually smaller than the original collection kernel, their numbers more than compensate. Test this hypothesis by considering the spontaneous breakup of a 0.2 cm radius water drop into:
(a) 2 equal-sized fragments.
(b) 5 equal-sized fragments.
(c) 20 equal-sized fragments.

7.2* Consider two drops of radius 10 μm and 15 μm initially.
(a) What is the collection kernel for the larger drop collecting the smaller droplet?
(b) Now consider that the larger drop collects one of the smaller droplets. What is the collection kernel for the larger drop collecting additional smaller droplets?

7.3* Show that the rate of change of radius (dr/dt) for a large drop growing by collecting much smaller droplets is proportional to the fall velocity of the large drop, assuming the collection efficiency is roughly constant and the fall velocity of the small droplets can be ignored.

7.4** Consider the 15 μm radius (collector) drop and 10 μm radius (collected) droplets in Problem 7.2.
(a) What is the probability that the collector drop will collect one or more droplets in 1000 seconds?
(b) What is the probability that the collector drop will collect two or more droplets in 1000 seconds?

7.5** Consider that the 15 μm and 10 μm radius drops in Problem 7.2 are growing in an updraft (each under the same ambient conditions). What is the collection kernel when the 15 μm radius drop has grown to 20 μm radius?

7.6** Show that the variance of the number of drops of mass m at time t under the pure-stochastic model is given by (7.13).

Hint: The variance is defined as

$$V[n] = E[n^2] - (E[n])^2 .$$

7.7** Show that the general probability that a drop will have collected only one droplet in time t is given by (7.12). Use the recursive relationship (7.8) with collection kernels Γ_1 for the first collection event and Γ_2 for the second.

7.8*** Show that (7.12) reduces to (7.9) for $\Gamma_1 = \Gamma_2$.

REFERENCES

Beard, K.V., 1976, 'Terminal velocity and shape of cloud and precipitation drops aloft,' *J. Atmos. Sci.* **33**, 851.

Beard, K.V., and Chuang, C., 1987, 'A new model for the equilibrium shape of raindrops,' *J. Atmos. Sci.* **44**, 1509.

Beard, K.V., and Ochs, H.T., III, 1983, 'Measured collection efficiencies for cloud drops,' *J. Atmos. Sci.* **40**, 146.

Beard, K.V., and Pruppacher, H.R., 1971, 'A wind tunnel investigation of collection kernels for small water drops in air,' *Quart. J. R. Meteorol. Soc.* **97**, 242.

Bradley, S.G., and Stow, C.D., 1978a, 'Collisions between liquid drops. I. The behavior of large water drops colliding during free fall,' *Phil. Trans. R. Soc. London* **A287**, 635.

Bradley, S.G., and Stow, C.D., 1978b, 'Collisions between liquid drops. II. An investigation of air film stability and drainage between colliding, electrically charge drops,' *Phil. Trans. R. Soc. London* **A287**, 654.

Bradley, S.G., and Stow, C.D., 1978c, 'Collisions between liquid drops. III. A parametric method for the investigation of the stability of large liquid drops,' *Phil. Trans. R. Soc. London* **A287**, 664.

Bradley, S.G., and Stow, C.D., 1984, 'Comments on 'Collision, coalescence and breakup of raindrops. Parts I and II',' *J. Atmos. Sci.* **41**, 701.

Brazier-Smith, P.R., Jennings, S.G., and Latham, J., 1972, 'The interaction of falling water drops: coalescence,' *Proc. R. Soc. London Ser A* **326**, 393.

Cooper, W.A., and Baumgardner, D., 1989, 'Comment on 'The collection kernel for two falling drops subjected to random perturbations in a turbulent air flow: A stochastic model',' *J. Atmos. Sci.* **46**, 1165.

Davis, M.H., 1972, 'Collisions of small cloud droplets: Gas kinetic effects,' *J. Atmos. Sci.* **29**, 911.

de Almeida, F.C., 1977, 'Collision efficiency, collision angle and impact velocity of hydrodynamically interacting cloud drops: A numerical study,' *J. Atmos. Sci.* **34**, 1286.

de Almeida, F.C., 1979, 'The effects of small-scale turbulent motions on the growth of a cloud droplet spectrum,' *J. Atmos. Sci.* **36**, 1557.

Gillespie, D.T., 1975, 'Three models for the coalescence growth of cloud drops,' *J. Atmos. Sci.* **32**, 600.

Lamb, H. 1945, *Hydrodynamics*. Dover Publications.

Langmuir, I., 1948, 'The production of rain by a chain reaction in cumulus clouds at temperatures above freezing,' *J. Meteorol.* **5**, 175.

Low, T.B., and List, R., 1982a, 'Collision, coalescence and breakup of raindrops. Part I: Experimentally established coalescence efficiencies and fragment size distributions in breakup,' *J. Atmos. Sci.* **39**, 1591.

Low, T.B., and List, R., 1982b, 'Collision, coalescence and breakup of raindrops. Part II: Parameterization of fragment size distributions,' *J. Atmos. Sci.* **39**, 1607.

Mason, B.J., 1971, *The physics of clouds*. Oxford: Clarendon Press.

McTaggart-Cowan, J.D., and List, R., 1975, 'Collision and breakup of water drops at terminal velocity,' *J. Atmos. Sci.* **32**, 1401.

Ochs, H.T., III, Czys, R.R., and Beard, K.V., 1986, 'Laboratory measurements of coalescence efficiencies for small precipitation drops,' *J. Atmos. Sci.* **43**, 225.

Pruppacher, H.R., and Klett, J.D., 1978, *Microphysics of clouds and precipitation*. Dordrecht: Reidel.

Reuter, G.W., de Villiers, R., and Yavin, Y., 1988, 'The collection kernel for two falling drops subject to random perturbations in a turbulent air flow: A stochastic model,' *J. Atmos. Sci.* **45**, 765.

Rogers, J.R., and Davis, R.H., 1990, 'The effects of van der Waals attractions on cloud droplet growth by coalescence,' *J. Atmos. Sci.* **47**, 1075.

Sartor, J.D., and Abbott, C.E., 1968, 'Charge transfer between uncharged water drops in free fall in an electric field,' *J. Geophys. Res.* **73**, 6415.

Schlamp, R.J., Grover, S.N., Pruppacher, H.R., and Hamielec, A.E., 1976, 'A numerical investigation of the effect of electric charges and vertical external electric fields on the collision efficiency of cloud drops,' *J. Atmos. Sci.* **33**, 1747.

Telford, J.W., 1955, 'A new aspect of coalescence theory,' *J. Meteorol.* **12**, 436.

Valdez, M.P., and Young, K.C., 1985, 'Number fluxes in equilibrium raindrop populations: A Markov chain analysis,' *J. Atmos. Sci.* **42**, 1024.

Woods, J.D., and Mason, B.J., 1964, 'Experimental determination of collection efficiencies for small drops in air,' *Quart. J. R. Meteorol. Soc.* **90**, 373.

Woods, J.D., and Mason, B.J., 1965, 'The wake capture of water drops in air,' *Quart. J. R. Meteorol. Soc.* **91**, 35.

Young, K.C., 1975, 'The evolution of drop spectra due to condensation, coalescence and breakup,' *J. Atmos. Sci.* **32**, 965.

8

GROWTH OF ICE PARTICLES BY COLLECTION

8.1. Introduction

Ice particles may grow by collecting water drops or by collecting other ice particles. The former process is formally termed *accretion* although it is more commonly referred to as *riming*. The process of collecting other ice particles is generally referred to as *aggregation*. The collection of a small ice crystal by a large water drop is often referred to as freezing since the process is similar to the freezing of a water drop due to the action of a freezing nucleus.

The determination of collection rates for ice particles is a much more difficult task than for water drops. Ice particles generally do not fall along a vertical path. Ice crystals may flutter or skitter with a considerable horizontal component to their motion; hailstones tend to gyrate as they fall, producing a tumbling motion. Thus, it is conceptually difficult to define a 'swept' volume or even a 'fall velocity.'

Most treatments attempt to force the accretion and aggregation processes into the rigid framework developed for coalescence. These generally assume a uniform fall velocity and a well-behaved collection efficiency. In reality, the fall velocity of a crystal varies as its orientation changes and the swept volume should consider the horizontal component of the particle's motion as well as its vertical component.

The accretion and aggregation processes proceed in a stochastic manner in the same fashion that was discussed in the previous chapter for coalescence. The interpretation of the collection equations presented in this chapter is intended to be the same as for coalescence in this regard.

8.2. The growth of rimed crystals and graupeln

As an ice crystal grows by diffusion, it may reach a size at which it starts to collect water drops. The supercooled water drops collide with the ice crystal and freeze. If the crystal and the water drops are not much colder than 0°C, the drop will tend to spread out before it freezes completely, conforming to the local shape of the ice crystal. If the crystal and drop are considerably colder than 0°C, the drop tends to retain its shape and freeze where it strikes. This leads to a more porous structure whose density is a function of the drop size, the accretion rate, and the ambient temperature.

8.2.1. *Nomenclature of rimed particles*

Accretion or riming proceeds in three distinct stages. As long as the basic crystal habit can be identified, the accreting particle is referred to as a *rimed crystal*. Once the accretion process has proceeded to obscure the crystal's habit, the particle is referred to as a graupel[1] (see Fig. 8.1). Such a graupel may grow to a size at which it is no longer able to freeze all of the water it accretes. At this stage, it is usually referred to as hail, and its density tends to approach the bulk density for ice.

Frozen water drops may also undergo accretion, growing into graupeln or hail. If the frozen drop is quite small, it may grow by diffusion to become an ice crystal and then accrete water to become a graupel. Large, frozen drops may grow directly as hail without intervening stages.

These particle classifications are based on the manner in which particles form rather than on the basis of their external appearance. Thus, a rimed snowflake may resemble an irregular graupel and, indeed, both are termed *snow pellets* in the *Glossary of Meteorology* (Huschke, 1959). However, the processes whereby these two types of particles have been formed are quite different, and represent quite different in-cloud conditions. Thus, for most purposes in the field of cloud physics, particle classification by the manner in which the particle was formed is preferred.

The distinctions between the various types of ice particles is not always clear cut and the transitions from one type of particle to another are usually gradual. A cross-section of a hailstone may show a center which consists mostly of crystals of small grain sizes, surrounded by a region of much larger grain sizes. The center would appear to be a graupel but it is very difficult to assign a size to the 'embryo' and to determine at what size the graupel embryo 'became' a hailstone.

8.2.2. *The collection equation for accreting particles*

Ice particles are considered to collect water drops in much the same manner as water drops collect other water drops. The swept volume is defined in terms of the fall velocity of the particle and its cross-section perpendicular to the fall axis. The collection equation is similar in form to (7.6),

$$\frac{dN_s}{dt} = AE_{iw}n_w\left|v_i - v_w\right|, \tag{8.1}$$

where dN_s/dt refers to the rate at which water drops of a particular size are collected by the ice particle whose cross-sectional area normal to its fall axis is A. The mutual collection efficiency is E_{iw}, the concentration of water drops is n_w, and v_i and v_w refer to the fall velocities of the ice particle and water drops, respectively.

The cross-sectional area (A) for a given crystal depends on its orientation as it falls. Plate crystals generally fall with their a-axes horizontal; A for plate

[1] The plural of graupel is graupeln.

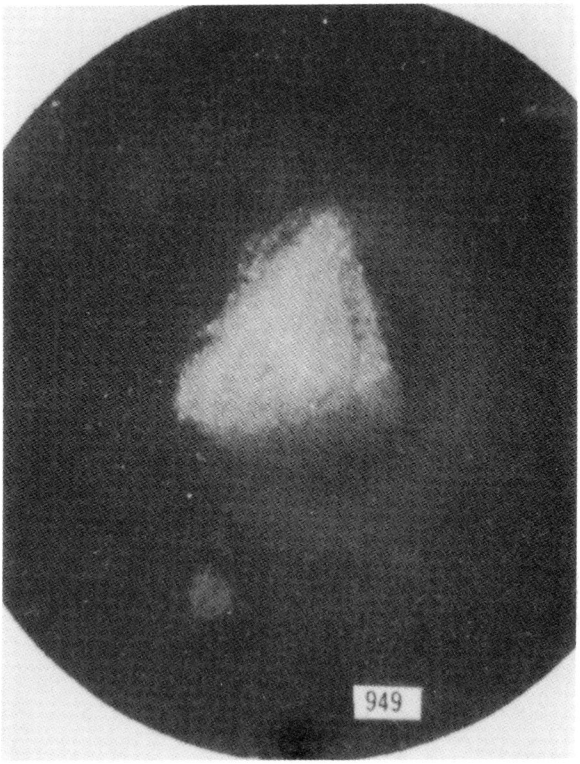

FIG. 8.1. Example of a conical graupel (R4c). From Nakaya (1954). Reprinted by permission of the publishers. From *Snow Crystals: Natural & Artificial*, Ukichio Nakaya, Cambridge, Mass.: Harvard University Press, Copyright © 1954 by the President and Fellows of Harvard College.

crystals can be approximated as πa^2. Columnar crystals tend to fall with their c-axis horizontal; A for columns is $4ac$. These fall orientations tend to increase the length of the minor axis of the crystal as it grows by accretion, i.e., plates become thicker and columns become fatter.

Cross-sectional areas for graupeln and hailstones are reasonably well approximated by πr_i^2 where r_i is the radius of the circumscribed sphere. For conical graupeln, the radius of the base is more applicable. Tumbling motions for falling hailstones (Knight and Knight, 1970) and graupeln (Heymsfield and Kajikawa, 1987) suggest the radius of the circumscribed sphere is often most appropriate.

(i) Terminal fall velocities for ice particles. Calculation of the growth of rimed ice particles, from the onset of riming through the growth of hail, requires defining the fall velocities of ice particles over a wide range of shapes and densities. In this section, the fall velocities of unrimed ice crystals will be examined first, followed by an examination of the fall velocities for graupeln and hailstones.

The term *fall velocity* applied to ice particles generally refers to the mean vertical (fall) velocity, ignoring the horizontal and irregular components of the particle's motion. A plate crystal, for example, will tend to fall vertically as long as its a-axes are perfectly horizontal. If the crystal is tilted slightly, it no longer lies flat and tends to slip horizontally. This motion tends to produce a lift force, which arrests the sideways motion of the crystal while changing its attack angle. It will then tend to slip horizontally in the opposite direction and repeat the process. Thus, the actual vertical component of the crystal's motion tends to change as the crystal changes its orientation, fluttering downwards much as a falling leaf does.

Columns likewise tend to fall with a horizontal component to their motion. A falling column tends to rotate, creating a sideways force which deflects the falling column from a vertical trajectory. As it falls, the rotation rate tends to increase until balanced by viscous forces. Thus, a columnar crystal will tend to fall at a constant angle to the vertical.

The irregular nature of the ice deposits in graupeln and hailstones tends to produce a more erratic falling behavior. This is due to the asymmetrical distribution of mass and the surface roughness, which may result in spinning and other gyrating motions, helical falling motions, and tumbling. Since these particles tend to be growing rapidly, the mass asymmetry and surface roughness are constantly changing, resulting in erratic changes in their fall behavior. These types of falling behavior are described by Harimaya (1976) and Pflaum et al. (1978).

The fall velocity of many particles is frequently described in terms of the Davies number for irregular objects,

$$N_{\mathrm{Da}} = \frac{8 m r_{\mathrm{c}}^2 g \rho_{\mathrm{air}}}{\eta^2 A},$$

where m is the mass of the particle, r_{c} is a characteristic semi-dimension such as the radius, and A is the cross-sectional area normal to the (vertical) fall axis. For plate crystals, r_{c} is the equivalent hexagonal radius $0.62\,A^{1/2}$; for columns, $r_{\mathrm{c}} = a$; for conical graupeln, r_{c} is the radius of the circumscribed circle about the base; for lump graupeln, r_{c} is the radius of the circumscribed sphere.

Heymsfield and Kajikawa (1987) provide empirical fits for the Reynolds number for a wide variety of planar crystals. They define the 'porosity' as the ratio of the actual cross-sectional area of the crystal to the cross-sectional area of a hexagon of the same dimensions. Crystals with porosities less than 0.6 include stellar (P1d), dendrite (P1e), and fernlike (P1f) crystals as well as crystal types P2b, P2c, P4b, P6c, P6d, P7a, and P7b. For these crystal types, the Reynolds number can be determined from the following:

$$N_{\mathrm{Re}} = 0.0684\, N_{\mathrm{Da}}^{0.999} \quad \text{for} \quad N_{\mathrm{Da}} < 1.47$$

$$N_{\mathrm{Re}} = 0.0715\, N_{\mathrm{Da}}^{0.883} \quad \text{for } 1.47 \le N_{\mathrm{Da}} < 66.1$$

$$N_{\mathrm{Re}} = 0.1759\, N_{\mathrm{Da}}^{0.6682} \quad \text{for } 66.1 \le N_{\mathrm{Da}}$$

For the remaining planar crystals with porosities $\geqslant 0.6$, the Reynolds number can be calculated from the following:

$$
\begin{aligned}
N_{\mathrm{Re}} &= 0.0684\, N_{\mathrm{Da}}^{0.999} && \text{for} && N_{\mathrm{Da}} < 1.47 \\
N_{\mathrm{Re}} &= 0.0715\, N_{\mathrm{Da}}^{0.883} && \text{for} && 1.47 \le N_{\mathrm{Da}} < 80.9 \\
N_{\mathrm{Re}} &= 0.1640\, N_{\mathrm{Da}}^{0.694} && \text{for} && 80.9 \le N_{\mathrm{Da}} < 395.4 \\
N_{\mathrm{Re}} &= 0.2140\, N_{\mathrm{Da}}^{0.6495} && \text{for} && 395.4 \le N_{\mathrm{Da}}
\end{aligned}
$$

The Reynolds numbers for columnar crystals may be derived from the tank experiments conducted by Jayaweera and Cottis (1969). Their data may be fitted by

$$
\log_{10} N_{\mathrm{Re}} = A \log_{10} N_{\mathrm{Da}} - B - C\,(\log_{10} N_{\mathrm{Da}})^2 ,
$$

where

$$
\begin{aligned}
A &= 0.240\, \Gamma^{-1/2} + 0.85 \\
B &= 0.478\, \Gamma^{-1/2} + 0.757 \\
C &= 0.032\, \Gamma^{-1/2} + 0.040
\end{aligned}
$$

and Γ is the aspect ratio. This relationship is valid over the range of Davies number from 10 to 10^4.

Heymsfield and Kajikawa (1987) also provide empirical fits for the Reynolds number for both conical and lump graupeln. For conical graupeln, the Reynolds number can be calculated from the following:

$$
\begin{aligned}
N_{\mathrm{Re}} &= 0.0431\, N_{\mathrm{Da}}^{0.782} && \text{for} && N_{\mathrm{Da}} < 5.47 \times 10^4 \\
N_{\mathrm{Re}} &= 0.2416\, N_{\mathrm{Da}}^{0.624} && \text{for} && 5.47 \times 10^4 \le N_{\mathrm{Da}} < 2.27 \times 10^5 \\
N_{\mathrm{Re}} &= 1.4039\, N_{\mathrm{Da}}^{0.481} && \text{for} && 2.27 \times 10^5 \le N_{\mathrm{Da}}
\end{aligned}
$$

and for lump graupeln, the Reynolds number can be calculated from:

$$
\begin{aligned}
N_{\mathrm{Re}} &= 0.0688\, N_{\mathrm{Da}}^{0.769} && \text{for} && N_{\mathrm{Da}} < 1.09 \times 10^4 \\
N_{\mathrm{Re}} &= 0.3470\, N_{\mathrm{Da}}^{0.595} && \text{for} && 1.09 \times 10^4 \le N_{\mathrm{Da}} < 6.58 \times 10^5 \\
N_{\mathrm{Re}} &= 3.6184\, N_{\mathrm{Da}}^{0.420} && \text{for} && 6.58 \times 10^5 \le N_{\mathrm{Da}}
\end{aligned}
$$

Since the plots are fairly similar for unrimed ice crystals and graupeln, a simple linear interpolation for lightly to moderately rimed crystals should suffice to provide the Reynolds number for intermediate stages of riming.

Rasmussen and Heymsfield (1987) provide Reynolds number relationships for rough-surfaced hailstones based on empirical fits to observed fall velocities as follows:

$$N_{Re} = 0.4487\, N_{Da}^{0.5536} \qquad \text{for } 1.83 \times 10^3 \leq N_{Da} < 3.46 \times 10^8$$

$$N_{Re} = \left(\frac{N_{Da}}{0.6}\right)^{1/2} \qquad \text{for } 3.46 \times 10^8 \leq N_{Da}$$

For large hailstones, the effects of skin friction drag are minimal and most of the resistance is form drag. Thus, sufficiently large rough or smooth hailstones have nearly the same fall velocities. This may not be true for smaller ice particles, as suggested by Johnson (1987). Smooth hailstones of smaller sizes may be expected to fall faster than rough hailstones of the same mass and density.

Böhm (1989) presents a general empirical relation for the fall velocities of ice particles. He modifies the Davies number by the fourth root of the inverse of porosity, as defined by Heymsfield and Kajikawa in the preceding,

$$N_{Da} = \frac{8mg\rho_{air}}{\pi\eta^2}\left(\frac{A_c}{A}\right)^{1/4},$$

where A_c is the circumscribed cross-sectional area. The Reynolds number may be calculated using

$$N_{Re} = 8.5\left[(1 + 0.1519\, N_{Da}^{1/2})^{1/2} - 1\right]^2.$$

Fall velocities calculated using this equation generally are within 10% of the values calculated using the fits provided by Heymsfield and Kajikawa (1987)

(ii) Collection efficiencies for ice particles. Whether or not an ice crystal will collect water drops of a given size depends on the size and fall velocity of the ice crystal and the size of the water drops. Small water drops approaching an ice crystal will tend to evaporate and follow the streamlines around the falling crystal. Thus, droplets smaller than 1 μm radius are not collected by ice particles, although they may add mass to the crystal as they evaporate and enhance the vapor gradient in the vicinity of the crystal, as discussed in Chapter 6. Droplets larger than 1 μm but smaller than 2.5 μm in radius may be collected, but usually spread out on the surface and do not retain an identifiable shape indicative of riming (Harimaya, 1976).

Pitter and Pruppacher (1974) find theoretically that small drops will be carried around a falling plate, as will large drops which have about the same fall velocity as the plate. They find that plates of radius smaller than 150 μm do not collect water drops of any size, and larger plates tend to collect only a certain

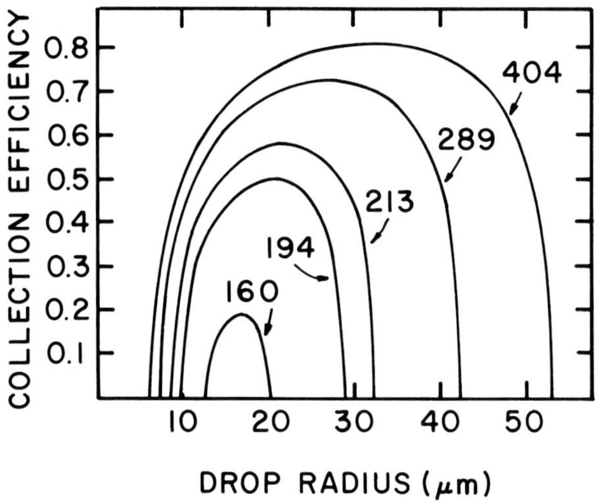

Fig. 8.2. Collection efficiencies for thin plates collecting water drops based on numerical simulations by Pitter and Pruppacher (1974) as a function of the drop radius. Plate radii are 160 μm, 194 μm, 213 μm, 289 μm, and 404 μm. *Journal of Atmospheric Sciences*, **31**, 2. American Meteorological Society. Reproduced with permission.

intermediate size range of drops, as shown in Fig. 8.2. These collection efficiencies can be approximated as

$$\log_{10}(1 - E_{\mathrm{iw}}) = 4.0\left[0.51 - \left(\frac{v_{\mathrm{w}}}{v_{\mathrm{i}}}\right)^{1/2}\right]^2 - 0.28\log(a_{\mathrm{i}} - 0.013) - 1.7\,,$$

where E_{iw} is limited to positive values. The fall velocities for the drops (v_{w}) and the ice crystal (v_{i}) are assumed to be in cm s^{-1} and the plate radius (a_{i}) is also in centimeters.

Pitter and Pruppacher find that the picture of collection presented in Fig. 7.1, in which all drops lying within the grazing trajectory are collected, does not always apply when dealing with plates. They find that droplets whose sizes are near the limits for accretion may be collected from a circular annulus, as depicted in Fig. 8.3. The range of drop sizes for which such annular collection occurs depends on the size of the plate.

Figure 8.4 depicts a rimed plate which shows the tendency for plate crystals to collect water drops near their perimeter. In addition, the corners of the plates tend to collect a disproportionate share of the droplets, leading to rather elongated protuberances with relatively few frozen drops in between. Once such a protuberance is initiated, by virtue of its location and the irregular falling motion of the crystal, it would tend to shield the regions in between, accentuating this pattern.

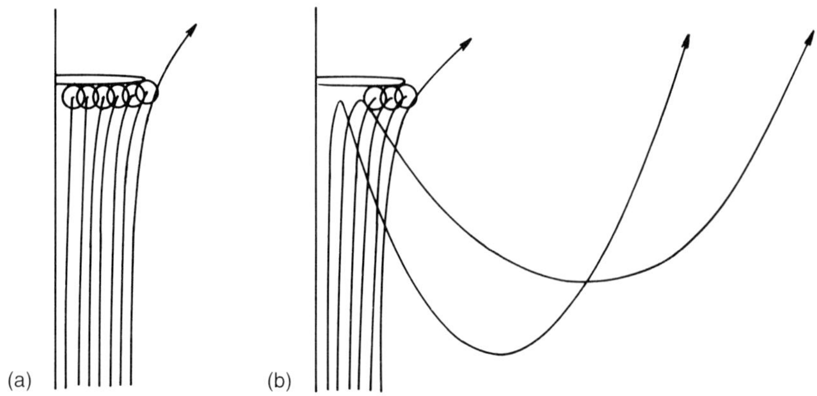

Fig. 8.3. Trajectories of drops interacting with a falling plate ($a_i = 404\ \mu m$) as described by Pitter and Pruppacher (1974). In (a), the drops have radius $52\ \mu m$; in (b) the drops have radius $52.5\ \mu m$. Note the annular nature of the collection in (b). *Journal of Atmospheric Sciences*, **31**, 2. American Meteorological Society. Reproduced with permission.

Theoretical collision efficiencies for simple columnar ice crystals are shown in Fig. 8.5, based on Schlamp et al. (1975). Since columnar ice crystals tend to rime uniformly along their c-axis,[2] the effects of the length of the column are small and most of the variation in collection efficiencies may be expected to result from variations in the column radius (a). For columns thinner than $75\ \mu m$ radius, only drops within a range of sizes are collected, as noted for plate crystals.

Observations of the onset of riming on plate crystals (P1a) by Harimaya (1975) are in good agreement with the theoretical treatment with an observed minimum radius of $150\ \mu m$. The observed minimum radius for plates with broad branches (P1c) was $350\ \mu m$; the minimum radius for dendrites (P1e) was $450\ \mu m$. Ono (1969) observed a minimum radius of $30\ \mu m$ required for riming on columns, in good agreement with the theoretical treatments.

Harimaya (1975) notes a minimum accreted drop radius near $7\ \mu m$ for plates; Ono (1969) notes a minimum accreted drop radius of $10\ \mu m$ for dendrites. Harimaya also notes a maximum accreted drop size of roughly $40\ \mu m$, although this may reflect the droplet sizes available for accretion.

Lew and Pruppacher (1983) studied the collection of small columnar ice crystals by larger water drops (between $100\ \mu m$ and $600\ \mu m$ radius) and found the collection efficiencies were slightly less than geometrical (0.9 to 1.0) for nearly isometric columns. The collection efficiency was noted to decrease as the aspect ratio increased, and was also noted to increase with decreasing pressure.

The collection efficiencies for graupeln and hailstones collecting much smaller water drops has been assumed to be close to unity. Relatively little

[2] Figures 12 and 13 from Zikmunda and Vali (1972) and figure 11 from Ono (1969) provide visual evidence of this.

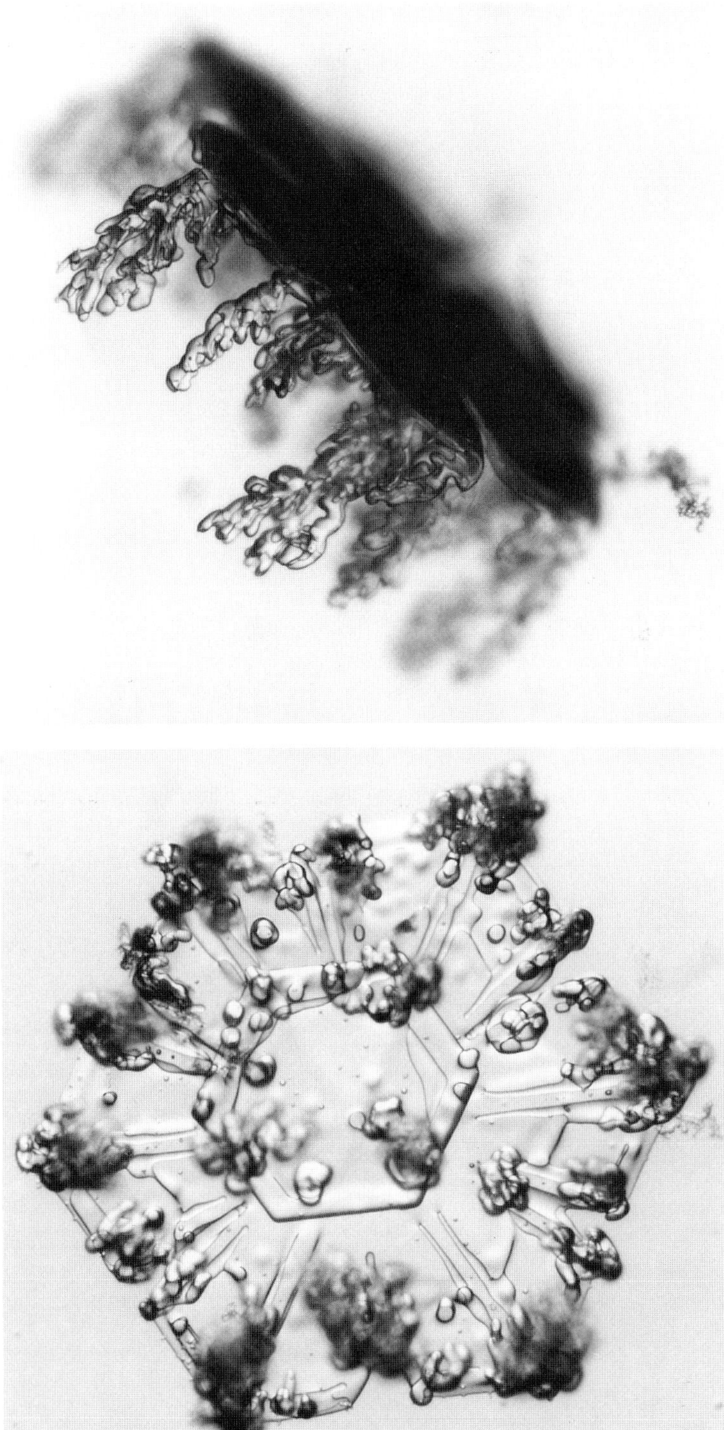

(a)

(b)

FIG. 8.4. Rime deposits on a plate crystal (magnification ×110). (a) Side view; (b) bottom view. Courtesy of C.A. Knight.

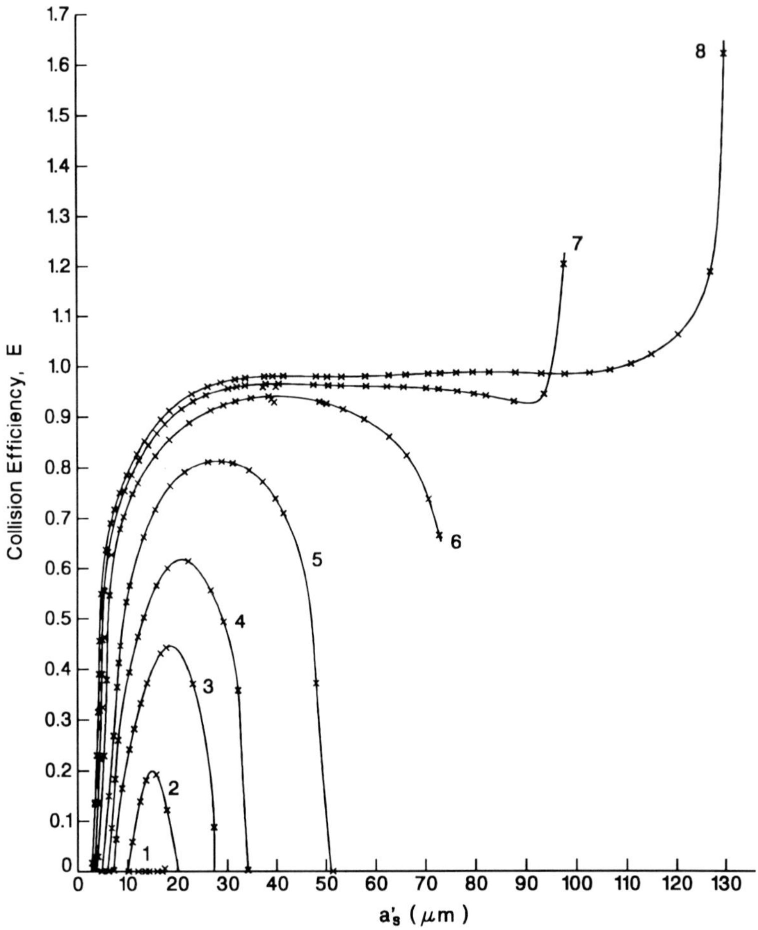

FIG. 8.5. Collection efficiencies for columns collecting water drops based on numerical simulations by Schlamp et al. (1975) as a function of the drop radius. The smallest column (1) with radius (a) of 23.5 μm and aspect ratio (Γ) of 1.43 does not collect drops of any size. Size parameters are: (2), a = 32.7 μm, Γ = 1.43; (3), a = 36.6 μm, Γ = 1.54; (4), a = 41.5 μm, Γ = 1.67; (5), a = 53.4 μm, Γ = 2.22; (6), a = 77.2 μm, Γ = 3.33; (7), a = 106.7 μm, Γ = 5.0; (8), a = 146.4 μm, Γ = 8.33. $T = -8°C$, P = 800 mbar. From Schlamp et al. (1975). *Journal of Atmospheric Sciences*, **32**, 12. American Meteorological Society. Reproduced with permission.

experimental or theoretical work has been done in this regard. The results of Macklin and Bailey (1968) suggest the collection efficiencies for hailstones may be considerably less than 1. However, they were only able to evaluate net collection, i.e., collection minus shedding. Pflaum and Pruppacher (1979) provide collection kernels for graupeln collecting 6 μm and 10 μm radius droplets. Their collection efficiencies range from 0.7 for 100 μm radius graupel to 0.87 for 600 μm radius graupeln. Most studies requiring collection efficiencies for graupeln and hailstones rely on extrapolation of the results of Beard and Grover (1974) for raindrops.

(iii) The density of rimed ice. The density of rimed ice that accretes on an object is determined by the relative rapidity of the freezing process versus the tendency of the colliding drops to spread out on the surface. If the drops are large or impact with a high velocity, they tend to spread out on impact, filling in any gaps in the ice structure. If the ice surface is fairly cold and the drops are small and impact with a low velocity, they tend to freeze where they strike, building up a porous ice structure of low density. Thus, the density of the ice structure may be expected to be a function of the surface temperature of the accreting particle, and the size and impact velocity of the accreting drops.

Macklin (1962) conducted extensive studies of the density of accreted ice for temperatures below $-5°C$ and found that the density could be described by the term

$$X = \frac{r_w v_{im}}{-T_s},$$

where v_{im} is the impact velocity and T_s is the surface temperature (°C). More recently, Heymsfield and Pflaum (1985) updated Macklin's original empirical fit as

$$\rho_{ac} = 0.30\, X^{0.44} \qquad \text{for } T_s \le -5°\text{C}$$

$$\rho_{ac} = \exp\,(-0.03115 - 1.7030\, X + 0.9116\, X^2 - 0.1224\, X^3)$$
$$\text{for } X < -1.60 \text{ and } T_s > -5°\text{C}$$

with limitations to the density range of 0.1 to 0.91 g cm^{-3}. Buser and Aufdermaur (1973) determine a minimum density of 0.17 g cm^{-3} for graupel based on random attachment of the droplets, although they acknowledge that protuberances could shield regions and result in still lower densities. Rasmussen and Heymsfield (1985) provide a set of empirical equations that can be used to determine the appropriate impact velocities.

8.3. The growth of hail

As an ice particle grows by accretion, its temperature must be sufficiently elevated to dissipate the latent heat of fusion as the supercooled water drops freeze on its surface. As the particle grows, the rate at which it collects water drops increases faster than its ability to dissipate the latent heat released due to the freezing drops. The conditions that define the growth of an ice particle which freezes all the drops it collects and whose surface temperature is 0°C are termed the Schumann–Ludlam limit. Once the growing ice particle exceeds the Schumann–Ludlam limit, it is termed a hailstone. The water that cannot be frozen may either be incorporated into a matrix of ice and water (spongy ice) or it may be shed.

8.3.1. *Calculation of accreting ice particle temperature*

The steady state temperature of an accreting ice particle is determined by the balance between the rate at which latent heat is released by the freezing of collected water drops and the rate at which heat can be lost by conduction and sublimation. The sensible and latent heat fluxes are similar to (5.2) and (5.5),

$$\left.\frac{dQ}{dt}\right|_{\text{sens}} = 4\pi r_i f_2^* K(T_i - T_\infty) \tag{8.2}$$

and

$$\left.\frac{dQ}{dt}\right|_{\text{lat}} = L_s \frac{dm}{dt} = 4\pi r_i f_2 DL_s (\rho_{s,i} - \rho_\infty) \tag{8.3}$$

where the subscript i refers to the ice particle, which is assumed to be approximately spherical.

The heat released by the freezing of the collected water drops may be obtained by summing (8.1) over all drops sizes (j) as

$$\left.\frac{dQ}{dt}\right|_{\text{accr}} = \pi r_i^2 L_f \sum_j E_{ij} \chi_j (v_i - v_j) \tag{8.4}$$

where χ_j represents the liquid water content contained in the jth category of drops. Note that the latent heat of fusion (L_f) as used here refers to the net heat resulting from warming the drops from their wet-bulb temperature[3] to 0°C, freezing the water at 0°C (333.69 J g^{-1}) and cooling the ice down to T_i. The formulation of (8.4) assumes that the water drops are all at the same temperature.

The steady state drop temperature follows from (8.2), (8.3), and (8.4) in a manner similar to the derivation of (5.8) as

$$T_i = \frac{\dfrac{L_f}{4} \displaystyle\sum_j E_{ij} \chi_j + f_2^* KT_\infty + f_2 DL_s (\rho_\infty - b)}{f_2^* K + sf_2 DL_s}, \tag{8.5}$$

where the terms s and b are based on a linear approximation of the saturation vapor density tables for ice, $\rho_{s,i} = sT_i + b$.

The ventilation coefficients for mass and heat transfer are provided by Rasmussen and Heymsfield (1987) as

$$f_2 = 0.78 + 0.308\, N_{Sc}^{1/3} N_{Re}^{1/2}$$

$$f_2^* = 0.78 + 0.308\, N_{Pr}^{1/3} N_{Re}^{1/2},$$

[3] The wet-bulb temperature for the water drops is usually assumed to be the ambient temperature.

where the Prandtl number (N_{Pr}) and the Schmidt number (N_{Sc}) are defined in Appendix 2.

Examination of (8.2) and (8.3) suggests that the sensible and latent heat fluxes away from the growing ice particle are proportional to the radius times the ventilation factor. The ventilation factor is proportional to the square root of the Reynolds number, which in turn is proportional to $r_x v_x$. Since v_x is roughly proportional to $r^{1/2}$, the sensible and latent heat losses are roughly proportional to $r^{7/4}$. Similarly, the release of latent heat due to freezing of collected water drops is roughly proportional to $r^{5/2}$, assuming negligible fall velocities for the water drops. Since for constant ambient conditions the release of latent heat by freezing increases more rapidly with radius than the rate at which sensible and latent heats may be conducted away from the ice particle, the temperature of the ice particle may be expected to increase as the particle grows.

8.3.2. *The Schumann–Ludlam limit*

The set of limiting conditions that define the maximum rate at which the ice particle may accrete water drops and still freeze all of the accreted water was first described by Schumann (1938) and later modified by Ludlam (1958). The defining conditions for the Schumann–Ludlam limit may be obtained by assuming the ice particle is at 0°C in (8.2) and (8.3),

$$\pi r_i^2 L_f \sum_j E_{ij}\, \chi_j (v_i - v_j) = 4\pi r_i \left[f_2^* K(T_o - T_\infty) + f_2 D L_s (\rho_{s,o} - \rho_\infty) \right]$$

where the subscript o refers to conditions at 0°C.

Hailstone growth is often simplified by assuming that the fall velocities of the water drops are fairly small and can be neglected and that the collection efficiency (E_{iw}) is unity. Under these assumptions, the defining conditions for the Schumann–Ludlam limit reduce to

$$\chi = \frac{4}{r_i L_f v_i} \left[K f_2^* (T_o - T_\infty) + D f_2 L_s (\rho_{s,o} - \rho_\infty) \right] \tag{8.6}$$

where χ is the liquid water content in units of mass per unit volume.

Figure 8.6 provides solutions for (8.6), defining the Schumann–Ludlam limit in terms of the particle radius, the ambient temperature, and the liquid water content. This figure may be interpreted in three ways. Given an ambient temperature and particle radius, the corresponding liquid water content represents the maximum that will permit all accreted water to freeze. Given a liquid water content and ambient temperature, the corresponding particle size is the largest that permits all accreted water to freeze.

Given typical liquid water contents in the hail growth zone of cumulonimbi as between 3 and 7 g m^{-3} and ambient temperatures colder than −20°C, Fig. 8.6 suggests that ice particles larger than 0.5 cm radius should be classified as

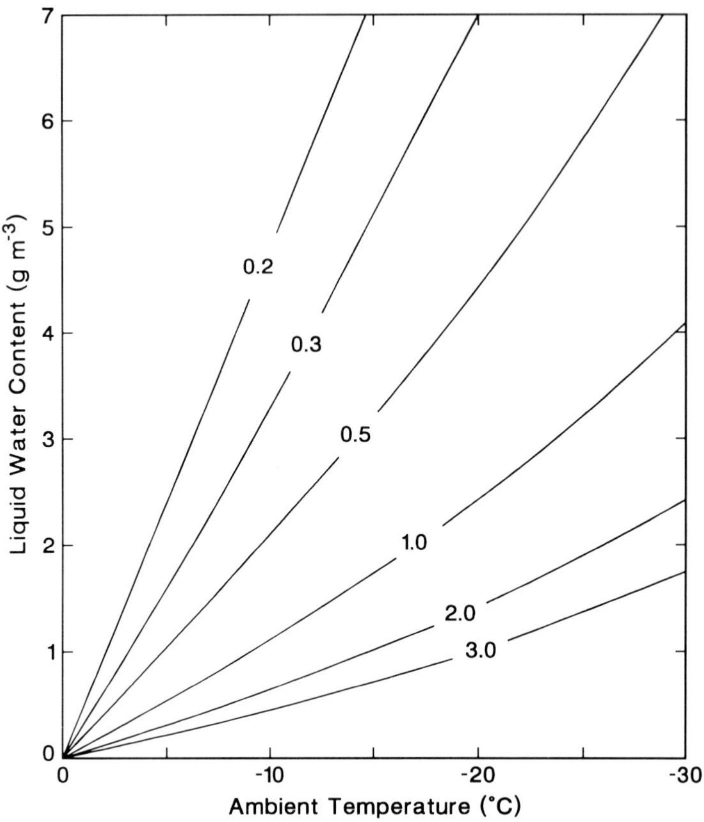

Fig. 8.6. Schumann–Ludlam limit as calculated from (8.6) for P = 700 mbar. Isopleths of particle radius (cm) are shown as functions of the ambient temperature and liquid water content.

hailstones. These values should only be taken as a rough guide to particle classification.

The preceding treatment assumes the hailstone has a uniform temperature. Lesins and List (1986) find a transition regime between ice particles growing with a completely dry surface (dry regime) and ice particles which incorporate the unfrozen water into a 'spongy ice' deposit (spongy regime without shedding) which they term the 'moist' regime. In the moist regime, the equatorial portions of the ice particle appears to have a thin film of water while the polar regions tend to appear dry. These regimes are shown in Fig. 8.7.

Johnson and Rasmussen (1992) suggest that once a graupel reaches the Schumann–Ludlam limit and begins spongy growth, its surface will become smoother, reducing its drag coefficient and increasing its fall velocity. This would have the effect of reducing the liquid water content required to maintain spongy growth. Thus, once an ice particle reaches sizes larger than the Schumann–Ludlam limit, it would not be likely to resume dry growth unless the liquid water content decreased significantly.

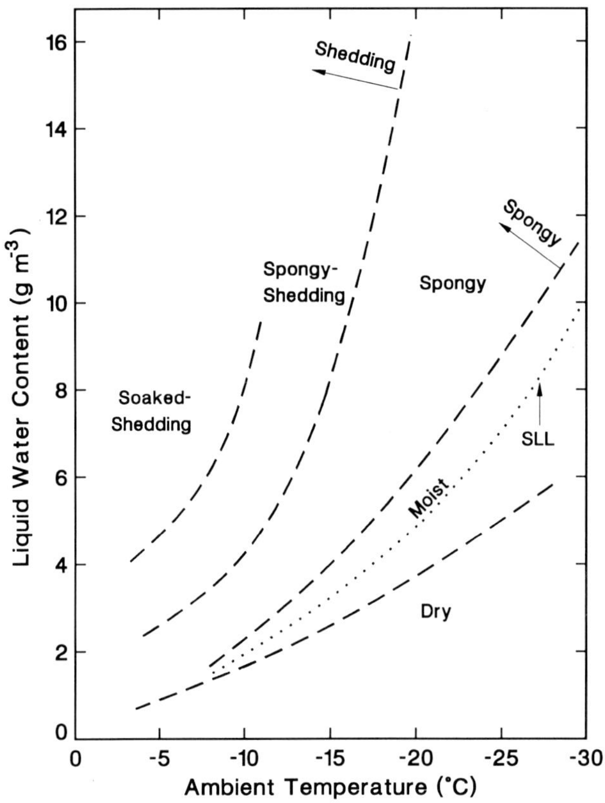

FIG. 8.7. The five observed hailstone growth regimes as a function of liquid water content and ambient temperature. The dotted line represents the theoretical Schumann–Ludlam limit. After Lesins and List (1986). *Journal of Atmospheric Sciences*, **43**, 23. American Meteorological Society. Reproduced with permission.

8.3.3. *Spongy ice growth and shedding*

As the ice particle grows to sizes beyond the Schumann–Ludlam limit, it enters a regime of spongy ice growth in which the unfrozen water is incorporated into an ice–water matrix. As the ice particle grows, the ice fraction decreases. Once the ice fraction decreases below 0.7, drops may be shed from the upper portions of the particle, directly enter the wake of the particle. Lesins and List (1986) refer to this regime as 'spongy-shedding.' Upon continued growth, the particle reaches its minimum ice fraction of 0.5 and all accreted water in excess of that required to maintain an ice fraction of 0.5 is shed.

Observations reported by List (1990) suggest that the surface temperature of growing hailstones in the shedding regimes may be −2°C or −3°C. List attributes this to the water skin, which is on the order of 0.03 cm thick, and the requirement that a temperature gradient be present within this skin to permit the observed growth of ice. List concludes that heat transfer through the water skin is by molecular conduction. Knight (1991) proposes that the observation is due to rapid cooling of the surface resulting from the intense ventilation, but that circulation within the water skin maintains a fairly uniform temperature near 0°C within the layer. List suggests it is erroneous to assume any portion of the surface of an accreting ice particle is at 0°C. These observations raise additional questions regarding the validity of equations such as (8.6), which assume that the hailstone has a surface temperature of 0°C.

8.4. The growth of snowflakes

The process of ice crystals or clusters of ice crystals collecting other ice crystals is termed aggregation. The clusters of ice crystals are sometimes termed aggregates although more commonly they are referred to as snowflakes. Since the fall velocities for snowflakes are not appreciably different than the fall velocities for the individual crystals that comprise them, the precipitation rate is not significantly enhanced by the aggregation process directly. However, when an aggregate falls below the melting level, the drops that form are typically larger and fall faster than the drops that would form from the melting of individual ice crystals.

The formation of aggregates involves the collision between two ice crystals followed by their 'sticking' together. As previously noted, ice crystals (and snowflakes) have a large horizontal component to their falling motion and a certain degree of randomness to their motion as well. Thus, a simple treatment of collection based on mean fall velocities is inadequate for describing aggregation processes. In this section, collisions and subsequent collection between ice crystals and aggregates will be discussed, followed by a discussion of snowflake densities and fall velocities.

8.4.1. *Collisions between ice crystals*

Since snowflakes of a given mass have a spectrum of fall speeds, two snowflakes of the same mass can collide. Thus, in defining a collection kernel for aggregation, it is necessary to incorporate a probabilistic description of randomly changing fall speeds. Passarelli and Srivastava (1979) describe the collection kernel for aggregation between particles of mass m_1 and m_2 as

$$K(m_1,m_2) = \pi \int_0^\infty \int_0^\infty (r_1 + r_2)^2 \, |v_1 - v_2| \, p\{v_1/m_1\} \, p\{v_2/m_2\} \, dv_1 \, dv_2 \,, \tag{8.7}$$

where $p\{v_1/m_1\}$ and $p\{v_2/m_2\}$ represent the probabilities that particles of mass m_1 and m_2 will have fall speeds of v_1 and v_2, respectively. The aggregation efficiency has been assumed to be unity.

Equation (8.7) may be interpreted in either of two ways. The fall speed distribution may arise solely from differences in snowflake density for flakes of a given mass, i.e., smaller fall speeds represent less dense snowflakes. Alternately, the fall speed of a snowflake of given mass and density may fluctuate more or less randomly, i.e., any given snowflake may possess a range of fall speeds. Under the former model, r_1 and r_2 vary with fall speed and must remain inside the integrals. Under the latter model, r_1 and r_2 may be taken outside the integrals and evaluated using the particle masses m_1 and m_2. Sasyō and Matsuo (1985) provide this relationship as

$$\pi(r_1 + r_2)^2 = 0.012^{-2/3} \left(m_1^{1/3} + m_2^{1/3}\right)^2 .$$

Passarelli and Srivastava (1979) assume that the probability distributions for the fall speeds are rectangular (uniform) between $\bar{v}(1 - X$ and $\bar{v}(1 + X)$, where X may range from 0 to 1. They partition the collection kernel K into K_1 and K_2 contributions. K_1 may be considered as the 'kernel-weighted' average particle in the fall speed distribution for m_1 particles collecting the kernel-weighted average particle in the fall speed distribution for m_2 particles. For $X <\sim 0.25$, K_1 is within ~5% of the standard collection kernel, which assumes a one-to-one relationship between size and fall speed. K_2 reflects the effects of the distribution of fall speeds.

Based on observations of the fluctuations in snowflake fall speeds, Passarelli and Srivastava estimate that $X \approx 0.05$ and that the maximum effect of the distribution of fall speeds is to increase the collection kernel by about 10%. The collection kernel for aggregation and the partitions K_1 and K_2 are shown in Fig. 8.8. Note that the K_1 contribution is virtually identical to the standard kernel. Increases in the collection kernel due to fluctuations in the fall speeds are greatest for the largest particles.

Is there a maximum size to which snowflakes can grow? The relative lack of cohesion between the aggregated ice crystals suggests that collisional breakup is possible for sufficiently large snowflakes. Based on airborne observations, Lo and Passarelli (1982) suggest collisional breakup may act to limit the maximum sizes of snowflakes.

8.4.2. *The sticking probability*

In nature, the largest snowflakes are observed when the air temperature is just below freezing. Figure 8.9 summarizes an extensive series of observations reported by Hobbs et al. (1974) relating the maximum sizes of snowflakes with the ambient temperature where they were collected. From this, it appears that aggregation is unimportant at temperatures colder than −20°C, presumably because ice crystals do not stick together when they collide at these temperatures.

Hosler et al. (1957) conducted a series of laboratory experiments in which they determined the force required to separate two ice spheres following momentary contact. Experiments were conducted at ice saturation and under 'dry' conditions, during which the ice spheres were subliming. As shown in Fig.

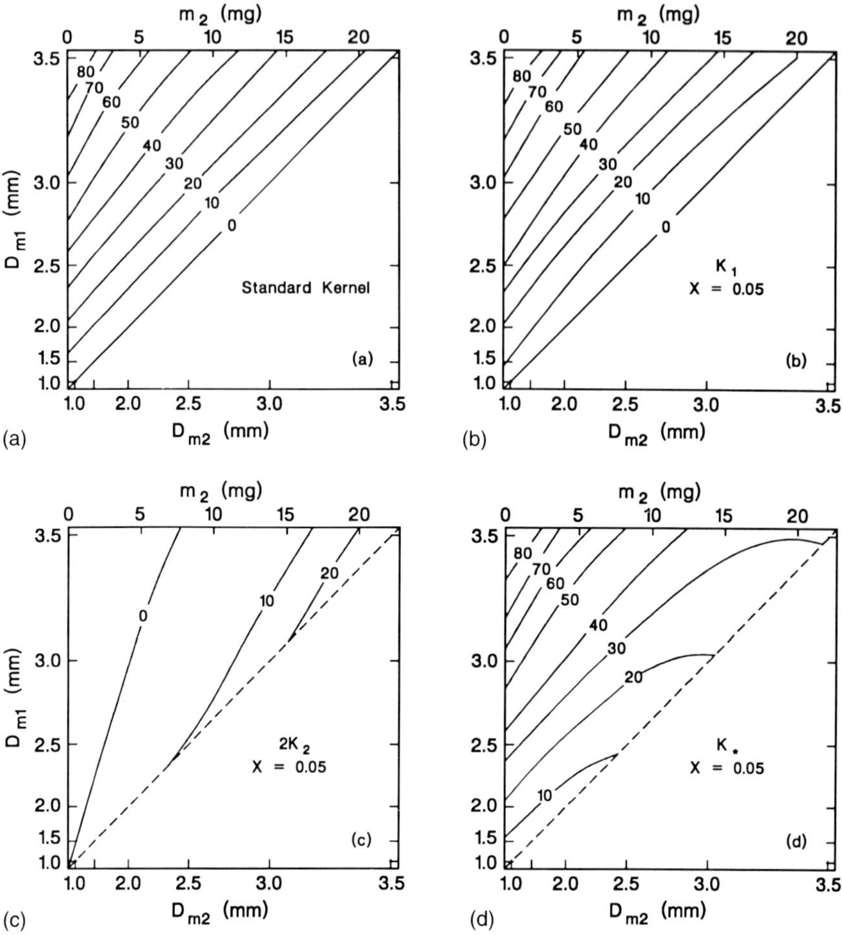

Fig. 8.8. The modified collection kernel for aggregation (d) is partitioned into K_1 representing 'kernel-weighted' average particles (b) and K_2 representing the effects of fluctuations in the fall speeds (c), as discussed in the text. The standard kernel represents the collection kernel for $X = 0$ (a). The D_{mx} scales represent equivalent melted diameter. After Passarelli and Srivastava (1979). *Journal of Atmospheric Sciences*, **36**, 3. American Meteorological Society. Reproduced with permission.

8.10, the force required to separate the spheres is markedly greater under conditions near ice saturation (and presumably above ice saturation).

When two ice particles come into contact, they tend to form a neck of ice between them, binding them together. This process is termed *sintering*. The more rapid the sintering process or the stronger the bond formed, the more likely it is that the two colliding particles will remain together. Hobbs (1965) discusses several mechanisms proposed to explain sintering. Sintering arising from the presence of a quasi-liquid layer on the ice would not explain observations of sintering below −12°C. Sintering through surface diffusion of water molecules is too slow by four orders of magnitude.

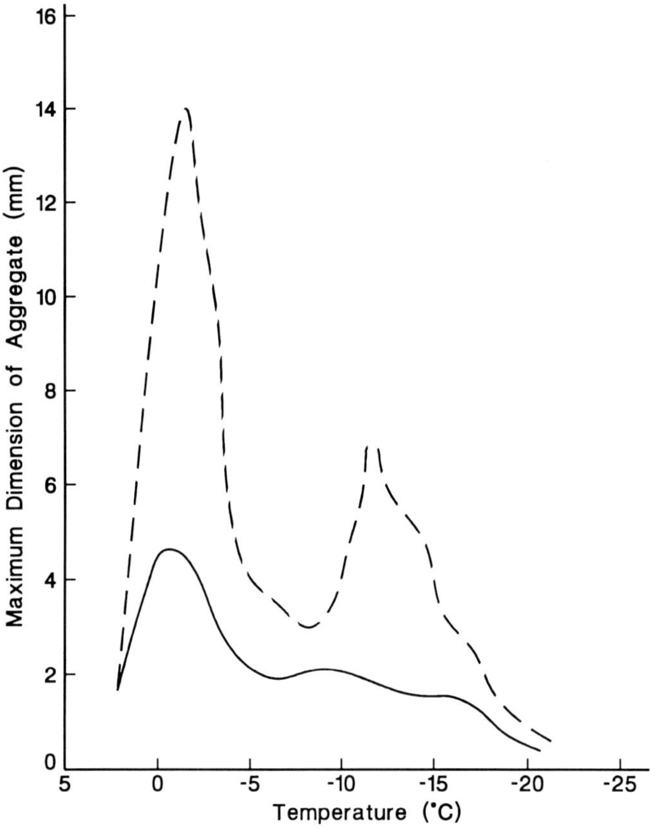

FIG. 8.9. Maximum dimensions of aggregates as a function of ambient temperature based on ground and aircraft observations by Hobbs et al. (1974). The solid curve represents the median of the observed aggregate sizes; the broken curve represents the largest aggregates observed. *J. Geophys. Res.* (1974) **79**.) Reproduced with permission.

Hobbs (1965) demonstrates that sublimation of ice from convex portions of the ice surface and deposition in the highly concave junction between the particles is able to account for the observed sintering rates. This explains why sintering is highly dependent on the ambient ice saturation ratio. Since the saturation vapor density decreases with decreasing temperature, the difference in vapor density over the convex and concave surfaces must also decrease with decreasing temperature. Thus, the sintering rate also decreases with decreasing temperature.

The secondary maximum in the maximum aggregate sizes in the region between $-12°C$ and $-15°C$ as shown in Fig. 8.9 suggests that dendrites and other branched ice crystals may 'stick' as a result of interlocking their branches rather than by sintering. Thus, the sticking probability is determined both by the sintering rate, which is temperature dependent, and by the crystal habit. The probabilities that two colliding dendrites will stick are not known.

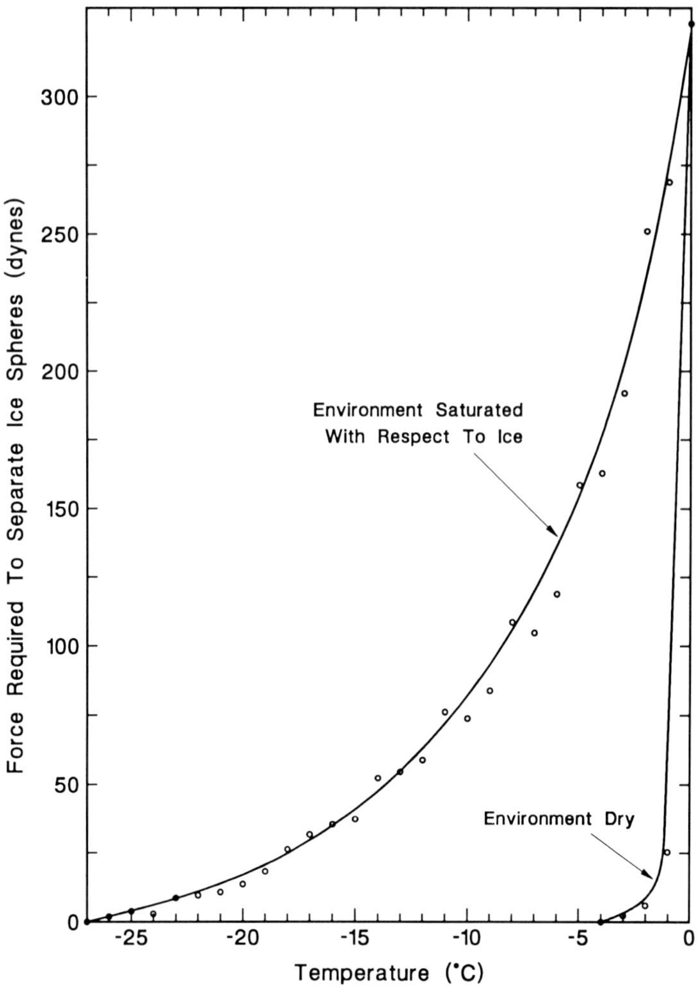

Fig. 8.10. Mean force required to separate two ice spheres following momentary contact as a function of temperature. After Hosler et al. (1957). *Journal of Meteorology*, **15**, 5. American Meteorological Society. Reproduced with permission.

8.4.3. *Snowflake densities and fall speeds*

The relatively open structure of snowflakes results in extraordinarily low bulk densities when considering their volume as that of a circumscribed sphere about the flake. Magono and Nakamura (1965) relate snowflake densities as a function of the snowflake diameter for aggregates of dendrites, as shown in Fig. 8.11. Passarelli and Srivastava (1979) provide a fit to their data as

$$\bar{\rho} = 0.0099\, r_f^{-0.6}, \tag{8.8}$$

where the average density ($\bar{\rho}$) is in g cm^{-3} and the radius of the snowflake (r_f) is in centimeters.

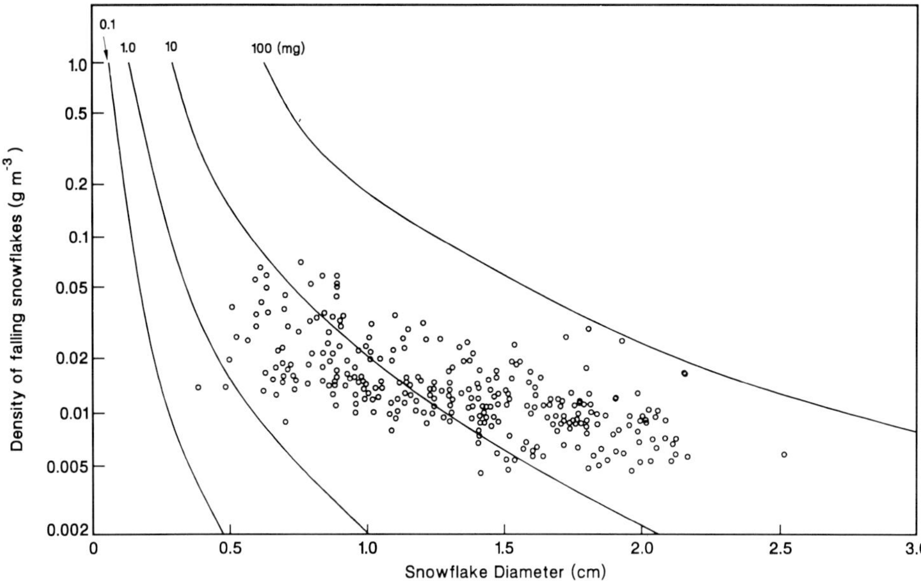

FIG. 8.11. Snowflake (dendrite aggregates) densities as a function of their diameter. The solid lines represent constant mass (mg), assuming spherical symmetry. From Passarelli and Srivastava (1979). *Journal of Atmospheric Sciences*, **436**, 3. American Meteorological Society. Reproduced with permission.

Passarelli and Srivastava (1979) present a composite of observations of snowflake fall velocities as a function of snowflake diameter, shown in Fig. 8.12. The solid curve in Fig. 8.12 represents the empirical relation

$$\bar{v}_f = 124\, r_f^{0.2}, \tag{8.9}$$

where the fall speed of the snowflake (v_f) is in $\mathrm{cm\,sec^{-1}}$. For a more direct solution to (8.7), the size versus density and fall speed versus density relationships can be solved from (8.8) and (8.9) (see Problem 8.1) as

$$\bar{v}_f = 162\, m^{0.083} \tag{8.10}$$

where the mass of the snowflake (m) is in grams.

Observations of the fluctuations of the falling motions of ice crystals and small snowflakes are provided by Kajikawa (1989). He notes that the standard deviation of the vertical fall speed is less than $2\,\mathrm{cm\,s^{-1}}$ which is roughly 3% to 5% of the mean vertical fall speed. The standard deviation of the horizontal speed ranges from 1 to $8\,\mathrm{cm\,s^{-1}}$ and is on the same order as the mean horizontal speed. Kajikawa (1982) notes that snowflakes tend to fall with a spiral motion resulting from asymmetric aggregation.

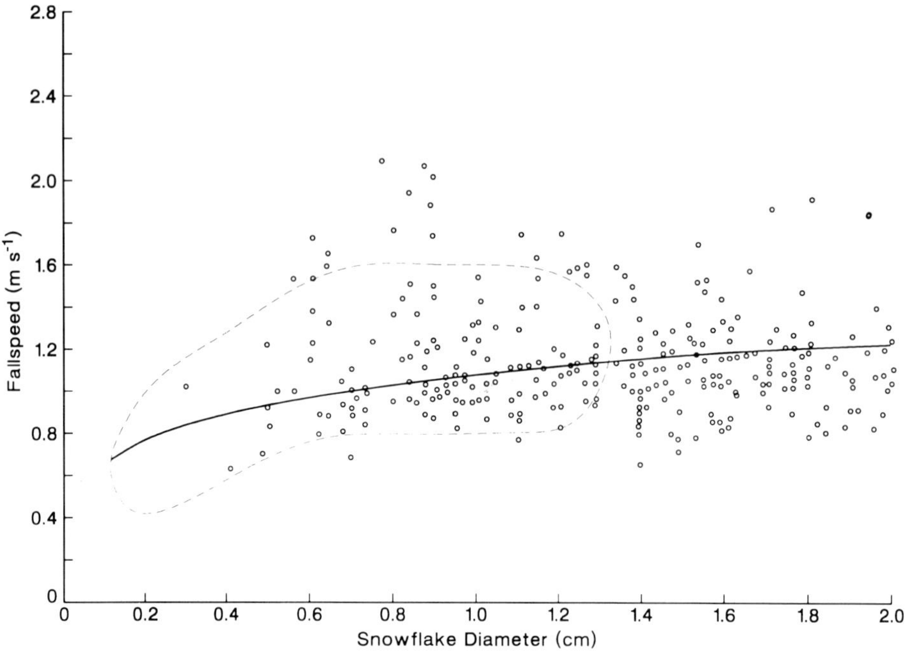

Fig. 8.12. Snowflake fall speeds as a function of their diameter. Points represent data from Magono and Nakamura (1965); the envelope represents data from Locatelli and Hobbs (1974). After Passarelli and Srivastava (1979). *Journal of Atmospheric Sciences*, **36**, 3. American Meteorological Society. Reproduced with permission.

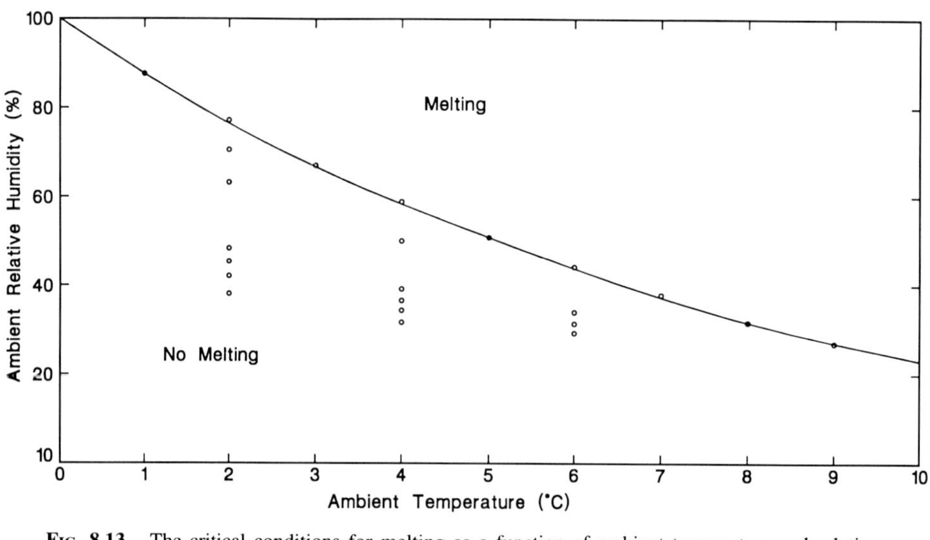

Fig. 8.13. The critical conditions for melting as a function of ambient temperature and relative humidity at 700 mbar as calculated based on (8.11). Points represent experimental observations showing no melting from Rasmussen and Pruppacher (1982). *Journal of Atmospheric Sciences*, **39**, 1. American Meteorological Society. Reproduced with permission.

8.5. Melting of ice particles

Ice particles falling in the atmosphere often encounter temperatures warmer than 0°C. The melting rate for such particles is determined by the rates at which sensible and latent heats are conducted to the particle. If the ice particle is subliming, latent heat is being conducted away from the particle. If the latent heat being conducted away from the ice particle is greater than the sensible heat being conducted to the particle, then the ice particle temperature would be below 0°C and no melting would occur. This critical condition for melting for an ambient pressure of 700 mbar is shown in Fig. 8.13.

Ice particles in the melting region shown in Fig. 8.13 may be expected to accumulate liquid water. If the ice particle were solid initially, liquid water would be expected to accumulate on its surface. Thus, continued melting of the ice particle requires that sensible heat be conducted to the liquid surface of the particle and then through the water shell to the ice interior. Mason (1956) assumes that the liquid coating is symmetrically distributed over the surface of the ice particle. The melting rate given by Mason is

$$\begin{aligned} L_f \frac{dm}{dt}\bigg|_{\text{melt}} &= \frac{4\pi r_w r_i K_w T_x}{r_w - r_i} \\ &= 4\pi r_w \left[K f_2^* (T_\infty - T_x) + L_v f_2 (\rho_\infty - \rho_{s,x})\right], \end{aligned} \tag{8.11}$$

where r_w and r_i represent the radii of the outer water shell and the inner ice core, respectively, and K_w is the conductivity of liquid water. Note that the conduction to the melting particle is based on a surface temperature (T_x) that is above 0°C.

These theoretical melting rates tend to overestimate the time required to melt an ice particle. Rasmussen and Pruppacher (1982) observed the time required to melt frozen drops in a vertical wind tunnel and found the observed times were typically 30% to 40% lower than the times calculated using (8.11). They note that melting must depend on shedding of melt water, the distribution of melt water around the ice core, internal circulations, and the bulk density and structure of the original ice particle.

The melting of a large ice sphere proceeds in stages as shown in Fig. 8.14. As ice melts, the melt water tends to collect in a torus about the particle equator, shedding drops. As the ice particle shrinks, the water torus tends to move upstream and the sizes of the shed drops increase. With continued shrinking, the shedding rate decreases and the melt water forms a stable raindrop around the ice core. Internal circulation commences, which enhances melting of the ice core. Note the asymmetric position of the ice core within the water drop.

The more complete theoretical treatments of the melting of spherical ice particles by Rasmussen et al. (1984a,b) include consideration of the asymmetric position of the ice core and internal circulations. They find the melting rates for ice particles smaller than 500 μm radius agree well with the theoretical treatment

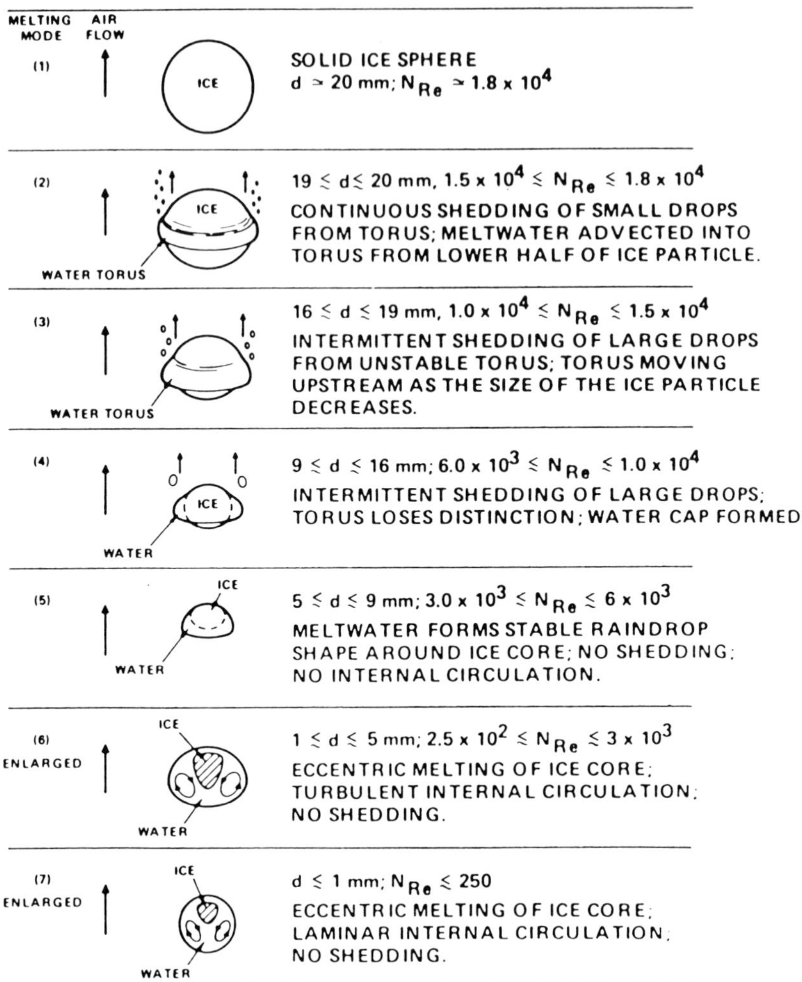

Fig. 8.14. Schematic of the melting modes. From Rasmussen et al. (1984b). *Journal of Atmospheric Sciences*, **41**, 3. American Meteorological Society. Reproduced with permission.

including laminar internal convection, provided the surface is smooth. For irregular and rough particles, the ventilation factor has to be increased by a factor of 2.4 to account for the observed melting rates.

For spherical ice particles between 500 μm and 2.5 mm radius, melting rates agree well with the theory, assuming that the meltwater is well mixed, i.e., the water surface of the particle (T_x) is also at 0°C. For larger ice particles, up to 4.5 mm radius, no internal circulations are noted and agreement with the theoretical treatments is good. Meltwater is not shed for sizes up to 4.5 mm radius.

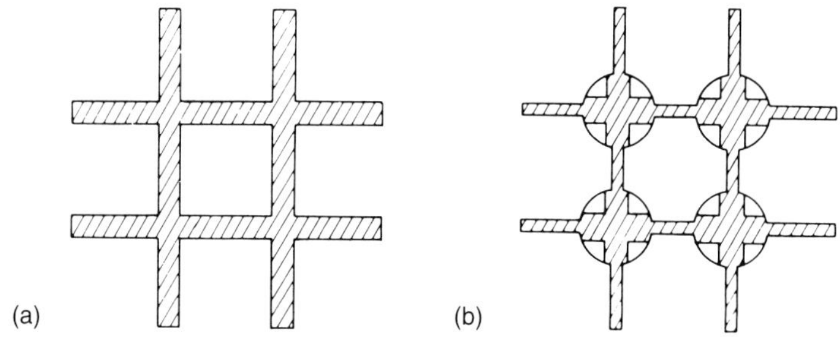

FIG. 8.15. Schematic representation of the melting behavior of an idealized aggregate. (a) The aggregate before the onset of melting; (b) the collection of the melt water at the ice branch links and the preferential melting of the ice structure directly exposed to air. From Knight (1979). *Journal of Atmospheric Sciences*, **36**, 6. American Meteorological Society. Reproduced with permission.

For larger spherical ice particles, Macklin (1963) provides an equation similar to (8.11) with a numerical factor intended to account for the effects of hailstone oblateness on the heat transfer. Macklin notes that the melting is more pronounced for more oblate hailstones, i.e, the numerical factor should be greater than 1. Rasmussen et al. (1984b) find this numerical factor must be set to 0.76 in order to provide agreement with their observations. This is considered to be a result of shedding for ice particles larger than radius 4.5 mm.

The melting behavior of ice crystals has been described by Knight (1979), who collected melting ice crystals in hexane. He noted that columnar ice crystals form a water film which thickens non-uniformly. As melting proceeds, the remaining ice column protrudes from a growing water drop. The exposed portions of the column are noted to melt more rapidly than those portions inside the water drop. The melt water is pulled by capillary action into the drop (or drops). Planar crystals were observe to melt with no tendency for the melt water to contract into drops.

Knight (1979) also observed the melting of aggregates. In these, the melt water was noted to fill interior corners (when they exist), acting through surface tension to tend to pull or collapse the aggregate into a more compact form. The ice framework tends to keep the aggregate extended. Again, the melting is more rapid for those portions of the ice structure directly exposed to the air. This concept is illustrated in Fig. 8.15. It is suggested that the preferentially thinning of the ice structure where exposed to the air may lead to breakup of the aggregate.

A theoretical treatment of the melting behavior of snowflakes is given by Mitra et al. (1990) based on (8.11). Since the melt water is considered to migrate to the interior of the snowflake, the surface temperature (T_x) may be considered to be 0°C. They assumed the aggregates were oblate spheroids whose axial ratio varied linearly from 0.3 for dry snowflakes to 1.0 for the resulting drop as the fraction of melt water (m_w/m_f) increased from 0 to 1. The density of the melting snowflake was also assumed to vary linearly with the fraction of melt water, from

$0.02\,g\,cm^{-3}$ for dry flakes, to $1.0\,g\,cm^{-3}$ for the water drop. Since they also observed the melting behavior of the snowflakes in a vertical wind tunnel, they were able to measure the fall velocity directly as the snowflakes melted. They note good agreement between their observations and the theoretical melting rates.

Problems

8.1* Derive the fall velocity versus mass relation for snowflakes (8.10) from the density versus size (8.8) and fall speed versus size (8.9) relationships.

8.2* Calculate the collection efficiency (according to (8.1)) for a columnar ice crystal which is falling along a trajectory inclined 30° from the vertical. The crystal dimensions are: a = 30 μm and c = 100 μm. Assume the crystal falls with its c-axis horizontal and that it collects all drops within the volume it traverses.

8.3* Compare the Schumann–Ludlam limit for a 1 cm radius spherical hailstone at an ambient temperature of −15°C for pressures of 500 mbar, 700 mbar, and 1000 mbar. Assume the water drops have negligible fall velocities and the collection efficiency is 1.

8.4* Calculate the critical conditions for melting at P = 1000 mbar for temperatures between 0°C and −10°C. Compare these results with Fig. 8.13.

8.5** Consider a 0.5 cm radius ice sphere, growing in a cloud of 10 μm radius water drops (concentration of $300\,cm^{-3}$). The ambient temperature is −15°C and the pressure is 700 mbar. You may assume the cloud is at water saturation and use a collection efficiency of 1.

(a) What is the temperature of the ice sphere?

(b) What fraction of the accreted water is lost by sublimation from the ice particle?

Computer problem

8.6** Consider a 1.0 cm radius (spherical) ice particle, falling in an isohumal environment with a lapse rate of $8°C\,km^{-1}$. Assume the freezing level is at 600 mbar and that the particle has a uniform temperature of 0°C (until all of the ice is melted).

(a) How far would the ice particle fall before melting completely in an environment at 60% relative humidity?

(b) How far would the ice particle fall before melting completely at 60% humidity in the presence of a rainwater content of $2\,g\,m^{-3}$? Assume (for simplicity) that the rain drops are all falling at mass-weighted fall velocity of $7\,m\,s^{-1}$ with a size of 0.1 cm radius and they are collected with a collection efficiency of 1.0.

REFERENCES

Beard, K.V., and Grover, S.N., 1974, 'Numerical collision efficiencies for small raindrops colliding with micron size particles,' *J. Atmos. Sci.* **31**, 543.

Böhm, H.P., 1989, 'A general equation for the terminal fall speed of solid hydrometeors,' *J. Atmos. Sci.*, **46**, 2419–27.

Buser, O. and Aufdermaur, A.N., 1973, 'The density of rime on cylinders,' *Quart. J. R. Meteorol. Soc.* **99**, 388.

Harimaya, T., 1975, 'The riming properties of snow crystals,' *J. Meteorol. Soc. Jpn* **53**, 384.

Harimaya, T., 1976, 'The embryo and formation of graupel,' *J. Meteorol. Soc. Jpn* **54**, 42.

Heymsfield, A.J., and Kajikawa, M., 1987, 'An improved approach to calculating terminal velocities of plate-like crystals and graupel,' *J. Atmos. Sci.* **44**, 1088.

Heymsfield, A.J., and Pflaum, J.C., 1985, 'A quantitative assessment of the accuracy of techniques for calculating graupel growth,' *J. Atmos. Sci.* **42**, 2264.

Hobbs, P.V., 1965, 'The aggregation of ice particles in clouds and fogs at low temperatures,' *J. Atmos. Sci.* **22**, 296.

Hobbs, P.V., Chang, S., and Locatelli, J.D., 1974, 'The dimensions and aggregation of ice crystals in natural clouds,' *J. Geophys. Res.* **79**, 2199.

Hosler, C.L., Jensen, D.C., and Goldschlak, L., 1957, 'On the aggregation of ice crystals to form snow,' *J. Meteorol.* **14**, 415.

Huschke, R.E. (Ed.), 1959, *Glossary of meteorology.* Boston: American Meteorological Society.

Jayaweera, K.O.L.F., and Cottis, R.E., 1969, 'Fall velocities of plate-like and columnar ice crystals,' *Quart. J. R. Meteorol. Soc.* **95**, 703.

Johnson, D.B., 1987, 'On the relative efficiency of coalescence and riming,' *J. Atmos. Sci.* **44**, 1671.

Johnson,D.B., and Rasmussen, R.M., 1992, 'Hail growth hyteresis,' *J. Atmos. Sci.* **49**, 2525–2532.

Kajikawa, M., 1982, 'Observation of the falling motion of early snowflakes. Part I. Relationship between the free-fall pattern and the number and shape of component snow crystals,' *J. Meteorol. Soc. Jpn* **60**, 797.

Kajikawa, M., 1989, 'Observation of the falling motion of early snowflakes. Part II. On the variation of falling velocity,' *J. Meteorol. Soc. Jpn* **67**, 731.

Knight, C.A., 1979, 'Observations of the morphology of melting snow,' *J. Atmos. Sci.* **36**, 1123.

Knight, C.A., 1991, 'Comments on 'Physics of supercooling of thin water skins covering gyrating hailstones',' *J. Atmos. Sci.* **48**, 217.

Knight, C.A., and Knight, N.C., 1970a, 'The falling behavior of hailstones,' *J. Atmos. Sci.* **27**, 672.

Lesins, G.B., and List, R., 1986, 'Sponginess and drop shedding of gyrating hailstones in a pressure-controlled icing wind tunnel,' *J. Atmos. Sci.* **43**, 2813.

Lew, J.K., and Pruppacher, H.R., 1983, 'A theoretical determination of the capture efficiency of small columnar ice crystals by large cloud drops,' *J. Atmos. Sci.* **40**, 139.

List, R., 1990, 'Physics of supercooling of thin water skins covering gyrating hailstones,' *J. Atmos. Sci.* **47**, 1919.

Lo, K.K., and Passarelli, R.E., Jr., 1982, 'The growth of snow in winter storms: An airborne observational study,' *J. Atmos. Sci.* **39**, 697.

Locatelli, J.D., and Hobbs, P.V., 1974, 'Fallspeeds and masses of solid precipitation particles,' *J. Geophys. Res.* **79**, 2185.

Ludlam, F.H., 1958, 'The hail problem,' *Nubila* **1**, 13.

Macklin, W.C., 1962, 'The density and structure of ice formed by accretion,' *Quart. J. R. Meteorol. Soc.* **88**, 30.

Macklin, W.C., 1963, 'Heat transfer from hailstones,' *Quart. J. R. Meteorol. Soc.* **89**, 360.

Macklin, W.C., and Bailey, I.H., 1968, 'The collection efficiencies of hailstones,' *Quart. J. R. Meteorol. Soc.* **94**, 393.

Magono, C., and Nakamura, T., 1965, 'Aerodynamic studies of falling snow flakes,' *J. Meteorol. Soc. Jpn* **43**, 139.

Mason, B.J., 1956, 'On the melting of hailstones,' *Quart. J. R. Meteorol. Soc.* **82**, 209.

Mitra, S.K., Vohl, O., Ahr, M., and Pruppacher, H.R., 1990, 'A wind tunnel and theoretical study of the melting behavior of atmospheric ice particles. IV: Experiment and theory for snowflakes,' *J. Atmos. Sci.* **47**, 584.

Nakaya, U., 1954, *Snow crystals natural and artificial*. Cambridge, MA: Harvard University Press.

Ono, A., 1969, 'The shape and riming properties of ice crystals in natural clouds,' *J. Atmos. Sci.* **26**, 138.

Ono, A., 1970, 'Growth mode of ice crystals in natural clouds,' *J. Atmos. Sci.* **27**, 649.

Passarelli, R.E., Jr., and Srivastava, R.C., 1979, 'A new aspect of snowflake aggregation theory,' *J. Atmos. Sci.* **36**, 484.

Pflaum, J.C., and Pruppacher, H.R., 1979, 'A wind tunnel investigation of the growth of graupel initiated from frozen drops,' *J. Atmos. Sci.* **36**, 680.

Pflaum, J.C., Martin, J.J., and Pruppacher, H.R., 1978, 'A wind tunnel investigation of the hydrodynamic behavior of growing, freely falling graupel,' *Quart. J. R. Meteorol. Soc.* **104**, 179.

Pitter, R.L., and Pruppacher, H.R., 1974, 'A numerical investigation of collision efficiencies of simple ice plates colliding with supercooled water drops,' *J. Atmos. Sci.* **31**, 551.

Rasmussen, R.M., and Heymsfield, A.J., 1985, 'A generalized form for impact velocities used to determine graupel accretional densities,' *J. Atmos. Sci.* **42**, 2275.

Rasmussen, R.M., and Heymsfield, A.J., 1987, 'Melting and shedding of graupel and hail. Part I: Model physics,' *J. Atmos. Sci.* **44**, 2754.

Rasmussen, R., and H.R. Pruppacher, 1982, 'A wind tunnel and theoretical study of the melting behavior of atmospheric ice particles. I. A wind tunnel study of frozen drops of radius $< 500\,\mu m$,' *J. Atmos. Sci.* **39**, 152.

Rasmussen, R.M., Levizzani, V., and Pruppacher, H.R., 1984a, 'A wind tunnel and theoretical study of the melting of atmospheric ice particles. II. A theoretical study for frozen drops of radius $<500\,\mu m$,' *J. Atmos. Sci.* **41**, 381.

Rasmussen, R.M., Levizzani, V., and Pruppacher, H.R., 1984b, 'A wind tunnel and theoretical study of the melting of atmospheric ice particles. III. Experiment and theory for spherical ice particles of radius $>500\,\mu m$,' *J. Atmos. Sci.* **41**, 374.

Sasyō, Y., and Matsuo, T., 1985, 'Effects of the variations of falling velocities of snowflakes on their aggregation,' *J. Meteorol. Soc. Jpn* **63**, 249.

Schlamp, R.J., Pruppacher, H.R., and Hamielec, A.E., 1975, 'A numerical investigation of the efficiency with which simple columnar ice crystals collide with supercooled water drops,' *J. Atmos. Sci.* **32**, 2330.

Schumann, T.E.W., 1938, 'The theory of hailstone formation,' *Quart. J. R. Meteorol. Soc.* **64**, 3.

Zikmunda, J., and Vali, G., 1972, 'Fall patterns and fall velocities of rimed ice crystals,' *J. Atmos. Sci.* **29**, 1334.

9

FORMATION OF PRECIPITATION BY COALESCENCE

9.1. Introduction

The formation of precipitation by coalescence is directly related to the cloud droplet spectrum resulting from the activation of cloud condensation nuclei (CCN) in the updraft. The cloud droplet spectrum is defined by the interactions between the nucleation and diffusion processes. Once the cloud droplet spectrum has been established, the growth of droplets by diffusion enhances the probability of collisions (and coalescences) between droplets, thereby forcing the growth of raindrops. As the coalescence process proceeds, collisional breakup shapes the spectrum, resulting in an approximately exponential raindrop spectrum.

The conversion of a spectrum of many small droplets to one containing relatively few large drops results in 'decontrol' of the supersaturation with respect to water. The numerous tiny cloud droplets are able to extract the water vapor being made available for condensation in the updraft at relatively low supersaturations. However, the relatively few large drops that make up the raindrop spectrum are much less able to extract this excess water vapor, allowing the water supersaturation to rise. Possible consequences of this rise in water supersaturation will be discussed.

Evaporation of precipitation falling into entrained (drier) air results in cooling and sinking air. Maintenance of the downdraft is dependent on a continued supply of small droplets. These droplets are provided by drop breakup and their rate of production is roughly proportional to the square of the rainfall rate.

9.2. The formation of a water cloud

The activation of a particular population of CCN in an updraft is dependent on the updraft velocity and the ambient temperature and pressure. The number of droplets formed increases as the number of available CCN increases. Likewise, the stronger the updraft, the higher the supersaturation can rise and the more droplets will be formed as is evident from an inspection of Fig. 3.15. The dependencies on ambient temperature and pressure are less obvious.

The interplay between the nucleation and diffusion processes is sufficiently complex to require a numerical model for solution. The essential aspects of such a model are discussed in the following sections, as are some of the results that may be obtained.

9.2.1. *Activation of CCN in an updraft*

As a parcel of air rises, both the actual vapor pressure and the saturation vapor pressure decrease, as discussed in Chapter 1. The actual vapor pressure decreases as the parcel expands and its total pressure decreases. The saturation vapor pressure is a function of temperature and decreases due to the cooling of the parcel as it rises. Since the saturation vapor pressure decreases more rapidly than the actual vapor pressure, the relative humidity increases as the parcel rises.

When the relative humidity exceeds a value close to 78%, hygroscopic aerosol containing ammonium sulfate or sodium chloride begin to deliquesce, forming solution droplets. These solution droplets are termed 'haze' particles. As the parcel rises and the relative humidity continues to increase, these haze particles continue to grow, becoming more dilute. For typical updrafts, most haze particles are able to maintain a radius close to the equilibrium radius as defined by the Köhler equation (3.53).

As the relative humidity increases above water saturation, the critical size for some of the hygroscopic aerosol may be exceeded. These CCN then are in the region of Fig. 3.13, in which the thermodynamic potential decreases as the droplets grow. These CCN are then said to have been 'activated' and the droplets can grow to sizes limited by the available time and water vapor.

The rate of condensation of water vapor on the growing particles increases with time. Droplets larger than their critical size will continue to remove more and more of the available water vapor as their size increases, as the ambient supersaturation rises, and as their numbers increase. Larger droplets have higher condensation rates due to the decrease in their equilibrium saturation ratio with increasing size (above their critical size) and due to the increase in radius, as indicated by (5.2). Subcritical nuclei also remove water vapor as they follow their Köhler curves to larger sizes. Eventually, the extraction of water vapor by the growing droplets and the unactivated CCN equals the rate at which water vapor is being made available in the rising parcel. This defines a peak supersaturation and, above this level the supersaturation decreases. Once the peak supersaturation has been reached, the cloud droplet spectrum is defined.

(i) Governing equations. As the saturation ratio in an updraft increases below the cloud base,[1] the CCN tend to increase in size, following their Köhler curve (3.53). However, the rate at which a given CCN can grow is limited by the rate at which water vapor can diffuse to the growing particle. Thus, the growth of the particle is governed by the diffusion equation (4.2), in which the saturation vapor density over water at the drop temperature is replaced by the equilibrium saturation ratio as defined by the Köhler equation 3.53 for the given masses of salt and water.

Solution of the diffusion equation requires solution for the particle temperature. Using the equilibrium saturation ratio (S_k) as defined by the Köhler

[1] The cloud base nominally occurs when the saturation ratio equals 1, although this is not quite true.

equation and following the derivation of (4.8), the particle temperature is given by

$$T_x = \frac{DL_v f_1 (\rho_\infty - S_k b) + K f_1^* T_\infty}{DL_v f_1 S_k s + K f_1^*},$$

where the terms s and b are the parameters for the linear approximation to the saturation vapor density of water.

The solution of the Köhler equation (3.53) requires evaluation of the term $\partial i / \partial m$ based on (3.54). The molality M is defined as

$$M = \frac{m_s / M_s}{m/1000},$$

where m and m_s represent the masses of water and salt, respectively, and M_s is the molecular weight of the salt. From this,

$$\frac{\partial i}{\partial m} = \frac{1}{m}\left\{0.1844 + 0.015862 \ln\left[\frac{1000 m_s}{m M_s}\right]\right\}.$$

(ii) A case study. Young (1992) describes a model of the activation of CCN in an updraft based on the complete version of the Köhler equation (3.53). Twenty-five categories of particles are followed with ammonium sulphate masses ranging from 10^{-18} g to 10^{-12} g. The number of nuclei corresponding to each category was derived from Fig. 3.19 for the typical continental CCN activity spectrum. Nuclei concentrations ranged from 6.3 cm^{-3} for the largest size to 191.5 cm^{-3} for the smallest size. The model was initiated at 95% relative humidity, assuming all of the nuclei were in equilibrium with their Köhler curve at that humidity.

The height (time) history of this population of CCN in a 2 m sec^{-1} updraft, initiated at 10°C and 900 mbar is shown in Fig. 9.1. The CCN can be grouped into three categories as follows.

The smallest CCN with critical supersaturations greater than the peak supersaturation grow as the saturation ratio increases, reaching their maximum size at the level of the peak supersaturation, and shrink above that level.

The middle size range of the CCN tend to follow their Köhler curve as the saturation ratio increases until the ambient supersaturation exceeds their critical supersaturation. Above this level, the particles grow rapidly to radii larger than 1 μm (4×10^{-12} g). This rapid increase in growth rate is more pronounced for the smaller of the CCN that are activated.

The largest CCN grow throughout the ascent, although their critical size is so large that they never become 'activated' in the sense that they experience a rapid increase in their growth rate. The broken curve in Fig. 9.1 indicates the height at which the various CCN reach the mass corresponding to their Köhler curve maximum (critical mass). The largest CCN have not reached their critical mass even well above the level at which the peak supersaturation is reached. The

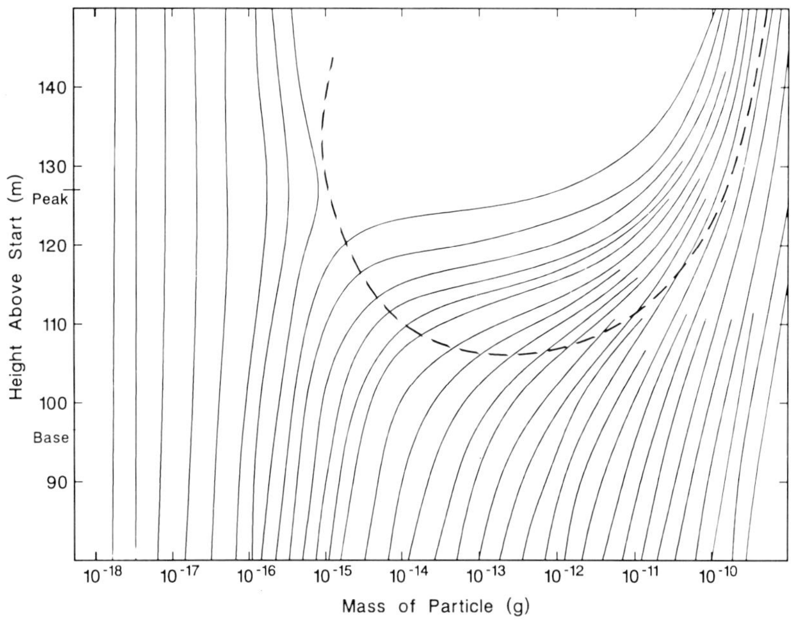

Fig. 9.1. The activation of a population of CCN in an updraft of $2\,\mathrm{m\,s^{-1}}$. Particle mass (salt plus water) is shown as a function of height. The peak supersaturation is 1.14% at a height of 127 m above the start. The level at which the saturation ratio equals 1.0 is 97 m (base). At the starting height, CCN were assumed to be in equilibrium at $S = 0.95$. Above the broken line, the particles are larger than their critical size. Note the sharp distinction between those CCN which become activated and those which fail to do so.

critical mass for the largest category of CCN is 2.0×10^{-8} g, corresponding to a radius of 16.8 μm. This size is large enough for the particle to participate in coalescence growth. Thus, the very largest CCN may never become activated prior to being collected by (or collecting) other drops.

(iii) Dependence on environmental parameters. The number of CCN activated in an updraft depends on the population of CCN, the updraft velocity, and the ambient temperature and pressure. The dependence of the number of CCN activated on the first three of these variables is shown in Fig. 9.2. The dependence on the number of CCN available is self-evident. The increase in the number of CCN activated as the updraft velocity is increased results from the higher peak supersaturations attained. Since there is less time available for growth in stronger updrafts, less of the excess water vapor is extracted by the growing particles, resulting in higher peak supersaturations as shown in Fig. 9.3. Thus, more of the CCN become activated.

The dependence on temperature is considerably more complex. The equilibrium saturation ratio from (3.53) is inversely related to temperature and directly related to the surface tension. Since the surface tension increases as temperature decreases, both of these effects yield a larger equilibrium saturation

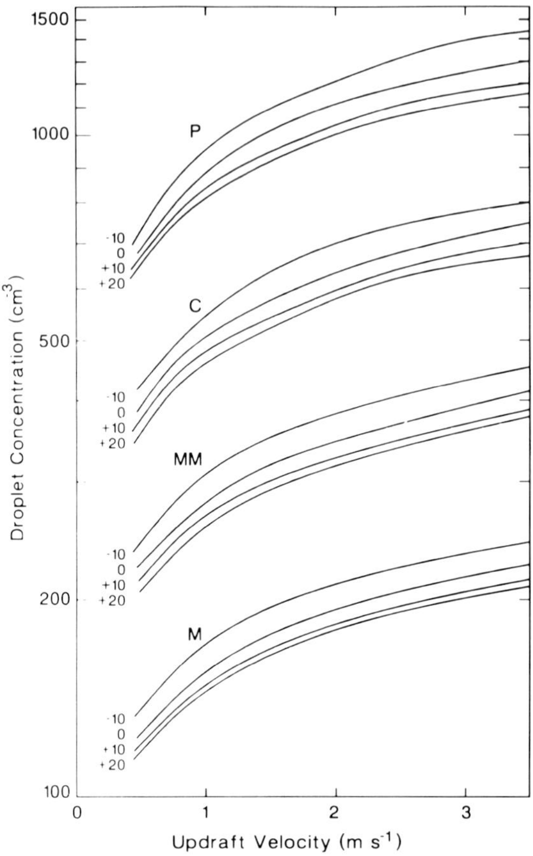

FIG. 9.2. Numbers of CCN activated as functions of the updraft velocity, initial temperature, and the CCN population. For simplicity, the CCN activity spectra have the form $n = n_o S^{0.5}$ where n_o is 150 cm^{-3} for the maritime (M) spectrum, 300 cm^{-3} for the modified maritime (MM) spectrum, 600 cm^{-3} for the continental (C) spectrum, and 1200 cm^{-3} for the polluted (P) spectrum.

ratio for colder temperatures for a particle containing given masses of salt and water.[2] The critical supersaturation is also higher for colder temperatures. Since the critical supersaturation is higher, one would expect fewer CCN to be activated. Yet this is not the case.

The growth rate of droplets by diffusion is reduced at colder temperatures since the diffusivity of water vapor in air is less and, more importantly, the saturation vapor density is less. For a given supersaturation, the difference in vapor density in (4.2) is smaller at colder temperatures. Thus, the rate at which the newly activated CCN and large CCN can extract excess water vapor is substantially reduced at colder temperatures. In addition, the particle mass for a given equilibrium saturation ratio is smaller at colder temperatures. These effects

[2] This results in a smaller equilibrium particle mass for a given saturation ratio at colder temperatures.

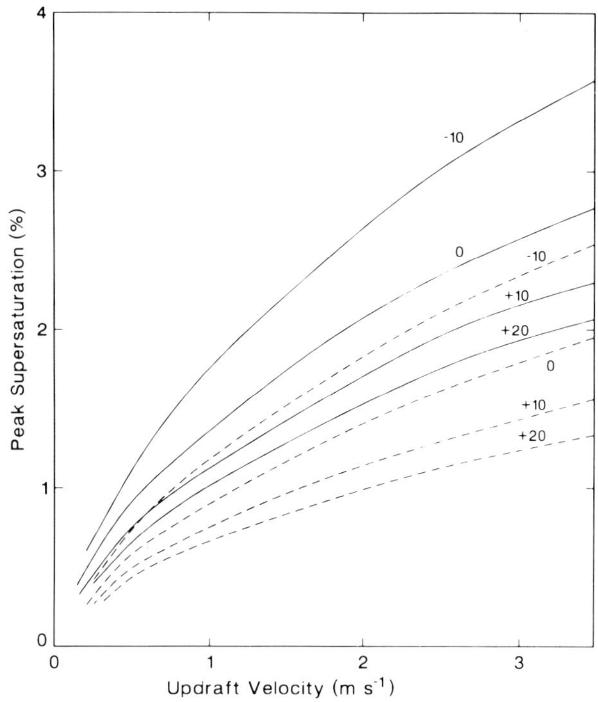

Fig. 9.3. Peak supersaturations for (—) maritime and (- - -) continental CCN spectra as a function of the updraft velocity and initial temperature.

increase the peak supersaturation to a greater extent than the colder temperatures increase the critical supersaturation. Thus, more CCN are activated at colder temperatures.

The dependence on pressure results from the increase in the diffusivity of water vapor with decreasing pressure. This promotes more rapid growth for the newly activated CCN as well as for the large CCN that are still smaller than their critical size. Pressure has little affect on the sizes attained by the smaller, subcritical CCN since they tend to be close to their equilibrium size anyway. The more rapid removal of the excess water vapor slightly reduces the peak supersaturation and thereby decreases the number of CCN activated as pressure decreases. For the case depicted in Fig. 9.1, at pressures of 900 mbar, 700 mbar, and 500 mbar, the numbers of CCN activated are $582\,cm^{-3}$, $578\,cm^{-3}$, and $573\,cm^{-3}$, respectively.

9.2.2. *Comparison between observed and modeled drop spectra*

Several field studies have examined the relationship between the population of CCN and the numbers of droplets found in the bases of clouds formed in an updraft. Fitzgerald (1972) collected samples of the air and measured the updraft

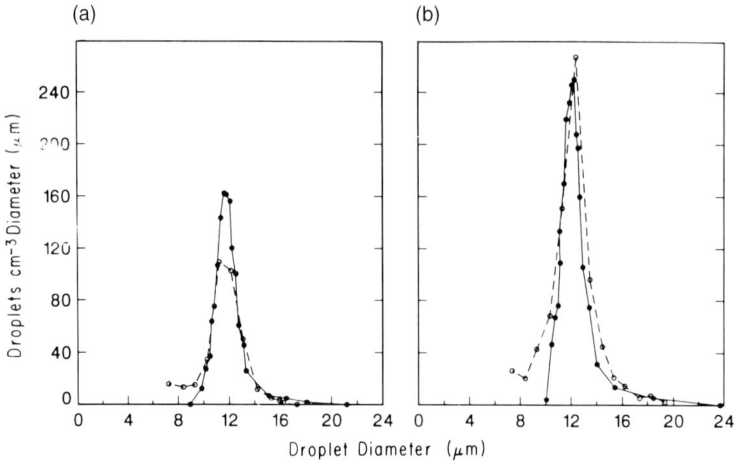

Fig. 9.4. Comparison of the observed and computed droplet size distributions. (a) 1430 LST 24 July 1970, Bemidji MN, cloud base temperature of 5.6°C, pressure of 789 mbar, with an updraft velocity of 0.58 m s^{-1}. (- - -), Observed distribution: $\bar{d}$ = 11.79 μm, $\sigma/\bar{d}$ = 0.153, concentration = 361 cm^{-1}; (—), computed distribution: $\bar{d}$ = 12.41 μm, $\sigma/\bar{d}$ = 0.171, concentration = 381 cm^{-1}. (b) 1600 LST 28 July 1970, Bemidji MN, cloud base temperature of 16.5°C, pressure of 840 mbar, with an updraft velocity of 1.25 m s^{-1}. (- - -), Observed distribution: $\bar{d}$ = 12.16 μm, $\sigma/\bar{d}$ = 0.162, concentration = 745 cm^{-1}; (—), computed distribution: $\bar{d}$ = 12.65 μm, $\sigma/\bar{d}$ = 0.154, concentration = 543 cm^{-1}. From Fitzgerald (1972). © J.W. Fitzgerald. Reproduced with permission.

velocity below the cloud base, and then sampled the droplet population 200 to 300 meters above the cloud base. The CCN activity spectrum was later determined using a thermal gradient diffusion chamber.

A numerical model similar to that described in § 9.2.1 was used to compute the droplet size distribution for comparison with the observed droplet distributions.[3] Results for two of these comparisons are shown in Fig. 9.4. In general, there is reasonable agreement between the observed and computed total concentrations and droplet sizes, although the computed spectra tend to be narrower than the observed spectra.

Figure 9.5 summarizes the comparisons reported by Fitzgerald (1972) and those by Twomey and Warner (1967). Twomey and Warner did not measure the updraft velocity but assumed a value of 3 m s^{-1} for all their observations. Fitzgerald's results show a correlation coefficient of 89% between the observed and computer droplet concentrations, with a slope of 1.25 indicating a tendency for the computed concentrations to be lower than the observed concentrations. The correlation coefficient for Twomey and Warner's results is 77% but the slope of their observations is close to 1 (0.97), suggesting the 3 m s^{-1} updraft velocity was chosen to obtain approximate overall agreement.

[3] It should be noted that the CCN salt masses were derived from the measured activity spectrum using the conventional form of the Köhler equation (3.55). This yields a larger salt mass for a given critical supersaturation than would be determined using (3.53). Since the conventional form of the Köhler equation is also used in the model, these effects tend to cancel.

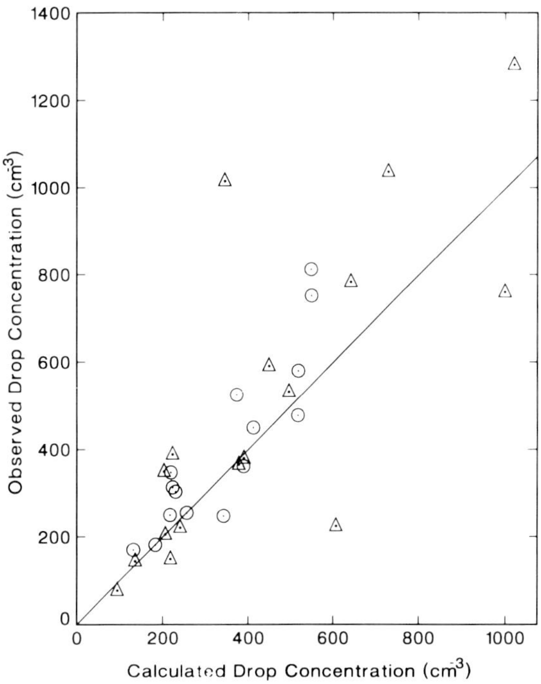

Fig. 9.5. Comparison of observed and computed drop concentrations. Data are taken from Fitzgerald (1972) and Twomey and Warner (1967). Correlation coefficients and slopes are 89% and 1.25 for the (⊙) Fitzgerald results; 77% and 0.97 for the (△) Twomey and Warner results.

9.2.3. *Effects of entrainment and mixing*

Broadening of the droplet spectrum due to the effects of competition was discussed in Chapter 3. In this section, broadening of the droplet spectrum due to the effects of entrainment and mixing are considered. Entrainment generally refers to bringing volumes of cloud-free, relatively dry air into the cloud through a boundary that is considered to be relatively well defined. The entrained parcel may be considered to be 'engulfed' by the cloud and often maintains its identity.

Mixing may involve exchange of air and particles between cloudy air and cloud-free air, or it may involve the exchange between different parcels of cloudy air. Mixing may alter the parcels involved to such an extent that they are indistinguishable from each other, or the parcels may retain their separate identities.

Traditionally, entrainment has been viewed as a horizontal movement of air through the vertical boundaries of the cloud. Squires (1958) was the first to suggest that evaporation-driven downdrafts could originate at the cloud top. Paluch (1979) used the equivalent potential temperature to determine that the majority of the air comprising the downdrafts and weak updrafts originated at the cloud top. Cloud top entrainment is associated with evaporative cooling, initiating sinking air. This sinking air then mixes with and dilutes nearby updrafts.

How does this entrainment (or mixing) process modify the cloud droplet spectrum? One conceptual approach is to mix the cloud-free and cloudy air, evaporating all of the drops to some extent until the mixed parcel reaches saturation, i.e., each drop is responsible for bringing some of the entrained air to saturation. An alternate approach is to evaporate a fraction of the drops completely such that saturation is attained, leaving the remainder of the drops unchanged.

The former approach is often termed *homogeneous mixing*; the latter approach is often termed *inhomogeneous mixing* although the term *entity mixing* is more descriptive. Telford et al. (1984) define an entity as a region of continuing identity in the turbulent volume. Thus, entity mixing involves the exchange of particles (as well as water vapor) between two entities while the entities retain their separate identities.

Baker and Latham (1979) and Baker et al. (1980) argue that since the time constant for evaporation is much smaller than the time constant for mixing on scales longer than 1 m, entity mixing is probably the dominant mode of mixing within the cloud. Telford et al. (1984) suggest that mixing may be homogeneous on small scales but entity mixing is prevalent on the cloud scale.

Paluch and Knight (1984) examine entity mixing at higher resolution by modeling the sedimentation of droplets into entrained air. As the larger droplets fall into dry air below, they evaporate and their fall speed decreases. This allows higher droplets to catch up with the lower droplets. As the former evaporate, their fall speed also decreases. Thus, relatively high concentrations of much smaller droplets accumulate in the region of the boundary between the cloudy and entrained air, as shown in Fig. 9.6. This process tends to broaden the average droplet spectrum in the parcel.

Telford et al. (1984) suggest that sinking air from cloud top entrainment can receive droplets from adjacent updrafts via entity mixing. This helps maintain the evaporative cooling necessary to sustain the downdraft and reduces the number of droplets in the updraft. The reduced number of droplets in the updraft are less able to extract the water vapor excess above saturation, resulting in an increase in the supersaturation. Baker and Latham (1979) suggest this will activate additional CCN and the remaining large droplets will grow faster due to the higher supersaturation. Paluch and Knight (1984) attribute their observations of sharp peaks in the concentrations of very small droplets far above the cloud base to this process.

Can the entrainment and mixing processes result in droplets larger than would be expected in an adiabatic (parcel) ascent? Modeling studies by Bower and Choularton (1988) suggest that the reduction in droplet concentrations in regions subject to entity mixing results in supersaturations higher than that which would be attained in an adiabatic parcel, despite activation of additional CCN.

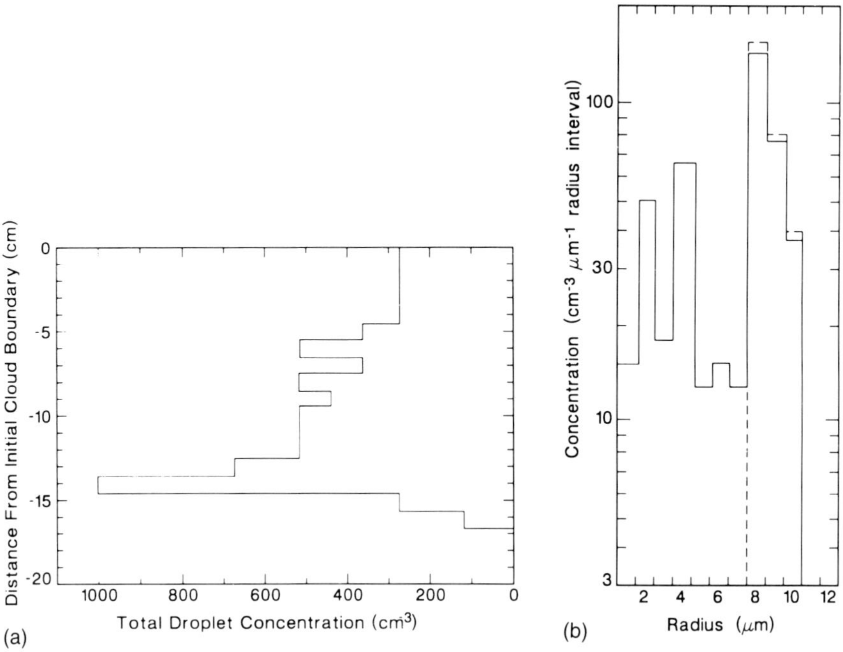

Fig. 9.6. The effects of entity mixing on the droplet spectrum in the region of the boundary between the entrained (dry) and cloudy air. (a) Total droplet concentration as droplets sediment into the entrained air. (b) Droplet spectrum averaged from 0 to −16 cm. The broken lines indicate the original droplet spectrum. From Paluch and Knight (1984). *Journal of Atmospheric Sciences*, **41**, 11. American Meteorological Society. Reproduced with permission

Based on observations of cloud droplet spectra and updraft velocities, they conclude that the supersaturation is not closely tied to the updraft velocity. Favored droplets that avoid evaporation are noted to grow to sizes larger than if they had grown in an adiabatic parcel. The average droplet however, survives on the order of 1 to 4 minutes. Paluch and Knight (1986) fail to find evidence of spectral broadening toward the larger sizes for continental cumuli whereas Telford (1987) cites evidence for production of larger drops by entity mixing in marine stratocumulus.

9.3. Initiation of the coalescence process

Continued growth of the droplets increases both their sizes and their fall velocities. Although the droplet spectrum displayed as a function of drop radius narrows as the droplets grow by condensation, the drop spectrum displayed as a function of the collection kernel broadens. As the collection kernel spectrum broadens, the likelihood of drop–drop collection events increases. Given sufficient condensation growth, the production of raindrops by the coalescence mechanism commences. This is associated with a rapid conversion from a log-normal droplet spectrum to a quasi-exponential drop spectrum.

9.3.1. *A pure-stochastic model of coalescence*

The initiation of the coalescence process is strongly dependent on the first few collection events. With each collection event, the collection kernel increases, thereby increasing the probability of another collection event. The probability of a collection event given a collection kernel Γ and small droplet concentration n_s is given by (7.7),

$$\Pr(0,t) = \exp(-n_s \Gamma t) .$$

Given a probability Pr, which may be obtained from a random number generator, the elapsed time to the next collection event is

$$t = \frac{\ln(1 - \Pr)}{-n_s \Gamma} . \tag{9.1}$$

To illustrate the use of (9.1), consider the interaction of a single collector drop with a monodispersed droplet spectrum. Assume that the collector drop has twice the mass of the droplets making up the population of collected droplets. Figure 9.7a defines isopleths of collection probability as a function of collected droplet radius assuming a fixed liquid water content of $2\,\mathrm{g\,m^{-3}}$. The collection kernel is calculated using collection efficiencies from Fig. 7.6 and fall velocities are calculated using the treatment in § 7.2.1.

Figure 9.7a makes the tacit assumption that the collector drop is formed by random coalescence between two of the collected droplets. For collector drops of radius 12.6 μm,[4] the average time required for 1% of the collector drops to undergo another collection event is roughly 40 s; for drops of radius 25.2 μm, this time is only 4 s. This suggests that for parcels at the same height in comparable clouds,[5] the cloud with fewer droplets is more likely to form rain by coalescence. Since maritime CCN spectra tend to produce fewer droplets than continental CCN spectra, it is to be expected that maritime clouds are more likely to produce rain by coalescence than continental clouds.

Figure 9.7b illustrates the increased probability that a collector drop will undergo a subsequent collection event. For example, the 12.6 μm radius collector drop has a 1% chance of collecting another droplet in 40 s. However, once the 12.6 μm radius 'double drop' has collected a third drop,[6] the probability that it will collect a fourth drop in 40 s has increased to 3%.

A pure stochastic model for coalescence could be developed based on a series of figures such as Fig. 9.7. Computationally, it is more efficient to generate a table of collection kernels or $n_s \Gamma$ to assist in the solution of (9.1) as shown in Table 9.1. Given such a table, it is relatively straightforward to generate a series of drop growth histories.

[4] This corresponds to a 'double drop' with the mass of 10 μm radius collected droplets.
[5] This assumes that the liquid water content is the same for both clouds.
[6] Its radius is now 14.4 μm.

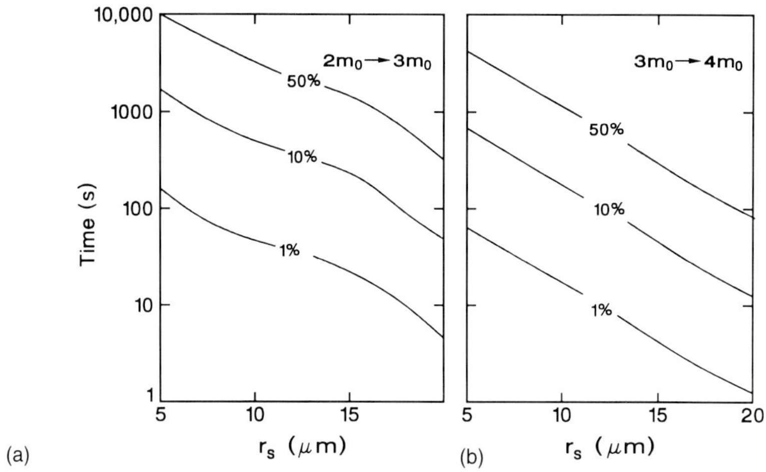

FIG. 9.7. The probability that a given collector drop will experience a collection event in the time specified. The liquid water content is assumed to be fixed at 2 g m^{-3}. The numbers of collected droplets range from 3820 cm^{-3} for 5 μm radius to 59.7 cm^{-3} for 20 μm radius droplets. (a) The probabilities that a 'double drop' will collect another drop as a function of time. (b) The probability that a 'triple drop' will collect another drop.

Consider for example, a simple program to generate the growth history of a randomly selected 'double' drop in a population of 15 μm droplets present in a concentration of 150 cm^{-3}. The collection kernels for the first three collection events are given in Table 9.1. The stochastic process that governs the growth of the drop is simulated using a random number generator (RNG). The time between successive collection events is determined from the collection kernel using (9.1).

Suppose the RNG returns a value of 0.4 for the first collection event ($k = 1$). From Table 9.1, $n_s\Gamma = 1.23 \times 10^{-3}\,s^{-1}$ and, from (9.1), $t = 415$ s. Thus, the collector drop remains at 18.9 μm radius until 415 s, when its radius increases to

Table 9.1. Parameters for use in the pure-stochastic model. The collected droplets of 15 μm radius are present in a concentration of 150 cm^{-3} with a fall speed of 2.82 cm s^{-1}. The event refers to the kth collection, i.e., $k = 1$ is for the double drop collecting a third droplet, which is the first collection starting with a double drop.

Event	Collector drop radius (μm)	Fall speed (cm s^{-1})	Collection efficiency (%)	Collection kernel (Γ) ($cm^3\,s^{-1}$)	$n_s\Gamma$ (s^{-1})
1	18.9	4.45	45	8.23×10^{-6}	1.23×10^{-3}
2	21.6	5.79	80	3.48×10^{-5}	5.22×10^{-3}
3	23.8	6.96	88	6.48×10^{-5}	9.72×10^{-3}

21.6 μm. Suppose for the second collection event, the RNG returns 0.8. Now the collector drop 'waits' another 308 s for its next collection. The collector drop remains at 21.6 μm radius from 415 s to 723 s, when its radius increases to 23.8 μm. For the third collection event, suppose the RNG returns a value of 0.85. The time to the next collection event is 195 s, at which time the collector drop grows to 25.6 μm radius. Problem 9.4 extends this concept to examine 100 collector drops, some of which may grow to raindrop sizes within 15 minutes.

The preceding is a highly simplified example designed to illustrate he growth of drops by stochastic coalescence. Cloud droplet spectra are not monodispersed but rather have a range of sizes. Consider the observed droplet spectrum shown in Fig. 9.4b. Collection events may occur involving any pair of droplets. Clearly, the likelihood of a collection event is larger for a greater disparity in the fall speeds of the droplets, for larger geometrical cross-section (larger drops), and for greater mutual collection efficiencies.

The number of interactions between any pair of drops r_i and r_j is proportional to their collection kernel times the concentrations of drops of the two sizes, n_i and n_j. Although the collection kernel is large for drop pairs involving the extreme upper and lower ends of the size spectrum, the scarcity of drops of extreme sizes results in relatively few collection events. At the other extreme, as r_i approaches r_j, the collection kernel approaches 0 and the number of collection events also approaches 0, regardless of the droplet concentrations.

Figure 9.8 presents the collection rate between droplet pairs for a normally distributed droplet spectrum with mean radius of 11.8 μm and standard deviation of 2 μm having 150 droplets cm^{-3}. The greatest number of interactions occur between droplets about one standard deviation larger than the mean radius and droplets slightly smaller than the mean radius. This combination provides a sufficiently large collection kernel and ample numbers of droplets at these sizes. The total interaction rate for this spectrum is 0.0074 s^{-1} cm^{-3} or about seven collection events every second in a volume of one liter.

9.3.2. *Effects of condensation on the collection kernel*

As the droplet spectrum grows by condensation in an updraft, the increased droplet sizes yield an increase in the geometric cross-section, an increase in the collection efficiency, and an increase in the fall speeds between corresponding portions of the spectrum. These effects are shown in Table 9.2 for a pair of droplets representing the most active interactions in Fig. 9.8 (droplet radii of 14.6 μm and 11 μm). The initial droplet spectrum is 'grown' by apportioning the additional mass among all of the droplets in proportion to their radius, thereby simulating diffusion growth without resort to solving (5.2).

As the droplets grow, their size disparity decreases. However, the increase in their respective sizes increases their fall speeds, with the net result of increasing the difference in their fall speeds by roughly 12% as the liquid water content (LWC) is increased from 1.0 g m^{-3} to 3.5 g m^{-3}. Their mutual

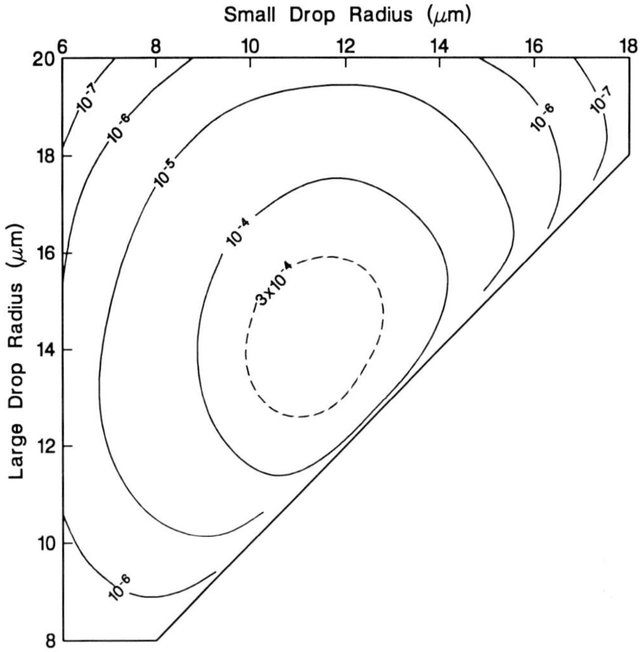

FIG. 9.8. Interaction rates between drop pairs. The droplet spectrum is normally distributed with a mean droplet radius of 11.8 μm and a standard deviation of 2 μm. The total droplet concentration is 150 cm^{-3} giving a liquid water content of 1.12 g m^{-3}. Isopleths are the interaction rate expressed as interactions s^{-1} μm^{-2} cm^{-3}. The interaction rate integrated over the entire spectrum is 0.00741 s^{-1} cm^{-3}.

Table 9.2. Interaction rates as a function of liquid water content, based on the spectrum used in Fig. 9.8 with large and small drop radii of 14.6 μm and 11 μm, respectively. Droplet concentrations are obtained by integrating the droplet spectrum from 14.4 μm to 14.8 μm and from 10.8 μm to 11.2 μm, yielding 4.5 cm^{-3} and 11.0 cm^{-3} for the large and small droplets, respectively. Drop sizes are adjusted for liquid water content (LWC) by apportioning the mass proportional to drop radius. Collection efficiencies (E_{LS}) are taken from Fig. 7.6. Fall speeds are calculated assuming P = 800 mbar and T = 10°C.

	Drop radii		Fall speeds			
LWC ($g\ m^{-3}$)	Large (μm)	Small (μm)	Large ($cm\ s^{-1}$)	Small ($cm\ s^{-1}$)	E_{LS} (%)	Events ($m^{-3}\ s^{-1}$)
1.0	14.2	10.5	2.47	1.35	6.0	64
1.5	15.7	12.4	3.00	1.87	8.4	116
2.0	16.9	13.8	3.48	2.33	9.1	154
2.5	18.0	15.0	3.93	2.75	10.5	211
3.0	19.0	16.1	4.36	3.13	15.0	351
3.5	19.8	17.0	4.76	3.50	18.5	494

geometrical cross-section is more than doubled, and their mutual collection efficiency is more than tripled for the same increase in LWC. Thus, the nearly eight-fold increase in their mutual interaction rate results largely from the increased geometrical cross-section and increased mutual collection efficiency.

Twomey (1976) suggests that the onset of the coalescence process in clouds takes place preferentially in relatively small regions of high LWC due to this strong dependence of the drop–drop interaction rate on LWC. Rokicki and Young (1978) use a quasi-stochastic model of drop growth by coalescence to examine the time required to produce a 20 dBZ radar echo for different entrainment rates. For a $3\,\mathrm{m}^{-1}$ updraft and a cloud base temperature of 10°C and using average entrainment rates[7] based on Warner (1970), the radar reflectivity factor (Z) after 25 minutes was −13 dBZ, as shown in Fig. 9.9. For the same updraft with no entrainment, the radar reflectivity factor after 25 minutes was 32 dBZ. Thus, roughly 30,000 more raindrops per unit volume were produced in the adiabatic parcel than in the average (entraining) parcel. As Twomey (1976) suggests, even if only 1% of the cloud volume exhibits adiabatic liquid water contents, these small volumes can produce more raindrops than the remainder of the cloud altogether. Nelson (1971) reached similar conclusions, comparing his model results with the observed development of rain by coalescence.

9.3.3. *Time required to produce precipitation*

The radar reflectivity factor (Z) is convenient for describing the development of precipitation since Z is proportional to the mass of the individual particles squared, i.e., for the same liquid water content, the drop spectrum with fewer but larger particles will have the higher Z. The radar reflectivity factor is defined as

$$Z = \sum_{i} n_i D_i^6$$

where D_i is the drop diameter for the ith size category. The radar reflectivity factor is usually expressed as decibels of Z (10 log Z) or dBZ. A radar reflectivity factor of 20 dBZ is commonly considered a threshold value for the development of precipitation.

Young (1975) describes a quasi-stochastic model of coalescence that incorporates activation of CCN, diffusional growth solving (5.2) and (5.8), and drop breakup based on (7.17). The model is a parcel model which permits homogeneous mixing (entrainment) and iteratively solves for the saturation ratio. This model, and its extension described by Valdez and Young (1985) will be used in this and the following two sections.

The time required to produce precipitation by coalescence varies with the CCN spectrum, the cloud base temperature, and the updraft velocity. As noted

[7] The model assumes homogeneous mixing.

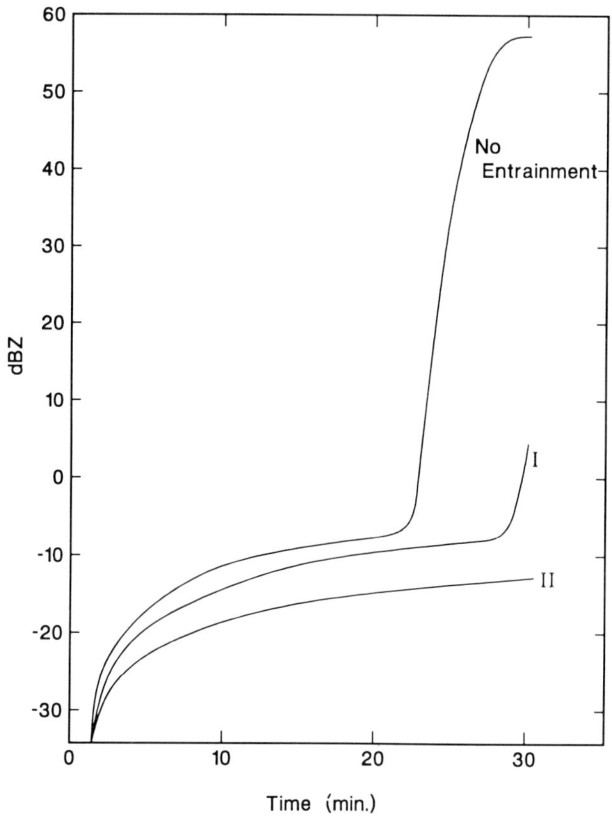

FIG. 9.9. Radar reflectivity factor (dBZ) as a function of time for different entrainment rates. The no entrainment curve represents an adiabatic ascent; curve II represents the average entrainment rate reported by Warner (1970); curve I is one-half of the entrainment rate for II. The cloud base temperature is 10°C with an updraft velocity of $3\,\mathrm{m\,s^{-1}}$. From Rokicki and Young (1978). *Journal of Atmospheric Sciences*, **17**, 6. American Meteorological Society. Reproduced with permission

previously, droplet spectra evolved from maritime CCN tend to have fewer but larger droplets and higher rates of collection than droplet spectra evolved from continental CCN. This is demonstrated in Fig. 9.10, comparing maritime, continental, and extreme continental CCN spectra. The least time required to produce a 20 dBZ reflectivity factor (17.8 min) is for the maritime CCN spectrum whereas the extreme continental spectrum required 26.5 min. It should be noted that the rate of increase in the reflectivity factor is about $12\,\mathrm{dBZ\,min^{-1}}$, which is comparable to the increase in radar first echoes summarized by Battan (1973).

The cloud base temperature affects the amount of liquid water at a given level, which in turn strongly affects the collection rate. Since the warmer cloud base temperatures provide higher LWC, the time required to produce a 20 dBZ radar echo is least for the warmest clouds. This is shown in Fig. 9.11 for cloud base temperatures ranging from -5°C to +24°C. The least time (28.3 min) was

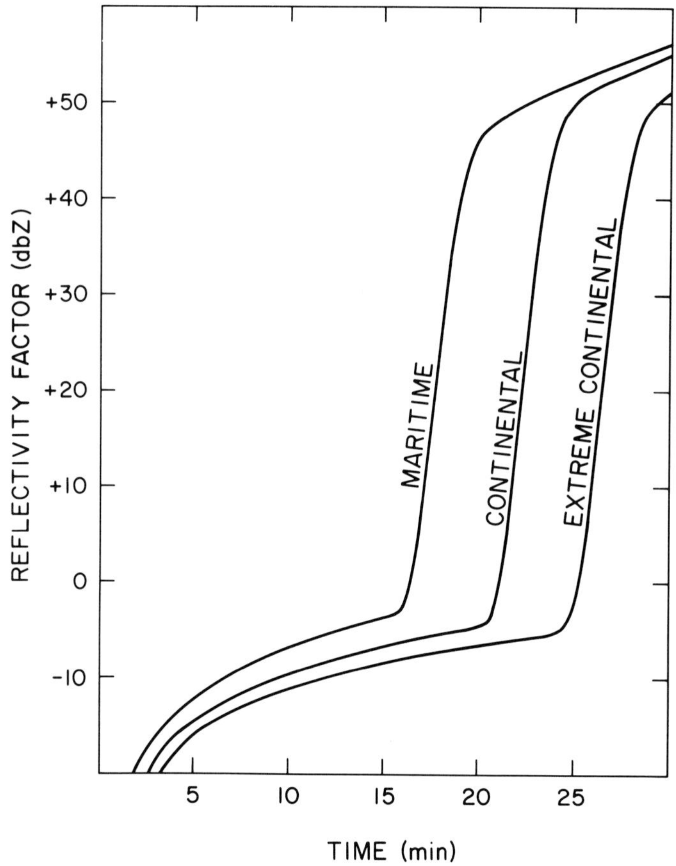

FIG. 9.10. Radar reflectivity factor (dBZ) as a function of time for different CCN spectra. The maritime CCN spectrum is $n = 291S^{0.46}$, yielding 247 droplets cm^{-3}; the continental CCN spectrum is $n = 710S^{0.55}$, yielding 457 droplets cm^{-3}; the extreme continental CCN spectrum is $n = 1700S^{0.82}$, yielding 670 droplets cm^{-3}. Cloud base temperature is 15°C with an updraft velocity of $3\,m\,s^{-1}$.

required for the +24°C cloud base; the 0°C cloud base required 42.1 min and the −5°C cloud base did not produce a 20 dBZ radar echo within 50 min. Since the number of droplets formed in each case was roughly the same, the influence of the higher LWC for the warmer clouds is clear.

The effects of updraft velocity are shown in Fig. 9.12 for updrafts ranging from $0.5\,m\,s^{-1}$ to $3\,m\,s^{-1}$. Although the weaker updrafts produce a 20 dBZ radar echo at a lower height (1.60 km), more time is required (53.3 min). The strongest updraft requires the least time (22.3 min) but the 20 dBZ radar echo occurs much higher in the updraft (4.02 km). Even though the stronger updrafts form more droplets ($457\,cm^{-3}$ versus $249\,cm^{-3}$), the forced condensational growth more than compensates for the smaller initial sizes.

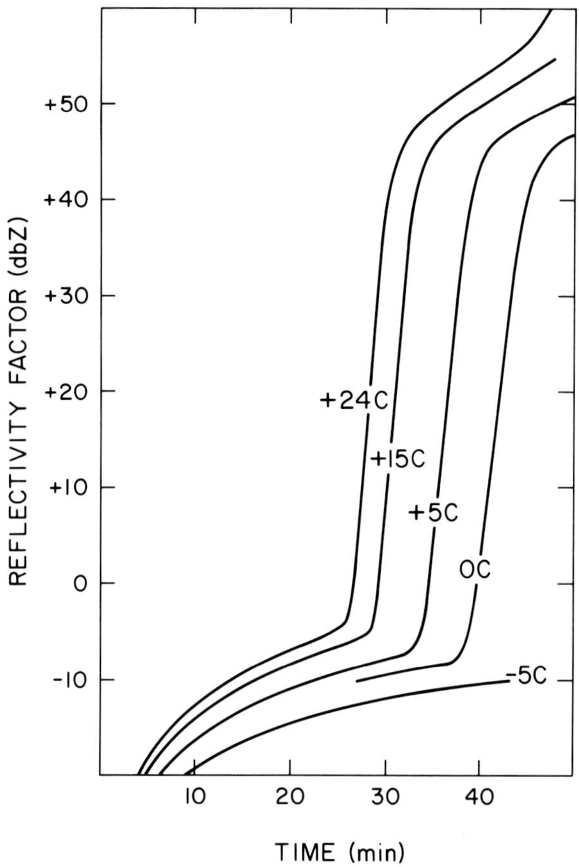

Fig. 9.11. Radar reflectivity factor (dBZ) as a function of time for different cloud base temperatures. The model is based on a maritime CCN spectrum with an updraft velocity of 1 m s^{-1}. A total of 169 droplets cm^{-3} were formed for the +24°C cloud base simulation; 202 droplets cm^{-3} were formed for the −5°C cloud base simulation.

9.4. Development of an exponential raindrop spectrum

The conversion from a log-normal cloud droplet size distribution to an approximately exponential raindrop distribution results from the collection of small droplets by large drops formed by coalescence. As the population of large drops increases, interactions between the large drops become more important. Many of these interactions between larger drops result in loss of mass due to breakup processes. This effectively limits the maximum radius of raindrops to about 2.5 mm and redistributes mass over the drop spectrum, resulting in an approximately exponential raindrop spectrum.

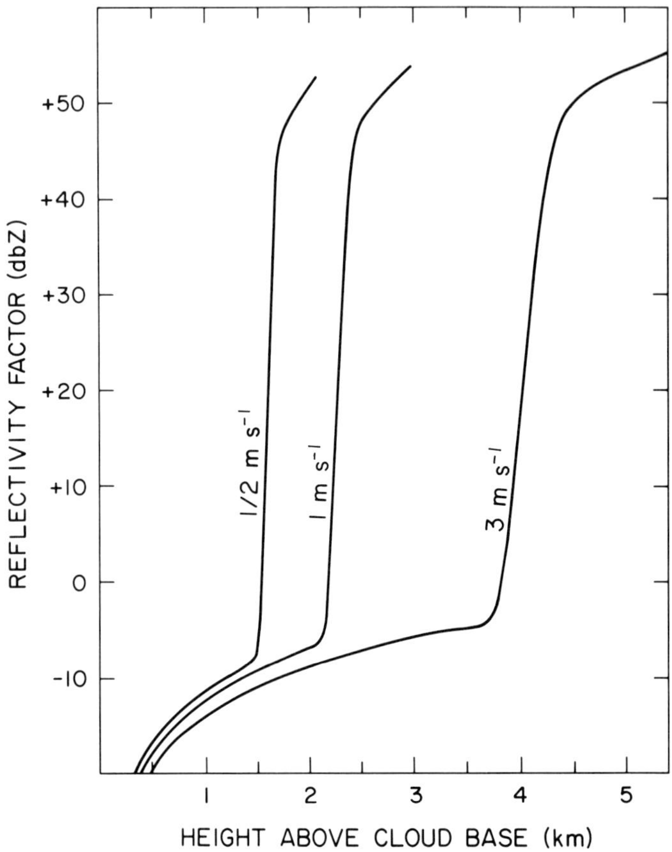

Fig. 9.12. Radar reflectivity factor (dBZ) as a function of time for different updraft velocities. The model is based on a continental CCN spectrum with a cloud base temperature of 15°C. The 0.5 m s^{-1}, 1.0 m s^{-1}, and 3.0 m s^{-1} updrafts yield droplet concentrations of 249 cm^{-3}, 322 cm^{-3}, and 457 cm^{-3}, and times to 20 dBZ reflectivity factor of 53.3 min, 38.0 min, and 22.3 min, respectively.

Exponential fits to raindrop spectra are termed Marshall–Palmer distributions after Marshall and Palmer (1948) who described observed raindrop distributions as exponentials of the form

$$n = n_o \exp(-\Lambda r)\,, \tag{9.2}$$

where $\Lambda = 82R^{-0.21}$, the rainfall rate (R) is in mm h^{-1} and the radius (r) is in centimeters.

9.4.1. *Mass transfer rates between drops of different sizes*

Following the development of a relatively small number of large drops through a stochastic coalescence process, the liquid water mass is rapidly transferred from the small cloud droplets to the raindrops, as shown in Fig. 9.13. At 16 min,

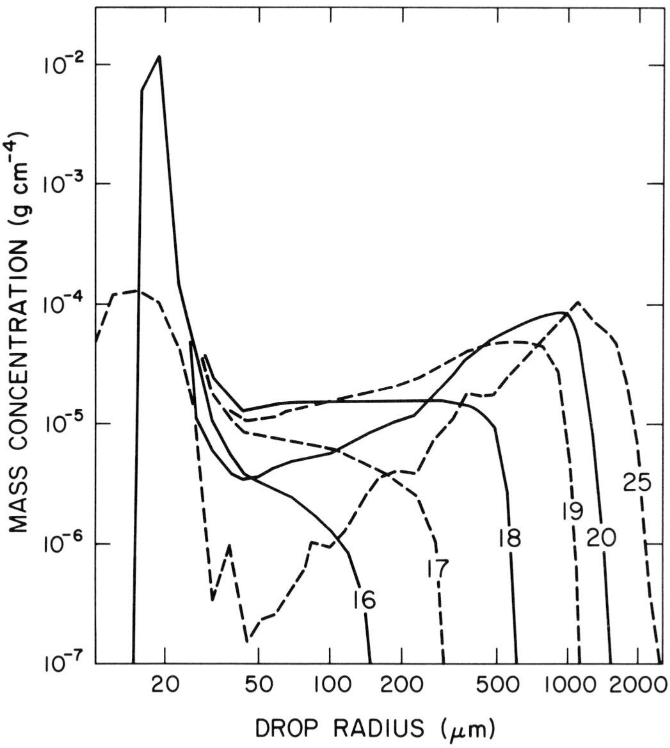

Fig. 9.13. Time evolution of the distribution of water mass in a cloud undergoing coalescence conversion. The simulation is for a maritime cloud with a cloud base temperature of 15°C and an updraft of 3 m s^{-1}. Note the rapid shift in the mass concentration from predominantly cloud droplets (<50 μm radius) to predominantly raindrops (>200 μm radius). From Young (1974).

the mass distribution is typical of cloud droplet spectra with a tail extending out to 150 μm radius. Three minutes later, the mass spectrum is bimodal with a pronounced peak near 500 μm radius. This shift from a cloud droplet spectrum to a raindrop spectrum via coalescence will be termed *coalescence conversion.*

The coalescence conversion eventually produces a steady state, quasi-exponential spectrum as shown in Fig. 9.14 for simulations using four different rainfall rates. The corresponding Marshall–Palmer (1948) distributions are shown for comparison. There is a tendency for the computed spectra to follow the Marshall–Palmer distributions. However, prominent peaks are evident at 0.12 mm, 0.44 mm, and 1.0 mm radius (Valdez and Young, 1985). Young (1975) points out that the exponential nature of the raindrop spectrum is a result of collisional breakup; an exponential spectrum does not result if the only breakup mode is disintegration of large (2.5 mm radius) drops.

An examination of the rate at which smaller drops are collected by larger drops over all possible drop–drop interactions is shown in Fig. 9.15. Figure 9.15a is based on the time period from 15 to 20 minutes, which represents the major

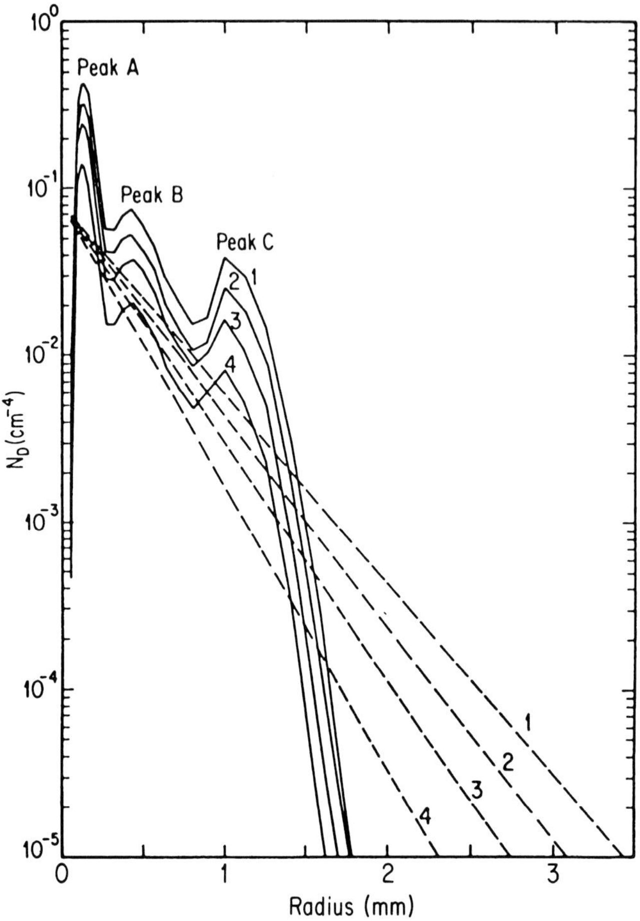

Fig. 9.14. Equilibrium drop size spectra after coalescence conversion. The logarithms of the number of drops after 30 min are plotted as a function of drop radius. The units for the ordinate are drops per unit volume per unit distance on the abscissa. The broken lines represent the corresponding Marshall–Palmer (1948) distributions. (1), LWC = 8.539 g m^{-3}, R = 207.5 mm h^{-1}; (2), LWC = 5.572 g m^{-3}, R = 134.5 mm h^{-1}; (3), LWC = 3.471 g m^{-3}, R = 81.5 mm h^{-1}; (4), LWC = 1.766 g m^{-3}, R = 41.0 mm h^{-1}; From Valdez and Young (1985). *Journal of Atmospheric Sciences*, **42**, 10. American Meteorological Society. Reproduced with permission

portion of the coalescence conversion. The predominant mass transfers are due to larger drops (70 μm to 1.0 mm radius) collecting cloud droplets (20 μm radius). This suggests that interactions between raindrops have a very small effect on the development of rain drops, i.e., simulations such as that described for Problem 9.4 reasonably reflect the initial growth of raindrops.

Once the steady state exponential spectrum has been established and most of the cloud droplets have been removed, the mass transfer rates for large drops collecting cloud droplets is greatly reduced, as shown in Fig. 9.15b. Still, there is an appreciable transfer of mass resulting from the collection of cloud droplets.

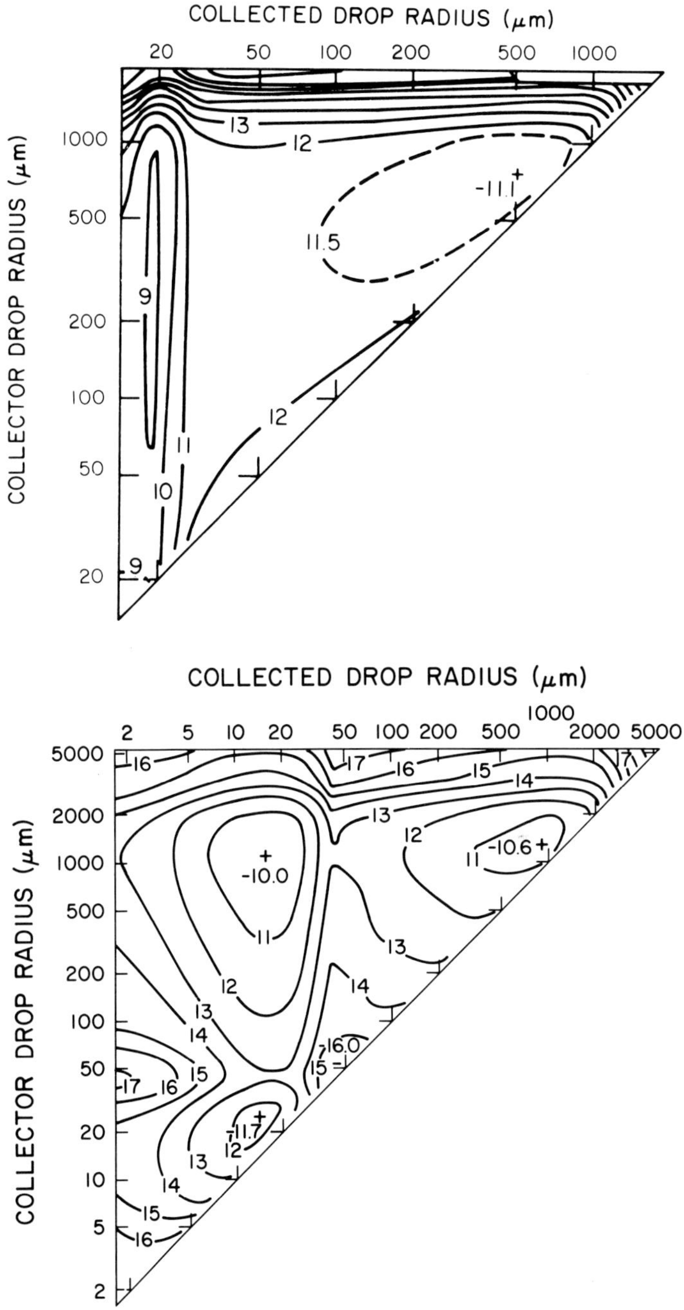

(a)

(b)

FIG. 9.15. Mass transfer rates between drops. The mass transfer from the smaller drops to the larger drops integrated over a 5 min period is shown. (a) The mass transfer between 15 and 20 min; (b) the mass transfer between 25 and 30 min. Isopleths represent the negative of the logarithm of the total mass transferred per cm^{-3} and per cm radius on each axis ($g\,cm^{-5}$). From Young (1974).

Clearly, mechanisms must be operating to provide a continuing supply of cloud droplets. In addition, it should be noted that the secondary maximum in mass transfer for 1.0 mm radius drops collecting somewhat smaller drops cannot result in a continued evolution to even larger drop sizes due to the steady state nature of the drop spectrum.

The mutual interactions between millimeter-sized drops generally result in filament breakup and the production of satellite droplets. Satellite production provides fragments that are usually larger than 50 μm radius. Thus, the relatively high mass transfer rates for millimeter-sized drops collecting drops between 70 μm and 200 μm represents the collection of previously produced satellite drops.

The production of droplets smaller than 30 μm radius results from the rise in water supersaturation discussed in § 9.4.2. The rise in water supersaturation activates additional CCN which grow rapidly by diffusion to sizes at which they are readily collected by the large drops. The half-life of newly activated CCN in this regime is less than 10 seconds. The removal of the droplets resulting from the newly activated CCN permits the supersaturation to rise still higher, activating additional CCN. Thus, the transfer of water vapor to the large drops occurs via an intermediate step involving the rapid diffusional growth of newly activated CCN and their subsequent collection by the large drops.

9.4.2. *Shaping the drop spectrum through births, deaths, and migration*

Three distinct portions of the drop spectrum may be identified based on the processes that form and remove drops. Drops smaller than 30 μm radius are formed in updrafts due to activation of CCN and rapid growth of the droplets and in downdrafts by evaporation from larger sizes. Drop fragments resulting from breakup processes are larger than this size and do not contribute directly to the population of droplets smaller than 30 μm radius.

Drops between 30 μm and 0.5 mm radius are formed primarily from breakup processes and are removed through collection by larger drops. Drops in this size range rarely grow significantly by collecting smaller droplets but remain at roughly the same size throughout their relatively short lifetime.

Drops larger than 0.5 mm radius have relatively long lifetimes and tend to meander up and down the size spectrum. Collisions with much smaller drops results in a slight growth; collisions with drops of comparable size often results in filament breakup and the loss of a small amount of mass. Since the number of much smaller drops is considerably greater than the number of comparable sized drops, drops in this size range tend to exhibit slow growth punctuated by sporadic breakup events that cause a sharp decrease in size. Only rarely does a drop undergo disk or bag breakup and lose its identity.

Valdez and Young (1985) describe the processes shaping the drop spectrum in terms of the births and deaths of drops and their migration to larger or smaller sizes. This may be expressed as

$$F_{\mathrm{b,i}} + F_{\mathrm{i}\uparrow,\mathrm{i}} - F_{\mathrm{e}\downarrow,\mathrm{i}} + F_{\mathrm{i}\downarrow,\mathrm{i}} - F_{\mathrm{e}\uparrow,\mathrm{i}} = F_{\mathrm{d,i}} \tag{9.3}$$

where $F_{b,i}$ and $F_{d,i}$ are the birth and death fluxes, $F_{i\uparrow,i}$ and $F_{i\downarrow,i}$ are the influxes from smaller and larger sizes, and $F_{e\downarrow,i}$ and $F_{e\uparrow,i}$ are the effluxes to smaller and larger sizes.

Drops in the size range from 30 μm to 0.5 mm may be described by the 'landed' model, in which the influx and efflux of drops is small compared to the birth and death fluxes. In this case, (9.2) reduces to

$$F_{b,i} = F_{d,i} = N_i \mathrm{Pr}_{d,i} = \frac{N_i}{L_i} \tag{9.4}$$

where $\mathrm{Pr}_{d,i}$ is the probability that a given drop will be removed in one second, N_i is the concentration of drops of size r_i, and L_i is the expected lifetime of the drops. This suggests that in the landed model, the equilibrium concentration of drops of a given size (N_i) is the product of the birth flux and the expected lifetime.

The birth flux is shown in Fig. 9.16 for a rainfall rate of 134.3 mm h^{-1}. This shows that the landed model is valid for radii smaller than 0.8 mm since the birth and death fluxes are virtually identical. The two smaller peaks evident in the computed spectra shown in Fig. 9.14 result from peaks in the birth flux rather than from peaks in the expected lifetime.

The expected lifetimes as a function of drop size for different rainfall rates are shown in Fig. 9.17. Note that the expected lifetime increases with increasing size for drops smaller than 1.0 mm radius. The slope of the equilibrium drop concentration in Fig. 9.14 is the result of a nearly exponential decrease in the birth flux coupled with an exponential increase in the expected lifetime. Thus, the slope of the drop concentration is less negative than the slope of the birth flux, as evident from (9.3).

For the landed model, there is little difference between the expected lifetime of the drops and their expected residence time within that size category. For drops larger than 0.4 mm radius, the residence times are less than the expected lifetime, indicating that the drops tend to migrate to another size category before losing their identity.

A 'nomadic' model must be used to describe the equilibrium shape of the drop spectrum for radii larger than 0.8 mm. Drops tend to accumulate in regions of the drop spectrum for which the net flux to larger sizes $F_{n,i} = F_{e\uparrow,i} - F_{i\downarrow,i}$ decreases as drop size increases (radii from 0.7 mm to 1.0 mm in Fig. 9.16). Thus, the death flux must be greater than the birth flux to maintain a balance between the creation of drops (births plus immigration) and the removal of drops (deaths plus emigration). The death flux increases since N_i increases, resulting in a peak near 0.5 mm radius.

9.4.3. *Relating computed spectra to observations*

A variety of observations of drop spectra are examined by Steiner and Waldvogel (1987) for the existence of peaks in the distributions, and compared to several

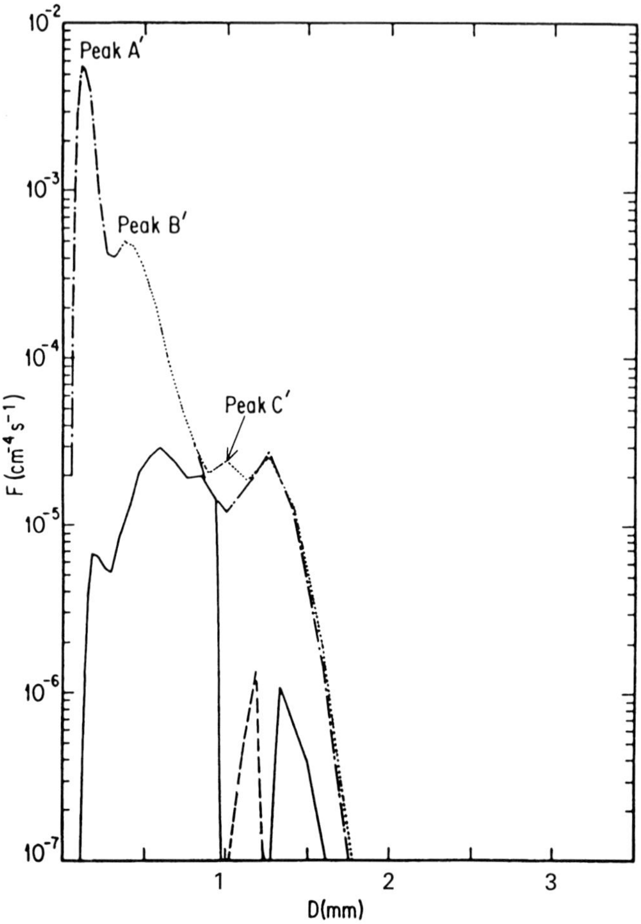

Fig. 9.16. (- · - · -) Birth flux ($F_{b,i}$), (······) death flux ($F_{d,i}$), and net (—), (- - -) positive and negative growth fluxes ($F_{n,i}$) for a rainfall rate of 134.3 mm h^{-1}. LWC = 5.572 g m^{-3}. The discontinuity in the birth and death fluxes near peak B' is meant to indicate that the two fluxes are virtually the same for radii <0.08 cm. From Valdez and Young (1985). *Journal of Atmospheric Sciences*, **42**, 10. American Meteorological Society. Reproduced with permission.

model results. They find good evidence for peaks at radii of 0.35 mm, 0.55 mm, and 0.95 mm, and some evidence for a peak at 1.6 mm, although they note that their observed spectra probably do not represent equilibrium spectra. The peak at 0.95 mm is reproduced by several models; the models appear to merge the peaks at 0.35 mm and 0.55 mm. None of the models reproduced the peak at 1.6 mm. The disdrometers used to obtain the drop spectra were unable to resolve drop sizes smaller than 0.2 mm.

The model peaks at 0.12 mm and near 0.4 mm result from maxima in the birth flux. Brown (1988) shows that the smaller peak is caused by fragments resulting from filament breakup. The peak near 0.4 mm appears to be a

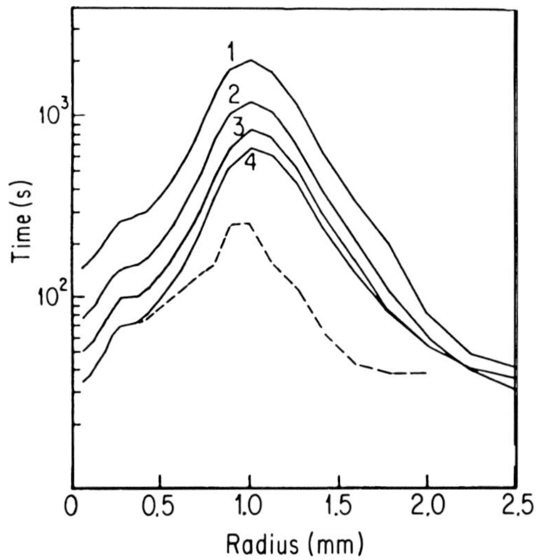

FIG. 9.17. Expected lifetimes for four equilibrium drop size distributions. Residence times for a rainfall rate of (---) 207.5 mm h^{-1} are shown for comparison. (1), R = 41.0 mm h^{-1}; (2), R = 81.5 mm h^{-1}; (3), R = 134.5 mm h^{-1}; (4), R = 207.5 mm h^{-1}. From Valdez and Young (1985). *Journal of Atmospheric Sciences*, **42**, 10. American Meteorological Society. Reproduced with permission

composite, resulting from sheet breakup producing fragments near 0.32 mm radius and disk breakup producing fragments near 0.55 mm radius.

List and McFarquhar (1990) model the evolution of the drop spectrum in a one-dimensional model using a pulse input of rain rather than a steady input, simulating convective precipitation. Due to different sedimentation rates for the different sized drops, instantaneous spectra do not exhibit the three-peaked distribution. The three-peaked nature of the distribution is shown to result from integrating the spectrum over time.

9.5. The rise of water supersaturation in updrafts

The water supersaturation in an updraft represents a balance between the rate at which water vapor is being made available for condensation due to adiabatic cooling and the rate at which the water vapor in excess of saturation can be extracted by the growing particles.[8] For a given amount of liquid water, dispersal among numerous, small droplets provides more efficient removal of the excess water vapor than does its dispersal among fewer, large drops. Thus, as precipitation particles appear in an updraft, the supersaturation with respect to water may be expected to increase in conjunction with the removal of the small droplets by coalescence.

The rise of water supersaturation in an updraft may exceed the peak supersaturation achieved just above the cloud base during the activation of CCN.

[8] This neglects changes in the excess water vapor required to change the supersaturation.

This would activate additional CCN that were too small to be activated near the cloud base. The activation of additional CCN might be expected to limit the rise of supersaturation. However, the short lifetime of these newly activated droplets limits their ability to restrict the rise in water supersaturation. Even the entrainment of additional CCN only reduces the rate at which the water supersaturation rises.

9.5.1. *Derivation of the steady state supersaturation*

The derivation of a steady state supersaturation[9] is made under the assumption that the rate at which excess water vapor is being made available for condensation in the updraft is exactly balanced by the rate at which the excess water vapor is being extracted by the growing particles. The actual supersaturation may be expected to lag behind its steady state value. The earliest derivation of the steady state supersaturation is due to Squires (1952). In the following, a derivation of the steady state supersaturation as a function of the updraft velocity and the drop spectrum will be provided.

The derivation of the steady state supersaturation is based on three equations. These are the hydrostatic equation

$$\frac{R_d T\, dP}{P} = -g\, dZ\,, \tag{9.5}$$

the pseudo-adiabatic equation

$$-g\, dZ = c_p\, dT + L_v\, dw_s\,, \tag{9.6}$$

and the Clausius–Clapeyron equation

$$dT = \frac{R_v T^2}{L_v - R_v T}\,\frac{d\rho_s}{\rho_s}. \tag{9.7}$$

Equation (9.6) assumes that L_v is constant; (9.7) differs from the conventional expression of the Clausius–Clapeyron in being expressed in terms of ρ_s rather than e_s.

It is necessary to express the pseudo-adiabatic equation (9.6) in terms of ρ_s rather than w_s in order to relate the growth of the particles ($d\rho_s/dt$) to the updraft velocity (dZ/dt). The saturation mixing ratio (w_s) may be approximated as ρ_s/ρ_a where the density of air is approximated as

$$\rho_a \approx \frac{P}{R_d T}.$$

[9] The supersaturation is steady-state in the sense that its value balances the rate at which water vapor is being made available for condensation in the updraft. In actuality, the derivation assumes that the vapor density excess over water saturation is constant.

Now, the change in saturation mixing ratio can be expressed as

$$
\begin{aligned}
dw_s &= \frac{1}{\rho_a}\, d\rho_s - \frac{\rho_s}{\rho_d^2}\, d\rho_a \\
&= \frac{R_d T}{P}\, d\rho_s - \frac{\rho_a R_d T}{P^2}\, dP + \frac{\rho_s R_d}{P}\, dT .
\end{aligned}
$$

Substituting from the hydrostatic equation (9.5) for dP and from the Clausius–Clapeyron equation (9.7) for dT, the change in saturation mixing ratio becomes

$$
dw_s = \frac{\rho_s g}{P}\, dZ + \left[\frac{R_d T}{P} + \frac{R_d}{P} \left(\frac{R_v T^2}{L_v - R_v T} \right) \right] d\rho_s .
$$

Now, the pseudo-adiabatic equation can be expressed in terms of $d\rho_s$ and dZ as

$$
-g \left(1 + \frac{L_v \rho_s}{P} \right) \frac{dZ}{dt} = \frac{T}{L_v - R_v T} \left(\frac{c_p R_v T}{\rho_s} + \frac{L_v^2 R_d}{P} \right) \frac{d\rho_s}{dt} . \tag{9.8}
$$

The rate at which water vapor is being extracted from the updraft is obtained by summing the condensation rate over all of the drops using the combined growth equation (5.11),

$$
\frac{d\rho_s}{dt} = - \sum n \frac{dm}{dt} = \frac{-4\pi (S - 1) \sum nr}{\dfrac{L_v^2}{K R_v T^2} + \dfrac{1}{d\rho_s}} . \tag{9.9}
$$

Now setting the rate that water vapor is being made available for condensation in the updraft (9.8) and the rate that water vapor is being extracted by condensation (9.9) and letting $U = dZ/dt$,

$$
\frac{g U (L_v - R_v T) \left(1 + \dfrac{L_v \rho_s}{P} \right)}{T \left[\dfrac{c_p R_v T}{\rho_s} + \dfrac{L_v^2 R_d}{P} \right]} = \frac{4\pi (S - 1) \sum nr}{\dfrac{L_v^2}{K R_v T^2} + \dfrac{1}{D \rho_s}} .
$$

Solving for the steady state supersaturation $(S - 1)$

$$
S - 1 = \frac{A U}{\sum nr} \tag{9.10}
$$

where

$$A = \frac{g(L_v - R_v T)}{4\pi T}\left[\frac{\left(1 + \frac{L_v \rho_s}{P}\right)\left(\frac{L_v^2}{K R_v T^2} + \frac{1}{D \rho_s}\right)}{\frac{c_p R_v T}{\rho_s} + \frac{L_v^2 R_d}{P}}\right].$$

Values for A as a function of temperature and pressure are provided in Fig. 9.18 for convenience in computing the steady state supersaturation. The temperature dependence of A is relatively small although it does exhibit a marked decrease with decreasing pressure.

9.5.2. *Estimates of water supersaturations in updrafts*

As a rough estimate of the magnitude of the water supersaturations that may be expected in updrafts, consider the case of a 3 m s^{-1} updraft with 1000 drops cm^{-3} with a mean radius of 10 μm. From Fig. 9.18, $A = 2.13 \times 10^{-6}$ s cm^{-3} for a temperature of 0°C and a pressure of 700 mbar. The steady state supersaturation is 0.06%. This value is quite reasonable for a mature cloud droplet spectrum based on the results presented for the activation of a population of CCN in § 9.2.

Now consider a cloud with the same water content but with only one drop per cm^3. The term Σnr is now 0.01 cm^{-2} compared to 1.0 cm^{-2} for the previous case. The steady state supersaturation is 6.4%. Clearly, this would result in the activation of additional CCN.

The relation between the total drop concentration (N) and the supersaturation as a function of time in a 3 m s^{-1} updraft is shown in Fig. 9.19. At 15 minutes, the drop concentration is 245 cm^{-3}. The supersaturation calculated from (9.10) is 0.20%;[10] the corresponding model supersaturation is 0.08%. The onset of coalescence conversion around 17 minutes results in a sharp decrease in the total drop concentration to 35 cm^{-3} at 20 minutes. The calculated supersaturation is now 1.59% compared to the model value of 0.84%. At 25 minutes with a drop concentration of 24 cm^{-3}, the calculated and model supersaturations are 4.29% and 3.69%, respectively.

The sharp rise in the drop concentration in Fig. 9.19 just after 20 minutes indicates the activation of additional CCN. These drops are quickly removed by the large drops present. The total number of drops decreases again as the supersaturation rises, since this forces the newly activated droplets to grow more quickly to sizes where they can be collected.

The rate of collection of the newly activated droplets by the large drops can be estimated as follows. At 25 minutes, the model drop distribution has 3.2 drops

[10] These values are calculated using the could base temperature and pressure. Since A decreases as pressure decreases with time, the calculated supersaturation should be less than these values.

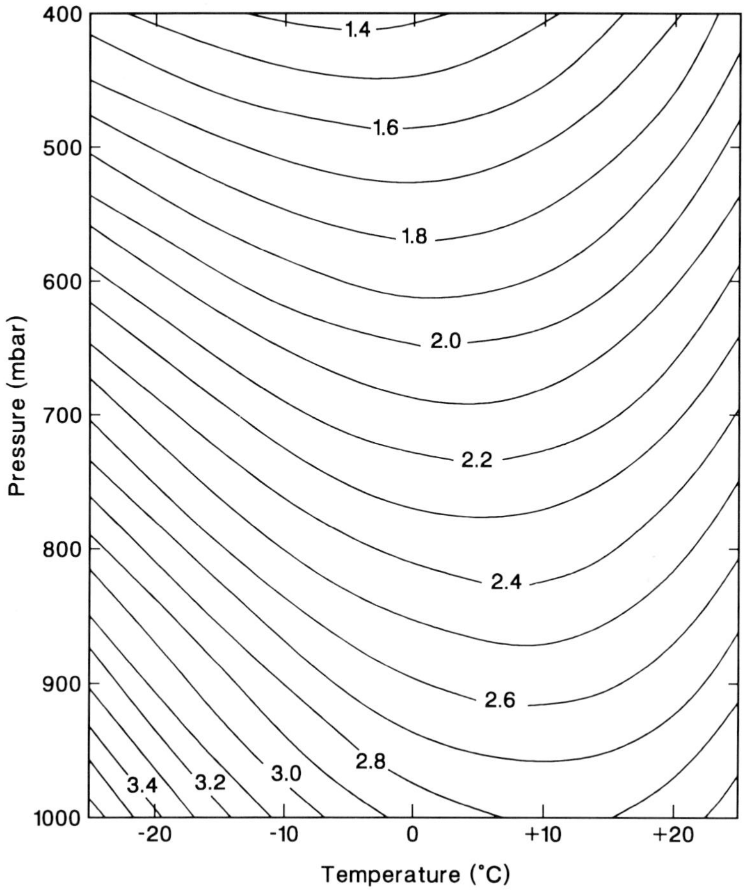

FIG. 9.18. Values for the coefficient A used in computing the steady state supersaturation as a function of temperature and pressure. Values for A are in units of 10^{-6} s cm^{-3} as used in (9.9).

ℓ^{-1} larger than 1 mm radius. If one assumes a mean radius of 1.5 mm for these drops with a fall speed of 9 m s^{-1} and a collection efficiency of 50% for 5 μm radius droplets, the sweep rate per drop is 31.8 cm^3 s^{-1}. Since the total concentration is 0.0032 cm^{-3}, the fractional sweep rate is 0.102 s^{-1}, or roughly 10% of the volume is swept every second. When considering the more numerous, smaller raindrops, it is clear that the removal of newly activated droplets is quite rapid, allowing the water supersaturation to rise at nearly 1% per minute in this case.

Graupeln and other large ice particles may be expected to have a similar affect on the water saturation. The ice supersaturation corresponding to water saturation increases roughly 1% for every 1°C decrease in temperature. Thus, at −10°C at water saturation, the ice supersaturation is roughly 10%. In a vigorous updraft of 5 m s^{-1}, this supersaturation could be achieved within 8 minutes,

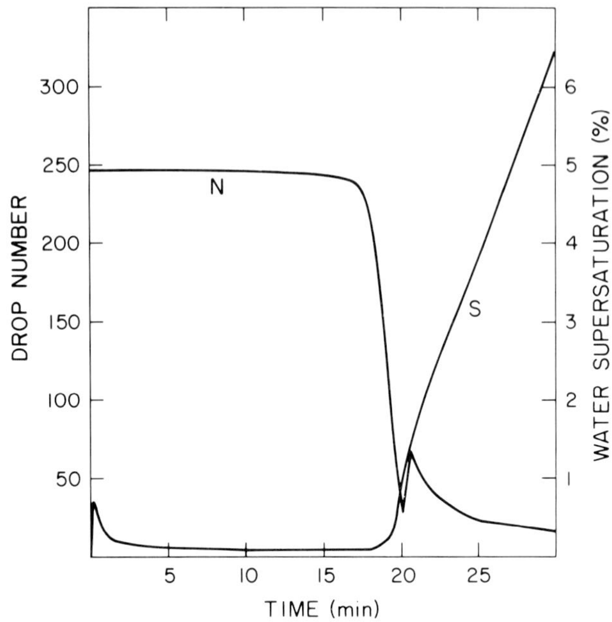

Fig. 9.19. The relation between total drop concentration (N) and water supersaturation is shown as a function of time (height) in a $3\,\mathrm{m\,s^{-1}}$ updraft. The cloud base temperature is 15°C; the CCN activity spectrum is $n = 291S^{0.46}$. From Young (1974).

starting from ice saturation. Generally, in a mixed phase cloud, the saturation ratio is closer to water saturation than to ice saturation, permitting significant water supersaturations to occur much more quickly than 8 minutes.

9.5.3. *Possible consequences of high water supersaturations*

There are two possible consequences of high water supersaturations that have significant implications for the evolution of the cloud. When the supersaturation increases, this represents water vapor that has not been condensed, associated with a decrease in the amount of latent heat that has been released. Second is the effect of high water supersaturations on the activation of ice nuclei.

The relationship between supersaturation and the release of latent heat was first noted by McDonald (1962) with respect to the activation of CCN near the cloud base. McDonald suggested that the rise in supersaturation near the cloud base during the initial formation of cloud droplets results in a lapse rate intermediate between the dry and pseudo-adiabatic lapse rates as the supersaturation increases. Above the level of peak supersaturation, McDonald hypothesized that the additional release of latent heat results in a cooling rate less than the pseudo-adiabatic rate, and could even produce a slight warming (negative lapse rate) as the parcel rose.

Das (1969) derives the lapse rate for updrafts and downdrafts in the presence of condensation–evaporation centers, extending this concept to ice particles as well as water drops. Das shows that the actual lapse rate could vary from negative for very high concentrations of very small droplets under certain conditions, to very close to a dry adiabatic lapse rate for very low concentrations of larger drops.

This variation of the lapse rate in response to changes in the particle size distribution may be expected to produce a modulation of the updraft velocity. The presence of large numbers of precipitation particles would allow the water supersaturation to rise, thereby reducing the latent heat release, and decreasing the buoyancy of the updraft. Conversely, a reduction in the numbers of precipitation particles should allow the supersaturation to decrease as newly activated CCN are permitted to grow, releasing additional latent heat, and increasing the buoyancy of the updraft.

Vertically-pointing Doppler radar studies of a hailstorm by Battan (1975) are suggestive of such a relationship between precipitation and modulations of the updraft. Time versus height profiles of updraft velocity and radar reflectivity factor (Z) are shown in Fig. 9.20. Updraft cores are shown by broken lines. The correspondence between acceleration of the updraft above zones of decreasing Z, and deceleration of the updraft above zones of increasing Z appears to be reasonably good. The two-dimensional nature of the cross-section does not properly reflect the three-dimensional nature of the updraft and these observations are merely suggestive. The observed rapid accelerations and decelerations of the vertical velocities are consistent with this explanation. The unloading of the updraft as precipitation falls out of the parcel would have a similar but smaller effect.

The presence of regions of high water supersaturation may also be expected to enhance the formation of ice crystals by condensation freezing nucleation. As discussed in § 4.6.3, provided the ambient temperatures are colder than the threshold for condensation freezing, the rise in ambient water supersaturation could result in rapid formation of high concentrations of ice crystals at temperatures much warmer than could be accounted for through deposition nucleation, or at rates much higher than could be accounted for by contact freezing nucleation.

9.6. The role of coalescence and breakup in rainshafts

A simple calculation can suffice to show that most raindrops would survive a fall from cloud base to the ground if evaporation were the only mechanism for decreasing the drop sizes (Problem 5.10). If the raindrop concentration is reasonably high, collisional breakup is sufficient to dominate evaporation as the main mechanism for mass loss by large drops. The production of small drops by the breakup process enhances the evaporation process in downdrafts in a manner analogous to the rise of water supersaturations in updrafts.

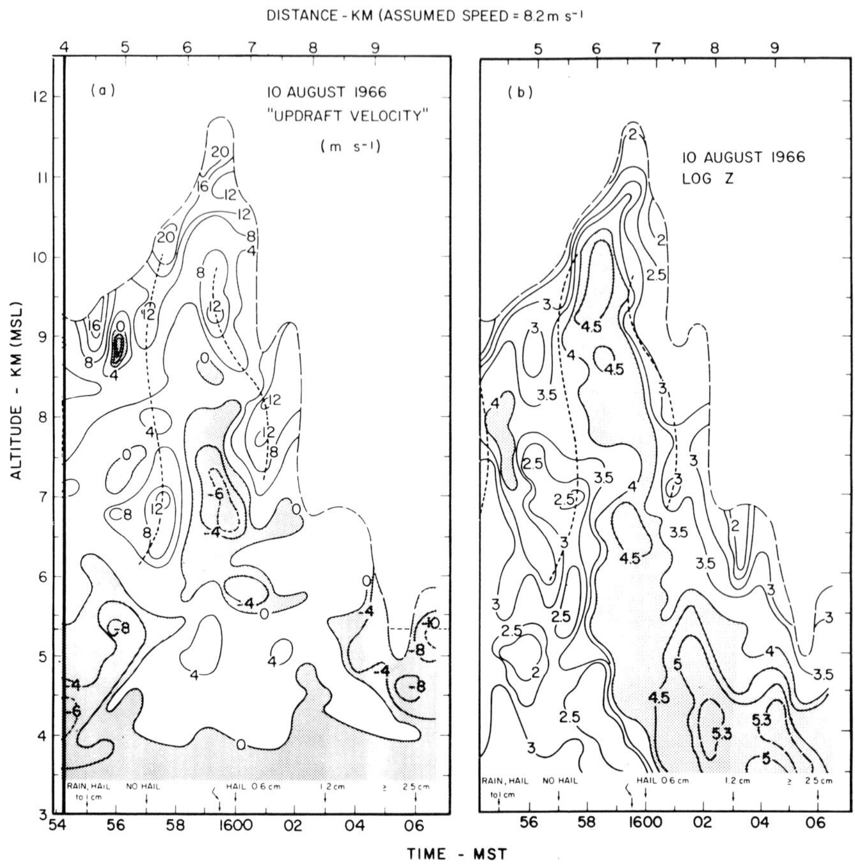

Fig. 9.20. Profiles of (a) updraft velocity and (b) reflectivity factor from a vertically pointing Doppler radar. The updraft velocities are estimated from the measured Doppler velocity; the radar reflectivity (Z in $mm^6\ m^{-3}$) is given as log Z. Updraft cores are indicated by dashed lines. Note the associations between decreasing Z and acceleration of the updrafts (and vice versa). From Battan (1975). *Journal of Atmospheric Sciences*, **14**, 2. American Meteorological Society. Reproduced with permission

9.6.1. *Evaporative cooling in downdrafts*

The initiation and maintenance of a downdraft generally results from mass loading and evaporative cooling. Studies such as those by Brown (1979) and Leary (1980) demonstrated the role of evaporative cooling, but parameterized the evolution of the drop spectrum by assuming it to be exponential (Marshall–Palmer). Leary notes that higher rainfall rates are associated with cooler and moister downdrafts. Emanuel (1981) extends these concepts using a thermodynamic approach, assuming that the evaporation is of cloud droplets rather than of raindrops.

Srivastava (1985, 1987) presents a simple model of evaporatively driven downdrafts, ignoring the effects of coalescence and breakup. He notes that strong downdrafts occur when the environmental lapse rate is close to dry adiabatic ($\Gamma_{env} > 9.5°C\,km^{-1}$) and are more likely when there is a high rainwater content. The calculated relative humidities in the simulated downdrafts were noted to be substantially lower than observed values.

Consider the mass lost due to evaporation from a 1 mm radius drop in a fall of 100 m in an environment at 50% relative humidity and 0°C. The drop's fall speed is $706\,cm\,s^{-1}$ and the ventilation factor is 8.426. Under these conditions, (5.11) may be written

$$\frac{dm}{dt} = 6.964 \times 10^{-5}\, r\,(S - 1)$$

and, expressing radius in terms of mass,

$$\frac{dm}{m^{1/3}} = 4.320 \times 10^{-5}\,(S - 1)\,dt = 1.5\,(m^{2/3} - m_o^{2/3})\,.$$

From this, the fractional mass lost in 100 m of fall is 1.8%.

Now consider the mass lost due to breakup processes during a fall over the same height. These calculations are summarized in Table 9.3 for a Marshall–Palmer (1948) size distribution with a rainfall rate of100 $mm\,h^{-1}$. The fall speeds are calculated using the formulation in § 7.2.1; the collision efficiencies are assumed to be geometric. The breakup probabilities are taken from Fig. 7.10; the fractional mass loss per breakup event is based on Brazier-Smith et al. (1973) for filament breakup as

$$\frac{m_{sat}}{m_L + m_S} = \frac{0.12\gamma^3}{(1 + \gamma^3)^2}, \tag{9.11}$$

where $\gamma = r_L/r_S$.

Table 9.3. Mass lost from a 1 mm radius raindrop due to breakup processes in a fall of 100 m. Drop concentrations are based on a Marshall–Palmer distribution with a rainfall rate of 100 $mm\,h^{-1}$. Falls speeds are based on the treatment in § 7.2.1; the number of collisions is based on (7.6). The breakup probability is taken from Fig. 7.10. The mass lost per drop is based on (9.10).

Radius (mm)	Number (ℓ^{-1})	Fall speed ($cm\,s^{-1}$)	Collisions	Breakup probability (%)	Mass lost per drop (%)	Mass lost total (%)
1.5 to 2.0	0.38	919	0.263	90	1.7	0.4
1.2 to 1.5	1.23	819	0.328	68	2.6	0.6
1.0 to 1.2	2.63	747	0.212	35	2.9	0.2
0.8 to 1.0	4.91	661	0.356	40	2.9	0.4
0.6 to 0.8	9.16	559	1.736	65	2.3	2.6
0.4 to 0.6	17.09	432	4.700	80	1.2	4.5
0.3 to 0.4	26.95	307	8.743	80	0.5	3.5
0.2 to 0.3	36.82	217	12.550	65	0.2	1.6

The fractional mass lost according to Table 9.3 is 13.8%, more than seven times the mass lost due to evaporation. It should be noted this is an underestimate of the mass lost due to drop breakup processes since it only considers filament breakup and does not quite cover the entire size range that participates in drop breakup. The fractional mass loss for a given drop due to breakup processes is roughly proportional to the rainfall rate. This implies that the rate of production of small drop fragments is roughly proportional to the square of the rainfall rate.

Tzivion et al. (1989) and Feingold et al. (1991) include the effects of drop coalescence and breakup in a model of a rainshaft. Figure 9.21 shows the shift in drop sizes from the larger to smaller drops as evaporation of the smaller drops reduces the ability of the large drops to maintain growth rates approximately equal to the loss of mass due to breakup. Note that the smaller drop sizes could not have fallen from the top of the rainshaft in the allotted time, i.e., these drops have been produced by the breakup process. Tzivion et al. (1989) note that starting with an initially narrow spectrum of smaller drop sizes results in spectral broadening through coalescence, as would be expected.

Feingold et al. (1991) compare simulations both with and without coalescence and breakup processes operating. For an initial liquid water content of $2\,g\,m^{-3}$ and an exponential raindrop spectrum, the maximum downdrafts are

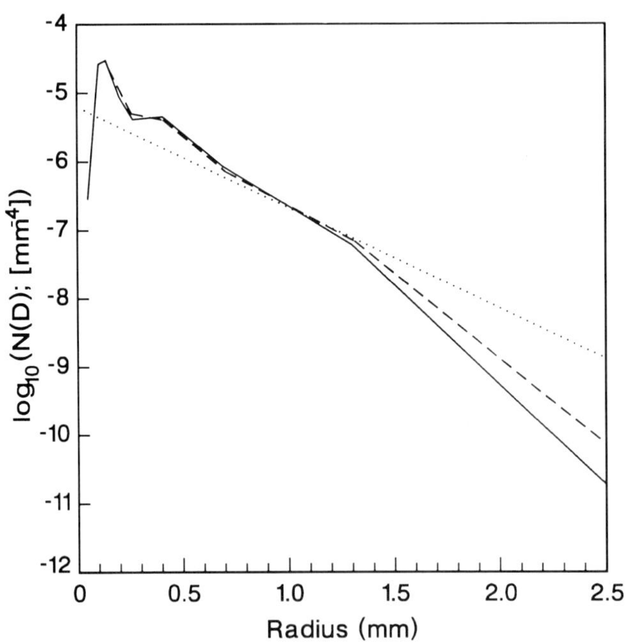

Fig. 9.21. Evolution of a drop spectrum in a subsaturated rainshaft including the effects of coalescence and breakup. The spectrum at the top of the rainshaft is Marshall–Palmer with a rainfall rate of $110\,mm\,h^{-1}$. (······), 0 km; (- - -), 1 km; (–), 2 km; $R = 110\,mm\,h^{-1}$; $t = 24$ min. From Tzivion et al. (1989). *Journal of Atmospheric Sciences*, **46**, 21. American Meteorological Society. Reproduced with permission

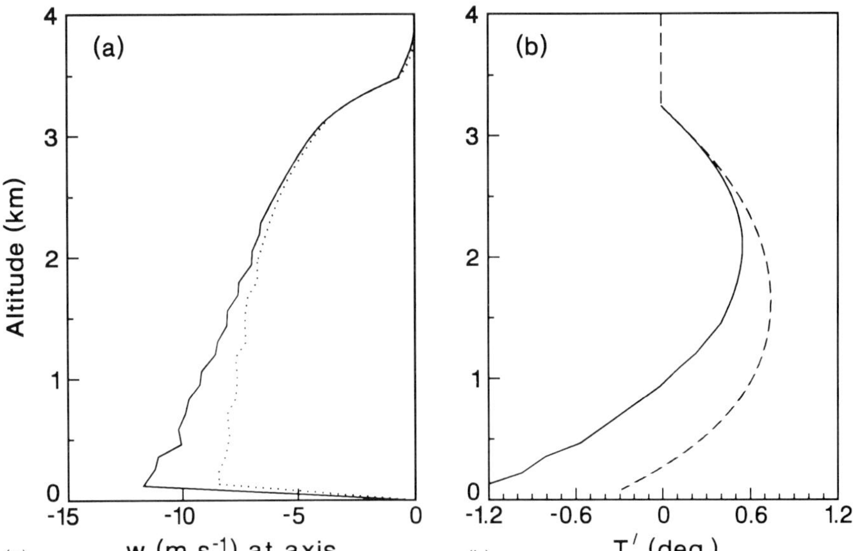

(a) (b)

Fig. 9.22. The effects of coalescence and breakup on (a) the velocity and (b) the temperature of a downdraft; (—) with, and (······), (- - -) without coalescence and breakup. t = 20 min. The environmental lapse rate is 9.0°C km^{-1} with the relative humidity increasing from 15% at altitude = 0 km to 100% at 3.2 km. The initial liquid water content is 2 g m^{-3} distributed as a Marshall–Palmer spectrum. From Feingold et al. (1991). *Journal of Atmospheric Sciences*, **48**, 12. American Meteorological Society. Reproduced with permission

roughly 50% stronger and the downdraft temperatures as much as 1°C colder when coalescence and breakup processes are included (see Fig. 9.22). They find for 2 g m^{-3}, the initiation of the downdraft is due to mass loading but its intensification is due to evaporative cooling. For higher liquid water contents, the mass loading becomes dominant and the effects of coalescence/breakup are smaller. For lower liquid water contents, mass loading is less important but the enhancement in the evaporation rate due to breakup is substantially reduced. Thus, for lower liquid water contents, the effects of coalescence/breakup are smaller and penetrating downdrafts are less likely.

9.6.2. *Formation of virga and scud clouds*

The role of breakup processes in reducing drop sizes in rainshafts is strongly dependent on the rainfall rate, whereas the evaporation of drops is relatively insensitive to the rainfall rate. For very low rainfall rates, breakup processes will be negligible and most of the raindrops will survive to reach the ground. As higher rainfall rates are considered, the rate of reduction of the drop mass through breakup processes becomes greater. This suggests that virga are associated with higher rainfall rates.

Figure 7.10 suggests that drops larger than 500 μm radius are more likely to lose mass through breakup processes. Once the drop radius becomes smaller than 200 μm, further reductions in size through drop breakup are unlikely and evaporation is responsible for the remaining size reduction. As the drop becomes smaller, it falls more slowly while its change of radius (dr/dt) increases. Thus, a plot of drop size as a function of height shows relatively little change until near the end of the drop's fall (Problem 9.3).

The formation of scud clouds (stratus fractus or cumulus fractus) in rainshafts may also be attributed to the breakup processes operating in rainshafts. Since rainshafts tend to be regions of sinking rather than rising air, formation of cloud cannot be attributed to adiabatic cooling. The evaporation from falling raindrops cannot produce supersaturated conditions unless the drop is considerably warmer than its environment.[11] The ragged appearance of scud clouds suggests that these are regions in which the humidity is close to saturation, allowing the slow evaporation and survival of breakup fragments. Breakup fragments would tend to accumulate in these regions, eventually reaching a balance between fragment production and removal by the falling raindrops.

Problems

9.1* Consider two drops with radii of 12 μm and 18 μm. Determine their mutual collection kernel (7.1).

Consider that the two drops grow in an updraft. Determine their mutual collection kernel when the larger drop has grown to 22 μm radius.

9.2** Consider a water cloud consisting of 700 drops cm^{-3} with a mean radius of 10 μm, with an updraft velocity of $2\,m\,s^{-1}$. What is the steady state supersaturation? Assume an ambient temperature of 0°C and a pressure of 700 mbar.

Consider this cloud undergoing coalescence conversion to a Marshall–Palmer size spectrum with the same liquid water content. Assume there are no drops smaller than 50 μm radius and the maximum radius is 2.5 mm. What is the steady state supersaturation now?

Computer problems

9.3** Modify the program used in Problem 5.10 to generate a plot of drop size as a function of height.

9.4** Consider the growth of 1000 individual drops as a stochastic process using a Monte Carlo model, as outlined in the text. Consider a cloud of 15 μm radius droplets in a concentration of 150 cm^{-3}. Start each drop as a 'double drop' at $t = 0$ and generate its growth history through 15 minutes.

(a) Plot the growth history (number of collections versus time) for the most favored drop (largest after 15 minutes), the second, the fifth, and the tenth most favored drops.

[11] Very small volumes in the vicinity of drops may experience supersaturated conditions due to the positive curvature of the saturation vapor density versus temperature curve.

(b) What fraction of the 1000 drops do not collect any drops? What fraction collect two or more drops? How do these compare with the predictions of the quasi-stochastic model?

(c) Repeat the simulations using drop concentrations of 120, 130, 140, and 160 drops cm^{-3}. Compare the maximum drop sizes after 15 minutes for each concentration. Plot the number of drops that undergo 100 collections and 1000 collections as a function of drop concentration.

(d) Repeat the simulations using drop radii of 14 μm and 16 μm. Compare the maximum drop sizes after 15 minutes for each drop size (14 μm, 15 μm, and 16 μm radius). Plot the number of drops that undergo 100 collections and 1000 collections as a function of initial drop size.

REFERENCES

Baker, M.B., and Latham, J., 1979, 'The evolution of droplet spectra and the rate of production of embryonic raindrops in small cumulus clouds,' *J. Atmos. Sci.* **36**, 1612.

Baker, M.B., Corbin, R.G., and Latham, J., 1980, 'The influence of entrainment on the evolution of cloud droplet spectra,' *Quart. J. R. Meteorol. Soc.* **106**, 581.

Battan, L.J., 1973, *Radar observation of the atmosphere*. Chicago: University of Chicago Press.

Battan, L.J., 1975, 'Doppler radar observations of a hailstorm,' *J. Appl. Meteorol.* **14**, 98.

Bower, K.N., and Choularton, T.W., 1988, 'The effects of entrainment on the growth of droplets in continental cumulus clouds,' *Quart. J. R. Meteorol. Soc.* **114**, 1411.

Brazier-Smith, P.R., Jennings, S.G., and Latham, J., 1973, 'Raindrop interactions and rainfall rates within clouds,' *Quart. J. R. Meteorol. Soc.* **99**, 260.

Brown, J.M., 1979, 'Mesoscale unsaturated downdrafts driven by rainfall evaporation: A numerical study,' *J. Atmos. Sci.* **36**, 313.

Brown, P.S., Jr., 1988, 'The effects of filament, sheet, and disk breakup upon the drop spectrum,' *J. Atmos. Sci.* **45**, 712.

Das, P., 1969, 'The thermodynamic equation in cumulus dynamics,' *J. Atmos. Sci.* **26**, 399.

Emanuel, K.A., 1981, 'A similarity theory for unsaturated downdrafts within clouds,' *J. Atmos. Sci.* **38**, 1541.

Feingold, G., Levin, Z., and Tzivion, S., 1991, 'The evolution of raindrop spectra. Part III. Downdraft generation in an axisymmetrical rainshaft model,' *J. Atmos. Sci.* **48**, 315.

Fitzgerald, J.W., 1972, *A study of the initial phase of cloud droplet growth by condensation*. Ph.D. Thesis, University of Chicago.

Leary, C.A., 1980, 'Temperature and humidity profiles in mesoscale unsaturated downdrafts,' *J. Atmos. Sci.* **37**, 1005.

List, R., and McFarquhar, G.M., 1990, 'The evolution of three-peak raindrop size distributions in one-dimensional shaft models,' *J. Atmos. Sci.* **47**, 2996.

Marshall, J.S., and Palmer, W. McK., 1948, 'The distribution of raindrops with size,' *J. Meteorol.* **5**, 165.

McDonald, J.E., 1962, 'A note on anomalous cooling rates in clouds,' *J. Atmos. Sci.* **19**, 309.

Nelson, L.D., 1971, 'A numerical study on the initiation of warm rain,' *J. Atmos. Sci.* **28**, 752.

Paluch, I.R., 1979, 'The entrainment mechanism in Colorado cumuli,' *J. Atmos. Sci.* **36**, 2467.

Paluch, I.R., and Knight, C.A., 1984, 'Mixing and the evolution of cloud droplet size spectra in a vigorous continental cumulus,' *J. Atmos. Sci.* **41**, 1801.

Paluch, I.R., and Knight, C.A., 1986, 'Does mixing promote cloud droplet growth? *J. Atmos. Sci.* **43**, 1994.

Rokicki, M.L., and Young, K.C. 1978, 'The initiation of precipitation in updrafts,' *J. Appl. Meteorol.* **17**, 745.

Squires, P., 1952, 'The growth of cloud drops by condensation,' *Aust. J. Sci. Res. Ser. A* **5**, 59.

Squires, P. 1958, 'Penetrative downdrafts in cumuli,' *Tellus* **10**, 381.

Srivastava, R.C., 1985, 'A simple model of evaporatively driven downdraft: Application to microburst downdraft,' *J. Atmos. Sci.* **42**, 1004.

Srivastava, R.C., 1987, 'A model of intense downdrafts driven by the melting and evaporation of precipitation,' *J. Atmos. Sci.* **44**, 1752.

Steiner, M., and Waldvogel, A., 1987, 'Peaks in raindrop size distributions,' *J. Atmos. Sci.* **44**, 3127.

Telford, J.W., 1987, 'Comments on 'Does mixing promote cloud droplet growth?'' *J. Atmos. Sci.* **44**, 2352.

Telford, J.W., Keck, T.S., and Chai, S.K., 1984, 'Entrainment at cloud tops and droplet spectra,' *J. Atmos. Sci.* **41**, 3170.

Twomey, S., 1976, 'The effects of fluctuations in liquid water content on the evolution of large drops by coalescence,' *J. Atmos. Sci.* **33**, 720.

Twomey, S., and Warner, J., 1967, 'Comparison of measurements of cloud droplets and cloud nuclei,' *J. Atmos. Sci.* **24**, 702.

Tzivion, S., Feingold, G., and Levin, Z., 1989, 'The evolution of raindrop spectra. Part II. Collisional collection/breakup and evaporation in a rainshaft,' *J. Atmos. Sci.* **46**, 3312.

Valdez, M.P., and Young, K.C., 1985, 'Number fluxes in equilibrium raindrop populations: A Markov chain analysis,' *J. Atmos. Sci.* **42**, 1024.

Warner, J., 1970, 'On steady-state one-dimensional models of cumulus convection,' *J. Atmos. Sci.* **27**, 1035.

Young, K.C., 1974, 'The evolution of drop spectra through condensation, coalescence and breakup,' *Proc. Int. Conf. Cloud Physics 1974. American Meteorological Society,* p. 95.

Young, K.C., 1975, 'The evolution of drop spectra due to condensation, coalescence and breakup,' *J. Atmos. Sci.* **32**, 965.

Young, K.C., 1993, 'Effects of simplifications of the Köhler equation on the activation of CCN in an updraft,' *J. Atmos. Sci.* (in press).

10

FORMATION OF PRECIPITATION BY THE ICE CRYSTAL MECHANISM

10.1. Introduction

The formation of precipitation by the ice crystal mechanism involves converting some of the supercooled water droplets in a cloud to ice particles that will grow large enough to fall out of the cloud and reach the ground without evaporating entirely. As noted in Chapter 1, the relative concentration of ice particles compared to water droplets is crucial in determining the ultimate sizes of the ice particles. The requirement for an epitaxial substrate for nucleation of ice, limits the number of potential ice nuclei compared to the potential number of cloud condensation nuclei (CCN). Thus, the concentrations of ice particles in clouds with temperatures warmer than $\sim -20°C$ are generally orders of magnitude less than the concentrations of water droplets. Under these conditions, the relatively few ice particles that form can grow to be orders of magnitude more massive than the original cloud droplets.

Once ice particles have been introduced into a supercooled water cloud, the ice water content increases at the expense of the liquid water content (LWC). This conversion of a supercooled water cloud to an ice cloud is termed *glaciation*. Glaciation proceeds via the diffusional growth of ice crystals, the collection of water drops by the ice particles (accretion), and by the collection of small ice crystals by water drops (coalescence-freezing).

The formation of precipitation by the ice crystal mechanism occurs in parallel with the formation of precipitation by the coalescence process. Interactions between these two precipitation mechanisms enhance the development of precipitation, and it may be difficult to identify which mechanism is dominant. The development of precipitation by the ice crystal mechanism is examined for both stratiform and cumuliform clouds.

10.2. The glaciation process

The glaciation process begins with the introduction of ice particles into a supercooled water cloud. The simplest case is to assume that there are no vertical motions with an initial state of an all-water cloud at water saturation and a final state of an all-ice cloud at ice saturation. Under these assumptions, changes in the ambient and particle temperatures and the supersaturation can be computed as a function of time. In an updraft, the saturation ratio will never reach ice saturation, and the glaciation process is never really complete in the sense of a final state at ice saturation.

10.2.1. *The thermodynamics of glaciation*

Glaciation is associated with a release of latent heat, resulting in an increase in the ambient temperature. The latent heat release results from converting liquid water to ice and from changes in the water vapor density between the initial and final states. The initial state is assumed to be at water saturation at the initial temperature ($\rho_{s,w}(T_o)$) with an LWC designated by χ_L. The final state is assumed to be at ice saturation at the final temperature ($\rho_{s,i}(T_f)$) with an ice water content of χ_I.

The calculation of the total heat released is simplified by considering the process as isothermal at T_o[1] and then subtracting the heat required to change the temperature of the ice particles from T_o to T_f. The total heat release is equated to the heat required to change the ambient air temperature from T_o to T_f,

$$\begin{aligned}\Delta Q &= L_f + L_s[\rho_{s,w}(T_o) - \rho_{s,i}(T_f)] - c_i(T_f - T_o)\,[\chi_L + \rho_{s,w}(T_o) - \rho_{s,i}(T_f)] \\ &= \rho_a c_p(T_f - T_o)\,, \end{aligned} \tag{10.1}$$

where L_s and L_f refer to the latent heats of sublimation and fusion at T_o, and c_i and c_p are the specific heats for ice and air at constant pressure, respectively.

Under the assumption that $\rho_{s,i}$ is a linear function of T over the range from T_o to T_f,

$$\rho_{s,i}(T_f) = \rho_{s,i}(T_o) + b\Delta T,$$

where $b = \rho_{s,i}(T_o + 1) - \rho_{s,i}(T_o)$. Letting $\Delta T = T_f - T_o$ and $\Delta\rho = \rho_{s,w}(T_o) - \rho_{s,i}(T_o)$, (10.1) may be written

$$\Delta T[\rho_a c_p + L_s b + \chi_L + c_i\Delta\rho] - c_i b(\Delta T)^2 = L_f\chi_L + L_s\Delta\rho\,.$$

Dropping the $(\Delta T)^2$ term and solving for ΔT,

$$\Delta T = \frac{L_f\chi_L + L_s\Delta\rho}{\rho_a c_p + L_s b + c_i\chi_L + c_i\Delta\rho}. \tag{10.2}$$

For initial conditions of -10°C, the latent heats of sublimation and fusion are 2836.6 J g^{-1} and 311.9 J g^{-1}, respectively, and the specific heat of ice is 2.0348 J g^{-1} °C^{-1}. The saturation vapor density over water at -10°C is 2.358 g m^{-3} and the saturation vapor densities over ice at -10°C and -9°C are

[1] The change in vapour density still assumes ice saturation at the final temperature but the determination of the associated latent heat release between the initial and final vapor densities relies on the latent heat of sublimation at T_o.

$2.139\,\mathrm{g\,m^{-3}}$ and $2.328\,\mathrm{g\,m^{-3}}$, respectively. At 700 mbar and $-10°\mathrm{C}$, the density of air is $926.7\,\mathrm{g\,m^{-3}}$. For an initial LWC of $1\,\mathrm{g\,m^{-3}}$, the temperature rise on glaciation is 0.635°C. Note that the temperature change is independent of the path, i.e., the final temperature is the same whether the glaciation occurs primarily by diffusion or by accretion.

The total heat released is $591.30\,\mathrm{J\,m^{-3}}$ as calculated from $\rho_s c_p \Delta T$. Of this total, $311.92\,\mathrm{J\,m^{-3}}$ comes from converting 1 gram of water to ice, and $280.72\,\mathrm{J\,m^{-3}}$ results from the decrease in the ambient vapor density from $2.358\,\mathrm{g\,m^{-3}}$ to $2.259\,\mathrm{g\,m^{-3}}$. The remaining $-1.3\,\mathrm{J\,m^{-3}}$ results from changing the temperature of the ice phase from $-10°\mathrm{C}$ to $-9.365°\mathrm{C}$. In this example, the contributions from 'freezing' the water and from decreasing the water vapor density are similar in magnitude.

Figure 10.1 provides the temperature rise on glaciation at 700 mbar for a range of temperatures and LWCs. The broken line indicates glaciation occurring with vapor phase balance, i.e., the initial and final water vapor densities are the same and the contribution from changing the water vapor density is 0. This line also represents conditions for which the initial LWC and the final ice water content are the same. Above this line, the final vapor density is higher than the initial vapor density and the contributions from changes in the water vapor

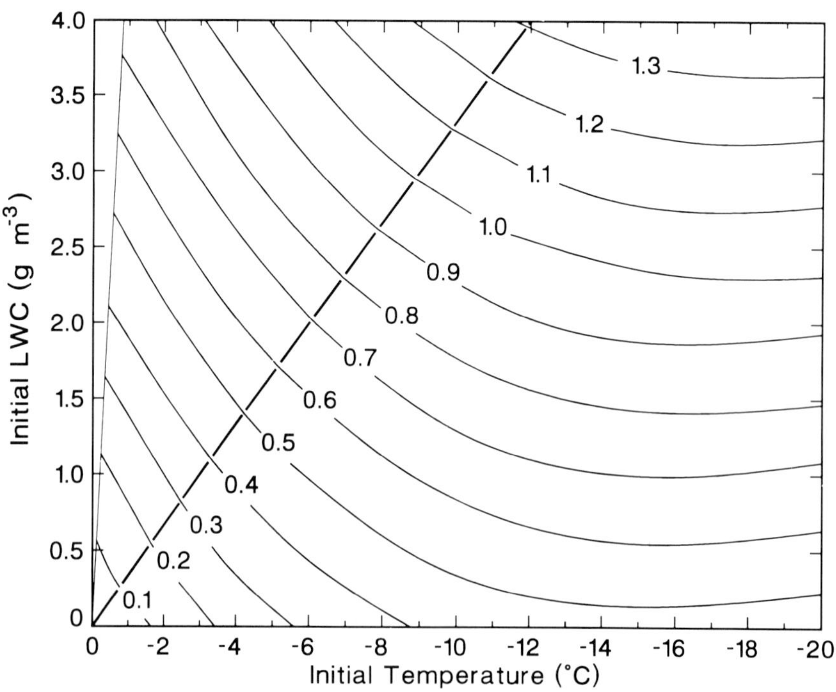

Fig. 10.1. The temperature rise associated with glaciation for an ambient pressure of 700 mbar. Isopleths of temperature rise (°C) are provided as a function of the initial temperature and liquid water content. The broken line indicates that glaciation occurs with vapor phase balance. The wedge-shaped region on the left delineates conditions which do not result in complete glaciation.

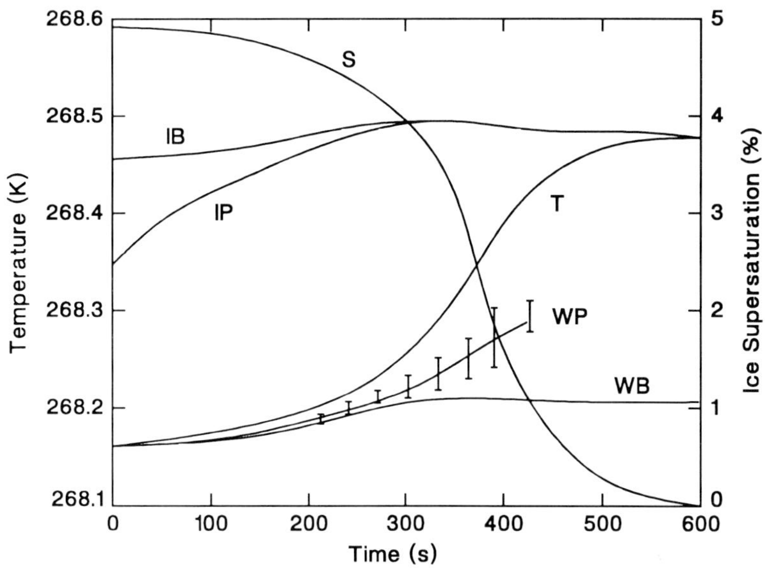

Fig. 10.2. Glaciation of a maritime cloud resulting from injection of 1000 ice crystals ℓ^{-1}. Ice supersaturation (S) and ambient (T), ice-bulb (IB), ice particle (IP), wet-bulb (WB), and water particle (WP) temperatures are plotted as a function of time for a cloud with no vertical motions. From Young (1974b).

density are negative. Below this line, conditions are as discussed in the above example. The small wedge-shaped region on the left represents a region in which the final temperature is 0°C and only partial glaciation occurs, i.e., the final cloud is mixed ice and water. Fukuta (1973) presents similar results for the glaciation process.

10.2.2. *Glaciation through diffusion and accretion*

In this section, the microphysical model described by Young (1974a) will be used to examine the glaciation process in more detail. This model calculates the saturation ratio and ambient temperature required to balance the mass and heat fluxes to and from the particles using an iterative technique. The mass and heat fluxes are determined from (5.2) and (5.6), which require solution of the particle temperature using (5.8) for water drops and (8.5) for ice particles (including accretion).

Figure 10.2 represents glaciation of a maritime cloud with 100 drops cm^{-3} and an LWC of $0.31\,\text{g}\,\text{m}^{-3}$ with no vertical motions. Initially, the cloud is at 4.9% supersaturation with respect to ice, which represents water saturation at the initial temperature of -5°C. Glaciation is initiated by the injection of 1000 crystals per liter.[2] Note that the water drops are able to maintain an ice

[2] This might be accomplished by seeding a cloud with dry ice for example.

supersaturation over 4% for more than 300 seconds and only when the water drops are nearly gone does the supersaturation drop rapidly. The temperature rise mirrors the decrease in supersaturation. It may be noted that the latent heat released due to 'freezing' the liquid water is $100.1\,\mathrm{J\,m^{-3}}$ whereas the latent heat released due to changes in the vapor density is $201.6\,\mathrm{J\,m^{-3}}$.

The wet-bulb and ice-bulb temperatures are shown for comparison with the calculated particle temperatures. Inclusion of kinetic effects (5.17), (5.18) result in particle temperatures that are in between the ambient temperature and the wet-bulb (or ice-bulb) temperatures. The vertical bars on the water drop temperatures represent the range of drop temperatures for the different sized drops used in this simulation.

Figure 10.3 represents glaciation of the same cloud but using 10 ice crystals ℓ^{-1} rather than 1000 ice crystals ℓ^{-1}. The time required for the glaciation process is roughly five times greater. The major changes in supersaturation and ambient temperature occur after all the water drops have evaporated. In addition, after 250 seconds, the ice particle temperature is warmer than the ice bulb temperature. This is a result of accretion, and indicates that accretion processes are much more important in this case than in the previous one.

Figure 10.4 shows the mass transfer rates by diffusion and by accretion as a function of time for a $3\,\mathrm{m\,s^{-1}}$ updraft. In these simulations, droplets are formed

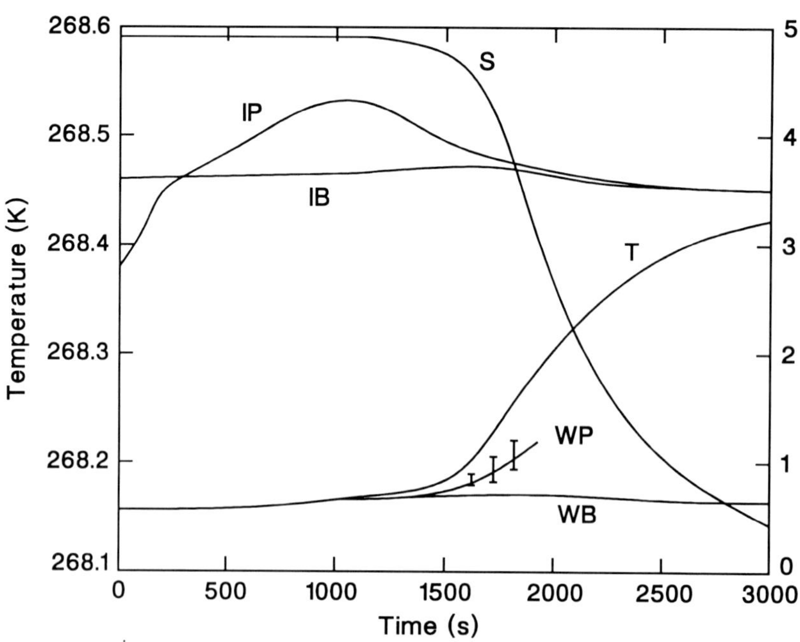

FIG. 10.3. Glaciation of a maritime cloud resulting from injection of 10 ice crystals ℓ^{-1}. Ice supersaturation and ambient and particle temperatures are shown as in Fig. 10.2. From Young (1974b).

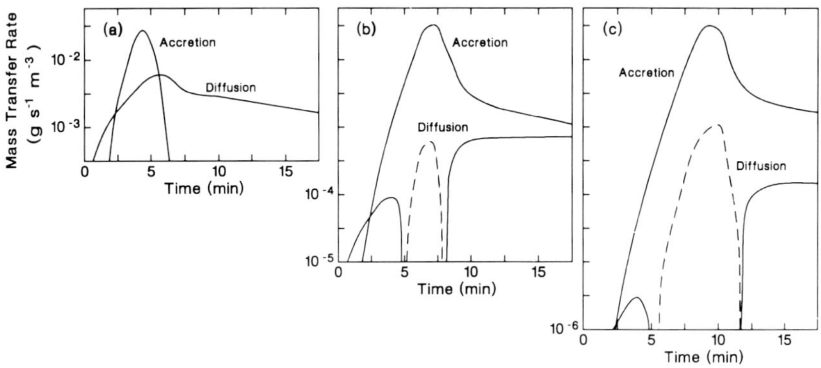

FIG. 10.4. Mass transfer rates to the ice phase by diffusion and accretion. Simulation is for a 3 m s^{-1} updraft, activating a continental CCN spectrum at a cloud base temperature of $+5°C$. Ice crystals were injected at 540 seconds, roughly corresponding to the $-5°C$ level. (a) 1000 crystals ℓ^{-1}, (b) 30 crystals ℓ^{-1}, (c) 1 crystal ℓ^{-1}. From Young (1974b).

based on a continental CCN spectrum ($n = 710S^{0.55}$) at a cloud base temperature of $+5°C$. Ice crystals are injected into the cloud (model) at 540 seconds, representing roughly the $-5°C$ level. For an ice crystal concentration of 1000 ℓ^{-1} (Fig. 10.4a), diffusion dominates for the first 2.5 minutes and after 6 minutes, although accretion accounts for 60% of the total mass transferred to the ice particles. Note that the diffusion transfer continues at a relatively high level well after the accretion rate has dropped off the scale. This reflects the extraction of water vapor through deposition on the ice particles as it is being made available in the updraft.

For an ice crystal concentration of 30 ℓ^{-1} (Fig. 10.4b), accretion clearly dominates, accounting for 99% of the total mass transfer. Between 5 and 8 minutes, the diffusion transfer rate is negative, reflecting ice particle temperatures warm enough such that the saturation vapor density over ice at the ice particle temperature is greater than the ambient vapor density.[3] In the latter stages, the accretion rate does not drop well below the diffusion rate but remains the dominant growth mechanism for the ice particles. This reflects the rise in water supersaturation, which results in continued activation and growth of CCN. Thus, most of the water vapor that is transferred to the growing ice particles is first condensed on small droplets, which are then collected by the ice particles.

For an ice crystal concentration of 1 ℓ^{-1} (Fig. 10.4c), the net diffusion transfer is negative, and diffusion is important only in producing ice particles large enough to undergo accretion, which is accomplished within 2.5 minutes. Figure 10.5 shows the relative total contributions to the ice particle growth by diffusion and accretion as a function of the ice particle concentration. For ice

[3] This implies that the ice particle temperature is greater than the ambient temperature. Note that the ice particle temperature can be greater than the ambient temperature and water vapor can still deposit on the ice particle, provided the ambient vapor density is higher than the vapor density at the particle surface.

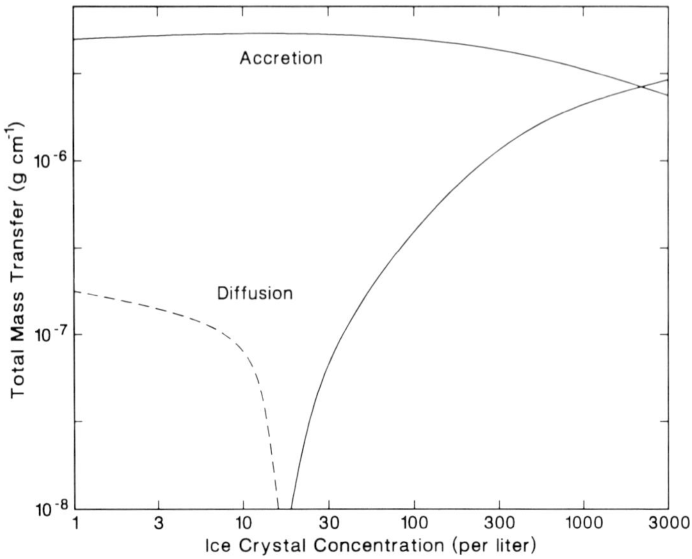

FIG. 10.5. Total mass transfer to the ice particles by diffusion and by accretion. The broken line indicates a negative transfer rate, i.e., there was a net loss of mass by diffusion.

particle concentrations less than $20\,\ell^{-1}$, the diffusion contribution is negative (net sublimation); for ice particle concentrations greater than $2000\,\ell^{-1}$, diffusion dominates over accretion.

10.2.3. *Interactions between the coalescence and ice crystal processes*

The ice crystal process cannot proceed without a concomitant coalescence process, although at colder temperatures and low LWCs, the coalescence process may not manifest itself before the ice crystal process becomes dominant. The growth of drops by coalescence may have a marked influence on the evolution of the ice crystal process, both as a 'short-cut' to producing larger ice particles more quickly and in enhancing the initiation of ice particles.

In Chapter 9, the presence of large drops in an updraft was shown to result in a rise in the water supersaturation, creating conditions favorable for condensation freezing nucleation. However, the rise in water supersaturation follows the coalescence conversion, i.e., precipitation has already formed in such a cloud before condensation freezing would be expected to occur. Thus, this mechanism would not be expected to *initiate* precipitation in a cloud, although it seems likely that it may account for rapid increases in ice particle concentrations in some clouds.

Contact freezing nucleation is also enhanced for large drops. In this case, drops larger than radius 100 μm are sufficient to produce reasonable contact freezing nucleation rates in an updraft, although ice particle production rates would probably be too slow to compete with the coalescence conversion. Again, contact freezing nucleation may explain high ice crystal concentrations at cloud boundaries where evaporation is important.

The primary interaction seems to be the concurrent development of large water drops via coalescence and the nucleation and growth of ice crystals via the ice crystal mechanism. As the water drops grow, their collection kernel increases, increasing their likelihood of collecting an ice crystal. Once the water drop collects an ice crystal, it freezes and continues growth as an ice particle, both through accretion and deposition. In such a case, if one sampled the precipitation particles at the time of the first appearance of a 20 dBZ radar echo, one would likely find the majority of the precipitation particles to be ice.

The time required to form a 20 dBZ radar echo for various updraft velocities is shown in Fig. 10.6 for a cloud base temperature of 15°C. The slowest updraft velocities develop a 20 dBZ echo at the warmest temperatures, although taking the longest time to do so. The development of precipitation is clearly via the coalescence mechanism since the precipitation particles at the time of the 20 dBZ echo are predominantly liquid water. As higher updraft velocities are considered, the temperatures at which the 20 dBZ echo appears are lower and the phase of the precipitation particles shifts from water to ice.

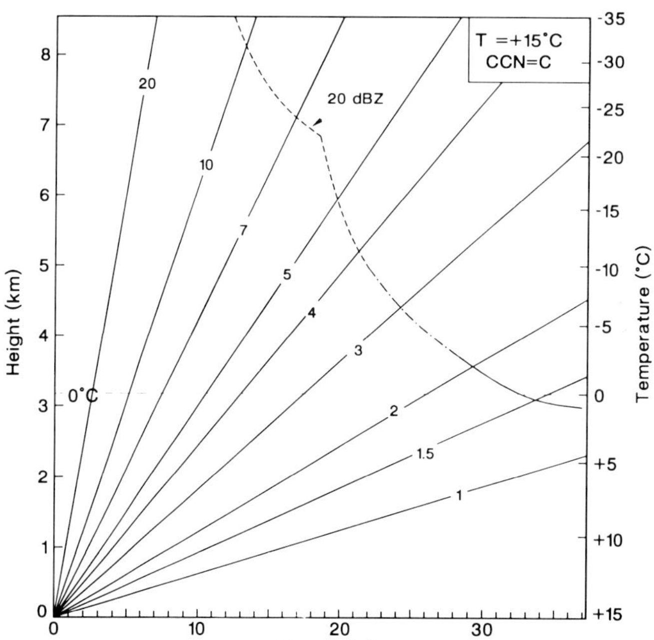

Fig. 10.6. Time required to develop a 20 dBZ radar echo for updraft velocities from 1 m s^{-1} to 20 m s^{-1}. The height of a rising parcel of air for a given updraft velocity can be traced as a function of time following its isopleth. The line designated 20 dBZ represents the time (height) that a radar echo of 20 dBZ is achieved. The solid portion indicates >90% of the precipitation particles were water; the broken portion indicates >90% of the particles were ice; the dot-dashed line represents comparable concentrations of ice and water particles.

The break in the plot of the time to a 20 dBZ radar echo in Fig. 10.6, corresponding to an updraft velocity of $6\,m\,s^{-1}$, reflects the shift from a coalescence driven mechanism to an ice crystal driven mechanism. For updraft velocities weaker than $6\,m\,s^{-1}$, the coalescence process produces large drops which, at colder temperatures, are more likely to freeze by collecting an ice crystal prior to the occurrence of a 20 dBZ echo. For updraft velocities between $4\,m\,s^{-1}$ and $6\,m\,s^{-1}$, virtually all of the drops formed by coalescence freeze before a 20 dBZ echo is achieved. For updraft velocities greater than $6\,m\,s^{-1}$, the ice crystals grow to precipitation sizes by diffusion and by collecting small water droplets.

Table 10.1 provides the time to a 20 dBZ radar echo and the temperature at which this occurs for a variety of cloud base temperatures and updraft velocities. Clearly, clouds with colder base temperatures are more likely to produce precipitation in the form of ice, although it is evident that the coalescence process strongly influences the development of precipitation even in relatively cold clouds.

Based on observations of the development of precipitation in summertime cumuli in southern Missouri, Braham (1964) concluded that large ice particles grow much faster than drops of comparable mass. He attributed this to the ice particles having a larger collection kernel due to their lower bulk densities, i.e.,

Table 10.1. The time required to produce a 20 dBZ radar echo due to ice particles and the temperature of the parcel at the time of the 20 dBZ echo. The upper number is the time to a 20 dBZ echo (min); the lower number is the corresponding temperature (°C). Italicized times reflect cases with comparable numbers of ice particles and water particles.

	U (m s^{-1})							
T_b (°C)	1	1.5	2	3	4	5	7	10
−5	26.4 −16.5	22.8 −19.1	20.5 −22.8	18.2 −29.8				
0	31.0 −12.0	25.6 −15.4	22.3 −18.1	18.5 −22.6	16.6 −27.6	15.3 −32.5		
5		*30.1* −11.5	*25.6* −14.0	20.6 −18.3	17.8 −22.0	16.2 −26.2	14.1 −35.0	
10		*36.7* −8.1	*30.5* −10.2	*24.3* −14.8	21.1 −19.6	18.1 −22.2	16.0 −28.1	12.8 −350
15		W[a] +0.1	W −2.0	*24.3* −6.6	*21.4* −11.2	*20.0* −16.2	16.8 −24.1	13.6 −31.8

[a] W = cases for which the precipitation particles were predominantly water drops.

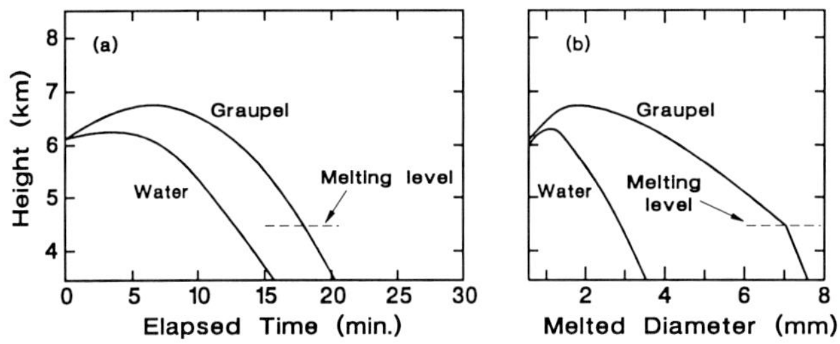

FIG. 10.7. Comparative growth of a water drop and a frozen drop. Particle growth trajectories are shown for a uniform updraft of $5\,m\,s^{-1}$ with a cloud base temperature of 20°C. Both particles are introduced at the −8°C level as 250 μm radius water drops with one allowed to freeze at the start of the calculations. From Johnson (1987). *Journal of Atmospheric Sciences*, **44**, 13. American Meteorological Society. Reproduced with permission.

a larger cross-section, and assuming comparable fall velocities between water drops and ice particles of the same mass.

Johnson (1987) used a numerical model to compare the growth of large water drops with that of frozen drops, assuming that the particles grow only by the collection of small cloud droplets. The simulations shown in Fig. 10.7 compare the growth of a 250 μm radius water drop with that of a solid ice particle with the same initial mass. The mass accreted on the ice particle was assumed to have a density of $0.5\,g\,cm^{-3}$. The ice particle grows at a faster rate than the water drop and, due to its smaller fall velocity, remains in the cloud longer. The resulting ice particle is ten times more massive than the water drop. It should be noted that most raindrops interact with comparable sized drops as well as with much smaller droplets and consequently undergo breakup as well as coalescence. This would tend to limit further the growth of the drops, increasing the disparity in the growth rates.

These interactions between the coalescence and ice crystal mechanisms should enhance the development and growth of precipitation to a considerable degree. Braham (1986) suggests that the term *coalescence-freezing* be used to describe these interactions. Clearly, the synergy between the two processes produces an outcome that is more than the sum of the two processes operating in isolation.

10.3. Formation of precipitation in stratiform clouds

The formation of precipitation in stably stratified clouds usually occurs within an extra-tropical cyclone or in association with orographic flow. Since temperatures generally decrease with height, ice crystals tend to form in the upper portions of the cloud system and grow as they fall through the cloud. As the ice particles fall through the 0°C isotherm, they melt, producing a radar feature known as a 'bright-band.' Below cloud base, evaporation serves to humidify the sub-cloud layers, reducing the mass lost to evaporation and resulting in relatively high precipitation efficiencies.

10.3.1. *Extra-tropical cyclones*

The dynamical processes that lead to the development of extra-tropical cyclones are examined in detail by Cotton and Anthes (1989). The most extensive region of precipitation in an extra-tropical cyclone is ahead of the warm front where warm, moist air from the southeast rises over the colder air at the surface. This tongue of warm, moist air is termed a *warm conveyor belt*,as shown in Fig. 10.8. If the air in the warm conveyor belt is conditionally unstable, the lifting may produce convective clouds; if the air in the warm conveyor belt is absolutely stable, the lifting will produce extensive regions of stratiform clouds. Convective clouds tend to form in the warm sector, particularly when surface heating

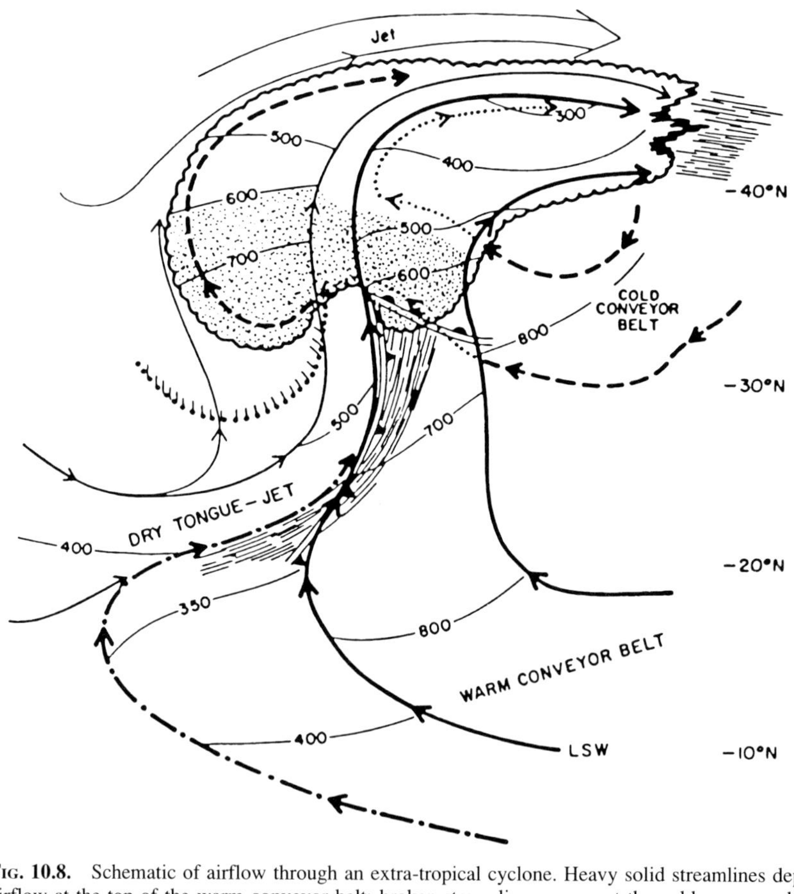

FIG. 10.8. Schematic of airflow through an extra-tropical cyclone. Heavy solid streamlines depict airflow at the top of the warm conveyor belt; broken streamlines represent the cold conveyor belt. Dot-dashed streamlines represent air originating at middle levels in the tropics; thin solid streamlines represent dry air that originates at upper levels west of the trough. The heights of the airstreams are shown in millibars. The region of dense upper- and middle-level cloud is represented by scalloping and sustained precipitation by stippling. Streaks indicate cirrus; the curved border of small dots with tails indicates the edge of the low level stratus. From Carlson (1980). *Monthly Weather Review*, **108**, 10. American Meteorological Society. Reproduced with permission.

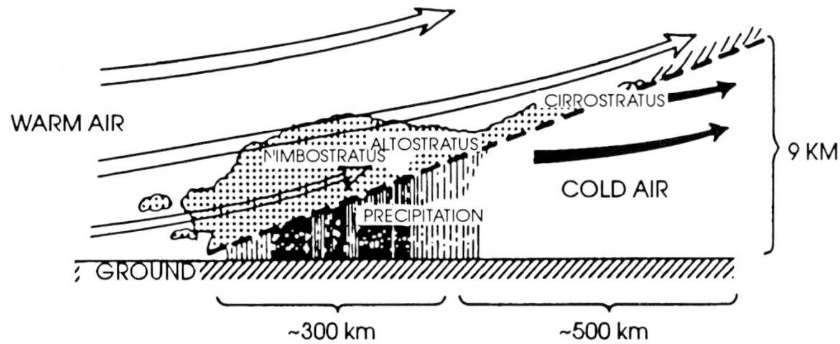

Fig. 10.9. Idealized cross-section through a warm front. The vertical scale is amplified by a factor of 30 compared to the horizontal scale. As the height of the frontal surface (broken line) increases, the clouds become thinner. Precipitation results only when the cloud is sufficiently deep. From Houze and Hobbs (1982). Reprinted by permission of Academic Press.

destablizes the air; stratiform clouds tend to form in the cold sector north of the warm front.

An idealized cross-section through a warm front is shown in Fig. 10.9. As the warm, moist air of the warm conveyor belt is forced to rise, a deep stratiform cloud is formed over the cold air, resulting in continuous precipitation. As the warm air is lifted further ahead of the surface front, the cloud bases are higher and the cloud depth decreases, leading to altostratus, cirrostratus, and cirrus.

Bands of more intense precipitation often occur within the precipitation shield associated with the warm front. These bands are convective, in contrast to the stratiform clouds making up most of the precipitation shield. Hobbs (1978) terms these *warm-frontal rainbands*. These bands appear to be associated with generating cells aloft, as shown in Fig. 10.10. Ice particles are nucleated and grow to precipitation sized particles in the generating cells through deposition and accretion. As they fall through the stratiform cloud below, they continue to grow primarily by deposition and aggregation.

Houze and Hobbs estimate that typically only 20% to 35% of the mass of the precipitation reaching the ground is derived from the generating cells (the seeder zone); the majority of the precipitation mass is derived from growth as the ice particles fall through the stratiform layers (the feeder zone), as shown in Fig. 10.11. They discuss several possible mechanisms for producing the instability associated with the generating cells.

The structure of the cold front is characterized by a region of more intense convection over the frontal boundary at the surface and rainbands similar to warm-frontal rainbands behind the front, as shown in Fig. 10.12. Hobbs (1978) terms the band of heavy precipitation associated with the cold front at the surface a 'narrow cold-frontal rainband' and the band just behind the cold front as a 'wide cold-frontal rainband.'

The narrow cold-frontal rainband is associated with a narrow (~5 km) updraft region with velocities of a few meters per second. The source of moisture is a low-level, southerly jet located just ahead of the front. The updraft is coupled

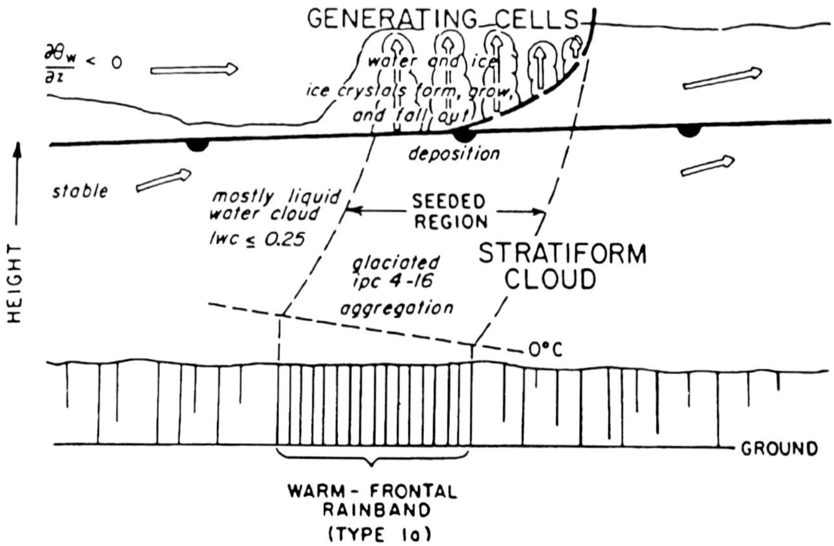

FIG. 10.10. Cross-section of a warm-frontal rainband. The air below the warm frontal boundary tends to be stable; the heavy broken line branching out from the front is a warm-frontal zone with convective ascent in the generating cells. Vertical hatching below the cloud base represents precipitation, with the spacing suggestive of the intensity. Ice particle concentrations (ipc) are ℓ^{-1}; cloud liquid water contents (lwc) are $\mathrm{g\,m^{-3}}$. The rainband is propagating from left to right. From Houze and Hobbs (1982). *Rev. Geophys. Space Phys.*, **16**. Reproduced with permission.

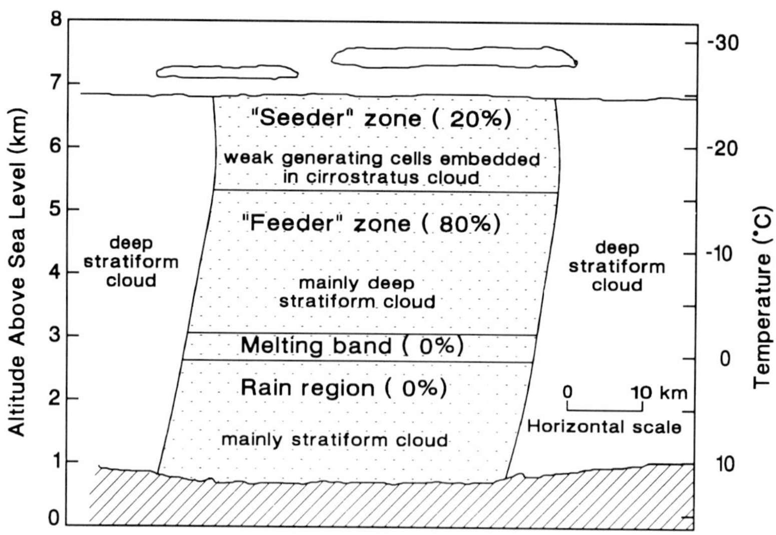

FIG. 10.11. Schematic cross-section of a wide cold-frontal rainband showing the seeder/feeder structure. Virtually all of the particle growth occurs above the melting band. From Hobbs et al. (1980). *Journal of Atmospheric Sciences*, **37**, 3. American Meteorological Society. Reproduced with permission.

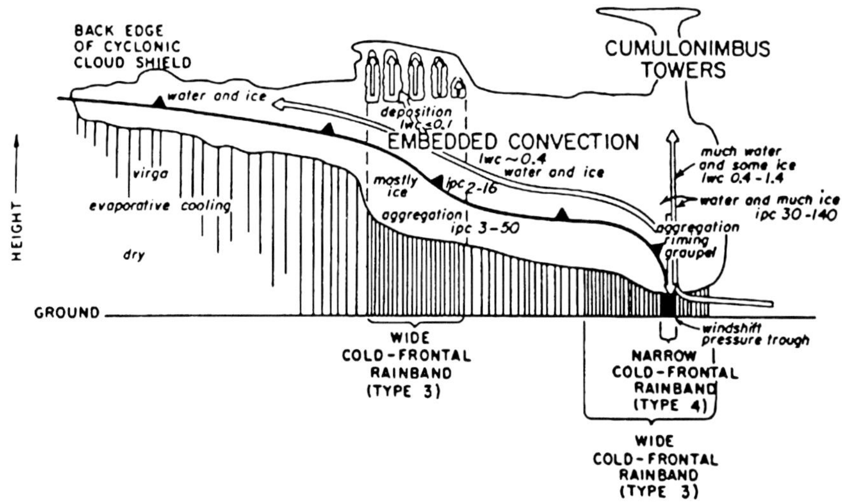

FIG. **10.12.** Cross-section of a cold front showing narrow and wide cold-frontal rainbands. Vertical hatching below the cloud base represents precipitation, with the spacing suggestive of the intensity. Open arrows represent airflow relative to the front. Ice particle concentrations (ipc) are ℓ^{-1}; cloud liquid water contents (lwc) are g m^{-3}. The motion of the front is from left to right. From Houze and Hobbs (1982). *Rev. Geophys. Space Phys.*, **16**. Reproduced with permission.

with a downdraft which is associated with heavy precipitation (~100 mm h^{-1}). The updraft typically contains large amounts of liquid water but relatively low ice particle concentrations. Accretion is the primary growth mechanism, with graupeln and even hail forming on occasion.

Precipitation efficiencies tend to be fairly high in extra-tropical cyclones because a large fraction of the condensate is converted into precipitation by the falling ice particles as they sweep out much of the liquid water available in the stratiform layers. Hobbs et al. (1980) estimate the precipitation efficiency by comparing the observed precipitation rate at the ground with the estimated condensation rate based on vertical motions estimated from Doppler radar observations of the horizontal winds and assuming mass continuity. This also assumes that the system is steady state, which is probably reasonable for these systems.

For a wide cold-frontal rainband just behind the front, Hobbs et al. estimated the precipitation efficiency as at least 80%. For the more convective narrow cold-frontal rainband, they estimated the precipitation efficiency as between 20% and 30%. Further behind the front, the precipitation efficiency for another wide cold-frontal rainband was estimated to be 20%. Figure 10.12 suggests that sub-cloud evaporation becomes more important with increasing distance behind the cold front. Hobbs et al. report radar observations suggesting that significant evaporation was taking place. Precipitation efficiencies for warm-frontal rainbands would be expected to be similar to those for the wide cold-frontal rainbands.

10.3.2. *Orographic clouds*

Precipitation from orographic clouds is also enhanced by the seeder/feeder mechanism. The residence time for particles formed in stable clouds that develop as air flows over a barrier is often too short to develop precipitation. The existence of a (higher) seeder cloud provides precipitation particles to the orographic cloud in much the same manner that the generating cells provide particles to the stratiform layers in frontal rainbands. In this case, the seeder cloud often arises due to large-scale ascent not associated with the orographic flow. Precipitation in the seeder cloud develops via the ice crystal mechanism and may reach the ground as light rain or snow.

The relation between the seeder and feeder clouds is shown in Fig. 10.13. The precipitation rate below the seeder cloud is P_o; without orographic enhancement, the precipitation rate at the ground is P_1. The formation of an orographic cloud as air is forced to rise over the barrier provides a supply of cloud droplets for further growth of the precipitation particles falling from the seeder cloud. The fall speed of the particles and the horizontal wind speed determine the length of time the particles remain in the orographic cloud, which, in turn, determines their ultimate size and the resulting precipitation rate (P_2). Note that the particles falling from the seeder cloud directly above the barrier crest follow a trajectory that passes through the evaporation zone downwind of the barrier, resulting in a lower precipitation rate.

The factors important to the orographic enhancement of precipitation include the mixing ratio of the air entering the orographic cloud, the slope of the barrier, the component of the wind perpendicular to the barrier, the precipitation rate from the seeder cloud, and the relative humidity of the layer between the seeder and feeder clouds. The mixing ratio of the low level air and the vertical motions resulting from impingement of the flow on the barrier determine the rate of production of liquid water in the feeder cloud. The precipitation rate from the

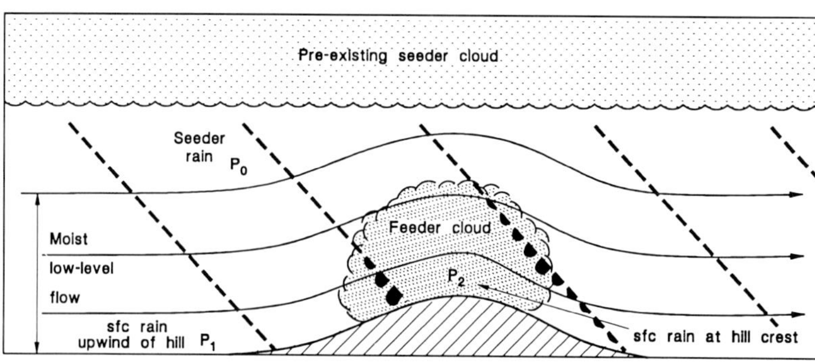

FIG. **10.13.** Conceptual model of the orographic enhancement of precipitation. Light precipitation (P_o) originates from the seeder cloud (thin broken lines) and reaches the ground with intensity P_1. Where the precipitation passes through the feeder cloud, particle growth by coalescence or accretion enhances the precipitation rate (heavy broken lines) to intensity P_2. From Browning (1979). *Global Atmos. Res. Programme* (1979) Ser. No. 23. © World Meteorological Organization. Reproduced with permission.

seeder cloud and the intervening evaporation determine the supply of particles to the feeder cloud.

The enhancement of precipitation due to orographic effects is much more efficient when the seeder particles are ice than when they are water. Choularton and Perry (1986) calculated the orographic enhancement by ice particles to be three times greater than by water drops when the feeder cloud contained relatively high concentrations of liquid water, promoting growth by accretion. They note that particle growth for a gradual ascent (long hills) was predominantly by deposition, and for a steep ascent (short hills) was predominantly by accretion. This is a result of the higher LWC for steeper ascents.

Chappell (1970) compares the condensation rates, the ice particle growth rates by deposition, and the observed precipitation rates near Climax, Colorado as a function of the 500 mbar temperature, as shown in Fig. 10.14. He assumed

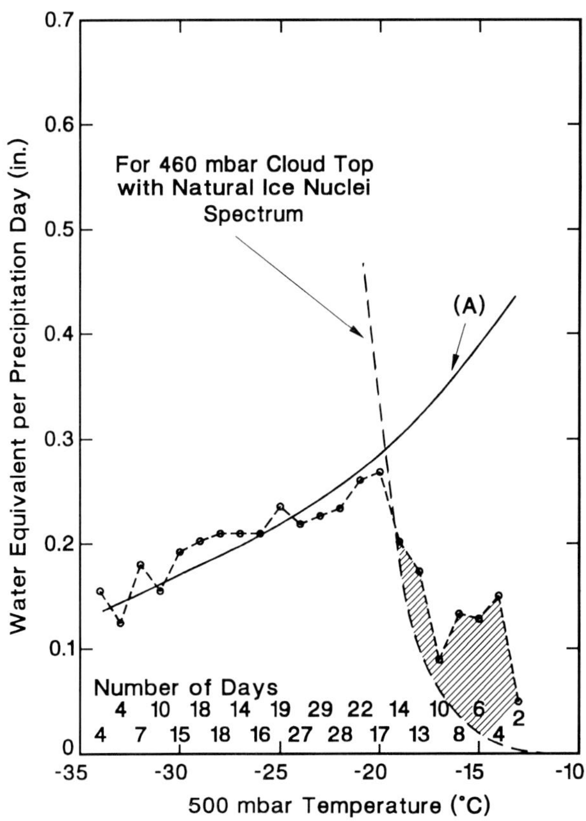

FIG. 10.14. The relation between (–) the rate of condensation of liquid water, (- - -) the growth rate of ice crystals, and (○) the observed precipitation rate near Climax, Colorado. The average condensation rate is based on 10.1 precipitation hours day^{-1} with an assumed updraft velocity of 11 cm s^{-1} at -20°C and adjusted for other temperatures according to variations in the horizontal wind speed. From Chappell (1970).

that the concentration of ice crystals would be determined from a standard ice nucleus activity spectrum and that growth would occur solely by deposition. At temperatures colder than −20°C, the condensation rate limits precipitation as the relatively high concentrations of ice crystals are easily able to deplete the liquid water. At warmer temperatures, precipitation is limited by the relative lack of ice crystals, implying relatively high LWCs in these clouds. Neglect of the accretion process probably underestimates the depletion of liquid water at temperatures warmer than than −20°C. The production of ice crystals via contact freezing nucleation, condensation freezing nucleation, or through secondary ice production would likewise be expected to enhance the precipitation rate.

Cotton et al. (1986) describe a two-dimensional, nonhydrostatic model of orographic precipitation applied to a case study in Colorado. They conclude that at temperatures warmer than −15°C, contact freezing nucleation is a significant source of ice particles and at temperatures colder than −10°C, deposition nucleation is a significant source. Aggregation was noted to produce a 30% increase in precipitation, in contrast to simulations that did not include aggregation.

Precipitation efficiencies in orographic clouds are dependent on the width of the barrier. Elliott and Hovind (1964) studied two mountain barriers in southern California and found precipitation efficiencies ranging from 17% to 27%. The highest efficiencies were noted for unstable conditions and for the wider barrier. Dirks (1973) noted precipitation efficiencies ranging from 25% to 80% for four cases in the Medicine Bow Mountains in Wyoming. Marwitz (1974) reported precipitation efficiencies of 62% for orographic clouds over the San Juan mountains in southwestern Colorado. Both the Dirks and Marwitz studies were for much wider barriers than those examined by Elliott and Hovind in California.

10.3.3. *The melting layer*

As ice particles fall into regions where the temperatures are above 0°C, they begin to melt and cool the surrounding air. In stratiform precipitation, this cooling associated with the melting ice particles can produce fairly deep isothermal layers. Just below the base of the isothermal layer, a shallow superadiabatic layer may result.

Figure 10.15 summarizes observations of the sizes and concentrations of ice particles through the melting layer made by Willis and Heymsfield (1989). The decrease in ice particle concentrations in the isothermal layer (shaded) is associated with an increase in the maximum particle size. The rapid growth of large aggregates is considered to result from increased differences in fall speeds resulting from differing degrees of melting, and from the enhancement of the sticking coefficient for temperatures near 0°C. Some of the large aggregates were noted to survive to the +5°C level.

A synthesis of observations obtained by Stewart et al. (1984) through the melting layer is presented in Fig. 10.16. The relationship between ice particle concentrations and maximum ice particle sizes mirrors that illustrated in Fig.

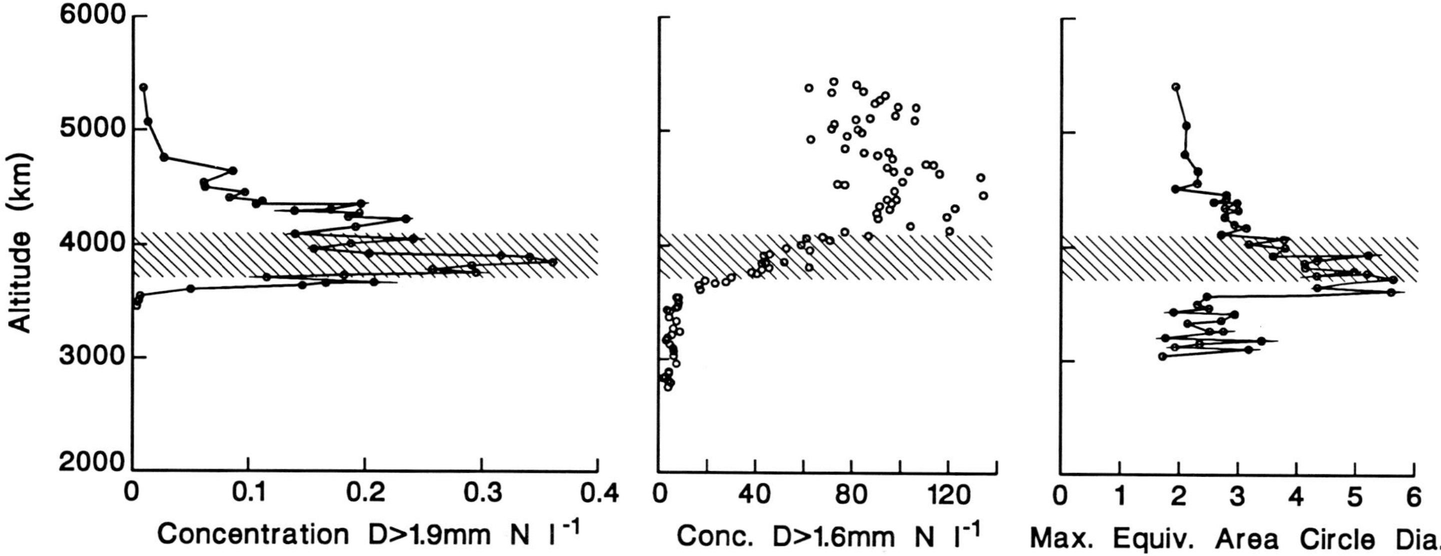

Fig. 10.15. Profiles of ice particle concentrations and maximum sizes through the melting layer. The shaded region represents the isothermal layer. (a) Concentration of particles with an equivalent circle diameter >1.9 mm; (b) concentration of particles with an equivalent circle diameter <1.6 mm; (c) maximum equivalent circle diameter in each 10-second data summation. From Willis and Heymsfield (1989). *Journal of Atmospheric Sciences*, **46**, 13. American Meteorological Society. Reproduced with permission.

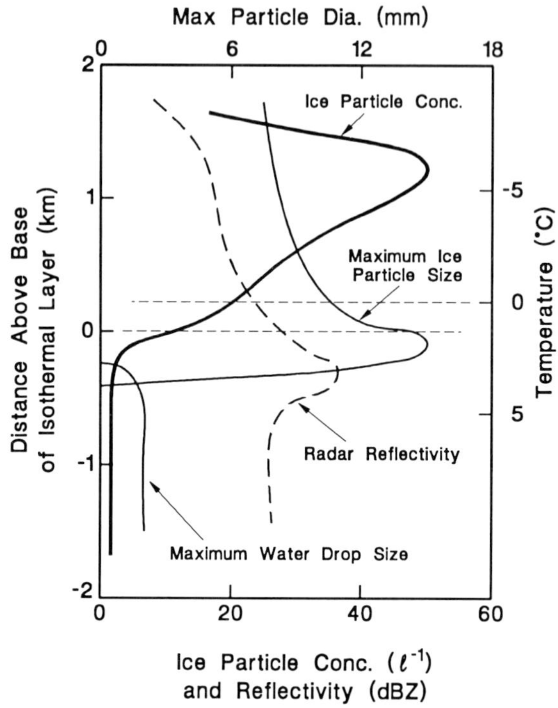

FIG. 10.16. Schematic profile through a typical melting layer. The isothermal layer is located between the horizontal broken lines. Particle diameters are shown for the largest ice particles and water drops as a function of height. The largest ice particles occur just below the isothermal zone; the region of maximum radar reflectivity occurs just below the region of largest ice particles. From Stewart et al. (1984). *Journal of Atmospheric Sciences*, **41**, 22. American Meteorological Society. Reproduced with permission.

10.15. The maximum in radar reflectivity occurs below the level of the maximum ice particle sizes. This is attributed to the increase in reflectivity as the large ice aggregates begin to melt and scatter more as water particles, increasing the reflectivity by roughly a factor of five over ice particles of the same size. As the large aggregates collapse to smaller sizes on melting, their fall speeds increase. Since the mass flux through any layer is roughly constant, the LWCs below the melting layer are considerably smaller than the corresponding ice water contents above the melting layer. The reduction in concentration, as well as the reduction in particle sizes, results in lower radar reflectivities.

The rate at which air cools due to the melting of falling precipitation may be derived following Wexler et al. (1954). If the air is not saturated, all of the heat used to melt the ice particles comes from cooling the air,

$$-L_{\mathrm{f}} \frac{dm}{dt}\bigg|_{\mathrm{melt}} = c_{\mathrm{p}} \rho_{\mathrm{a}} \frac{dT}{dt}, \tag{10.3}$$

where ρ_a is the density of air. Note that (10.3) reflects only the sensible heat exchanges, and that latent heat exchanges are also occurring. From (8.11), the melting rate is

$$-L_f \left. \frac{dm}{dt} \right|_{\text{melt}} = 4\pi r_i [K f_2^* (T_\infty - T_o) + L_v f_2 (\rho_\infty - \rho_{s,o})] ,$$

assuming that the ice particles are at 0°C.

As precipitation continues to fall through the layer, the ambient vapor density in the layer will increase. If the ambient vapor exceeds the saturation value at 0°C, when the ambient temperature cools to the saturation point, further cooling is accompanied by the release of latent heat of vaporization due to the necessity of reducing ρ_s, assuming that the air is maintained at saturation over water. Equation (10.3) becomes

$$-L_f \left. \frac{dm}{dt} \right|_{\text{melt}} = c_p \rho_a \frac{dT}{dt} + L_v \frac{d\rho_s}{dt} .$$

From (9.6), $d\rho_s$ may be written in terms of dT. Solving for the rate of change of temperature,

$$\frac{dT}{dt} = \frac{-L_f \left. \dfrac{dm}{dt} \right|_{\text{melt}}}{\left[c_p \rho_a + \dfrac{L_v \rho_s (L_v - R_v T)}{R_v T^2} \right]} , \tag{10.4}$$

where ρ_s refers to the saturation vapor density at the ambient temperature. Note that L_v is in reference to the ambient temperature whereas L_f is in reference to 0°C. Under saturated conditions, the cooling rate is roughly half of the cooling rate under unsaturated conditions.

10.4. Formation of precipitation in convective clouds

Convective clouds tend to be organized into cells, which evolve through a growing stage, a mature stage, and a dissipating stage. Precipitation is initiated in the growing stage and, indeed, the downdrafts that characterize the mature stage result from the evaporation of precipitation in air entrained at or near the cloud top. Growing particles tend to rise in the updraft until they either fall back through the updraft or transit to weaker updrafts or downdrafts. As a result of evaporation in the downdrafts, the precipitation efficiencies for convective systems tend to be considerably lower than those for stratiform systems.

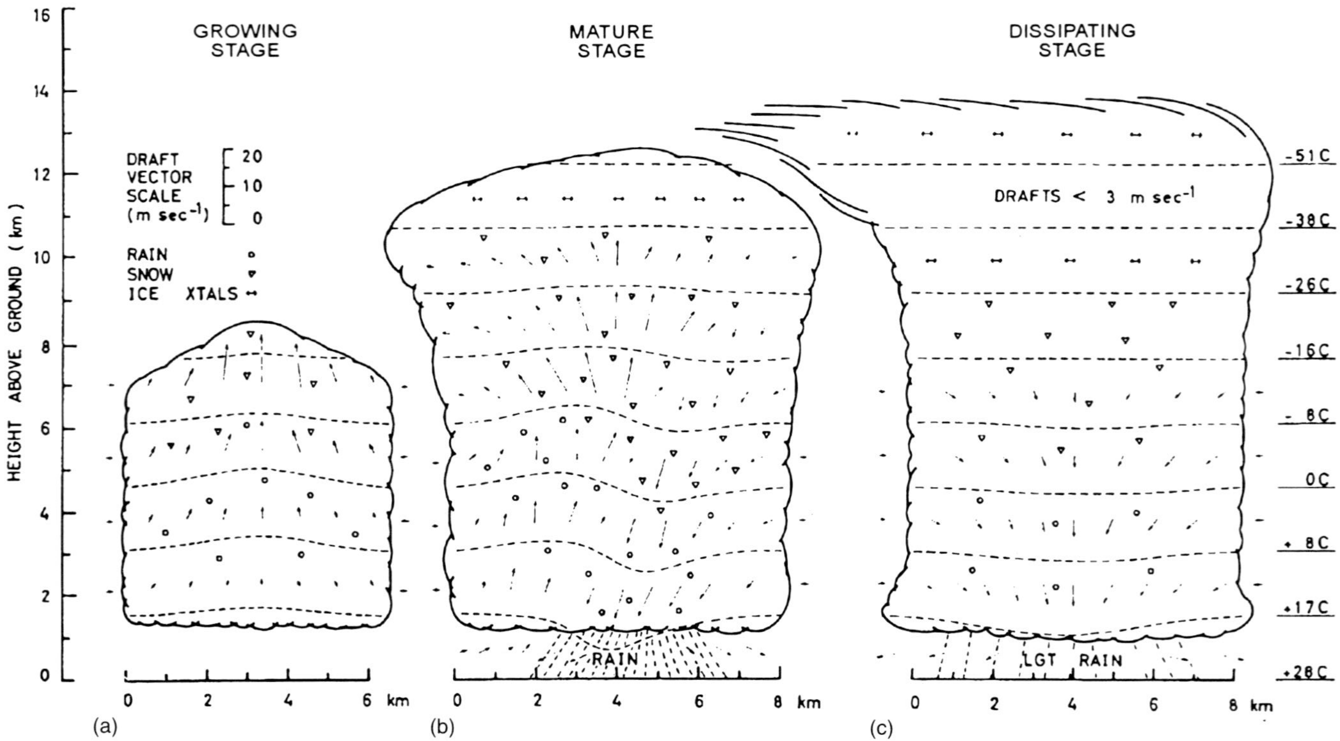

Fig. 10.17. Idealized cross-section of the three stages in the life cycle of a typical thunderstorm cell. (a) Growing stage characterized by updrafts and no precipitation; (b) mature stage characterized by both updrafts and downdrafts and heavy precipitation; (c) dissipating stage characterized by downdrafts and light precipitation. From Byers and Braham (1949). *Meteorological Monographs*, **14**, 36. American Meteorological Society. Reproduced with permission.

10.4.1. *Organization of precipitating convective systems*

The evolution of a thunderstorm cell through the growing, mature, and dissipating stages was first described by Byers and Braham (1949). The growing stage is considered to consist entirely of updraft with liquid particles below the 0°C level and a mixture of ice and liquid particles above that, as shown in Fig. 10.17a. Particles tend to be rising in this stage, having fall speeds less than the updraft velocity.

The mature stage shown in Fig. 10.17b is characterized by both updrafts and downdrafts. The initiation of downdrafts results from the development of precipitation and the cooling associated with precipitation evaporating into air entrained at or near the top of the cloud. Once the downdraft is initiated, precipitation particles are rapidly carried to the ground, resulting in heavy precipitation. High precipitation rates are associated with considerable mass loading leading to strong negative buoyancy. High precipitation rates are also associated with enhanced production of small drops through collisional breakup, enhancing the evaporative cooling of the downdraft.

The dissipating stage is defined by the disappearance of the updraft. In Fig. 10.17c, this stage is characterized as having weak downdrafts and light precipitation. Loss of the updraft may result from the spreading of the cold, downdraft air at the surface or it may result from particle loading and/or entrainment.

In midlatitudes, most convective systems consist of multiple cells. Daughter cells may develop along the outflow boundary of the downdraft air from a parent cell[4] or may develop as a result of some other highly localized forcing such as sea breeze convergence. Organized multicell storms tend to form in midlatitudes in an environment characterized by substantial wind shear. Figure 10.18 shows a typical multicell thunderstorm in northeast Colorado, with two cells in the growing stage and one each in the mature and dissipating stages. The daughter cells tend to form on the right front flank of the storm. Thus, the storm propagates to the right of the direction of the individual cell motion.

Thunderstorms that form in environments with weak wind shear tend to have nearly vertical updrafts. Precipitation particles rise while they grow until their fall speed equals the updraft velocity. This level is often referred to as their *balance level*. With further growth, the particle falls back down through the updraft. The maximum size of precipitation particles is determined by the time available for growth in the updraft and rarely exceeds a few millimeters in radius.

The tilted updraft in Fig. 10.18 is typical of thunderstorms that form in environments having moderate to strong vertical wind shear. In this case, precipitation particles can grow in the relatively weak updrafts and may be transported horizontally into stronger updrafts as they fall through the rising air. If the width and tilt of the updraft are suitable, the growing particle is roughly balanced in the updraft and moves nearly horizontally. The maximum size of a

[4] Outflow from more than one parent cell may be involved.

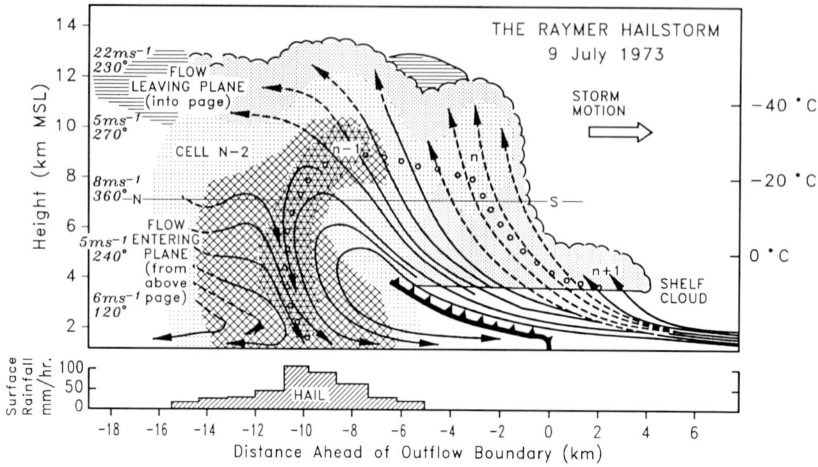

Fig. 10.18. Schematic diagram of the Raymer hailstorm described by Browning et al. (1976), in vertical cross-section along the storm's motion. Solid lines represent streamlines relative to the moving system; broken lines indicate flow in or out of the plane of the cross-section. Open circles represent the growth trajectory of a hypothetical hailstone. Shading represents radar reflectivities of 35 dBZ, 45 dBZ, and 50 dBZ. The cells marked n and n+1 are in the growing stage; the cell marked n−1 is in the mature stage; the cell marked n−2 is in the dissipating stage. From Browning et al. (1976). *Monthly Weather Review*, **104**, 5. American Meteorological Society. Reproduced with permission.

particle growing in a tilted updraft is determined by the vertical velocity of the maximum updraft encountered by the growing particle, as discussed in Chapter 12.

10.4.2. *Characteristics of the precipitation particles*

The coalescence and ice crystal mechanisms both contribute to the development of precipitation in the majority of convective clouds. Clouds with colder cloud base temperatures and higher updraft velocities tend to produce precipitation via the ice crystal mechanism whereas clouds with warmer cloud base temperatures and weaker updrafts tend to be driven by the coalescence mechanism. The limited time available for the growth of precipitation and the relatively high LWC in convective clouds imply that growth by collection will dominate growth by diffusion.

Heymsfield (1978) concludes that the dominant precipitation mechanism in cumulus congestus clouds in northeast Colorado is the accretional growth of small ice crystals, resulting in graupeln. Braham (1964) found that most of the precipitation particles in southern Missouri cumuli consisted of liquid drops and in some cases, frozen drops. Clearly, the warmer cloud base temperatures in Missouri favored the coalescence process, whereas the colder cloud bases in Colorado favored the accretional growth of ice particles.

The multicellular nature of midlatitude thunderstorms suggests that precipitation particles produced in an earlier cell may be transferred to a cell still in the growing stage. This recycled precipitation may be in the form of ice particles or as liquid particles, resulting from melting of ice particles. One potential source of recycled particles is precipitation from the anvil.

Heymsfield (1986) describes the formation of aggregates of ice particles in the anvil of a severe thunderstorm at temperatures between −25°C and −36°C. He cites high ice particle concentrations and the presence of a wide variety of crystal habits, including many with dendritic features, as permitting aggregation by mechanical linking rather than by sintering. Aggregates as large as 5 mm radius were observed, which could serve as a source of recycled particles.

Recycled precipitation particles can grow to centimeter-sized graupeln or hailstones much more quickly than particles which must grow from scratch. Heymsfield (1982) calculated the growth times required for 2.5 mm radius graupeln, and found frozen drops developed most rapidly, followed by graupeln, aggregates, and planar crystals. He postulates two pathways for the growth of large graupeln and hailstones, as shown in Fig. 10.19. The growth of aggregates (A) in decaying cloud towers or in the anvil is followed by sedimentation to

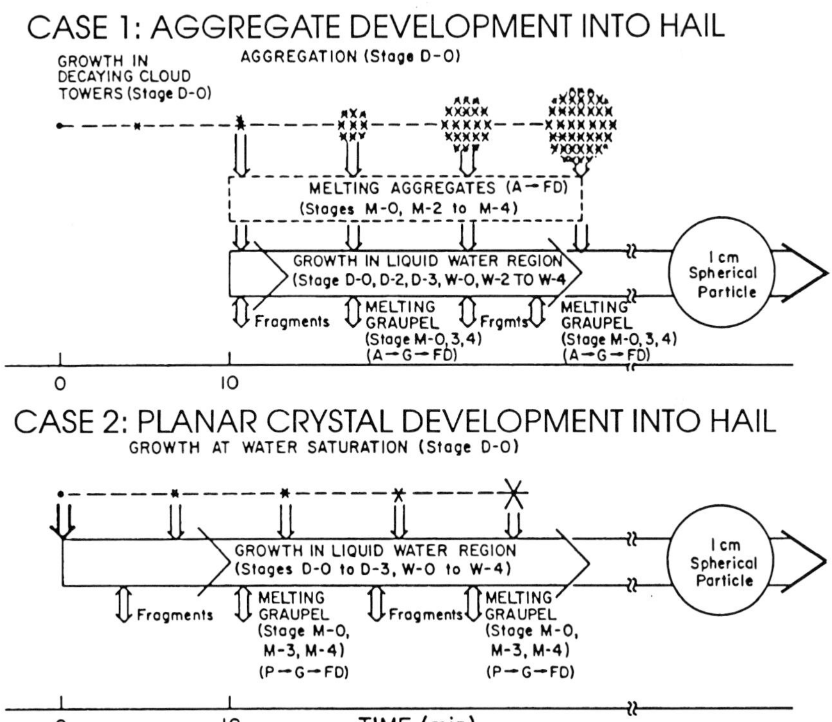

FIG. 10.19. Schematic diagram showing two possible pathways leading to the development of graupeln and hail in multicellular thunderstorms. Growth in the liquid water region occurs in the growing stage of a cell. In Case 2, the planar crystal originated in the cell; in Case 1, the aggregates originated in earlier cells and have been transferred into the growing cell. From Heymsfield (1982). *Journal of Atmospheric Sciences*, **39**, 12. American Meteorological Society. Reproduced with permission.

lower levels and melting. As the raindrop is carried up in a growing cell, it freezes (FD) and grows rapidly by accretion. Alternately, the initial growth of a planar ice crystal (P) is by deposition to sizes determined by the LWC before accretional growth dominates. The particle grows as a graupel (G) before reaching hailstone sizes.

10.4.3. *Moisture fluxes and precipitation efficiency*

Precipitation efficiencies for convective clouds are generally estimated from observations of the moisture influx in the lower portions of the cloud and from estimates of the precipitation at the ground. Since most precipitating convective systems are not steady state, it is necessary to integrate these fluxes over the life cycle of the cell.

Braham (1952) describes the moisture fluxes for a typical thunderstorm based on observations of Florida and Ohio thunderstorms made by the Thunderstorm Project. The moisture influx was estimated using a swarm of constant level balloons to determine convergence and radiosondes to determine the water vapor mixing ratio. The surface rainfall was estimated on the basis of a network of precipitation gauges.

Braham found the total influx of water vapor was 890 kton[5] of which 360 kton never condensed, leaving a total condensate of 530 kton. The total rainfall was 100 kton, resulting in a precipitation efficiency of 18%. Of the remaining condensate, 240 kton evaporated in the downdraft, leaving 190 kton that evaporated due to entrainment or remained as non-precipitating cloud.

Newton (1966) examined the moisture budgets for squall-line thunderstorms which he assumed were essentially steady state. He estimated the water vapor flux entering the updraft as 8.8 kton s^{-1}, of which 0.6 kton s^{-1} never condensed, leaving a total condensate flux of 8.2 kton s^{-1}. The precipitation flux based on measurements was 4.7 kton s^{-1} for a precipitation efficiency of 57%. Downdraft evaporation accounted for 3.6 kton s^{-1}, and 0.6 kton s^{-1} was exported to the anvil. The precipitation flux determined by subtracting the downdraft evaporation and the anvil export from the total condensate was 4.0 kton s^{-1} for a precipitation efficiency of 49%. This discrepancy gives an idea of the magnitude of the errors in estimating precipitation efficiencies using these methods. Auer and Marwitz (1968) found an average precipitation efficiency of 55% for High Plains hailstorms.

A precipitation efficiency of 15% was reported by Foote and Fankhauser (1973) for a supercell thunderstorm in a highly sheared environment. Strong vertical wind shear apparently increases the fraction of the condensate that is exported to the anvil, thereby reducing the overall precipitation efficiency. Figure 10.20 summarizes the relation between the precipitation efficiency of convective clouds and the vertical wind shear.

Rosenfeld and Gagin (1989) examine the factors that are correlated with the total rainfall from convective clouds in Israel and South Africa in which the ice

[5] Units are in metric tons $\times 10^3$ (kton), or 10^9 grams.

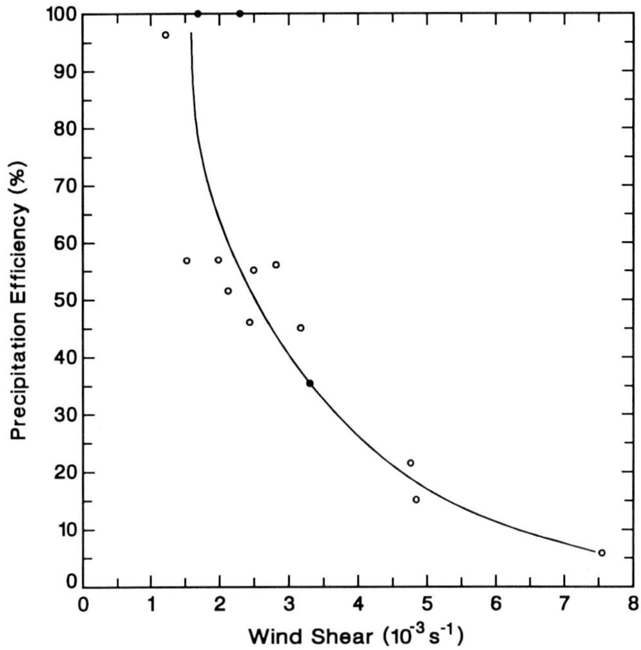

FIG. 10.20. Precipitation efficiency as a function of vertical wind shear in the layer from cloud base to cloud top. The different symbols represent different sources of precipitation efficiency observations. From Foote and Fankhauser (1973). *Journal of Applied Meteorology*, **12**, 8. American Meteorological Society. Reproduced with permission.

crystal mechanism is dominant. They find a 75% correlation between the height of the cloud above the 0°C level and total rain volume; an increase of 23% in the cloud height above the 0°C level is associated with doubling the total volume of rain. A 56% correlation was found between the number of radar reflectivity peaks with a cloud cluster and the total rain volume; most of the increase in rain volume was due to an increase in the area of precipitation, with increases in intensity and duration of less importance. Increasing the absolute humidity at cloud base was found to increase the intensity of the precipitation and the precipitation efficiency. Lower static stability was noted to reduce rain volume.

Problems

10.1* Consider a mixed phase cloud which contains small ice crystals in a concentration of $1\,\ell^{-1}$. Consider a large drop falling through such a cloud. If it collects one or more ice crystals, it will freeze. Assume the drop collects the ice crystals with a geometric collection efficiency of 1 and that the ice crystals are very small and have negligible fall velocities.

(a) What is the probability that a 0.5 mm radius drop will freeze in 100 seconds?

(b) What is the probability that a 1.0 mm radius drop will freeze in 100 seconds?

10.2** Consider a supercooled water cloud consisting of 7 μm radius droplets present in a concentration of 500 cm^{-3}. Assume an ambient temperature of −10°C and a pressure of 700 mbar. What concentration of lump graupel would just suffice to yield a graupel fall velocity of 1 $m\,s^{-1}$, assuming that all of the liquid water was converted to graupel?
Hint: Refer to Chapter 8.

10.3** Consider a supercooled water cloud at −10°C and 700 mbar, with a liquid water content of 0.5 $g\,m^{-3}$. Glaciate the very top of the cloud with 0.1 ℓ^{-1} of very small ice crystals. Assume all of the supercooled water in the top layer is converted into (roughly) spherical ice particles.

Allow the ice particles to grow by accretion as they fall through the water cloud. Calculate the precipitation rate ($mm\,h^{-1}$) out of the base of the cloud. You may assume the cloud droplets have 0 fall velocity and are collected by the ice particles with a collection efficiency of 1. You may assume the accreted ice density is 0.5 $g\,cm^{-3}$.

Repeat for ice particle concentrations of 1, 10, 100, 1000, and 1000 ℓ^{-1}.

Plot the precipitation rate as a function of ice particle concentration.

10.4** How long would it take to transform a 100 m thick layer with a pseudo-adiabatic lapse rate to an isothermal layer at 0°C by the melting of precipitation falling at a rate of 5 $mm\,h^{-1}$? Assume the ice particles melt completely before they exit the base of the 100 m thick layer. Assume the top of the layer initially is at 0°C and 800 mbar and that the layer is saturated with respect to water.

REFERENCES

Auer, A.H., Jr., and Marwitz, J.D., 1968, 'Estimates of air and moisture flux into hailstorms on the High Plains,' *J. Appl. Meteorol.* **7**, 196.

Braham, R.R., Jr., 1952, 'The water and energy budgets of the thunderstorm and their relation to thunderstorm development,' *J. Meteorol.* **9**, 227.

Braham, R.R., Jr., 1964, 'What is the role of ice in summer rainshowers? *J. Atmos. Sci.* **21**, 640.

Braham, R.R., Jr., 1986, 'The cloud physics of weather modification. Part I: Scientific basis,' *Wld Meteorol. Org. Bull.* **35**, 215.

Browning, K.A., 1979, 'Structure, mechanism and prediction of orographically enhanced rain in Britain,' *Global Atmos. Res. Prog.* Ser. 23, World Meteorological Organization, p. 88.

Browning, K.A., Fankhauser, J.C., Chalon, J.-P., Eccles, P.J., Strauch, R.C., Merrem, F.H., Musil, D.J., May, E.L., and Sand, W.R., 1976, 'Structure of an evolving hailstorm. Part V: Synthesis and implications for hail growth and hail suppression,' *Mon. Weath. Rev.* **104**, 603.

Byers, H.R., and Braham, R.R., Jr., 1949, *The thunderstorm.* Washington, DC: US Government Printing Office.

Carlson, T.N., 1980, 'Airflow through midlatitude cyclones and the common cloud pattern,' *Mon. Weath. Rev.* **108**, 1498.

Chappell, C.F., 1970, *Modification of cold orographic clouds.* Ph.D. Thesis, Colorado State University.

Choularton, T.W., and Perry, S.J., 1986, 'A model of the orographic enhancement of snowfall by the seeder–feeder mechanism,' *Quart. J. R. Meteorol. Soc.* **112**, 335.

Cotton, W.R., and Anthes, R.A., 1989, *Storm and cloud dynamics.* San Diego, CA: Academic Press.

Cotton, W.R., Tripoli, G.J., Rauber, R.M., and Mulvihill, E.A., 1986, 'Numerical simulation of the effects of varying ice crystal nucleation rates and aggregation processes on orographic snowfall,' *J. Climate Appl. Meteorol.* **25**, 1658.

Dirks, R.A., 1973, 'The precipitation efficiency of orographic clouds,' *J. Rech. Atmos.* **7**, 177.

Elliott, R.D., and Hovind, E.L., 1964, 'The water balance of orographic clouds,' *J. Appl. Meteorol.* **3**, 235.

Foote, G.B., and Fankhauser, J.C., 1973, 'Airflow and moisture budget beneath a northeast Colorado hailstorm,' *J. Appl. Meteorol.* **12**, 1330.

Fukuta, N., 1973, 'Thermodynamics of cloud glaciation,' *J. Atmos. Sci.* **30**, 1645.

Heymsfield, A.J., 1978, 'The characteristics of graupel particles in northeastern Colorado cumulus congestus clouds,' *J. Atmos. Sci.* **35**, 284.

Heymsfield, A.J., 1982, 'A comparative study of the rates of development of potential graupel and hail embryos in High Plains storms,' *J. Atmos. Sci.* **39**, 2867.

Heymsfield, A.J., 1986, 'Ice particle evolution in the anvil of a severe thunderstorm during CCOPE,' *J. Atmos. Sci.* **43**, 2463.

Hobbs, P.V., 1978; 'Organization and structure of clouds and precipitation on the mesoscale and microscale in cyclonic storms. *Rev. Geophys. Space Phys.*, 316, 741.

Hobbs, P.V., Matejka, T.J., Herzegh, P.H., Locatelli, J.D., and Houze, R.A., Jr., 1980, 'The mesoscale and microscale structure and organization of clouds and precipitation in midlatitude cyclones. I. Case study of a cold front,' *J. Atmos. Sci.* **37**, 568.

Houze, R.A., Jr., and Hobbs, P.V., 1982, 'Organization and structure of precipitating cloud systems,' *Adv. Geophys.* **24**, 225.

Johnson, D.B., 1987, 'On the relative efficiency of coalescence and riming,' *J. Atmos. Sci.* **44**, 1671.

Marwitz, J.D., 1974, 'An airflow case study over the San Juan mountains of Colorado,' *J. Appl. Meteorol.* **13**, 450.

Newton, C.W., 1966, 'Circulations in large sheared cumulonimbus,' *Tellus* **18**, 699.

Rosenfeld, D., and Gagin, A., 1989, 'Factors governing the total rainfall yield from continental convective clouds,' *J. Appl. Meteorol.* **28**, 1015.

Stewart, R.E., Marwitz, J.D., Pace, J.C., and Carbone, R.E., 1984, 'Characteristics through the melting layer of stratiform clouds,' *J. Atmos. Sci.* **41**, 3227.

Wexler, R., Reed, R.J., and Honig, J., 1954, 'Atmospheric cooling by melting snow,' *Bull. Am. Meteorol. Soc.* **35**, 48.

Willis, P.R., and Heymsfield, A.J., 1989, 'Structure of the melting layer in mesoscale convective system stratiform precipitation,' *J. Atmos. Sci.* **46**, 2008.

Young, K.C., 1974a, 'A numerical simulation of winter-time, orographic precipitation: Part I. Description of model microphysics and numerical techniques,' *J. Atmos. Sci.* **31**, 1735.

Young, K.C., 1974b, 'The conversion of a supercooled cloud to ice via contact nucleation and direct injection of ice crystals,' *Proc. Int. Conf. Cloud Physics 1974*. American Meteorological Society, p. 174.

11

THE MICROPHYSICAL BASES FOR MODIFICATION OF CLOUDS

11.1. Introduction

A wide variety of effects resulting from the modification of clouds have been postulated. These include precipitation enhancement and suppression, hail suppression, dispersal of fog, and lightning suppression. In some cases, the physical processes leading from the injection of seeding material to the desired effect are well understood. In other cases, one or more links in the chain of physical effects are poorly understood and the case for producing the desired seeding effect is weak.

Inadequate or improper dispersal of the seeding material may act to mask or negate any postulated seeding effect. Typically, the volumes that need to be seeded are on the order of cubic kilometers and the volume actually seeded may be only a small fraction of this. Problems also arise in delivering the seeding material to the appropriate region within the cloud to be seeded or even identifying the appropriate region to be seeded.

To overcome many of these difficulties, researchers have relied on statistical experiments to detect the effects of seeding operations. A statistical experiment attempts to compare randomly selected seeded and not-seeded events. Tests of significance can be applied to observed differences between the seeded and not-seeded events in an effort to determine whether the observed results could have occurred by chance. Due to the naturally large variability in most precipitating cloud systems, the seeding effect must be quite pronounced or the experiment must be run over a period of several years in order to conclude that seeding had an effect. Even then, without a coincident physical experiment, it is difficult to convince skeptics of a seeding effect, and it is generally not possible to transfer the results to another region.

11.2. Physical approaches to precipitation modification

Clearly, a physically valid concept designed to achieve any of the stated goals of a cloud modification activity must depend on a proper understanding of the natural processes and of the effects the treatment will have on these processes. A conceptual model such as, 'adding silver iodide nuclei to a cloud in order to increase the number of ice particles which will increase the rain' is not physically valid in this sense. In fact by itself, this is physically invalid because arbitrarily increasing the number of ice particles may ultimately result in a decrease in precipitation or even no precipitation at all.

Extending this concept to:

> uniformly dispersing silver iodide nuclei to enhance contact nucleation, thereby increasing the concentration of ice particles such that their growth by diffusion and accretion is not significantly reduced by competition and their ultimate size is such that they can fall to the ground without substantial evaporation losses, increasing precipitation

is a physically valid concept within our present knowledge. There are technological limits in regard to achieving the designated dispersal goals, and there are limits in our understanding of the production of ice particles by contact nucleation within clouds. There are also uncertainties in regard to the initial state of the cloud, and other processes may be operating which have been neglected. However, this concept is testable at each step, at least in principle.

The feasibility of any modification activity is also dependent on its cost to benefit ratio. If the cost exceeds the anticipated benefit, there isn't much point in carrying out the activity. Modification activities are usually assumed to take advantage of some inherent instability within the cloud, and rely on an initially small perturbation or modification becoming amplified in its effects. The addition of a few hundred grams of silver iodide (AgI) to a cloud to produce 100 kton of rain when and where it is needed for agricultural purposes would clearly be cost effective. Such initially small modifications will be termed *modification levers.*

Three principal types of modification levers have been proposed to alter precipitation processes in clouds. These are (i) modification of the ice crystal process, (ii) modification of the coalescence process, and (iii) modification of the dynamic processes within the cloud and its environment. Each of these will be examined in some detail in the following sections.

11.2.1. *Modification of the ice crystal process*

A wide variety of ways of modifying the ice crystal process have been proposed. The most viable of these is to enhance the concentration of ice crystals in clouds deemed deficient in naturally produced ice crystals. Since the efficiency of the ice crystal process is predicated on the proper balance between water droplet and ice crystal concentrations, it is clear that some optimal ice crystal concentration may be defined, which would be expected to vary from cloud to cloud.

(i) Modification levers. Modification of the ice crystal process may involve levers which operate at the level of ice nucleation, ice crystal growth, growth by accretion, or growth by aggregation. Conceivably, collection processes could be enhanced by charging the particles, or one could increase aggregation rates by

increasing the level of turbulence in the cloud. The growth of ice crystals by diffusion can be altered, and presumably reduced, by the adsorption of certain organic vapors. The means by which such levers could be employed is not clear and their cost-to-benefit ratio appears to be high.

The modification levers most commonly relied on either enhance the concentrations of active ice nuclei or directly inject additional ice particles into the cloud. Such methods of enhancing the concentrations of ice crystals within a cloud are termed *glaciogenic seeding*. The direct injection of ice particles is carried out by rapidly cooling a small volume of cloud air to temperatures colder than −40°C. The mechanism by which ice crystals are formed in this way is subject to some dispute. Vonnegut (1981) supports the homogeneous deposition hypothesis; Mason (1981) suggests homogeneous freezing of water droplets formed by homogeneous condensation. In either event, the numbers of ice crystals produced is too large to be explained by the activation of cloud condensation nuclei (CCN) in the high supersaturations caused by the rapid cooling followed by homogeneous freezing (see Problem 11.3).

The cooling of small volumes of cloud air to temperatures colder than −40°C is most commonly achieved by dropping pellets of dry ice[1] into the cloud. Alternately, rapid expansion resulting from the release of highly pressurized gas from an orifice can have a similar effect. Eadie and Mee (1963) estimate the production of ice crystals by falling dry ice pellets as roughly 10^{10} crystals per gram of dry ice based on laboratory experiments. At ambient temperatures warmer than −5°C, they note that the number of ice crystals produced decreases markedly as the fall velocity of the pellet increases. Holroyd et al. (1978) estimate the production of ice crystals by dry ice pellets falling through natural clouds as roughly 10^{11} crystals per gram.

Enhancement of the concentrations of active ice nuclei has most often been accomplished using silver iodide (AgI) or lead iodide (PbI_2). Several other inorganic and organic substances that have been found to produce enhanced concentrations of ice crystals are listed in Table 4.1. Dispersal of AgI is usually achieved by burning an acetone solution of AgI complexed with sodium iodide (NaI), potassium iodide (KI), or ammonium iodide (NH_4I), or by burning a pyrotechnic mixture containing silver iodate ($AgIO_3$). These methods are discussed in more detail by Ruskin and Scott (1974) and Dennis (1980).

Modification of the lattice structure of AgI by partial substitution of the silver or iodide ions has been undertaken in an effort to improve the fit between the AgI and ice lattices. An improved lattice fit would be expected to reduce the lattice strain and promote formation of ice at warmer temperatures. DeMott et al. (1983) compare the effectiveness of AgI aerosol formed by burning an AgI–NH_4I acetone solution with various concentrations of ammonium perchlorate (NH_4ClO_4). The resulting AgI–AgCl aerosol provides an equivalent number of ice crystals per gram compared with AgI, at temperatures roughly 1°C warmer, as shown in Fig. 11.1. Palanisamy et al. (1986) attribute the enhanced nucleating

[1] Dry ice is solid CO_2, which sublimes at −78°C at 1000 mbars.

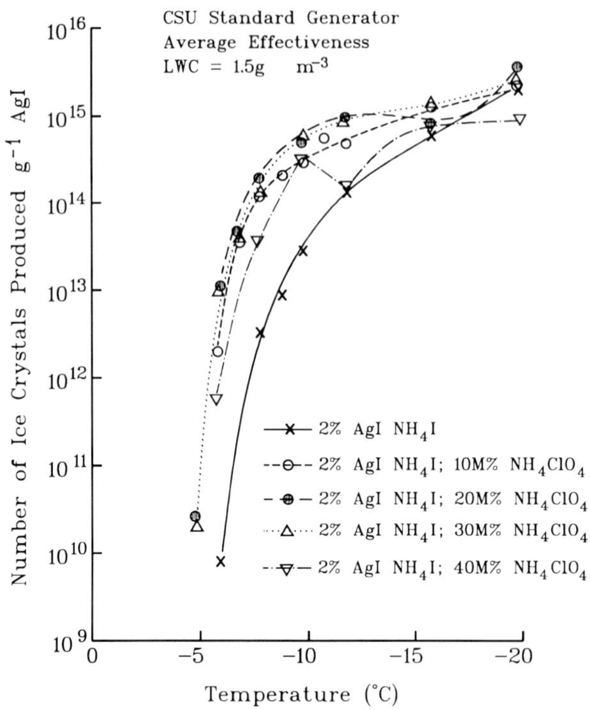

Fig. 11.1. Number of ice crystals produced per gram of AgI as a function of the mole fraction of NH_4ClO_4. The aerosols formed are a solid solution of AgI and AgCl in varying ratios (mole %). From DeMott et al. (1983). *Journal of Climate & Applied Meteorology*, **22**, 7. American Meteorological Society. Reproduced with permission.

ability of the AgI–AgCl aerosol to a small contraction in the AgI lattice, providing a better match with ice. Scott et al. (1989) describe substitution of bismuth and oxygen in the AgI lattice by burning a solution containing bismuth triiodide (BiI_3).

The ice forming ability of AgI in clouds appears to be predominantly *via* contact freezing nucleation (DeMott, 1991). The yields of ice crystals per gram of $AgI_{0.76}Cl_{0.24}$ for contact freezing and condensation freezing nucleation are shown in Fig. 11.2 based on experiments using the Colorado State Univeristy (CSU) isothermal cloud chamber. Condensation freezing nucleation was achieved by injecting the sample at 25°C at water saturation into the cloud chamber; this produced ~3×10^{14} crystals per gram within 1 minute. Contact freezing acts over a longer period of time, relying on the collection of the aerosol by Brownian motion and phoretic forces, with the number of ice crystals reaching 10^{14} per gram after ~20 minutes. DeMott suggests that the limited opportunities for activation *via* condensation freezing imply that contact freezing nucleation is the dominant mode of action for AgI in clouds. Finnegan and Pitter (1987) suggest that AgI ground generators located within an orographic (target) cloud can form ice crystals via condensation freezing nucleation. Feng and

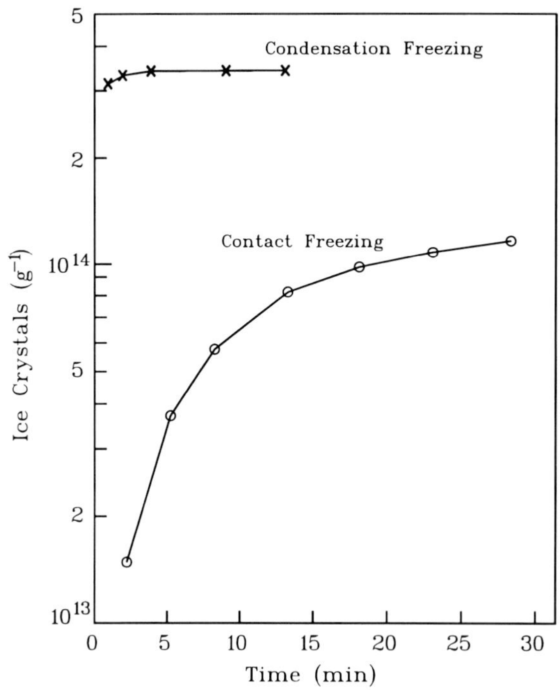

Fig. 11.2. The number of ice crystals per gram of $AgI_{0.76}AgCl_{0.24}$ resulting from contact freezing and condensation freezing nucleation at $-10°C$. The number of ice crystals formed by contact nucleation increases with time as the nuclei are collected by the water drops; virtually all of the ice crystals formed by condensation freezing are formed within the first minute. After DeMott (1991). *Journal of Applied Meteorology*, **30**, 9. American Meteorological Society. Reproduced with permission.

Finnegan (1989) claim the addition of sodium chloride (NaCl) to AgI·AgCl aerosol also promotes condensation freezing nucleation.

(ii) Optimal ice nucleus concentrations. If natural ice crystal concentrations are relatively low, increasing the concentration of ice crystals can increase the precipitation rate by converting a larger fraction of the cloud water to precipitation particles. As the concentration of ice crystals is increased, competition for the available liquid water increases, reducing the ultimate sizes of the ice particles. If this is carried to the extreme, the resulting ice particles will be too small to be able to fall to the ground before subliming (or melting and evaporating) completely. Thus, for a given cloud, there must be some optimal ice crystal concentration that will maximize the amount of precipitation from the cloud.

The optimal ice crystal concentration may be assessed using a parcel model and examining the time required to develop precipitation. Rokicki and Young (1978) simulate the development of precipitation in a 3 m s^{-1} updraft with a cloud base at +5°C for ice crystal concentrations from 1 ℓ^{-1} to 1000 ℓ^{-1}. For

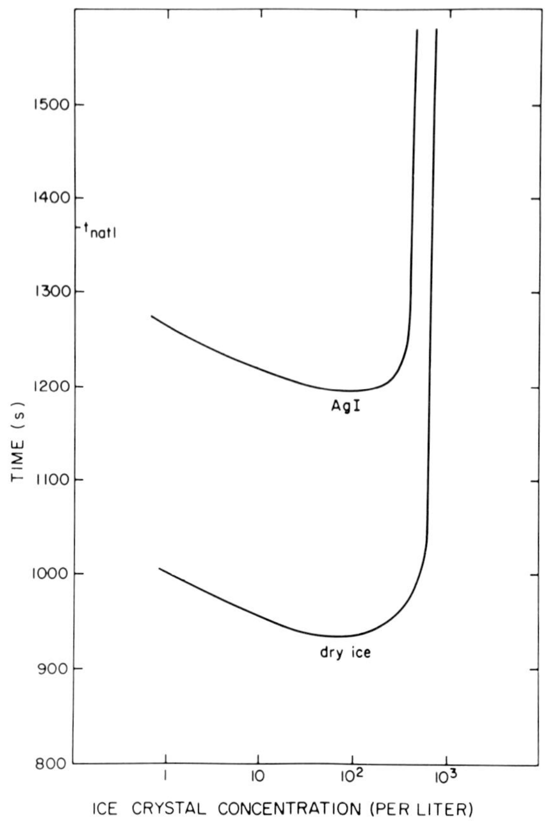

FIG. 11.3. The time required for formation of a 20 dBZ radar reflectivity as a function of ice crystal concentration. The crystal concentrations are those at the $-10°C$ level produced by the seeding and natural processes combined. t_{natl} refers to the time required to produce a 20 dBZ reflectivity without seeding. The cloud base is at 5°C with a constant updraft of 3 m s^{-1}. From Rokicki and Young (1978). *Journal of Applied Meteorology*, **17**, 6. American Meteorological Society. Reproduced with permission.

these conditions, Fig. 11.3 suggests that the optimal ice crystal concentration is roughly 100 crystals ℓ^{-1} for both direct injection of ice crystals (dry ice seeding) and for formation of ice crystals by AgI seeding. The direct injection of ice crystals is assumed to produce ice crystals at the $-2°C$ level; the AgI seeding continues to produce ice crystals up to the $-20°C$ level. The time required to produce precipitation via direct injection is about 280 seconds less than the time via AgI seeding, which, in turn, is about 175 seconds less than the natural case.

The reduction in time required to produce a 20 dBZ reflectivity may be considered to be a measure of the magnitude of the seeding effect. A relatively small reduction in the time required to produce precipitation probably will not result in a substantial change in the amount (or duration) of precipitation. A large reduction in the required time may differentiate between a non-precipitating

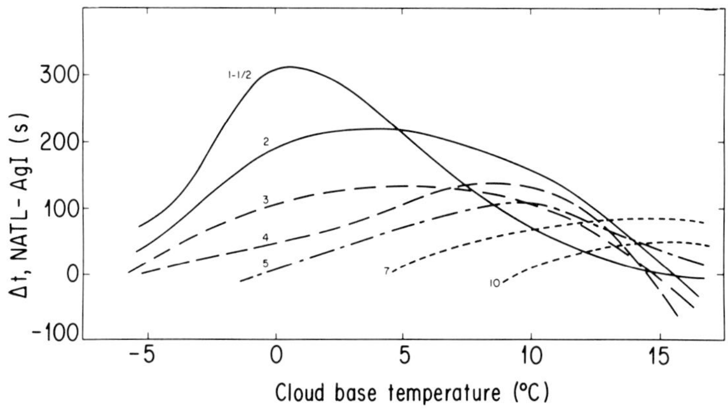

Fig. 11.4. The reduction in time required to produce a 20 dBZ radar reflectivity in a AgI seeded cloud compared to a natural cloud. Seeding rates were chosen to be close to the optimal seeding rate in each case. Updraft velocities are given in $m\,s^{-1}$. From Rokicki and Young (1978). *Journal of Applied Meteorology*, **17**, 6. American Meteorological Society. Reproduced with permission.

natural cloud and a precipitating seeded one, especially if the lifetime of the cloud or its depth are insufficient to produce precipitation naturally.

The reduction in time required to produce a 20 dBZ radar echo is shown in Fig. 11.4 as a function of the cloud base temperature and the updraft velocity. It is evident that AgI seeding would be expected to have its greatest effect for cloud base temperatures between 0°C and +5°C. At colder temperatures, AgI is effective but the natural process is also quite efficient. At warmer temperatures, the coalescence process is more rapid. Weaker updrafts tend to exhibit the greatest effect.

(iii) Windows of seeding opportunity. Clouds seedable using glaciogenic methods may be defined as those clouds which do not already have an efficient precipitation mechanism and whose temperatures are cold enough for formation of ice crystals by seeding, with lifetimes long enough for precipitation to develop. Hudak and List (1988) partition cumulus clouds in South Africa by air mass type and find only those clouds in maritime tropical air masses with cloud top temperatures warmer than −20°C are seedable, estimating a potential increase in precipitation efficiency from 2% to 15%. Continental tropical clouds are noted to possess ample concentrations of ice particles and maritime polar clouds with tops colder than −15°C are noted to have naturally efficient accretion processes. Maritime polar clouds with tops warmer than −15°C are found to have lifetimes too short to produce precipitation.

Rokicki and Young (1978) define the optimal AgI seeding rates as a function of the cloud base temperature and updraft velocity, as shown in Fig. 11.5. Region I is characterized by relatively cold cloud base temperatures and strong updrafts. Simulations suggest that precipitation would not develop via the ice crystal mechanism, regardless of seeding. Region III is characterized by relatively warm cloud bases and weaker updrafts, and relatively high optimal seeding rates with

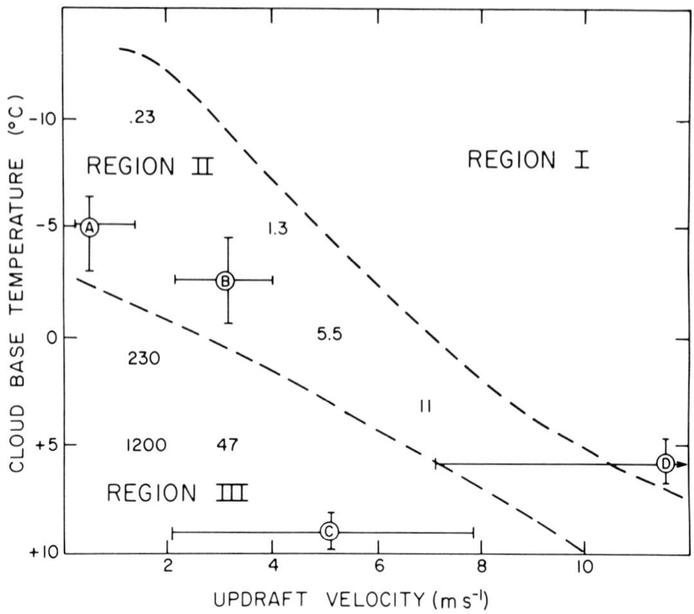

Fig. 11.5. Regions of differing response to AgI seeding. In Region I, no precipitation occurs, regardless of seeding; conditions in Region II suggest it is easy to overseed; conditions in Region III suggest it is difficult to overseed. Plotted numbers represent optimal seeding rates in g km^{-2} min^{-1}. Circled letters indicate the range of clouds seeded by various projects: (A) Climax I (Grant et al., 1971); (B) Great Lakes Project (Weickmann, 1974); (C) Israeli Project (Gagin, 1975) and (D) Santa Barbara Project (Elliott and Hovind, 1964). From Rokicki and Young (1978). *Journal of Applied Meteorology*, **17**, 6. American Meteorological Society. Reproduced with permission.

regard to producing a radar first echo. Optimal seeding rates in Region II are relatively low, suggesting that low seeding rates should be employed to avoid overseeding.

Combining Figs. 11.4 and 11.5, as shown in Fig. 11.6, the maximum AgI seeding effect occurs in Region III (difficult to overseed) for relatively weak updraft velocities and cloud base temperatures between 0°C and +10°C. At colder cloud base temperatures, the natural ice crystal process is able to achieve the relatively low optimal ice crystal concentrations required (Region II) and AgI seeding has little or no effect. At warmer cloud base temperatures, the coalescence mechanism is efficient and there are suggestions that glaciogenic seeding may reduce the liquid water available for accretional growth, thereby reducing total precipitation.

Of the seeding projects shown in Fig. 11.5, only the Climax and Israeli Projects have been suggested to exhibit increases in precipitation due to seeding.[2] Climax reported increases for the warmer clouds, consistent with Fig. 11.6. Clouds seeded during the Israeli Project lie mostly within the 100 s seeding effect isochrone.

[2] Questions regarding the positive seeding effect reported for the Climax Project have been raised and are discussed later in this chapter.

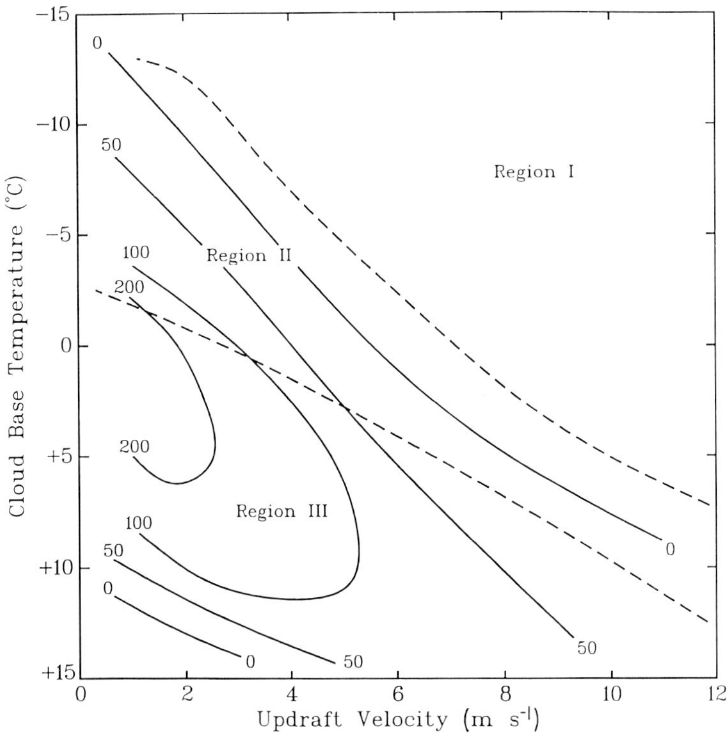

FIG. 11.6. Seedability isopleths as a function of cloud base temperature and updraft velocity. Isochrones of the reduction in time to produce a 20 dBZ radar reflectivity for optimally AgI-seeded updrafts compared to natural clouds are given in seconds. Regions of response to AgI seeding are as defined in Fig. 11.5.

A proper definition of the seeding window for a given cloud must consider the life cycle of the cloud and the variations in updraft velocity as the cloud evolves. The implicit assumption in Figs. 11.4 through 11.6 is that the cloud lifetime is sufficiently long to produce precipitation, and that the updraft remains fairly steady throughout the period of precipitation development. These simulations simply demonstrate that 'windows of opportunity' for precipitation enhancement *via* glaciogenic seeding exist for those clouds whose ice crystal and coalescence mechanisms are not optimal.

(iv) Dispersal of seeding material. The efficacy of the seeding application depends on adequate dispersal of the seeding material throughout the volume to be seeded. Silver iodide generators are point sources, whether they are carried by aircraft or ground-based. Seeding with dry ice pellets tends to produce a curtain of seeding material as the pellets fall from the aircraft. In either case, the seeding material or the small ice crystals produced by the seeding agent must be dispersed more or less uniformly throughout a volume which is much larger than the volume initially seeded.

A Gaussian plume model is often used to describe the dispersal of seeding material from ground-based generators. Dennis (1980) relates the concentration of seeding material (n) to the source strength (n_o), the mean horizontal wind speed (U) and the horizontal and vertical standard deviations of the plume (σ_y and σ_z) as

$$n = \frac{n_o}{8\pi U \sigma_y \sigma_z} \exp\left[-\frac{y^2}{2\sigma_y^2} - \frac{z^2}{\sigma_z^2}\right].$$

The standard deviations of the plume in the horizontal and vertical directions may be expected to be different, the latter being directly dependent on the static stability of the atmosphere. The standard deviation of the plume in the horizontal depends on the distance (time) downwind of the source. Within the first 10 to 15 minutes, σ_y varies as time cubed; decreasing to a linear dependence on time after about an hour.

The spread of the plume in the vertical is markedly dependent on the stability of the atmosphere and the topography. Ground-based seeding of convective clouds would be expected to be more effective under conditions of low static stability, allowing convective elements to carry the seeding material upward into the clouds. Attempts to seed orographic clouds from generators sited in valleys may be unsuccessful if the valley air is trapped under a temperature inversion. This suggests siting generators on higher ground, which often reduces the time available for the seeding plume to spread since there may be no suitable mountains far enough upwind to provide effective dispersal times.

Seeding with an AgI-generator from an aircraft flying perpendicular to the mean wind results in an expanding conical plume moving downwind. Dennis (1980) suggests that the turbulence induced by the aircraft results in an initial plume diameter of about 10 m, rapidly spreading within a few minutes to a few hundred meters. Subsequent spreading would be much slower. Repeated seeding passes would be required to produce a relatively continuous plume of seeding material. Project Whitetop utilized three seeding aircraft, each shuttling back and forth along its own 16 km track, to produce a relatively continuous seeding plume.

Seeding below a convective cloud base allows the seeding material to be carried upward into the cloud. The time available for the plume to spread before reaching the −5°C level is inversely related to the updraft velocity. Dennis estimates that turbulence within moderate updrafts would be sufficient to fill a typical convective updraft 500 m in diameter but in larger, more intense updrafts, as little as 1% to 2% of the updraft may be seeded.

There is a trade-off between the need to target a particular cloud and the need to disperse the seeding agent over a large volume. The further the generator site is from the target cloud, the more difficult it is to assure that the seeding material reaches its target. The closer to the target cloud the seeding agent is released, the less dispersed is the seeding material. The presence of stable layers between the generator site and the target cloud and/or the presence of strong wind shear may prevent the seeding material from reaching the target cloud. Proper dispersal of the seeding agent remains perhaps the greatest obstacle to the success of glaciogenic seeding.

(v) Extra-area effects. Extra-area effects may be defined as any modification of clouds (or precipitation) outside of the intended area of effect (target area). This may involve the persistence of the seeding material in the target area, carry over of the seeding material into the control area, or downwind effects affecting areas well outside the research area.

Persistence of a seeding effect has been postulated by direct action of AgI remaining in the target area and secondary effects resulting from changes in CCN production or changes in moisture or heat fluxes from ground wetted by presumed increases in precipitation (NRC, 1973). Gabriel et al. (1967) examined data from the Israeli project and found no evidence for persistence up to two days following the seeding. Bigg and Turton (1988) found evidence for an increase in precipitation for a variety of Australian seeding projects, starting about 9 to 10 days after seeding and lasting for about 9 days. Mather et al. (1990) find a similar anomaly commencing 11 to 12 days after seeding in the Nelspruit, South Africa area and lasting for 10 to 11 days. No physical hypothesis has been advanced to explain such anomalies.

Strong evidence supporting pooling of AgI aerosol under stable conditions during the Colorado River Basin Pilot Project conducted in southwestern Colorado has been given by Marwitz (1980) and Elliott et al. (1978). Marwitz shows that the early stages of a storm are characterized by low level stability, during which the low level flow is parallel to the orographic barrier. During this period, AgI aerosol remains in the valleys, while, aloft, air flows over the barrier. Elliott et al. show that AgI released on seeding days was trapped at low levels, moving westward along the barrier, to be released on the subsequent (control) day. They considered 18 control days to be contaminated by such a carry over effect. Neyman et al. (1969) find evidence for carry over effects at distances in excess of 160 km in Project Whitetop (negative seeding effect) and Grossversuch III (positive seeding effect).

Speculation of the effects of seeding extending well outside the research area have been raised almost since the initiation of modern weather modification activities. A variety of studies show both increases and decreases in precipitation, both downwind and upwind of the research area and at distances of 290 km or more. Many of these are summarized in NRC (1973).

11.2.2. *Modification of the coalescence process*

Modification of the coalescence process differs from modification of the ice crystal process in that one may expect continued increases in precipitation as the amount of seeding agent is increased, i.e., there is no 'optimal' seeding rate.[3] This is due to the drop breakup process, which serves to limit the growth of

[3] The numerical simulations described by Nelson (1971) suggest small reductions in precipitation intensity, and total amounts may result from high seeding rates.

individual raindrops while producing additional 'seeds' to spread the coalescence conversion process to other regions of the cloud. Provided the lifetime of the cloud is sufficiently long, a relatively small injection of seeding agent may be expected to accelerate development of precipitation which could then spread throughout the cloud.

(i) Modification levers. Addition of relatively large hygroscopic particles or water drops is considered the most feasible approach to modifying the coalescence process. Other approaches, such as enhancing coalescence between colliding drops through addition of a surfactant to reduce the surface tension of the drops or to increase the collision efficiency by placing electric charges on the drops, have been proposed. Electric charges on drops would also be expected to enhance coalescence between colliding drops. The means by which surfactants could be dispersed or how appropriate charges could be placed on drops have not been fully defined, nor does the cost-to-benefit ratio seem favorable (see Problem 11.6).

Addition of large hygroscopic particles or water drops would be expected to broaden the droplet spectrum and thereby enhance the development of precipitation by coalescence. As shown by Fig. 9.7, the likelihood of a drop collecting another droplet increases as the large drop size is increased. Since the time between collections tends to decrease as the drop grows, addition of drops somewhat larger than the existing droplets can markedly shorten the time required to grow drops to millimeter sizes, bypassing the need for the first few (slow) collection events. A more rapid production of raindrops by coalescence would be expected to produce a detectable radar 'first echo'[4] at a lower height in an updraft. This could result in precipitation in a cloud that would not otherwise precipitate, or could extend the duration of precipitation in a cloud that would produce precipitation naturally.

Hygroscopic materials have an advantage in seeding to enhance the coalescence process in that they deliquesce at relative humidities less than 100%. Sodium chloride (NaCl), for example, deliquesces at 78% humidity. If dry NaCl particles are dispersed in an updraft below a cumulus cloud, the particles will extract water vapor from their environment and become solution droplets. They would then grow in the same manner that other CCN grow below the cloud base. Although large hygroscopic seed particles would be expected to lag significantly behind their equilibrium radius, even a doubling of their dry radius by the cloud base would effectively multiply the mass of seeding material delivered to the cloud by nearly an order of magnitude (see Problem 11.7).

Sodium chloride or urea are commonly chosen as seeding materials to enhance the coalescence process. Silverman and Weinstein (1974) suggest that these two substances have similar efficiencies for use in clearing fogs, which implies that they would have similar effects on enhancing the coalescence process. Nelson and Silverman (1972) describe the use of microencapsulation techniques to provide a well-defined particle size spectrum with reduced

[4] A radar first echo is usually associated with a 20 dBZ reflectivity.

clumping and caking. Dennis and Koscielski (1972) describe the use of a mixture of finely ground NaCl and soapstone in aircraft seeding of convective clouds in South Dakota. Biwas et al. (1967) describe ground-based spraying of an NaCl solution to seed clouds in India. Urea may also be dispersed as a powder or as a slurry (Ruskin and Scott, 1974).

Braham et al. (1957) describe experiments in which cumulus clouds were seeded by dispersing large water drops from an aircraft. A large valve was used to disperse water, emptying a 400 gallon tank in 18 seconds for a dispersal rate of 1 kg m^{-1}. The median seed drop size was ~400 μm radius and seed drops were spread over a 15 m wide path. Seeding was carried out about 1000 m above the cloud base, allowing the large seed drops to grow as they fell through the cloud.

There is a trade-off between the seed particle size and the seeding effect per particle. Small seed particles do not possess a substantial advantage over the naturally occurring droplets in terms of coalescence growth, but many more particles can be obtained from a given mass of seeding agent. Large seed particles may grow rapidly by collection but relatively fewer large particles can be obtained from a given mass. In addition, larger particles have greater fall speeds and may fall from the cloud before they can affect a significant fraction of the cloud. This suggests there is an appropriate (optimal) seed particle size for a given cloud.

(ii) Hygroscopic versus AgI seeding effects. An estimate of the relative magnitudes of the effect of seeding with large drops (or hygroscopic materials) is provided by Rokicki and Young (1978) using a parcel model to determine the time required to produce a 20 dBZ radar echo (Fig. 11.7). Reductions in the time

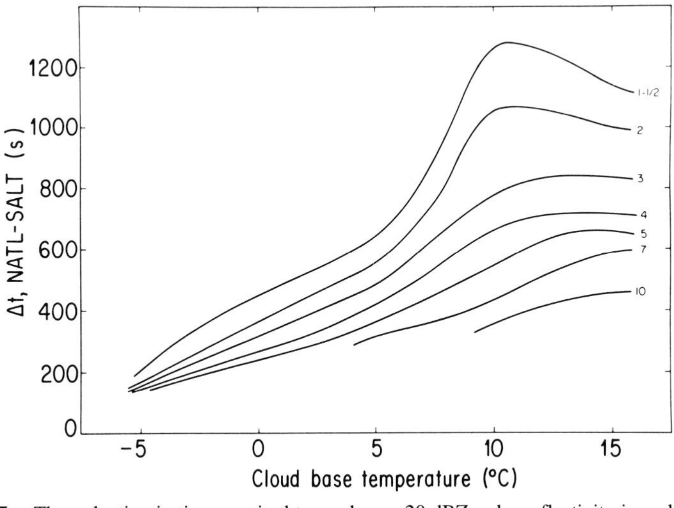

Fig. 11.7. The reduction in time required to produce a 20 dBZ radar reflectivity in a cloud seeded with large drops compared to a natural cloud. A seeding rate of 100 g km^{-2} min^{-1} is used with a seed drop size corresponding to drops having a fall speed of one-tenth of the updraft velocity (m s^{-1}). From Rokicki and Young (1978). *Journal of Applied Meteorology*, **17**, 6. American Meteorological Society. Reproduced with permission.

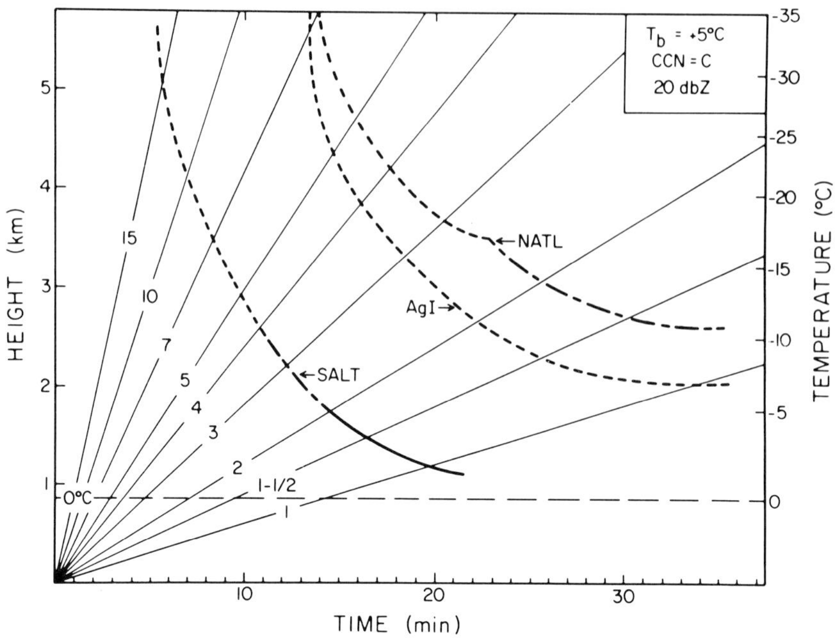

FIG. 11.8. Time required to reach a 20 dBZ reflectivity for natural, AgI-seeded, and large drop (salt) seeded cases for a cloud base temperature of +5°C. Seeding rates are as given for Figs. 11.4 and 11.7. Updraft velocities (m s^{-1}) are represented by the lines sloping upward to the right. Solid lines indicate that >90% of the large particles (>100 μm radius) are water; broken lines indicate >90% of the large particles are ice; dot-dashed are intermediate (mixed water and ice). From Rokicki and Young (1978). *Journal of Applied Meteorology*, **17**, 6. American Meteorological Society. Reproduced with permission.

to a 20 dBZ echo are noted for all conditions with the largest reductions (in time) for cloud base temperatures near +10°C and for the weakest updrafts. For clouds with base temperatures warmer than +10°C, the natural coalescence process is relatively efficient; for clouds with colder bases, the efficiency of the coalescence process whether seeded or natural is lower due to the lower liquid water content.

The relative efficiencies of seeding with large drops and seeding with AgI are shown in Fig. 11.8. The reduction in the height of the 20 dBZ 'first echo' is greater for large drop seeding than for AgI seeding in all cases. This finding is consistent with the results of Dennis and Koscielski (1972), who found that the average first echo height in AgI-seeded clouds was 2100 m above the cloud base whereas the average first echo height in salt-seeded clouds was 1600 m. The average first echo height in the unseeded clouds was 3350 m above the cloud base for a reduction of 1750 m in the salt-seeded clouds. Figure 11.7 suggests a reduction of 2000 m for the 2 m s^{-1} updraft with a cloud base temperature of +10°C.

(iii) Optimal seed drop size. The time required to produce a radar first echo by seeding with large drops is quite sensitive to the seed drop size. Figure 11.9

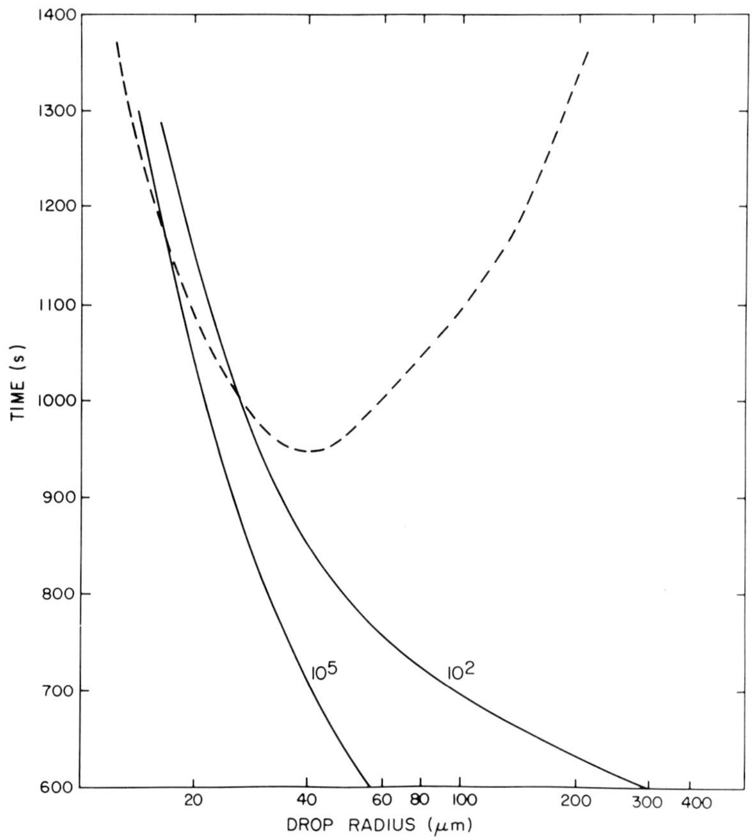

FIG. 11.9. Time required for a 20 dBZ first echo as a function of the seed drop radius. Solid lines represent seeding rates of 10^2 and 10^5 g km^{-2} min^{-1}. The broken line includes the estimated effects of sedimentation. The cloud base is +15°C with an updraft velocity of 3 m s^{-1}. From Rokicki and Young (1978). *Journal of Applied Meteorology*, **17**, 6. American Meteorological Society. Reproduced with permission.

shows the time required to produce a 20 dBZ radar echo for two different seeding rates. Larger seed drops tend to produce precipitation more quickly but would be expected to fall out of the parcel more rapidly (broken line). This serves to define an optimal seed drop size, which would be expected to vary depending on the updraft velocity and the liquid water content of the cloud.

The sensitivity of the seeding effect to the seed drop size suggests that relatively small changes in the seed drop size can have the same effect as changing the seeding rate by an order of magnitude or more. According to Fig. 11.9, an increase in the seed drop radius from 26 μm to 34 μm would permit a reduction in the seeding rate by a factor of 1000, with the same result. This suggests that the size spectrum of seed particles should be carefully controlled and optimized for each cloud. It appears likely that most of the seeding material used in field experiments to enhance the coalescence process was not dispersed in the optimal size range, and was thereby wasted.

Airborne dispersal of hygroscopic seeding material or large drops has been carried out both at the cloud base and at varying heights above the cloud base. Dispersing seeding material at the cloud base assumes that the material will be carried into the cloud by the updraft. The seed particles are assumed to grow as they rise, reaching a balance level near the top of the cloud, and continuing to grow as they fall back down through the cloud. If the seed particles are too large, they may never enter the cloud, or they may directly affect only the lower portions of the cloud. Relatively small seed particles would be expected to be carried to the upper portions of the cloud, where the vertical velocities weaken. Although small seed particles would remain in the cloud and continue to grow, somewhat larger seed particles would provide a more rapid development of precipitation.

Dispersing the seeding material at or near the top of the cloud would permit larger seed particles since they would not need to be lifted from the cloud base by the updraft but a greater mass of seeding material would be required. The more rapid development of precipitation using larger seed particles may permit enhancement of precipitation in clouds too shallow or too short-lived to produce precipitation when seeded from the cloud base. Both the updraft velocity and cloud depth would be expected to determine the optimal seed drop size.

11.2.3. *Modification of dynamical processes*

Glaciation of a supercooled cloud releases latent heat, resulting in an increase in the ambient temperature as given by (10.2). Since the glaciation usually does not occur with vapor phase balance, the final ice water content (IWC) is generally not the same as the original liquid water content (LWC). The contribution to the change in buoyancy of the parcel due to the change in the amount of condensate is generally much smaller than the contribution due to the change in temperature (Orville, 1986).

Increasing the buoyancy of the parcel would be expected to result in taller clouds. This would increase the total condensate, although the conversion to relatively small ice crystals may increase the amount of condensate that enters the anvil, offsetting the increase in total condensate. Woodley et al. (1982) postulate that the invigorated updraft would lead to pressure falls at the surface and increased inflow at low levels. Simpson (1980) suggests the acceleration of the updraft resulting from seeding would enhance entrainment of drier, environmental air into the cloud, resulting in more evaporation and stronger downdrafts. As the downdrafts spread out near the surface, they would promote growth of secondary towers. Woodley et al. consider that this would lead to cloud mergers, resulting in increased precipitation efficiency due to larger storms.

Orville (1986) points out that a stronger updraft may result in a net decrease in precipitation if the additional condensate injected into the anvil exceeds the additional condensate resulting from enhanced low-level convergence. Simpson (1980) notes that the enhanced mid-level entrainment could result in a detached or cutoff turret with little or no change in the precipitation. Numerical simulations by Levy and Cotton (1984) suggest that surface pressure decreases due to seeding would be negligible, although acceleration of the updraft, leading to enhanced entrainment at the seeding levels, would be significant. They point out that the seeding *must* enhance the precipitation at the ground in order to promote cloud growth and subsequent precipitation.

Seeding to enhance the updraft by conversion of the supercooled water to ice must proceed fairly rapidly in order to have the desired dynamical effect. Young (1975) simulates the glaciation process as a function of seeding rate for both AgI and dry ice seeding for several updraft profiles. Figure 11.10 shows the extent of glaciation as a function of height (as given by temperature) for AgI seeding rates up to $10\,g\,km^{-2}\,min^{-1}$ for an updraft which accelerates from $2\,m\,s^{-1}$ at cloud base (+9°C) to $8\,m\,s^{-1}$ at the -40°C level. Seeding rates of roughly $100\,g\,km^{-2}\,min^{-1}$ would be required to convert half of the liquid water

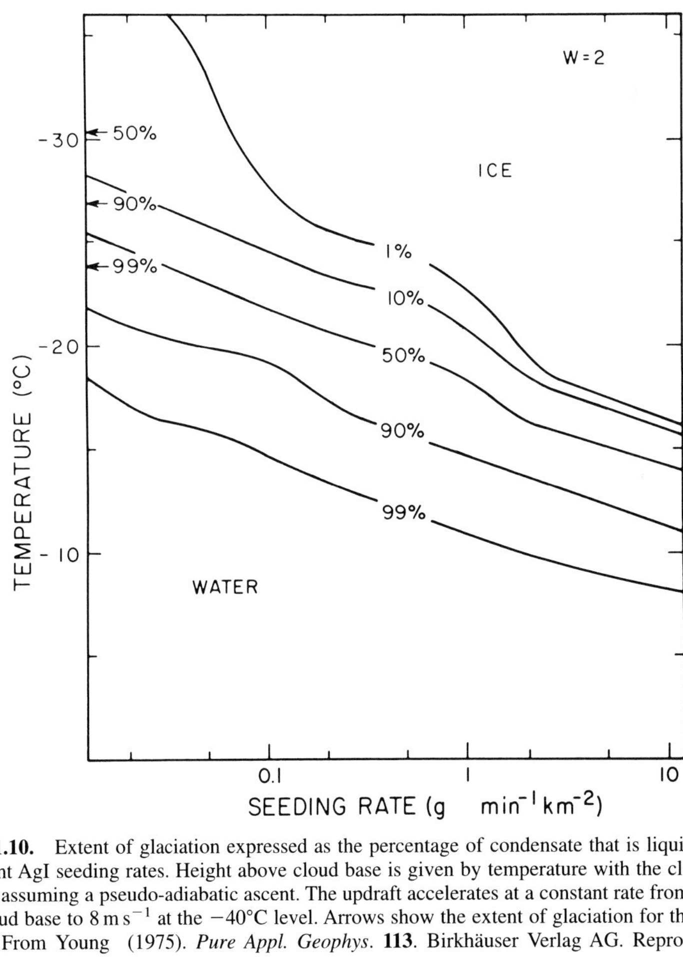

FIG. 11.10. Extent of glaciation expressed as the percentage of condensate that is liquid water for different AgI seeding rates. Height above cloud base is given by temperature with the cloud base at +9°C, assuming a pseudo-adiabatic ascent. The updraft accelerates at a constant rate from $2\,m\,s^{-1}$ at the cloud base to $8\,m\,s^{-1}$ at the -40°C level. Arrows show the extent of glaciation for the unseeded cloud. From Young (1975). *Pure Appl. Geophys.* **113**. Birkhäuser Verlag AG. Reproduced with permission.

to ice by the $-10°C$ level. These seeding rates appear to have been achieved in experiments described by Simpson et al. (1967), for example, who assumed the glaciation process was complete by the $-8°C$ level.

11.3. Statistical approaches to precipitation modification

This section is intended to provide only an introduction to statistical approaches to precipitation modification. The reader is referred to NRC (1973), Dennis (1980), or Flueck (1986) for more in-depth discussions. Here, the relation between physical and statistical experiments, the design of statistical experiments, and some evaluation methods will be discussed.

11.3.1. *Physical versus statistical approaches*

Statistical approaches to any weather modification activity should not be seen as unrelated to and independent of an understanding of the physical processes involved. The 'believability' of any 'positive' result from a statistical experiment is enhanced when the results are also physically plausible. Flueck (1986) argues that believability is enhanced when the conceptual model is clearly and succinctly stated at each step, when it is internally consistent, with no proposed 'miracles,' and when it is consistent with past accepted evidence. Likewise, the credibility of the statistical evidence depends on the quality of the design and its implementation, whether or not randomized trials were conducted and their level of significance, the quality and internal consistency of the data, and the integrity of the investigation.

Each approach has its advantages and disadvantages. The principal advantages of a physical approach are that an optimal seeding design is possible, the results are transferable to other regions, and the experiment can be conducted in a relatively short time which means it is less expensive. The main disadvantage to a physical experiment is that the atmosphere is too complex to permit exact predictability. Thus, there is always some uncertainty in the chain of physical events leading to the predicted effect. The longer and less well-founded the chain of events, the weaker is the physical link between the application and the desired effect.

A statistical experiment can proceed with limited knowledge of the physical processes involved, although a more complete understanding of the physical processes can lead to a better design of the statistical experiment. A statistical experiment tests only a given experimental configuration, e.g., use of a particular seeding agent at a particular seeding rate, without giving any idea of how to modify the configuration to optimize the seeding effect,[5] if any is detected. Quite often, the results obtained in one geographical region cannot be applied directly in a different region. If the effects of seeding are relatively small compared to the variability in precipitation, several years of experimentation may be required to establish a seeding effect with an acceptable level of confidence.

[5] *Post hoc* stratification can help to identify subgroups of the original group of 'go' units that may be more seedable than others.

11.3.2. *Design of experiments*

The natural variability of the atmosphere precludes identifying two identical situations, seeding in one case and keeping the other as a control. Thus, it is necessary to define a set of conditions such that the group of candidate seeding situations is relatively homogeneous, i.e., each situation would be expected to develop precipitation in pretty much the same way as any other situation in the group.

Flueck (1986) suggests that, at the broadest level, the experiment defines a general group within which the experiment will take place. For example, this may be the population of all wintertime clouds in the Sierra Nevada range in California. Once a target area is defined and a conceptual model is adopted, a subgroup is defined. This may be the set of orographic clouds that occur within the target area. The conceptual model may allow identification of 'seedable' clouds within this subgroup, e.g., all clouds with top temperatures warmer than -15°C. Within the seedable group, some clouds may not be accessible due to equipment malfunction or other operational difficulties, leaving a treatment group. Within this treatment group, clouds are randomly selected for seeding.

Three common experimental designs are (1) target-only, (2) fixed target–control, and (3) crossover (NRC, 1973), as shown in Fig. 11.11. The target-only design consists of a single target area which is seeded or not seeded on a random basis. The experimental unit may be a cloud, a day, or a storm. Units may be defined as 'go' or 'no go' depending on whether they are within the seedable group. The decision to seed or not seed is made randomly. The effect of seeding is based on the differences between the seeded and not-seeded units.

Evaluation of some non-randomized (operational) seeding projects has been made on the basis of considering all 'go' days during the operation of the project as seeded units and considering the historical record of all 'go' units prior to commencement of seeding as the not-seeded units. The difficulty here is that there is no assurance that the underlying population of 'go' days is stationary, *e.g.*, predominant storm types may vary from year to year.

The fixed target–control design employs a designated control area close to the target area which is presumed not to be affected by the seeding. Multiple control areas surrounding the target area can be used, selecting the particular upwind control area to be used based on the prevailing winds for that particular experimental unit (Fig. 11.11c). Seeding is done only in the target area and on a random basis; the control area is never seeded. Selection of the control area should maximize the correlation between the precipitation in the target and control areas, which generally implies that the target and control areas are nearby. However, the target and control areas must be far enough apart to minimize the possibility that seeding agent could affect the control area.

The crossover design also uses two areas (A and B) as in the fixed target–control design, but seeding is carried out in both areas. Seeding is carried out for all 'go' units but only one of the areas is seeded. The decision to seed A or B is made randomly. This design has the advantage that it reduces the number of

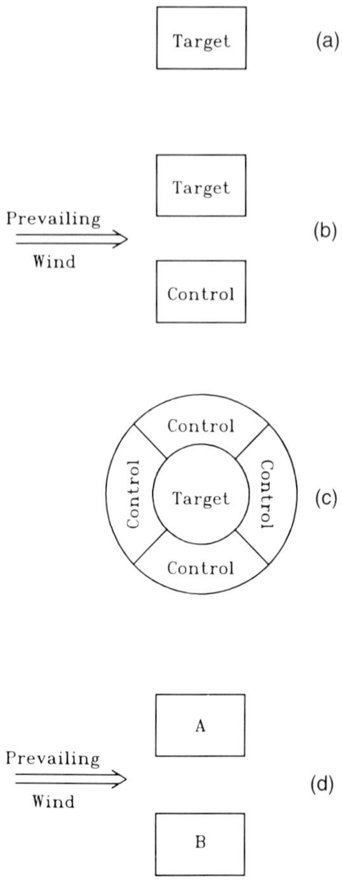

FIG. 11.11. Conceptual design of statistical experiments in weather modification. (a) Target-only design, (b) fixed target–control design, (c) fixed target–control design for variable wind directions, which defines the control area as upwind of the target area, and (d) crossover design.

experimental units required to reach a given level of significance for a given effect.

The crossover design can also be applied to seeding of individual clouds. In this case, two seemingly identical clouds are selected, usually on the same day in the same region. One of the clouds is selected for seeding on a random basis. Evaluation is based on the differences between the seed/no-seed pairs.

11.3.3. *Evaluation of experiments*

Different experimental designs require different methods of evaluation. In addition, the power of the statistical tests and the ability to resolve a seeding effect can be enhanced through use of covariates. The hazards of *post hoc* analysis and optional stopping also need to be considered.

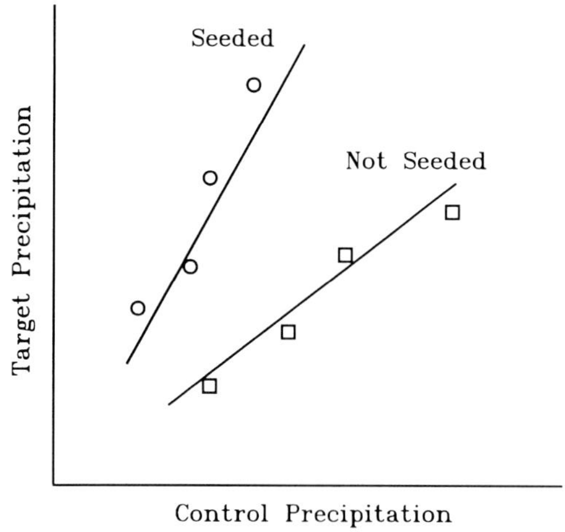

FIG. 11.12. Hypothetical evaluation of a fixed target–control design seeding experiment. The abscissa represents the precipitation observed in the control area; the ordinate represents that observed in the target area. The 'seeded' regression represents those units for which the target area was seeded; the 'not-seeded' regression represents those units for which the target area was not-seeded.

(i) Direct statistical evaluation. Target-only designs are often evaluated on the basis of the difference between the mean precipitation for the seeded clouds and the mean precipitation for the not-seeded clouds. If the precipitation values are approximately normally distributed, a t-test can be used. If not, non-parametric methods are suggested (e.g., Battan, 1966)

Evaluation of the fixed target–control design compares the regression lines between the target and control precipitation for the seeded and not-seeded units, as shown in Fig. 11.12. The slopes of the two regression lines are then compared (e.g., Brownlee, 1965). Evaluation of fixed target–control designs for operational seeding projects may be done using the not-seeded regression based on the historical record of 'go' units prior to commencement of the seeding operation.

Evaluation of fixed target–control experiments may also be based on the 'double ratio' (DR) defined as

$$DR = \left.\frac{S}{NS}\right]_t \left.\frac{NS}{S}\right]_c$$

where the S and NS refer to the mean precipitation on seeded and not-seeded days, respectively, and the subscripts t and c refer to the target and control areas, respectively. Evaluation of the statistical significance of the double ratio is done using Monte Carlo techniques assuming the null hypothesis that there is no effect of seeding.[6] This technique is described further for Problem 11.8.

[6] This is often termed 'rerandomization'

The evaluation of the crossover design is similar to that for the fixed target–control in that two regression lines are compared. However, in this case the two areas are designated A and B, and the regression lines are seed A and seed B.

(ii) Use of covariates. The function of the control area in the fixed target–control and crossover designs is to reduce the variability between experimental units. For example, a given experimental unit may experience above normal precipitation in both the target and control areas. This would produce a point toward the upper right in Fig. 11.12. Another unit with below normal precipitation in both areas would produce a point toward the lower left. In this way, the precipitation in the control area is used to improve the discrimination between seeded and not-seeded precipitation in the target area.

This concept can be extended using statistical or physical models to estimate (predict) the effect of seeding in the target area. Simpson et al. (1967) compare the predicted cloud top heights with those observed for 23 clouds (both seeded and unseeded) over the Caribbean, as shown in Fig. 11.13. Seeding was accomplished using dropable AgI-pyrotechnic flares in clouds with top temperatures between −5°C and −20°C. The model was used to predict both the natural and seeded cloud top heights. The abscissa in Fig. 11.13 is the seedability expressed as the difference between the seeded and not-seeded predicted cloud top heights. The ordinate is the difference between the observed and predicted heights.

The correlation between the observed and predicted cloud top heights for the seeded units in Fig. 11.13 is 0.973. The null hypothesis that the two regression lines (seeded and not-seeded) have the same slope can be rejected at the 99% level, i.e., seeded cloud grew taller as a result of the seeding. Analysis of the cloud top heights for the same set of clouds without using the model predicted heights would not suggest a difference between the seeded and not-seeded clouds significant at the 95% level.

(iii) Post hoc analysis. Once a seeding project has been completed, it is tempting to analyze the results in a variety of ways not anticipated at the start of the project. For example, the results from Climax I were stratified (partitioned) by 500 mbar temperature and 700 mbar equivalent potential temperature, and wind direction and speed after the composite results indicated no statistically significant change in the precipitation due to seeding. In the *post hoc* analysis, significant increases in precipitation were found for 500 mbar temperatures warmer than −20°C, 700 mbar equivalent potential temperatures between 308 K and 325 K, 700 mbar wind directions between 190° and 250°, and 700 mbar wind speeds between 12 m s^{-1} and 14 m s^{-1}. In addition, significant decreases in precipitation were found for 500 mbar temperatures colder than −25°C.

The difficulty with such *post hoc* analyses is in determining the appropriate level of statistical significance that should be applied to tests of a seeding effect. The appropriate level of significance is a function of the number of 'tests' performed and the independence of the tests. If one examines enough partitions, one is bound to find at least one result that is 'significant' at the 95% level. The

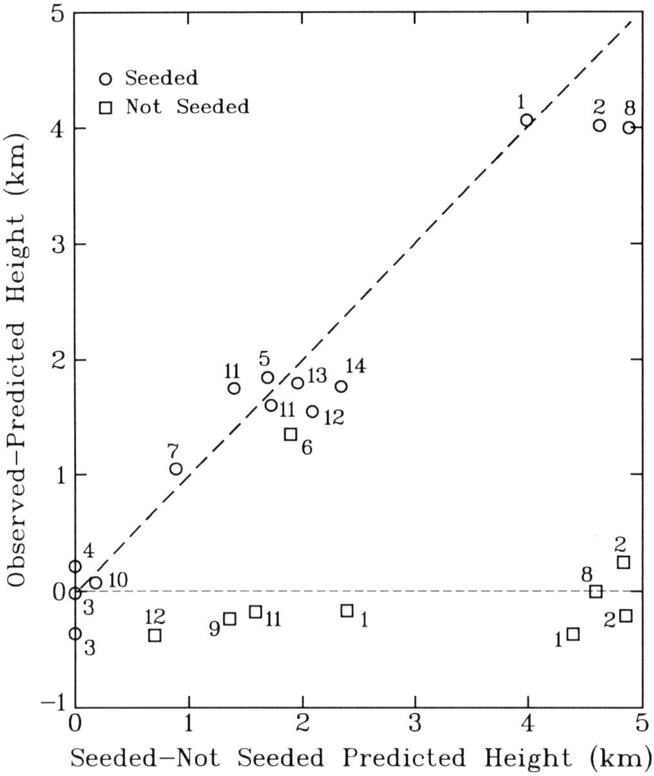

FIG. 11.13. Comparison of predicted and observed cloud top heights for seeded and not-seeded clouds. The abscissa represents the predicted difference in height due to seeding; the ordinate represents the difference between the observed height and the predicted height for the not-seeded cloud. The not-seeded clouds (squares) would be expected to scatter about the 0 ordinate (light broken line). The seeded cloud top heights (circles) scatter about a slope of 1. Numbers within the squares and circles represent groups of clouds on the same day with widths within a range of 300 m. After Simpson et al. (1967). *Journal of Atmospheric Sciences*, **24**, 5. American Meteorological Society. Reproduced with permission.

proper procedure following such *post hoc* analysis is to replicate (repeat) the experiment to test a new hypothesis, based on the *post hoc* analysis.

Mielke et al. (1981) describe the replication of the original Climax cloud seeding experiment. They found that Climax II confirmed the increases in precipitation cited above but failed to support the decrease in precipitation for the 500 mbar temperature below −25°C partition.[7] Climax II stands as a model that too few seeding projects have followed in this regard.

(iv) Optional stopping. Optional stopping refers to conducting a seeding experiment until a statistically significant result is found, and then stopping the experiment. Following any number of experiment units n, there exists a finite

[7] There is controversy regarding these findings, as discussed in §11.5.2.

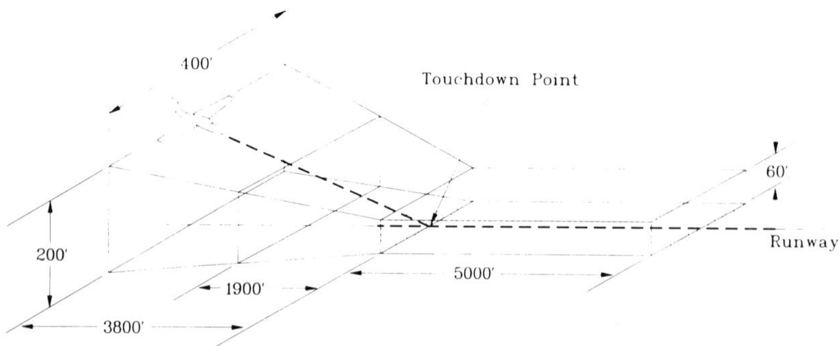

FIG. 11.14. Zones to be cleared by fog dispersal techniques at airports. The zones are termed the approach zone, the touchdown zone, and the roll-out zone. Typical dimensions of the volume to be cleared for each zone are given in feet. From Silverman and Weinstein (1974). Reprinted with permission.

(non-zero) probability that a significant result will first be attained. For a 95% level of significance, there is a probability of 19% of achieving a 'significant' result sometime within the first 10 trials. This probability rises to 37% after 100 trials and to 53% after 1000 trials. Obviously, this is a poor way to conduct a scientific investigation of cloud seeding.

11.4. Modification of fog

Fog may be categorized on a microphysical basis as (1) warm fog–water droplets at temperatures warmer than freezing, (b) supercooled fog–water droplets at temperatures below freezing, or (c) ice fog – comprised entirely of ice crystals. For ice fog, the only cure seems to be prevention. For both supercooled and warm fogs, the physical processes involved are reasonably well understood, the dynamical and microphysical structures are fairly well known, and the fogs are easily accessible by ground-based seeding techniques.

The objective in most fog modification operations is to improve visibility for aircraft landings (and takeoffs), as shown in Fig. 11.14. Thus, it is not necessary to dissipate the fog entirely but merely to improve the visibility enough to permit aircraft to land. Silverman and Weinstein (1974) give the daytime visibility (V)[8] as

$$V = \frac{3.912}{\pi \sum_{i} K_i n_i r_i^2}, \tag{11.1}$$

where n_i and r_i are the concentration and radius of the ith particle size category and K_i is the scattering efficiency of a fog particle of that size. For droplets, K_i has a value of approximately 2. From (11.1), it is evident that the greatest

[8] The visibility is defined as the distance at which a black target can just be detected against a horizon sky with a contrast threshold of 2%.

improvement in visibility would be expected from reducing the size of the droplets, although reductions in visibility could also be achieved by reducing droplet concentrations.

Reducing the concentration of fog droplets by filtering (screens, trees, etc.) or enhanced coagulation induced by ultrasonic sound waves or by application of surfactants may be possible, but the relatively low concentrations of fog droplets generally renders these approaches ineffective. Reduction in the fog particle sizes can be accomplished through evaporation. In supercooled fogs, introduction of ice crystals to exploit the ice crystal mechanism is the most effective method. In warm fogs, seeding with hygroscopic particles or direct heating may be employed. Since the costs of diverting planes at a major airport can be substantial[9], brute force methods can be cost effective.

11.4.1. *Supercooled fog*

The introduction of ice crystals into a cloud of supercooled droplets results in evaporation of the droplets as the ice crystals grow. For a fog droplet spectrum, this would be similar to the case shown in Fig. 10.2. The rapidity of the clearing is in direct relation to the concentration of ice crystals used. If the ice crystal concentrations are relatively low, the ice crystals would grow by diffusion to sizes at which they would settle out. If their concentrations are relatively high, aggregation would be expected to become more important in promoting fallout.

The release of latent heat through the glaciation process serves to enhance turbulence levels and promotes spreading of the ice crystals and hence the clearing effect. Silverman and Weinstein (1974) state that the lateral spread of the ice crystals is about $1\,\mathrm{m\,s^{-1}}$, and spreading of the cleared zone continues until the crystal concentration is significantly reduced by sedimentation. A single line of ice crystals can generally produce a cleared path 2 km wide.

Use of dry ice or ice nucleants to clear supercooled fog is now considered to be an operational technology. Silverman and Weinstein state that during the winter of 1969–1970, dry ice seeding operations at 13 airports in the north-central and north-western United States cost approximately $80,000 and resulted in savings of more than $900,000 in airline operating expenses.

11.4.2. *Warm fog*

Three methods have been used in attempts to clear warm fogs. Direct heating (thermokinetic approach) and hygroscopic seeding will be discussed in some detail. The use of helicopters or other means for mixing drier air above the fog, down into the fog is discussed by Silverman and Weinstein (1974). If the relative humidity in the air aloft is less than 90%, the mixture of fog air with drier air is below water saturation and fog droplets will evaporate to some extent.

[9] The cost of diverting a single jumbo jet is typically on the order of $50,000 and upwards.

Unlike the dissipation of supercooled fog, the treatment effects for warm fogs do not propagate laterally to any great extent. Thus, continued application is required, making the costs of dissipating warm fog considerably greater than those for dissipating supercooled fog.

(i) Thermokinetic approach. Direct heating of volumes of foggy air will cause the relative humidity to decrease, resulting in partial or complete evaporation of the fog droplets. At the same time, heating can increase the level of turbulence and, in shallow fogs, promote downward mixing of drier air. The minimum energy that must be supplied to evaporate the fog droplets contained in a given volume is the sum of the latent heat of vaporization required to evaporate the drops and the sensible heat required to raise the ambient air temperature sufficiently to accommodate the additional water vapor resulting from the evaporation of the fog droplets. Typically, the energy required is provided through the combustion of hydrocarbon fuels. Since the combustion products include water vapor, the ambient air temperature must be raised by some additional amount to accommodate this source of water vapor. The amount of energy required to evaporate a given fog decreases with increasing temperature, as shown in Fig. 11.15, due to the increasing slope of the saturation vapor density curve with increasing temperature.

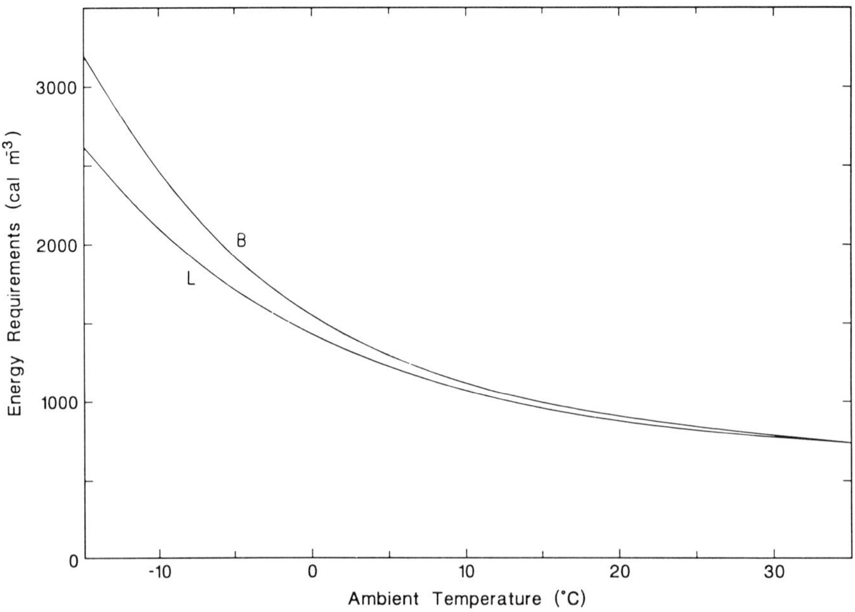

FIG. 11.15. Calculations of the energy required to evaporate 1 m^3 of fog with a liquid water content of 1 g m^{-3} at 1000 mbar as a function of the initial ambient temperature. L refers to use of laser energy; B to the burning of octane, which releases additional water vapor. The initial and final states are assumed to be at water saturation. From Young and Tomlinson (1993). American Meteorological Society. Reproduced with permission.

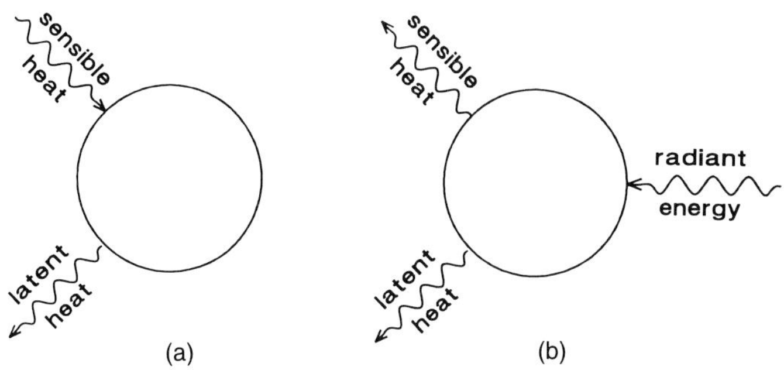

FIG. 11.16. Energy transfer for evaporating drops. (a) Heating the air, which results in a transfer of sensible heat to the drop and latent heat away from the drop; (b) heating the drops directly with a laser. In the latter case, the laser energy is absorbed by the drop and is dissipated by both the sensible and latent heat transfers away from the drop. From Young and Tomlinson (1993). American Meteorological Society. Reproduced with permission.

Several operational systems have been installed, including a FIDO[10] system at the Los Angeles International Airport (LAX) in 1949 and a Turboclair system at the Orly International Airport in Paris, France in 1970. The LAX system was abandoned in 1953; the Turboclair system at Orly was abandoned in the late 1970s. Presumably these systems were not found to be cost effective.

The use of lasers has been proposed as a more efficient means of clearing warm fog. Young and Tomlinson (1993) cite several advantages of using lasers rather than burning hydrocarbon fuels. Since the laser would not produce additional water vapor, the total energy required would be less, as shown in Fig. 11.15, although this factor is insignificant for temperatures warmer than 10°C. Lasers can be directed to illuminate the desired volume more precisely, thereby heating a smaller volume. At the proposed 10.6 μm wavelength, very little of the energy is absorbed by water vapor or other atmospheric gases. Most of the laser energy would be absorbed directly by the drops, i.e., only volumes containing fog would absorb and the strongest absorption would be in the most dense regions of the fog.

This latter feature provides an interesting microphysical problem. Heating the air to induce evaporation means that the sensible heat transfer is toward the drop while the latent heat transfer is in the opposite direction, as shown in Fig. 11.16. For drops absorbing laser energy directly, the energy absorbed from the laser beam is transferred to the ambient air by sensible and latent heat fluxes from the drop (both fluxes are in the same direction).

The increase in the ambient water vapor density due to the latent heat transfer is smaller than the increase of the saturation vapor density due to the increase in ambient temperature resulting from the transfer of sensible heat. This implies that the ambient relative humidity decreases with time, as shown in Fig. 11.17. Reduction in the relative humidity is important to retard further fog development and to maintain a limited amount of evaporation after the laser is

[10] Fog Intensive Dispersal Of.

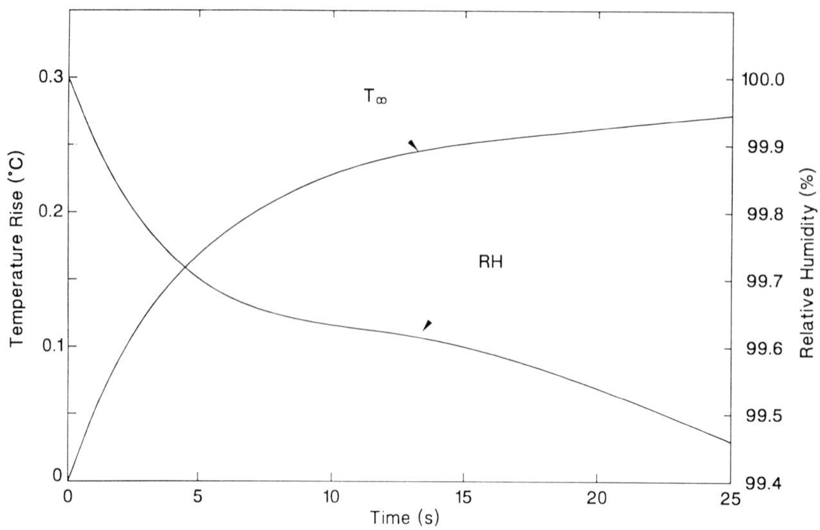

Fig. 11.17. Time evolution of ambient temperature and relative humidity in foggy air under laser illumination. Initial conditions are 20°C and 1000 mbar with a liquid water content of 0.194 g m^{-3}. The CO_2 laser intensity is 1 W cm^{-2}. The continued increase in temperature and decrease in relative humidity after 15 s is due to the slight absorption of laser energy by the water vapor. From Young and Tomlinson (1993). American Meteorological Society. Reproduced with permission.

turned off. The reduction in relative humidity is enhanced for higher power densities.

Since the laser energy is absorbed in proportion to radius squared[11] and the sensible and latent heat fluxes in proportion to the radius, the larger drops are warmer than the smaller drops. Assuming the drop temperature is quasi-steady, the energy absorbed can be equated to the sensible and latent heat fluxes away from the drop as

$$I\pi r^2 = 4\pi r(K\Delta T + DL_v \Delta\rho)\,. \tag{11.2}$$

Assuming that the environment is at water saturation, small values of $\Delta\rho$ can be estimated from the Clausius–Clapeyron equation by integrating

$$d\rho_s = \rho_s \frac{L_v - R_v T}{R_v T^2}\, dT\,.$$

from T_∞ to T_x, assuming ρ_s may be approximated by $\overline{\rho_s}$,

$$\Delta\rho = \overline{\rho_s}\left[\frac{L_v}{R_v}\left(\frac{1}{T_\infty} - \frac{1}{T_x}\right) - \ln\left(\frac{T_x}{T_\infty}\right)\right].$$

[11] For drops smaller than 8 μm radius, the absorption cross-section is proportional to radius cubed (see Young and Tomlinson, 1993).

For small ΔT, the logarithm term can be neglected and $\Delta\rho$ can be approximated as

$$\Delta\rho \approx \frac{\overline{\rho_s} L_v \Delta T}{R_v T_\infty^2}.$$

Now, (11.2) can be solved explicitly for ΔT as

$$\Delta T = \frac{Ir}{4}\left[K + \frac{DL_v^2 \, \overline{\rho_s}}{R_v T_\infty^2}\right]^{-1}. \tag{11.3}$$

Equation (11.3) demonstrates that the drop temperature elevation is proportional to radius.

The evaporation rate, dr/dt, is essentially independent of drop size for radii greater than 8 μm, and decreases with decreasing size for radii smaller than 8 μm (see Problem 11.4). Thus, the drop spectrum will narrow as evaporation proceeds. Since the visibility (11.1) is proportional to drop radius squared, this implies that visibility improvements would be more rapid than if traditional methods of heating the air were employed. In the latter case, the drop spectrum would broaden, and proportionately more of the energy would be used to evaporate the smaller drops rather than the larger drops.

(ii) Hygroscopic seeding approach. Hygroscopic seeding relies on the relatively low equilibrium saturation ratio over hygroscopic substances such as NaCl and urea to evaporate fog droplets. As noted by Silverman and Weinstein (1974), the clearing effectiveness of hygroscopic seeding is critically dependent on the seed particle size. If the seed particle sizes are very small, the fog droplets will be replaced with haze particles of about the same size, *i.e.*, the seed particles will remain suspended and keep the visibility low. If the seed particles are very large, they will fall through the fog too rapidly to evaporate a significant fraction of the liquid water. Silverman and Weinstein estimate the optimal seed particle size as approximately 15 μm radius.

Silverman and Weinstein point out that a candidate hygroscopic material for dissipating fogs should retain its hygroscopicity as it becomes a more and more dilute solution droplet. For example, calcium chloride ($CaCl_2$) deliquesces at 45% relative humidity but quickly loses its affinity for water vapor on dilution. Lithium hydroxide (LiOH) has a deliquescent point at a humidity of 83% but maintains a high growth rate as the solution droplets become more dilute. Silverman and Weinstein compare the fog clearing efficiencies for a variety of proposed hygroscopic materials and conclude that urea is the most effective non-corrosive material for operational warm fog dissipation. They note that microencapsulation would be necessary since urea is relatively friable and would produce large numbers of small fragments in handling if not encapsulated.

11.5. Modification of orographic clouds

The potential for seeding orographic clouds to enhance precipitation was first detailed by Bergeron (1949), who concluded these to be the only category of clouds with a significant glaciogenic seeding potential. Ludlam (1955) advanced a model of orographic precipitation processes, and estimated that increases of up to 100% of the normal seasonal snowfall would be possible in central Sweden. A wide variety of orographic cloud seeding projects have been carried out, many of which are summarized by Grant and Kahan (1974) and by Elliott (1986).

An estimate of the seedability of orographic clouds is provided by Hill (1980), based on observations of aircraft icing as a function of the cloud top temperature and maximum vertical motions in the cloud (Fig. 11.18). Clouds with tops colder than $-21°C$ are considered not seedable since they already possess ample concentrations of ice crystals. Clouds with warmer tops are considered seedable, although the yield is expected to be greater for stronger updrafts since the liquid water content is greater. Since the vertical velocity for a given barrier is directly related to the cross-barrier component of the horizontal wind speed (u_x) and the moisture flux over the barrier is also proportional to u_x, the seeding potential is roughly proportional to the square of the horizontal wind speed.

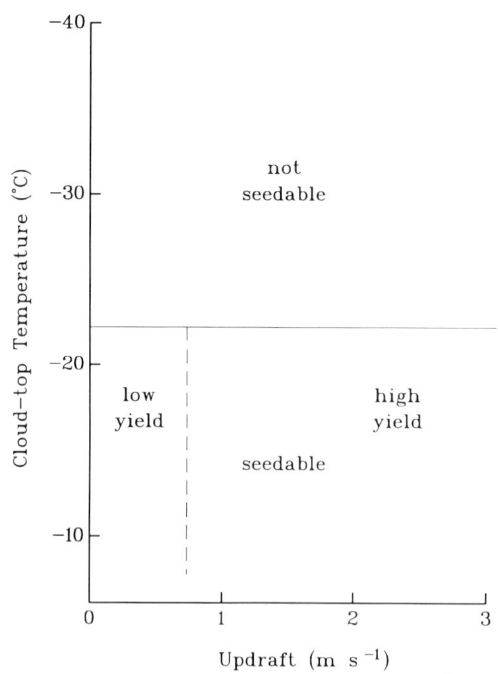

FIG. 11.18. Seedability of orographic clouds as a function of the cloud top temperature and average maximum vertical motion. After Hill (1980). *Journal of Applied Meteorology*, **19**, 12. American Meteorological Society. Reproduced with permission.

Rangno (1986) points out a number of shortcomings in our conceptual models of seeding orographic clouds. There is no widely accepted parameter that will predict the maximum ice particle concentrations in most clouds; some clouds exhibit a temperature-dependence while other clouds exhibit a dependence on the breadth of the cloud droplet spectrum. Ice particle formation appears to be enhanced near and downwind of the barrier crest by mechanisms not well understood. Better knowledge of the distribution and evolution of supercooled liquid water with the cloud and its effect on particle growth are needed. Considerable uncertainty exists on the optimal methods for seeding orographic clouds. These provide a focus for ongoing research efforts.

11.5.1. *Numerical modeling approach*

The model described by Ludlam (1955) is basically a calculation of the diffusional growth of a single ice crystal in a cloud assumed to be maintained at water saturation. Accretional growth of the ice particles was neglected on the basis of observations suggesting that few or no rimed particles occurred under the conditions being simulated. Ludlam found that a residence time of less than 1500 s resulted in 'blow-over', i.e., the crystals pass over the barrier crest into the evaporation zone to the lee of the barrier. Thus, the seeding material must be introduced far enough upwind of the barrier and at a height such that the precipitation would be targeted on the barrier.

The optimal number of ice crystals was estimated as that concentration sufficient to deposit water vapor at the rate at which the updraft provides water vapor for deposition. Ludlam assumed a plate crystal radius of 500 μm and estimated the optimal concentration as $20\,\ell^{-1}$.

Young (1973,1974a,b) describes the simulation of orographic precipitation under natural and AgI-seeded conditions using a complete microphysical approach in a kinematic framework. The model provides for ice nucleation *via* deposition and contact freezing nucleation, ice crystal growth with variable axial ratios and densities, and growth by accretion and aggregation. The ambient saturation ratio was calculated using an iterative technique to balance the changes in ambient vapor density and temperature with the requirements of the particles' growth. The model utilized nine levels, each initially 300 m thick, to simulate the growth and sedimentation of particles. The flow was assumed to be quasi-steady from west to east such that the distance and time axes were equivalent,[12] permitting calculations to be made once for each grid in two-dimensions.

The case chosen (Grand Junction CO rawinsonde for 28 January 1971) had a cloud top temperature of −11°C and saturated conditions for the top six levels. The upwind cloud was assumed to have a liquid water content (LWC) of $0.06\,g\,m^{-3}$. The water mass flux through the cloud unavailable for precipitation was defined for each streamline as the water vapor flux at ice saturation at the coldest temperature achieved for that streamline. The available mass flux is then

[12] In order to permit sedimentation, the levels were kept vertically aligned. Since the horizontal wind speed differed between levels, the time step for each level was inversely proportional to the wind speed.

the total mass flux minus the unavailable mass flux. The precipitation efficiency follows as the precipitation flux divided by the available mass flux. The precipitation efficiency calculated for the natural case was 0.04%, in good agreement with Fig. 10.14 for Climax CO.

Three cases with cloud top seeding using AgI are compared. In case A, seeding was simulated 40 km upwind of the barrier crest, with a concentration of AgI sufficient to produce 50 ℓ^{-1} of ice crystals by contact nucleation (CACN) at -4°C. In case B, the seeding rate was doubled to CACN = 100 ℓ^{-1}. In case C, seeding was simulated 45 km upwind with CACN = 50 ℓ^{-1}. Cloud top seeding was assumed to fill the uppermost model level completely. Vertical transport of AgI by turbulent mixing was simulated. A case of cloud base seeding was found to produce results virtually identical to the natural case.

Precipitation rates and efficiencies are shown in Table 11.1. Doubling the seeding rate from case A to case B reduced the precipitation efficiency slightly, a result of decreasing the average sedimentation velocity of the particles. Moving the seeding line further upwind also reduced the precipitation efficiency by increasing sub-cloud evaporation upwind of the barrier crest. Case A appears to be close to the optimal seeding rate. The ice crystal concentration in the uppermost layer for cases A and C was 771 ℓ^{-1} and for case B it was 1218 ℓ^{-1}. These figures are somewhat higher than the optimal ice crystal concentrations shown in Fig. 11.3 for a 3 m s^{-1} updraft, although the residence time in the orographic cloud is longer and aggregation processes are more important.

Figures 11.19 and 11.20 compare the liquid water contents for the natural and seeded case A. In the natural case, the peak LWC was 0.35 g m^{-3} over the barrier crest. In the seeded case, the upper portion of the cloud is completely glaciated and the maximum LWC is reduced to 0.22 g m^{-3}. This illustrates that a substantial portion of the LWC is not converted to ice particles, and is lost to evaporation on the lee side of the barrier crest, even under optimal seeding conditions.

The mass transfer rates to the ice phase by diffusion and by accretion are shown in Figs. 11.21 and 11.22. Diffusional growth of ice crystals is dominant just downwind from the seed line. The zone of maximum accretion is coincident with the zone of high LWC but upwind of the maximum LWC. Note the

Table 11.1. Total precipitation and precipitation efficiencies (PE) for orographic cloud simulations under natural and seeded conditions. Cloud top temperature is -11°C.

Case	CACN (ℓ^{-1})	Precipitation (g cm^{-1} s^{-1})	Precipitation (mton h^{-1} km^{-1})	PE (%)
Natural	None	0.06	20	0.04
A	50	30.26	10 850	19.5
B	100	29.47	10 550	19.0
C	50	29.79	10 700	19.2

Source: Young (1974b).

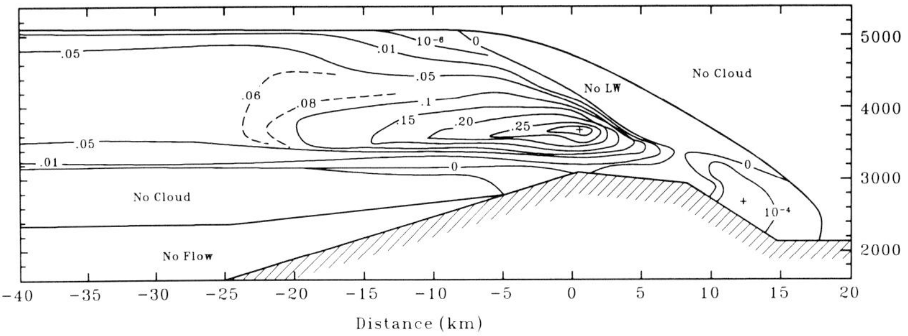

Fig. 11.19. Cross-section of the liquid water content (g m^{-3}) for the natural orographic cloud case. Flow is from west (left) to east (right). No flow region represents stable conditions beneath the westerly flow. From Young (1974b). *Journal of Atmospheric Sciences*, **31**, 7. American Meteorological Society. Reproduced with permission.

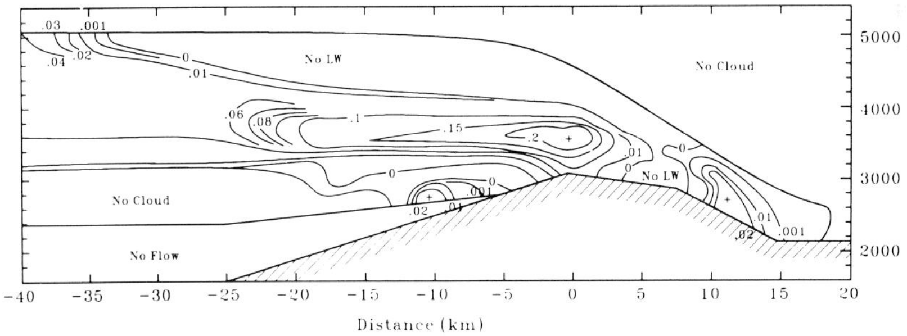

Fig. 11.20. Cross-section of the liquid water content (g m^{-3}) for the seeded case A. Seeding is at 40 km upwind of the barrier crest at cloud top with CACN = 50 ℓ^{-1}. From Young (1974b). *Journal of Atmospheric Sciences*, **31**, 7. American Meteorological Society. Reproduced with permission.

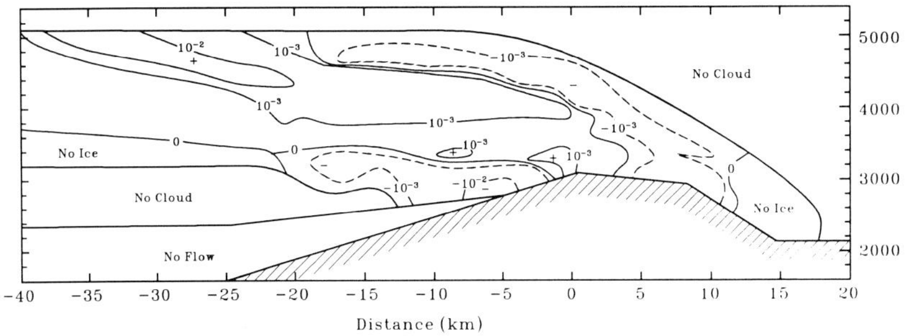

Fig. 11.21. Mass transfer rate to the ice phase by diffusion for the seeded case A. Mass transfer rates are in units of g m^{-3} per kilometer path length. Broken isopleths indicate sublimation. From Young (1974b). *Journal of Atmospheric Sciences*, **31**, 7. American Meteorological Society. Reproduced with permission.

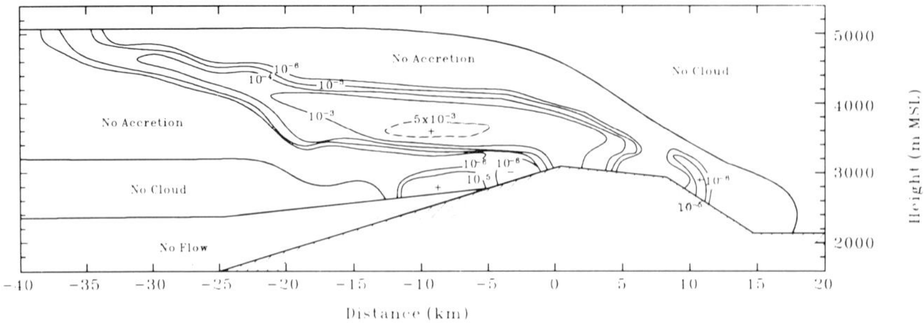

FIG. 11.22. Mass transfer rate to the ice phase by accretion for the seeded case A. Mass transfer rates are in units of g m^{-3} per kilometer path length. From Young (1974b). *Journal of Atmospheric Sciences*, **31**, 7. American Meteorological Society. Reproduced with permission.

dominance of accretion processes in depleting the zone of maximum LWC. However, even with this relatively high concentration of ice particles, the combined effects of diffusional and accretional growth are insufficient to extract water vapor at the rate it is being made available for condensation in the upward moving air upwind of the barrier crest.

Precipitation particles reaching the ground near the barrier crest are predominantly rimed ice particles and aggregates. For case A, 58% of the precipitation near the barrier crest is in the form of aggregates. For case B, a slightly higher fraction of the precipitation is in the form of aggregates, and slightly more of the precipitation falls downwind of the barrier crest.

11.5.2. *The Climax experiment*

The Climax field experiment consisted of two phases, as described by Grant (1986). Climax I was an exploratory experiment designed to evaluate the cloud seeding potential of wintertime orographic clouds in the central Colorado Rocky Mountains and was carried out between 1960 and 1965. Climax II was a confirmatory experiment intended to verify (or disprove) conclusions drawn from *post hoc* analysis of the results from Climax I and was conducted between 1965 and 1970.

The target area was located near the town of Climax CO at altitudes ranging from 3 km to 4 km MSL. The terrain features north–south oriented ridgelines with ground-based seeding conducted from various upwind (west) locations. The seeding was accomplished by burning a solution of AgI and NaI in acetone. The experimental design was target-only, although some later analyses incorporated a number of National Weather Service (NWS) stations as controls. No microphysical observations of the clouds were made.

The original analysis of the replication (Climax II) is provided by Mielke et al. (1971). They confirm the increased precipitation resulting from seeding for the 500 mbar temperature partition warmer than −21°C, for the 700 mbar equivalent potential temperature partition 308 K to 325 K, and for the 700 mbar

Table 11.2. Evaluation of the Climax I and II experiments. The double ratio for each of the four partitions noted as significant for both Climax I and II by Mielke et al. (1971) are given with 90% confidence limits. Target group mean precipitation is contrasted to the mean of eight NWS 'control' stations.

Partition	Climax I	Climax II
T 500 mbar −11°C to −20°C	1.520 0.683 to 1.475	1.089 0.692 to 1.398
θ_e 700 mbar 308 K to 325 K	1.220 0.721 to 1.387	1.160 0.786 to 1.274
700 mbar wind 190° to 250°	1.552 0.685 to 1.483	1.504 0.757 to 1.320
700 mbar wind 12 to 14 m s^{-1}	2.266 0.446 to 2.211	1.022 0.765 to 1.288

Source: Mielke et al. (1981).

wind partitions with directions from 190° to 250° and wind speeds from 12 m s^{-1} to 14 m s^{-1}.

A re-analysis of the Climax experiment is described by Mielke et al. (1981) considering possible regional differences in precipitation between the seeded and not-seeded days. Eight NWS precipitation stations were used as 'controls' and the average precipitation for the reporting stations on the 'go' days was used as the control precipitation. The double ratios and 90% confidence limits based on rerandomization are given in Table 11.2 for the four partitions that were found to be significant at the 5% level[13] by Mielke et al. (1971). Only the southwest 700 mbar wind partition is significant at the 5% level for both Climax I and II.

Further analysis of the Climax I and II experiments by Rangno and Hobbs (1987) refines the estimation of the 500 mbar temperatures in the target area and corrects some errors in the precipitation values used for the Climax recording precipitation gauge. Their conclusion is that neither Climax I nor Climax II exhibit any statistically significant increases in precipitation for the warm 500 mbar temperature partition.

Hobbs and Rangno (1979) criticize the Climax conceptual model on several accounts. They point out that the 500 mbar temperature is not a good indicator of the cloud top temperature. They also note that there is a poor correlation between ice nucleus concentrations and ice crystal concentrations in clouds, suggesting that ice crystals are more common in Colorado orographic clouds than assumed by the Climax experimenters.

[13] Since the seeding hypothesis was to increase precipitation, a one-sided test may be used. The upper bound of the 90% confidence limit is equivalent to the critical value for a one-sided test at a P-value of 5% level. The P-value represents the probability of incorrectly rejecting the null hypothesis that there is no seeding effect.

The Climax experiment is illustrative from several aspects. First, it demonstrates the need for replication and even independent verification by other groups of researchers. Second, it illustrates the value of peer review of the experiment, made possible through ready availability of the collected data. Third, it highlights the limitations of a (largely) statistical experiment.

11.5.3. *The Cascade project*

The Cascade project is almost at the other extreme from the Climax experiment. Here, the emphasis was on gathering dynamical and microphysical data for natural conditions and for conditions seeded with either AgI or dry ice. Numerical models were used to estimate the growth trajectories of particles resulting from cloud-top seeding and to predict the time and location at which the effects would be expected at the ground. Aircraft measurements of ice nuclei and the types and concentrations of cloud particles as well as visual observations of the cloud appearance were made, in addition to ground observations in the 'target' area of the types, concentrations, and degree of riming of ice crystals and the precipitation rate. These are detailed in Hobbs (1975a,b), and Hobbs and Radke (1975).

The immediate effects of glaciogenic seeding were frequently evident by the replacement of cloud bows and coronas with optical phenomena characteristic of ice crystals, such as haloes, parhelia, and sub-suns. Measurements of ice crystal concentrations prior to and subsequent to seeding demonstrated two order of magnitude increases in concentrations when the natural ice crystal concentration was less than $100\,\ell^{-1}$ but changes were relatively small when the natural ice crystal concentration exceeded $1000\,\ell^{-1}$. Seeding rates were relatively high in order to produce more easily detectable changes in the cloud and in the precipitation at the ground. Dry ice seeding rates of $300\,\mathrm{g\,km^{-1}}$ to $600\,\mathrm{g\,km^{-1}}$ and AgI (pyrotechnic) seeding rates from $100\,\mathrm{g\,km^{-1}}$ to $2500\,\mathrm{g\,km^{-1}}$ were employed. Seeding was typically carried out over a period of 1 to 2 hours.

At the ground, precipitation rates were measured with automatic weighing bucket and tipping bucket gauges. Snow was collected and weighed, as well as being analyzed for silver and for freezing nucleus activity. Replicas of precipitation particles were made every 15 minutes. Particle trajectory analysis based on field data was used to estimate the predicted period of effect (PPE) at each ground station. Analysis of the snow collected at the ground indicated a good correlation between silver contents above the detection threshold and the PPE.[14]

Changes in the intensity and type of precipitation at the ground associated with the seeding are shown in Fig. 11.23. Increases in the precipitation rate may be noted within the PPE.[15] These are associated with decreases in the amount of

[14] High silver concentrations do not necessarily imply that the AgI was responsible for the precipitation, since the AgI aerosol could have been scavenged by pre-existing ice particles.

[15] Uncertainties in the particle trajectories result in a longer PPE than the actual time span over which seeding was conducted.

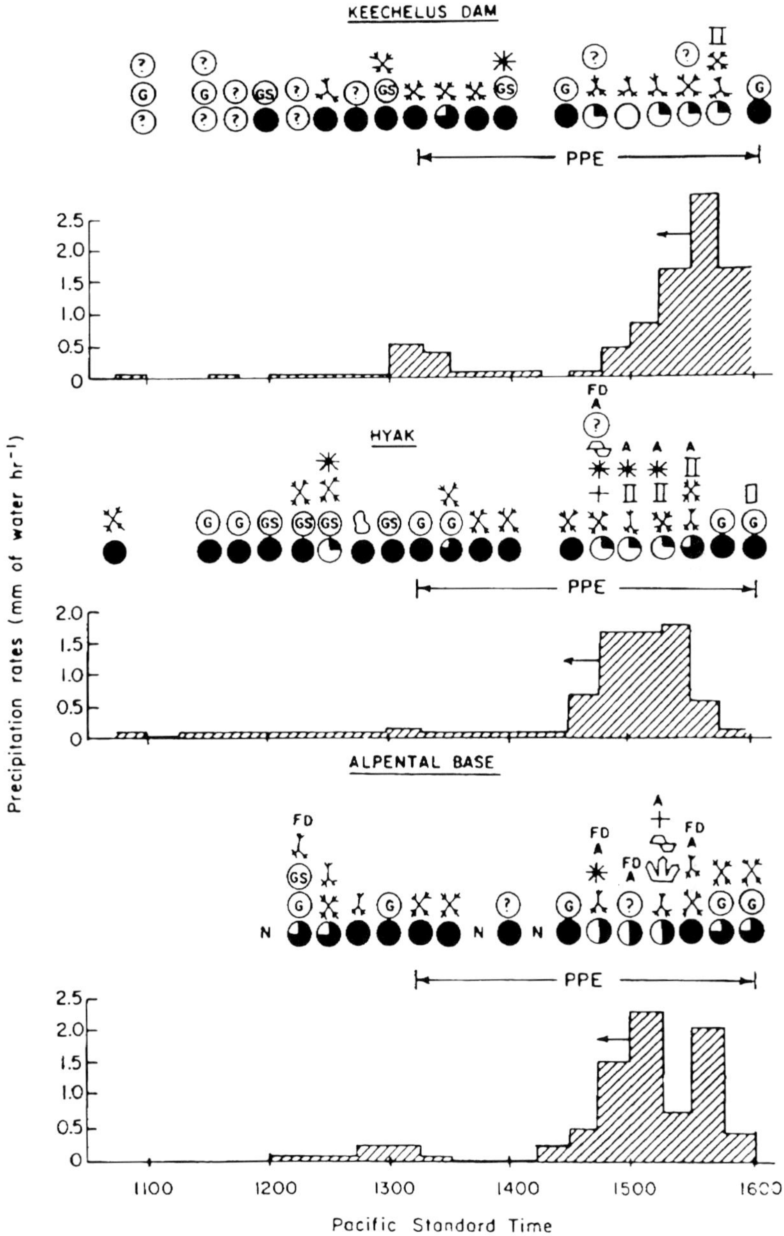

FIG. **11.23.** Dominant ice particle types, degrees of riming, and precipitation rates at Keechelus Dam, Hyak, and Alpental Base during seeding operations conducted as part of the Cascade project. PPE indicates the predicted period of seeding effect. Shaded circles reflect the degrees of riming from unrimed (open circle) to densely rimed (closed circle). A circled 'G' indicates graupel; a circled 'GS' indicates graupel-like snow; 'A' indicates aggregates. The remaining symbols are defined in Hobbs (1975a). From Hobbs and Radke (1975). *Journal of Applied Meteorology,* **14**, 5. American Meteorological Society. Reproduced with permission.

riming and the appearance of aggregates. Both of these changes would be expected to result from an increased concentration of ice crystals due to seeding. Note that the precipitation prior to the seeding consisted mostly of densely rimed particles including graupeln and graupel-like snow.

Of the twenty cases the aircraft was able to fly in the seeded clouds, the effects of seeding were detected in sixteen cases, with possible detection in the remaining four. For the same twenty cases, distinct changes in precipitation type as predicted were detected within the PPE on six occasions, and four other cases were judged as having demonstrated possible effects of the seeding. In this experiment, the evidence for an effect of cloud-top seeding on the precipitation at the ground is overwhelming. The physical chain of events is reasonably well understood and fairly well documented. However, from this limited set of observations, it is difficult to estimate the economic impact an operational seeding program would have.

11.5.4. *Other orographic seeding projects*

A variety of orographic seeding projects have been and are being conducted. Some of these are summarized by Elliott (1986). Super et al. (1988) and Super and Heimbach (1988) describe orographic cloud seeding experiments in the Bridger Range in Montana. Super et al. provide a good statement of their physical hypothesis, including testable aspects.

Proper targeting of the effects of airborne seeding is crucial to the success of the seeding efforts. King (1984) suggests that particle trajectories depend largely on the liquid water content of the cloud and the cloud depth, and are less dependent on the temperature structure of the cloud and details of the particle growth. The relative success of the Cascade project in identifying the seeding effect at the ground would appear to support these conclusions. A more sophisticated technique for targeting is provided by Rauber et al. (1988) for the Sierra Cooperative Pilot Project in California. Deshler et al. (1990) reproduced the findings of the Cascade project in the Sierra Nevada, tracing the effects of airborne seeding to changes in the precipitation types and rates at the ground. They note increases in precipitation rates from $0.1\,\mathrm{mm\,h^{-1}}$ to $1.0\,\mathrm{mm\,h^{-1}}$.

The Colorado River Basin Pilot Project conducted in the San Juan Mountains in southwestern Colorado found that ground-released AgI often would remain trapped near the surface due to low level stability. As a storm progressed, conditions would become more unstable near the surface, releasing the AgI to seed the cloud up to 12 hours after the generators were turned off. Cooper and Marwitz (1980) suggest these storms be categorized as having three stages, progressing from stable to neutral to unstable, as shown in Fig. 11.24. The stable stage is characterized by little vertical lifting, and consequently there is relatively little liquid water with little seeding potential. As the storm becomes more unstable, the amount of liquid water increases in concert with its seedability.

It is evident that conditions governing the development of precipitation and the seedability of orographic clouds vary widely from region to region and from

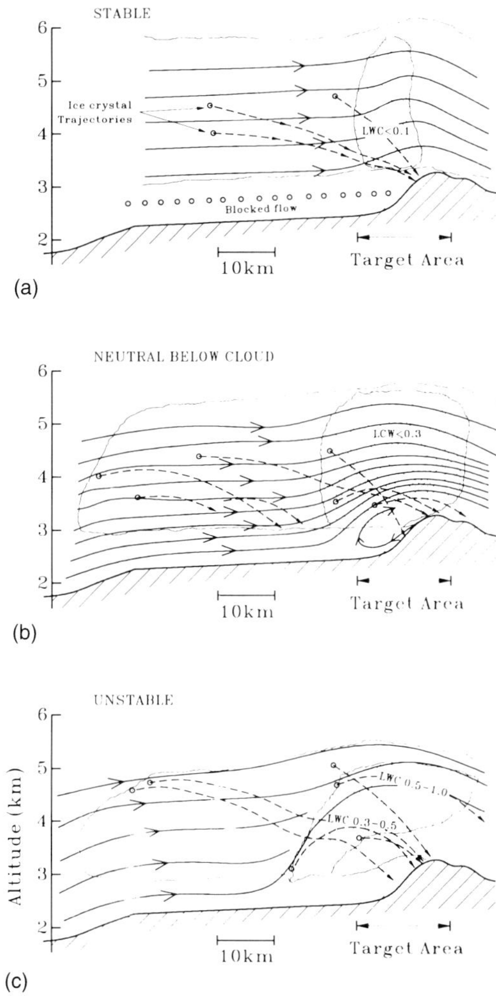

Fig. 11.24. Typical structure of orographic clouds over the San Juan Mountains during the (a) stable, (b) neutral, and (c) unstable stages of the storm. Airflow is from the left (southwest) at 15 m s^{-1}, shown by solid lines with arrows. Cloud boundaries are scalloped; shaded regions represent regions with liquid water. Some typical ice particle trajectories are shown as broken lines. From Cooper and Marwitz (1980). *Journal of Applied Meteorology*, **19**, 8. American Meteorological Society. Reproduced with permission.

time to time within a given storm. Woodley and Rosenfeld (1992) suggest that precipitation processes are quite efficient in most orographic clouds and that regions of liquid water are mostly found within 1 km of the barrier crest. Windows of seeding opportunity may exist within a storm but in general, seeding during the entire storm is not advised. It is clear that improved physical models of orographic precipitation are needed to insure success in seeding wintertime orographic clouds.

11.6. Modification of convective clouds

Convective clouds may be seeded to produce changes in their microphysical structure that lead directly to precipitation enhancement, or they may be seeded to promote glaciation and the rapid release of latent heat, to invigorate the updraft and produce larger, longer-lasting clouds which would produce more precipitation. The former of these is termed *static mode* seeding whereas the latter is termed *dynamic mode* seeding.

Ice nucleant seeding may be utilized for either static or dynamic mode seeding. Often it may be difficult to separate these two effects, although relatively low seeding rates are generally designed to promote changes in the microphysical structure of the cloud, whereas relatively high seeding rates are designed to promote changes in the dynamical structure of the cloud. Hygroscopic seeding would be expected to reflect static mode seeding, although one might also expect accompanying changes in the mass loading as well as changes in the sedimentation and evaporation in entrained air, leading to development of downdrafts, i.e., a dynamical effect.

11.6.1. *Glaciogenic seeding for microphysical effects*

Glaciogenic seeding for microphysical effects is based on the assumption that precipitation development in convective clouds is inefficient due to the lack of sufficient concentrations of ice particles. Frequent observations of liquid water at temperatures as cold as −20°C support this hypothesis. The presence of supercooled liquid water is commonly taken as evidence of seedability.

Precipitation formation through a dominant coalescence process may be expected in clouds with warmer cloud base temperatures due to their higher liquid water contents. The ice crystal mechanism would be expected to be dominant in relatively cold clouds with low liquid water contents. An intermediate range of clouds in which both processes are comparable may be expected. Thus, the seedability of convective clouds must hinge on both the natural coalescence and ice crystal mechanisms being relatively inefficient.

Silverman (1986) suggests that an active coalescence mechanism is associated with secondary ice crystal production, and that adding ice nucleants to such a cloud would result in overseeding and a reduction in precipitation.[16] This appears to be the case for Project Whitetop which reported significant decreases in precipitation. The Israeli Project appears to have found the 'seeding window' for convective clouds, having reported significant increases in precipitation from clouds with relatively low natural concentrations of ice crystals and an inefficient coalescence mechanism.

(i) Project Whitetop. Project Whitetop was conducted in southern Missouri during the summers of 1960 through 1964. Airborne seeding was conducted along a 50 km seedline, burning a solution of AgI and NaI in acetone to release

[16] The presence of large drops in an updraft would also be expected to enhance condensation freezing nucleation.

2700 g h^{-1} AgI. For each day, a seeding plume (target) was determined based on the winds, leaving the remainder of the research area, defined as a 100 km radius circle centered on the radar site, as the out-of-plume or control area.

Observations of the microphysical structure were obtained by aircraft penetrations of the convective clouds as part of Project Whitetop. Koenig (1963) found ice particle concentrations considerably higher than would be expected from standard ice nucleus activity spectra, and no apparent dependence of ice particle concentration on cloud top temperature was found. The size spectra for large ice particles were noted to resemble the drop size spectra quite closely. Of the 84 clouds studied, 57 developed precipitation but only one of these developed ice particles without the prior existence of large drops. Braham (1964) suggests that these clouds naturally produce relatively high concentrations of ice particles and, on occasion, may be naturally overseeded.

Results from Project Whitetop have been widely analyzed. The analyses described by Neyman et al. (1969) for example, show decreases in rain gauge precipitation of up to 50% within the seeding plume. Similar decreases in precipitation were noted in § 11.2.1, part (v), extending well beyond the research area. On the basis of these findings, it is likely that the control area was contaminated as well. Braham (1979) suggests that the decreases in precipitation resulted from overseeding.

Since these clouds tended to have relatively high natural ice crystal concentrations,[17] the high seeding rates employed in Project Whitetop probably introduced small ice crystals in concentrations large enough to compete for the supercooled water and reduce the accretional growth of the larger ice particles. This reduction in particle sizes would be expected to increase evaporation losses in the downdraft, as well as increase the mass of debris as the cloud dissipates.

Similar but non-significant decreases in precipitation were reported by Battan (1966) for AgI-seeding in convective clouds with an active coalescence process in Arizona. These findings suggest that glaciogenic seeding of convective clouds with relatively warm base temperatures to enhance precipitation without concomitant changes in the dynamical structure of the clouds (static seeding mode) is ineffective and probably will decrease rather than increase precipitation.

(ii) The Israeli project. The Israeli Project consists of three phases, with Israeli I conducted during the winters from 1961 through 1967, and Israeli II conducted from 1969 through 1975. Israeli III is ongoing. In all phases, seeding was done from aircraft flying at roughly cloud base altitude. The Israeli I project employed a randomized cross-over design with north and south areas. For the Israeli II project, a fixed target–control design was used, with the target area in northeastern Israel in the Lake Tiberias catchment and the control area upwind along the Mediterranean Sea.

[17] See § 9.5.3.

Microphysical observations of the clouds indicated that they are totally deficient of drops larger than 50 μm radius. Ice particle concentrations were found to have a clear dependence on cloud top temperature, with no evidence of secondary ice particle production. Typical cloud base temperatures were 8°C with updrafts in the range for a significant seeding potential according to Fig. 11.6.

Gagin (1986) presents the analysis of the 388 days of operations for the Israeli II project. These are summarized in Table 11.3, showing an increase of 45% in the mean precipitation for the −11°C to −21°C partition. This 45% increase in mean precipitation is associated with a 47% increase in the duration of precipitation (P value = 0.6%), suggesting that the intensity of the precipitation was relatively unaffected by the seeding. Increases in the frequency of precipitation were also noted but these were not significant at the 5% level.

An analysis of the seeding operations done in the south target area during the Israeli II project suggests that seeding resulted in a decrease in precipitation. Gabriel and Rosenfeld (1990) suggest that this may have resulted from enhanced natural ice nucleus concentrations from wind-blown soil from the desert regions in the southern part of Israel. Woodley and Rosenfeld (1992) suggest that large dust particles might initiate the coalescence in these clouds, leading to conditions similar to those found in Project Whitetop.

The findings of the Israeli II project confirm the previous results for the Israeli I project. Together with a relatively good understanding of the physical processes leading to precipitation in these clouds, this set of experiments perhaps best demonstrates the ability to increase precipitation through static mode glaciogenic seeding. These results also underscore the seedability requirement that both the ice crystal and the coalescence mechanisms must be relatively inefficient.

Table 11.3. Evaluation of the Israeli II results partitioned by cloud top temperature. Mean precipitation for the target and control areas (mm day^{-1}), the double ratio (DR) and significance of the double ratio as a P-value are provided for several overlapping cloud top temperature stratifications.

Cloud top temperature (°C)	Days		Seeded		Not seeded			
	S	NS	Target	Control	Target	Control	DR	P (%)
<−26	49	38	10.66	10.24	9.87	10.97	1.16	18.2
<−21	70	52	11.14	10.26	10.55	10.92	1.12	16.3
−21 to −26	21	14	12.25	10.30	12.41	10.78	1.03	43.9
>−11	27	21	9.85	8.06	7.12	7.87	1.35	19.1
>−15	45	38	9.51	7.82	6.88	7.85	1.39	5.5
−11 to −26	59	53	11.37	9.80	10.24	11.88	1.35	1.3
−15 to −26	41	36	12.41	10.94	10.31	11.96	1.32	5.1
−11 to −21	38	39	10.77	9.64	9.45	12.27	1.45	1.3
−15 to −21	20	22	12.57	11.61	8.97	12.70	1.53	5.3

Source: Gagin (1986).

(iii) The High Plains Experiment (HIPLEX). The HIPLEX was conducted during the summers of 1979 and 1980 in southeastern Montana. The primary objective was to document the various links in the physical hypothesis of static mode seeding in small, isolated cumulus congestus. An instrumented KingAir was used for the initial pass to determine the liquid water content and ice crystal concentrations at the −8°C level. If the cloud was considered a 'go' cloud, it was seeded with dry ice at the −10°C level by a Learjet. Typical cloud base temperatures were 5°C with cloud tops not colder than −12°C. A total of 20 clouds met the seeding criteria, of which 12 were actually seeded. The experimental design is discussed by Smith et al. (1984).

Figure 11.25 compares the development of precipitation in the seeded and not-seeded clouds as summarized by Cooper and Lawson (1984). The seeding resulted in a two order of magnitude increase in the ice crystal concentrations after 2 minutes, achieving concentrations from $1\,\ell^{-1}$ to $100\,\ell^{-1}$ in the seeded clouds compared to ice crystal concentrations less than $1\,\ell^{-1}$ in the not-seeded clouds. Many of the seeded clouds developed concentrations of rimed ice crystals after 5 minutes in excess of $2\,\ell^{-1}$. However, the development of significant concentrations of precipitation-sized ice particles was only slightly (not statistically significant) enhanced over the not-seeded clouds, as was the estimated rainfall at cloud base.

Cooper and Lawson attribute the lack of continued development of precipitation to the limited lifetimes of these clouds. They cite entrainment as the principal factor in the relatively short lifetimes. Isaac (1986) summarizes several observations of lifetimes of continental cumuli defined on the basis of a cloud top colder than −7°C and finds (top) lifetimes are typically 10 to 12 minutes. Roughly 5 to 8 minutes are needed for a cumulus cloud to reach the −7°C level from the initial condensation at cloud base. This suggests that clouds must last longer than 15 to 20 minutes, otherwise the time is insufficient to develop precipitation naturally or through glaciogenic seeding.

11.6.2. *Hygroscopic seeding*

The term *hygroscopic seeding* will be used to include seeding with hygroscopic materials, whether in dry form or in solution, as well as seeding with non-hygroscopic large particles including water drop seeding. The particular seeding agent used has relatively little effect on the enhancement of the coalescence process, although it would markedly affect the amount of seeding material that must be dispersed and the method by which the material is dispersed. The importance of the seed particle size has been stressed in § 11.2.2, part (v).

Relatively few randomized experiments have been conducted that permit an evaluation of hygroscopic seeding to enhance precipitation. The experiments on tropical cumuli seeded with large water drops in the Caribbean described by Braham et al. (1957) provide evidence of precipitation enhancement when seeding rates as high as $1\ \mathrm{kg\ m^{-1}}$ are used. They describe experiments conducted in the winter and fall of 1954 in which pairs of similar cumulus clouds were

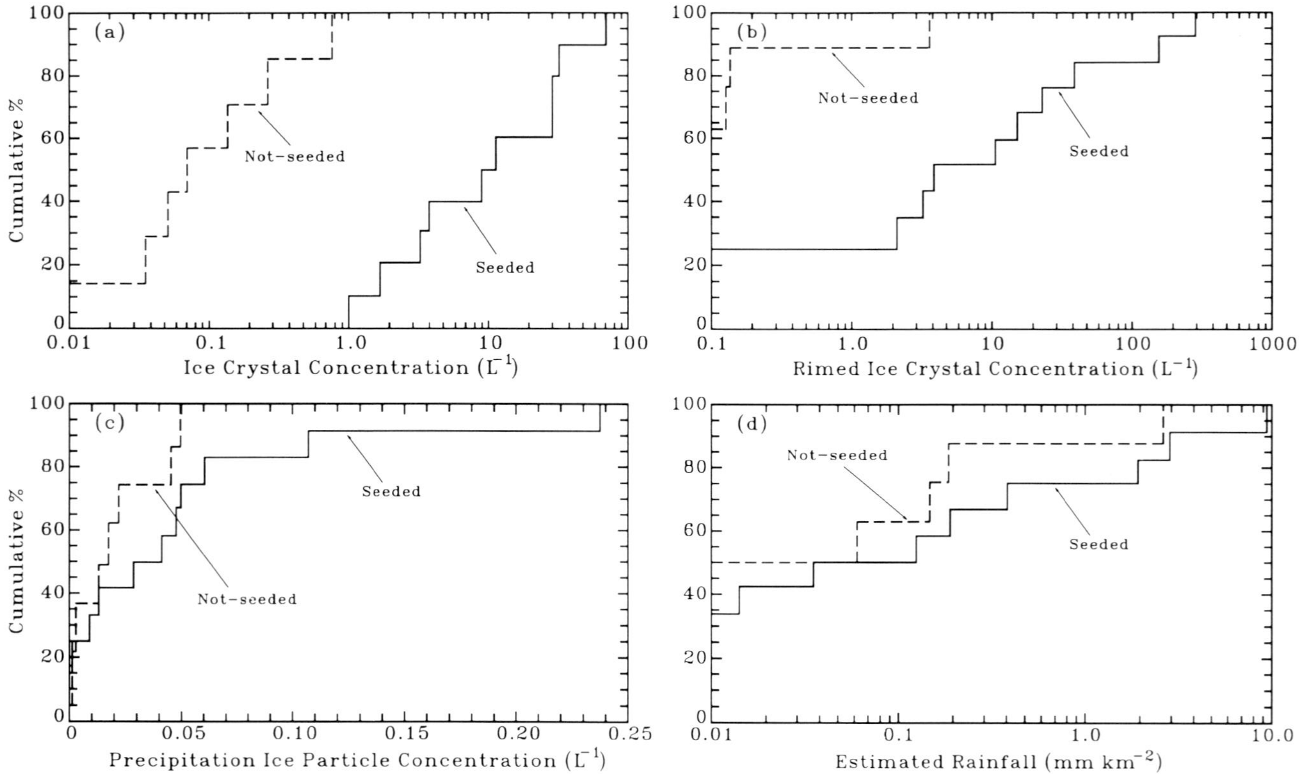

Fig. 11.25. Comparison of the development of precipitation in seeded and not seeded clouds in HIPLEX. (a) Ice crystal concentrations after 2 minutes. (b) Rimed ice crystal concentrations after 5 minutes. (c) Concentration of precipitating ice particles at least 600 μm in longest dimension. (d) Aircraft estimated rainfall. From Cooper and Lawson (1984). *Journal of Climate & Applied Meteorology*, **23**, 4. American Meteorological Society. Reproduced with permission.

Table 11.4. Formation of radar echoes in seeded and not-seeded tropical cumuli over the Caribbean Sea. Seeding was carried out with large water drops. 'E' indicates clouds that produced a radar echo; 'NE' indicates clouds that did not produce an echo. The probability that this result could have occurred by chance if there were no effect of seeding is 1.7%.

		Seeded clouds		
		E	**NE**	**Sums**
Not-seeded clouds	**E**	6	11	17
	NE	17	18	35
	Sums	22	24	46

Source: Braham et al. (1957).

selected, one of which was seeded at random. The appearance or non-appearance of a radar echo was used to evaluate the effects of seeding.

A total of 46 pairs of clouds were studied, with results as detailed in Table 11.4. Seeded clouds were more likely to develop radar echoes as not-seeded clouds. The probability that the distribution of radar echoes shown in Table 11.4 could have occurred by chance is 1.7%, i.e., the seeding was responsible for initiating precipitation in clouds that otherwise would not have developed precipitation.

Seeded clouds displayed a more rapid initiation of precipitation, averaging 6.4 minutes to produce a first echo compared to an average of 11.9 minutes for the not-seeded clouds. The distribution of radar echoes by time for the seeded and not-seeded clouds is shown in Fig. 11.26. After 10 minutes, 46% of the seeded clouds had produced an echo, compared to only 10% of the not-seeded ones. Braham et al. calculate that the probability that these differences in time to first echo could have occurred by chance is 0.03%.

Dennis and Koscielski (1972) compare the heights of first echoes in natural, AgI-seeded, and salt-seeded cumuli in South Dakota and find that the salt-seeded clouds produce radar first echoes an average of 1600 m above cloud base, compared to 2100 m for the AgI-seeded clouds, and 3350 m for the not-seeded clouds. The reduction in height of first echo for the salt-seeded clouds is significant beyond the 1% level.

Dennis and Koscielski also note that the height of the AgI-seeded cloud first echo is closely related to the temperature, whereas the height of the salt-seeded cloud first echo appears to be independent of temperature. This indicates initiation of the ice crystal mechanism when clouds were seeded by AgI, and the coalescence mechanism when they were seeded with salt. The average cloud base temperature for the salt-seeded clouds was 9°C.

Although both of these experiments suggest earlier initiation of precipitation in hygroscopic-seeded clouds, they do not demonstrate increases in the total rain volume from the cloud. It is possible that earlier initiation of precipitation could

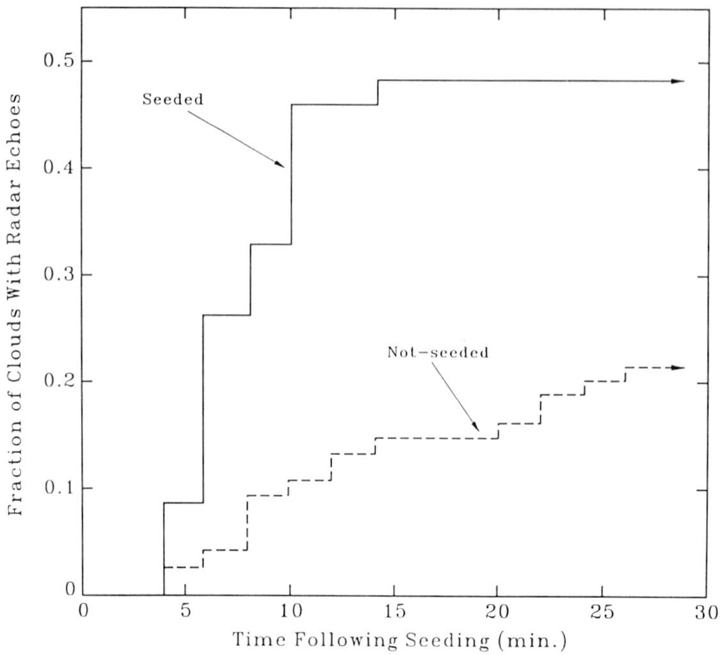

FIG. 11.26. Time to a radar first echo for cumulus clouds seeded with large water drops compared to not-seeded clouds. Twenty-two of the 46 seeded clouds produced a radar echo within 25 minutes, compared to 16 of the 76 not-seeded clouds. Data obtained from Braham et al. (1957) for tropical cumuli over the Caribbean Sea.

simply cause the cloud to rain out earlier than it would otherwise have done with no change in the total precipitation, or even a decrease if the formation of precipitation hastened the demise of the cloud before it reached the height (depth) it would have naturally. Earlier initiation of precipitation would be expected to produce precipitation from clouds whose lifetimes are too short or whose depth is insufficient to produce precipitation naturally.

11.6.3. *Glaciogenic seeding for dynamical effects*

The initial successes in seeding individual cumulus clouds for dynamical effects reported from Project Stormfury (Simpson et al., 1967) led to the Florida Area Cumulus Experiment (FACE) designed to evaluate the potential for enhancing precipitation by dynamic mode seeding. The conceptual model for a seeding effect on precipitation evolved as the FACE evolved, as discussed in § 11.2.3. The Texas component of the Southwest Cooperative Program was established based on the findings from the FACE. A review of dynamic mode seeding of cumulus clouds is provided by Orville (1986).

(i) Project Stormfury. As part of Project Stormfury, a dynamical mode seeding experiment was conducted during the summer of 1965 south of Puerto Rico over

the Caribbean Sea. Cumulus clouds were seeded with 1700 g AgI within a single turret to promote glaciation and stimulation of the updraft through the release of latent heat. A total of 14 clouds were seeded, with 9 clouds serving as controls. The seeded clouds grew roughly 1600 m higher than the comparable control clouds. These results are shown in Fig. 11.13.

Ruskin (1967) provides microphysical observations, documenting the changes within a seeded turret. The seeding aircraft penetrated near the cloud top at 7.2 km, with four research aircraft penetrating simultaneously at cloud base, 3.0 km, 5.2 km, and 5.8 km, respectively. The liquid water content observed on the seeding pass at the $-5°C$ level (5.2 km) was $1.9\,g\,m^{-3}$, decreasing to $0.9\,g\,m^{-3}$ on the subsequent pass five minutes later. This was accompanied by an increase in the small ice particle water content from $0.1\,g\,m^{-3}$ to $1.0\,g\,m^{-3}$ and an increase in the temperature of 1.5°C. The seeded portion of the cloud grew from 7.3 km to 11.0 km following the seeding, while no other cloud in the area grew above 8.2 km.

(ii) Florida Area Cumulus Experiment (FACE). The FACE was designed to evaluate the potential for enhancing precipitation on an area basis rather than from individual storms. Individual clouds were seeded using AgI-flares with the intention of creating longer lasting storms, as well as to promote mergers between adjacent clouds. The exploratory phase of the FACE was conducted from 1970 through 1976[18] (FACE-1) using 10 AgI-flares for each seeded cell; the confirmatory phase was conducted during 1979 and 1980 (FACE-2), using a variable number of flares. Statistical analyses were based on both floating and total targets.

The FACE-1 results are described by Woodley et al. (1982), who note significant increases in precipitation due to seeding for both the floating and total targets. The average 6-hour precipitation for the floating target on seeded days was 3.40 mm; the corresponding precipitation for the not-seeded days was 2.29 mm. The difference is significant at the 1% level. Time resolution of the floating target precipitation revealed no significant differences for the first 2 hours but marked increases in precipitation from the seeded clouds subsequent to 2 hours after seeding (Fig. 11.27).

Although the FACE-2 was designated as a confirmatory experiment, the amount of AgI delivered to a given cell varied, from the amount used in the FACE-1 to considerably less. Although the overall effect of the seeding on precipitation was not conclusive, when cells treated with nine or more flares were studied, significant increases in precipitation were noted. The seeded to not-seeded ratio of the precipitation volume for this partition was 2.60, significant at the 0.6% level (Gagin et al., 1986).

Woodley et al. also report that cells treated shortly after the initial appearance of a radar echo exhibited stronger responses to seeding than those treated later. The seeded to not-seeded ratio of the precipitation volume for the

[18] Experiments were not conducted in 1972 or 1974.

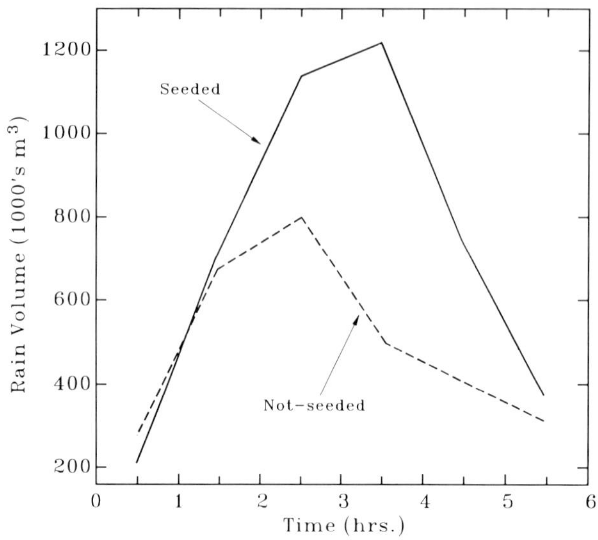

Fig. 11.27. Time history of precipitation from seeded and not-seeded clouds for the floating target for the FACE-1. Precipitation is expressed as the hourly rain volume in 10^3 m^3. Woodley, et al. (1982). *Journal of Applied Meteorology*, **21**, 2. American Meteorological Society. Reproduced with permission.

early treatment partition was 1.58, significant at the 4.4% level. When combined with the nine or more flares partition, the double ratio was 3.57, significant at the 0.5% level. These experiments suggest for an optimal dynamical effect, seeding should be conducted early in the life cycle of the cell with relatively massive amounts of AgI.

(iii) Southwest Cooperative Program/Texas. The Texas component of the Southwest Cooperative Program focused on small mesoscale convective clusters that would be produced by seeding within individual cells. The experimental units were 25 km radius circles, centered on the cell that qualified the unit for treatment. Results of this experiment for the summers of 1986 and 1987 showed greater than a two-fold increase in precipitation from seeded units (statistically significant at the 5% level) and merging was twice as frequent for seeded clouds (Rosenfeld and Woodley, 1989).

Rosenfeld and Woodley (1993) summarize results for 183 cells from 1986 through 1990 (90 seeded, 93 not-seeded) and note a 130% increase (2.3 fold) increase in precipitation. Differences in precipitation between the seeded and not-seeded clouds began to be evident after 10 to 15 minutes, unlike the FACE-2 clouds which did not exhibit differences between seeded and not-seeded clouds until more than 90 minutes after seeding.[19] Increases in cell height due to seeding were comparable to those in the FACE.

[19] This may be due to the higher time resolution of precipitation volume in the Texas experiment.

The Texas experiment represents a verification of the FACE results in a different geographical location. This may represent the first time that a seeding experiment demonstrating enhanced precipitation in one location has been successfully transferred to another location.

11.7. Inadvertent modification of cloud processes

Inadvertent modification of cloud processes is most often related to urban effects on cloud formation and the development of precipitation within those clouds. The Metropolitan Meteorological Experiment (METROMEX) was designed to study the effects of St. Louis MO on clouds and precipitation over the city and in the surrounding areas, relying on aircraft and ground-based physical measurements as well as statistical analyses. The METROMEX was conducted from 1971 through 1973; a summary of it may be found in Braham (1981).

Urban influences may be classified as dynamical or microphysical. Changnon et al. (1976) suggest cloud development over cities may be enhanced due to thermodynamic effects leading to greater instability and/or thermodynamic effects leading to low-level convergence over the city. They also suggest cities are sources for giant CCN, which may enhance the coalescence process. Summer precipitation increases of 25% found downwind of St. Louis have been attributed to these effects.

Braham (1981) notes that urban production of CCN broaden the cloud droplet spectrum, producing higher concentrations of small as well as large droplets. The reduction in mean droplet size associated with the increase in droplet concentrations would imply a lessened probability for an active coalescence process. Braham cites modeling studies that suggest that giant and ultragiant CCN more than compensate for the additional droplets and reduction in mean droplet size. He notes the importance of extending the CCN size spectrum used in modeling precipitation development to larger sizes, citing a need to simulate the growth of large particles present in concentrations much less than $0.1\,\ell^{-1}$, representing CCN larger than 30 μm radius.

Braham notes that the urban heat island promotes low level convergence and rising air over the city. When the incoming air is relatively unstable, stronger updrafts are produced. This effect is most pronounced during afternoons and in the summer (Changnon et al., 1991). Braham also suggests that an enhanced coalescence process in urban clouds leads to enhanced formation of ice. The glaciation associated with enhanced ice formation is postulated to invigorate the updrafts and produce urban clouds that are taller.[20]

Paper and pulp mills have been singled out as significant sources of CCN. Hobbs et al. (1970) estimated a production rate of 10^{19} CCN s^{-1} from a paper mill in Washington state. The CCN were noted to consist primarily of sodium sulfate (Na_2SO_4). Statistically significant increases in precipitation were found downwind of the paper mill. Mather (1991) reports an enhancement of the coalescence process downwind of a paper mill in South Africa.

[20] Ice nucleus concentrations are noted to be lower over and downwind from St. Louis.

Automobile exhaust has been associated with enhanced ice nucleus concentrations. Schaefer (1968) suggests that lead from combustion of leaded gasoline reacts with naturally occurring iodine to form lead iodide. Hobbs and Locatelli (1969) report enhanced ice nucleus concentrations downwind from a forest fire.

Clearly, the formation of clouds and development of precipitation within those clouds can be affected by anthropogenic influences. These influences may be dynamical as well as microphysical in their action.

Problems

11.1* Consider a warm fog with 50 drops cm^{-3} and a liquid water content of 0.21 $g\,m^{-3}$. If the droplet spectrum was monodispersed, what would the visibility be?

Now suppose that coagulation between droplets has resulted in halving the drop concentration. What is the improvement in visibility due to coagulation?

11.2* Determine the energy required to completely dissipate 1 m^3 of fog at +20°C and 1000 mbar with 1 $g\,m^{-3}$ of liquid water by using a laser.

11.3* How large a volume of air could be cooled from −10°C to −40°C by the sublimation of 1 g of dry ice. Assume a latent heat of sublimation of 398 $J\,g^{-1}$ and an ambient pressure of 700 mbar.

11.4* (a) Demonstrate that dr/dt for drops forced to evaporate by laser illumination is roughly constant for drops larger than 8 μm radius.
(b) Demonstrate that dr/dt under these conditions decreases with decreasing drop size for drops smaller than 8 μm radius.
Hint: Use (11.3).

11.5** Determine the energy required to completely dissipate 1 m^3 of fog at 20°C and 1000 mbar with 1 $g\,m^{-3}$ of liquid water by burning octane (C_8H_{18}). The heat of combustion for *n*-octane at 20°C and 1000 mbar is 5527 $kJ\,mol^{-1}$.

11.6** Some researchers have proposed using surface active agents to enhance the coalescence efficiency between water drops and thereby enhance the coalescence mechanism. What seeding rate (in units of $g\,s^{-1}$) would be required to coat all the cloud droplets with a monolayer of such a surfactant having a molecular weight of 200 $g\,mol^{-1}$ for an updraft area of 1 km^2 and an updraft velocity of 2 $m\,s^{-1}$?

You may assume a density of 1 $g\,cm^{-3}$ for the surfactant and that the molecules are spherical. You may also assume that the cloud has a liquid water content of 1 $g\,m^{-3}$ with a monodispersed droplet spectrum and a radius of 15 μm.

11.7** Consider seeding an updraft with 5 μm radius sodium chloride (NaCl) particles sufficiently below cloud base such that the NaCl particles will be in equilibrium with their Köhler curve at cloud base (+5°C and 700 mbar). What would be the radius of the resulting solution drop?

What is the enhancement factor in the seeding rate ($g\,s^{-1}$) attained using NaCl particles rather than simply seeding with water drops at cloud base?

Computer problem

11.8** The following data are the rain volumes for the target and control cells on both the seeded (S) and not-seeded (NS) days for FACE-2 (from Gagin et al., 1986). Daily mean rain volumes in $10^3\,m^3$ and the numbers of target and control cells are given for each day.

Date	S/NS	Target	Control
1979			
Aug 20	S	158.5 (18)	108.8 (269)
Aug 28	NS	151.9 (22)	114.9 (110)
Aug 29	NS	329.3 (21)	173.0 (151)
1980			
Jun 25	S	152.6 (6)	34.4 (15)
Jul 4	S	107.5 (5)	61.8 (70)
Jul 7	S	238.5 (14)	86.3 (59)
Jul 8	S	297.1 (19)	85.6 (194)
Jul 9	S	223.2 (35)	67.7 (64)
Jul 10	NS	165.7 (7)	109.7 (60)
Jul 12	NS	88.3 (21)	40.6 (47)
Jul 16	NS	256.1 (18)	120.1 (93)
Jul 29	NS	155.5 (12)	28.8 (8)
Aug 2	S	158.5 (25)	52.3 (179)
Aug 3	NS	115.5 (13)	87.0 (334)
Aug 4	S	82.7 (16)	76.6 (31)
Aug 18	S	257.1 (23)	93.9 (185)
Aug 26	NS	150.0 (8)	141.7 (43)
Aug 29	NS	113.3 (17)	66.7 (157)

(a) Determine the double ratio for the effect of seeding on precipitation for FACE-2. Remember to weight each day's mean rain volume by the number of cells.

(b) Determine the probability that this result could have been obtained by chance given the null hypothesis that seeding had no effect, using a Monte Carlo approach. For each of 1000 random trials, randomly select nine of the eighteen experimental days to be seeded days, leaving the remaining nine days to be not-seeded days. Calculate the double ratio as in (a). Count the number of trials for which the simulated double ratio is as extreme or more extreme than the observed double ratio, i.e., if the observed double ratio is >1, then count all simulated double ratios greater than or equal to the observed double ratio. Dividing by the number of trials gives the probability that the observed result could have been obtained by chance.

REFERENCES

Battan, L.J., 1966, 'Silver iodide seeding and rainfall from convective clouds,' *J. Appl. Meteorol.* **5**, 669.

Bergeron, T., 1949, 'The problem of artificial control of rainfall on the globe. I. General effects of ice nuclei in clouds,' *Tellus* **1**, 32.

Bigg, E.K., and Turton, E., 1988, 'Persistent effects of cloud seeding with silver iodide,' *J. Appl. Meteorol.* **27**, 505.

Biwas, K.R., Kapoor, R.K., and Kanuga, K.K., 1967, 'Cloud seeding experiment using common salt,' *J. Appl. Meteorol.* **6**, 914.

Braham, R.R., Jr., 1964, 'What is the role of ice in summer rain-showers?' *J. Atmos. Sci.* **21**, 640.

Braham, R.R., Jr., 1979, 'Field experimentation in weather modification,' *J. Am. Statist. Ass.* **74**, 57.

Braham, R.R., Jr., 1981, 'Urban precipitation processes,'x *Meteorol. Monogr.* **18**(40), 75.

Braham, R.R., Jr, Battan, L.J., and Byers, H.R., 1957, 'Artificial nucleation of cumulus clouds,' *Meteorol. Monogr.* **2**(11), 47.

Brownlee, K.A., 1965, *Statistical theory and methodology in science and engineering*, New York: John Wiley and Sons.

Changnon, S.A., Jr., Semonin, R.G., and Huff, F.A., 1976, 'A hypothesis for urban rainfall anomalies,' *J. Appl. Meteorol.* **15**, 544.

Changnon, S.A., Jr., Shealy, R.T., and Scott, R.W., 1991, 'Precipitation changes in fall, winter, and spring caused by St. Louis,' *J. Appl. Meteorol.* **30**, 126.

Cooper, W.A., and Lawson, R.P., 1984, 'Physical interpretation of results from the HIPLEX-1 experiment,' *J. Clim. Appl. Meteorol.* **23**, 523.

Cooper, W.A., and Marwitz, J.D., 1980, 'Winter storms over the San Juan Mountains. Part III: Seeding potential,' *J. Appl. Meteorol.* **19**, 942.

DeMott, P.J., 1991, 'Comments on 'The persistence of seeding effects in a winter orographic cloud seeded with silver iodide burned in acetone',' *J. Appl. Meteorol.* **30**, 903.

DeMott, P.J., Finnegan, W.G., and Grant, L.O., 1983, 'An application of chemical kinetic theory and methodology to characterize the ice nucleating properties of aerosols used for weather modification,' *J. Clim. Appl. Meteorol.* **22**, 1190.

Dennis, A.S., 1980, *Weather modification by cloud seeding*. New York: Academic Press.

Dennis, A.S., and Koscielski, A., 1972, 'Height and temperature of first echoes in unseeded and seeded convective clouds in South Dakota,' *J. Appl. Meteorol.* **11**, 994.

Deshler, T., Reynolds, D.W., and Huggins, A.W., 1990, 'Physical response of wintertime orographic clouds over the Sierra Nevada to airborne seeding using dry ice or silver iodide,' *J. Appl. Meteorol.* **29**, 288.

Eadie, W.J., and Mee, T.R., 1963, 'The effect of dry-ice pellet velocity on the generation of ice crystals,' *J. Appl. Meteorol.* **2**, 260.

Elliott, R.D., 1986, 'Review of wintertime orographic cloud seeding,' *Meteorol. Monogr.* **21**(43), 87.

Elliott, R.D., and Hovind, E.L., 1964, 'On convective bands within Pacific coast storms and their relation to storm structure,' *J. Appl. Meteorol.* **3**, 143.

Elliott, R.D., Shaffer, R.W., Court, A., and Hannaford, J., 1978, 'Randomized cloud seeding in the San Juan Mountains, Colorado,' *J. Appl. Meteorol.* **17**, 1298.

Feng, D., and Finnegan, W.G., 1989, 'An efficient, fast functioning nucleating agent – AgI(FA)AgCl–4NaCl,' *J. Weath. Modif.* **21**, 41.

Finnegan, W.G., and Pitter, R.L., 1987, 'Rapid ice nucleation by acetone–silver iodide generator aerosols,' *J. Weath. Modif.* **20**, 51.

Flueck, J., 1986, 'Principles and prescriptions for improved experimentation in precipitation augmentation research,' *Meteorol. Monogr.* **21**(43), 155.

Gabriel, K.R., and Rosenfeld, D., 1990, 'The second Israeli rainfall stimulation experiment: Analysis of precipitation on both targets,' *J. Appl. Meteorol.* **29**, 1055.

Gabriel, K.R., Avichai, Y., and Steinberg, R., 1967, 'A statistical investigation of persistence in the Israeli artificial rainfall stimulation experiment,' *J. Appl. Meteorol.* **6**, 323.

Gagin, A., 1975, 'The ice phase in winter continental cumulus clouds,' *J. Atmos. Sci.* **32**, 1604.

Gagin, A., 1986, 'Evaluation of 'static' and 'dynamic' seeding concepts through analyses of Israeli II and FACE-2 experiments,' *Meteorol. Monogr.* **21**(43), 63.

Gagin, A., Rosenfeld, D., Woodley, W.L., and Lopez, R.E., 1986, 'Results of seeding for dynamic effects on rain-cell properties in FACE-2,' *J. Clim. Appl. Meteorol.* **25**, 3.

Grant, L.O., 1986, 'Hypotheses for the Climax wintertime orographic cloud seeding experiments,' *Meteorol. Monogr.* **21**(43), 105.

Grant, L.O., and Kahan, A.M., 1974, 'Weather modification for augmenting orographic precipitation,' *Weather and climate modification* (W.N. Hess, Ed.). New York: John Wiley and Sons, p. 282.

Grant, L.O., Chappell, C.F., and Mielke P.W., Jr., 1971, 'The Climax experiment for seeding cold orographic clouds,' *Preprints Int. Conf. Weather Modification 1971*. American Meteorological Society, p. 78.

Hill, G.E., 1980, 'Seeding opportunity recognition in wintertime orographic clouds,' *J. Appl. Meteorol.* **19**, 1371.

Hobbs, P.V., 1975a, 'The nature of winter clouds and precipitation in the Cascade Mountains and their modification by artificial seeding. Part I: Natural conditions,' *J. Appl. Meteorol.* **14**, 783.

Hobbs, P.V., 1975b, 'The nature of winter clouds and precipitation in the Cascade Mountains and their modification by artificial seeding. Part III: Case studies of the effects of seeding,' *J. Appl. Meteorol.* **14**, 819.

Hobbs, P.V., and Locatelli, J.D., 1969, 'Ice nuclei from a natural forest fire,' *J. Appl. Meteorol.* **8**, 833.

Hobbs, P.V. and Radke, L.F., 1975, 'The nature of winter clouds and precipitation in the Cascade Mountains and their modification by artificial seeding. Part II: Techniques for the physical evaluation of seeding,' *J. Appl. Meteorol.* **14**, 805.

Hobbs, P.V., and Rangno, A.L., 1979, 'Comments on the Climax and Wolf Creek Pass cloud seeding experiments,' *J. Appl. Meteorol.* **18**, 1233.

Hobbs, P.V., Radke, L.F., and Shumway, S.E., 1970, 'Cloud condensation nuclei from industrial sources and their apparent influence on precipitation,' *J. Atmos. Sci.* **27**, 81.

Holroyd, E.W., Super, A.B., and Silverman, B.A., 1978, 'The practicability of dry ice for on-top seeding of convective clouds,' *J. Appl. Meteorol.* **17**, 49.

Hudak, D.R., and List, R., 1988, 'Precipitation development in natural and seeded clouds in southern Africa,' *J. Appl. Meteorol.* **27**, 734.

Isaac, G.A., 1986, 'Summer cumulus lifetime – Importance to static mode seeding,' *Meteorol. Monogr.* **21**(43), 25.

King, W.D., 1984, 'Seeding stratiform clouds: The effects of cloud parameters on aiming,' *J. Clim. Appl. Meteorol.* **23**, 1265.

Koenig, L.R., 1963, 'The glaciating behavior of small cumulonimbus clouds,' *J. Atmos. Sci.* **20**, 29.

Levy, G., and Cotton, W.R., 1984, 'A numerical investigation of mechanisms linking glaciation of the ice-phase to the boundary layer,' *J. Clim. Appl. Meteorol.* **23**, 1505.

Ludlam, F.H., 1955, 'Artificial snowfall from mountain clouds,' *Tellus* **7**, 277.

Marwitz, J.D., 1980, 'Winter storms over the San Juan Mountains. Part I: Dynamical processes,' *J. Appl. Meteorol.* **19**, 913.

Mason, B.J., 1981, 'The mechanisms of cloud seeding with dry ice,' *J. Weath. Modif.* **13**, 11.

Mather, G.K., 1991, 'Coalescence enhancement in large multi-cell storms caused by emissions from a Kraft paper mill,' *J. Appl. Meteorol.* **30**, 1134.

Mather, G.K., Bigg, E.K., and Renton, S., 1990, 'Apparent persistence effects in the Nelspruit area from silver iodide seeding for hail suppression,' *J. Appl. Meteorol.* **29**, 806.

Mielke, P.W., Jr. Grant, L.O., and Chappell, C.F., 1971, 'An independent replication of the Climax wintertime orographic cloud seeding experiment,' *J. Appl. Meteorol.* **10**, 1198.

Mielke, P.W., Jr., Brier, G.W., Grant, L.O., Mulvey, G.J., and Rosenzweig, P.N., 1981, 'A statistical reanalysis of the replicated Climax I and II wintertime orographic cloud seeding experiments,' *J. Appl. Meteorol.* **20**, 643.

Nelson, L.D., 1971, 'A numerical study on the initiation of warm rain,' *J. Atmos. Sci.* **28**, 752.

Nelson, L.D., and Silverman, B.A., 1972, 'Optimization of warm-cloud seeding agents by microencapsulation techniques,' *Mon. Weath. Rev.* **100**, 153.

Neyman, J., Scott, E.L., and Wells, M.A., 1969, 'Statistics in meteorology,' *Rev. Int. Statist. Inst.* **37**, 119.

NRC (National Research Council), Committee on Atmospheric Sciences, 1973, *Weather and climate modification, problems and progress*. Washington, DC: National Academy of Sciences.

Orville, H.D., 1986, 'A review of dynamic-mode seeding of summer cumuli,' *Meteorol. Monogr.* **21**(43), 43.

Palanisamy, M., Thangaraj, K., Gobinathan, R., and Ramasamy, P., 1986, 'X-ray diffraction and ice nucleation studies of AgI–AgCl solid solutions,' *J. Cryst. Growth* **79**, 1005.

Rangno, A.L., 1986, 'How good are our conceptual models of orographic cloud seeding?' *Meteorol. Monogr.* **21**(43), 115.

Rangno, A.L., and Hobbs, P.V., 1987, 'A reevaluation of the Climax cloud seeding experiment using NOAA published data,' *J. Clim. Appl. Meteorol.* **26**, 757.

Rauber, R.M., Elliott, R.D., Rhea, J.O., Huggins, A.W., and Reynolds, D.W., 1988, 'A diagnostic technique for targeting during airborne seeding experiments in wintertime storms over the Sierra Nevada,' *J. Appl. Meteorol.* **27**, 811.

Rokicki, M.L., and Young, K.C., 1978, 'The initiation of precipitation in updrafts,' *J. Appl. Meteorol.* **17**, 745.

Rosenfeld, D., and Woodley, W.L., 1989, 'Effects of cloud seeding in west Texas,' *J. Appl. Meteorol.* **28**, 1050.

Rosenfeld, D., and W. L. Woodley, 1993; 'Effects of cloud seeding in West Texas: Additional results and new insights', *J. Appl. Meteor.* (in press).

Ruskin, R.E., 1967, 'Measurements of water–ice budget changes at −5°C in AgI-seeded tropical cumulus,' *J. Appl. Meteorol.* **6**, 72.

Ruskin, R.E., and Scott, W.D., 1974, 'Weather modification instruments and their use,' *Weather and climate modification* (W.N. Hess, Ed.). New York: John Wiley and Sons, p. 136.

Schaefer, V.J., 1968, 'Ice nuclei from auto exhaust and organic vapors,' *J. Appl. Meteorol.* **7**, 148.

Scott, P.T., Finnegan, W.G., and Sinclair, P.C., 1989, 'Characterization of a modified hexagonal silver iodide nucleus aerosol,' *J. Appl. Meteorol.* **28**, 722.

Silverman, B.A., 1986, 'Static mode seeding of summer cumuli – A review,' *Meteorol. Monogr.* **21**(43), 7.

Silverman, B.A., and Weinstein, A.I., 1974, 'Fog,' *Weather and climate modification* (W.N. Hess, Ed.). New York: John Wiley and Sons, p. 355.

Simpson, J., 1980, 'Downdrafts as linkages in dynamic cumulus seeding effects,' *J. Appl. Meteorol.* **19**, 477.

Simpson, J., Brier, G.W., and Simpson, R.H., 1967, 'Stormfury cumulus seeding experiment 1965: Statistical analysis and main results,' *J. Atmos. Sci.* **24**, 508. Smith, P.L., Dennis, A.S., Silverman, B.A., Super, A.B., Holroyd, E.W., III, Cooper, W.A., Mielke, P.W., Jr., Berry, K.J., Orville, H.D., and Miller, J.R., Jr., 1984, 'HIPLEX-1: Experimental design and response variables,' *J. Clim. Appl. Meteorol.* **23**, 497.

Super, A.B., and Heimbach, J.A., Jr, 1988, 'Microphysical effects of wintertime cloud seeding with silver iodide over the Rocky Mountains. II. Observations over the Bridger Range, Montana,' *J. Appl. Meteorol.* **27**, 1152.

Super, A.B., Boe, B.A., Holroyd, E.W., III, and Heimbach, J.A., Jr., 1988, 'Microphysical effects of wintertime cloud seeding with silver iodide over the Rocky Mountains. I. Experimental design and instrumentation,' *J. Appl. Meteorol.* **27**, 1145.

Vonnegut, B., 1981, 'Misconception about cloud seeding with dry ice,' *J. Weath. Modif.* **13**, 9.

Weickmann, H.K., 1974, 'The mitigation of Great Lakes storms,' *Weather and climate modification* (W.N. Hess, Ed.). New York: John Wiley and Sons, p. 318.

Woodley, W.L., Jordan, J., Barnston, A., Simpson, J., Biondini, R., and Flueck, J., 1982, 'Rainfall results of the Florida Area Cumulus Experiment, 1970,' *J. Appl. Meteorol.* **21**, 139.

Young, K.C., 1973, *A numerical simulation of wintertime, orographic precipitation under natural and seeded conditions.* Ph.D. Thesis, The University of Chicago.

Young, K.C., 1974a, 'A numerical simulation of wintertime, orographic precipitation: Part I. Description of model microphysics and numerical techniques,' *J. Atmos. Sci.* **31**, 1735.

Young, K.C., 1974b, 'A numerical simulation of wintertime, orographic precipitation: Part II. Comparison of natural and AgI seeded conditions,' *J. Atmos. Sci.* **31**, 1749.

Young, K.C., 1975, 'Growth of the ice phase in strong cumulonimbus updrafts,' *Pure Appl. Geophys.* **113**, 1005.

Young, K.C., and Tomlinson, E., 1993, 'The microphysical effects of irradiating a fog with a 10.6 μm CO_2 laser,' *J. Atmos. Sci.* (in press).

12

THE FORMATION AND MODIFICATION OF HAIL

12.1. Introduction

Hail-producing thunderstorms may be classified as multi-cell or single-cell (supercell) thunderstorms. This distinction has important consequences for the development of hail in the storm. In a multi-cell thunderstorm, many of the embryos which eventually become hail develop in the growing stage which then evolves into a mature stage. Thus, the updraft velocity increases as the embryos grow, serving to balance the increasing fall speed of the hail embryos, keeping them at a relatively constant level. This serves to maintain the growing hailstones in a temperature and liquid water regime favorable for continued growth.

A supercell thunderstorm consists of a single, quasi-steady cell that maintains updraft and downdraft continuity over a period of several hours. Thus, the region of the cloud in which potential hail embryos develop and the region in which they grow to hailstone sizes are distinct and separate. Not only must the potential hailstones be transferred into regions of stronger updrafts as they grow, but the increase in the updraft velocities they encounter must closely match the increase in their fall speeds as they grow.

The region of weaker updrafts within which the embryos that may become hail are formed is referred to as the *embryo formation region* (EFR); the region of stronger updrafts in which some of the embryos may grow to become hail is referred to as the *hail growth zone* (HGZ); the region of weak updrafts and downdrafts in which the hail falls to the ground is referred to as the hail cascade. These are depicted in Fig. 12.1.

Only a relatively small fraction of the potential hail embryos that are formed in the EFR ever become hail. Most of these embryos remain small, and melt before reaching the ground as rain. Although the size and position of the potential hail embryos influences the probability that they will be 'selected' to become hail, random perturbations of the particle trajectory due to turbulence introduces a stochastic influence on the selection process.

Evaluation of hail suppression concepts is dependent on understanding the mechanisms that produce hail in thunderstorms. The most viable seeding hypotheses for hail suppression rely on seeding in the EFR. Thus, identification of the EFR is crucial to the success of attempts to suppress hail.

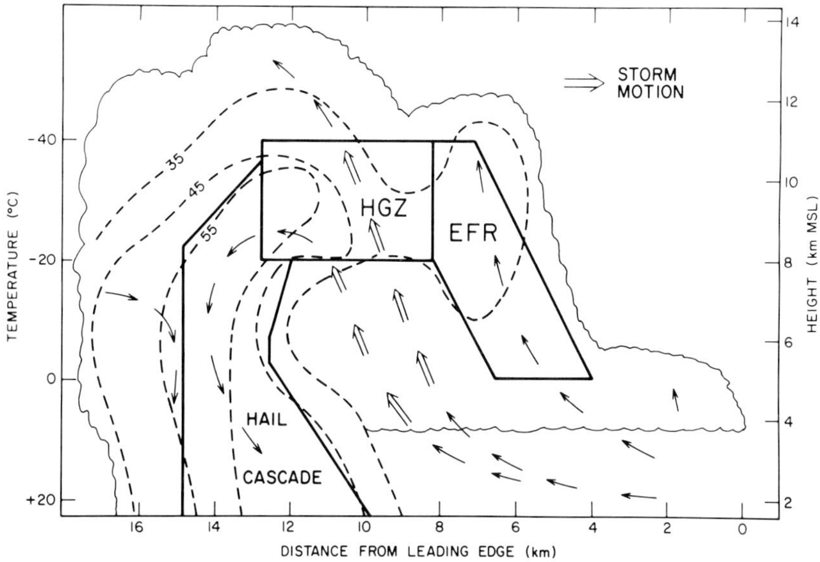

Fig. 12.1. Schematic diagram of a hailstorm showing the embryo formation region (EFR), the hail growth zone (HGZ), and the hail cascade. Single arrows represent weaker updrafts; double arrows represent stronger updrafts. Dashed isopleths indicate radar reflectivity in dBZ. The abscissa may be interpreted as a time axis as a multi-cell storm evolves. From Young (1977). *Meteorological Monographs*, **16**, 38. American Meteorological Society. Reproduced with permission.

12.2. Structure of hailstorms

Hailstorms generally form in regions of moderate to strong vertical wind shear. Typically, environmental wind speeds increase with height and the storm tends to move with the mean tropospheric wind. This results in the inflow of moist air at low levels, as shown in Fig. 12.1. As the inflow air rises, its inertia results in an updraft which is tilted upshear at low and middle levels and turning in the downshear direction at the upper levels, producing the anvil outflow (Browning, 1977).[1]

The tilted updraft serves to move particles horizontally through the updraft, limiting the time they can spend in the HGZ. This helps prevent the accumulation of precipitation that might otherwise weaken the updraft. It also serves to locate the EFR in the weaker updrafts downshear of the HGZ. The tilted updraft provides the mechanism to transfer growing particles from the EFR into the HGZ.

12.2.1. *Multi-cell thunderstorms*

Multi-cell thunderstorms consist of a temporal sequence of cells, each progressing through the growing, mature, and dissipating stages, as described in

[1] Downshear is in the same direction as the wind shear vector. In this case, the wind direction is to the right with wind speeds increasing with height. Hence, the wind shear vector is to the right and upshear is to the left.

Chapter 10. Figure 10.18 depicts the organization of cells in a typical multi-cellular thunderstorm. The shelf cloud delineates the region of updrafts; the rainshaft serves to delineate the downdraft region. Daughter cells tend to form on the right flank of the storm such that the storm propagates to the right of the mean wind.

Hail production in a multi-cell thunderstorm is restricted to the mature cells. Thus, the pattern of hail at the ground (hailswath) consists of several distinct 'hailstreaks.' Changnon (1977) notes that the average duration of hailswaths ranged from 45 minutes to 3 hours, whereas hailstreaks had an average duration of 10 minutes. Goyer (1977) estimates that hail cores are separated in time by 17.6 to 19.5 minutes, corresponding to the times between successive mature cells.

The three-dimensional pattern of airflow in a typical multi-cell thunderstorm is shown in Fig. 12.2. The updraft air shown by ribbons A and B rises sharply in the storm interior, leaving the storm at upper levels toward the northeast, forming the anvil outflow. Most of the environmental air at middle levels passes around the storm (ribbons labeled C) but some environmental air (ribbons D and E) enters the storm at middle levels and contributes to the downdraft.

In the mature cell, updraft velocities are sufficiently strong to preclude development of precipitation particles from the cloud base within the updraft core. Larger particles tend to be advected into the main updraft above the $-20°C$ level, forming the upper boundary of a weak echo region (WER) as seen by radar. This weak echo region is evident in Fig. 12.1. These large particles grow as they advect through the main updraft, as evidenced by the increase in radar reflectivity toward the upshear side of the HGZ.

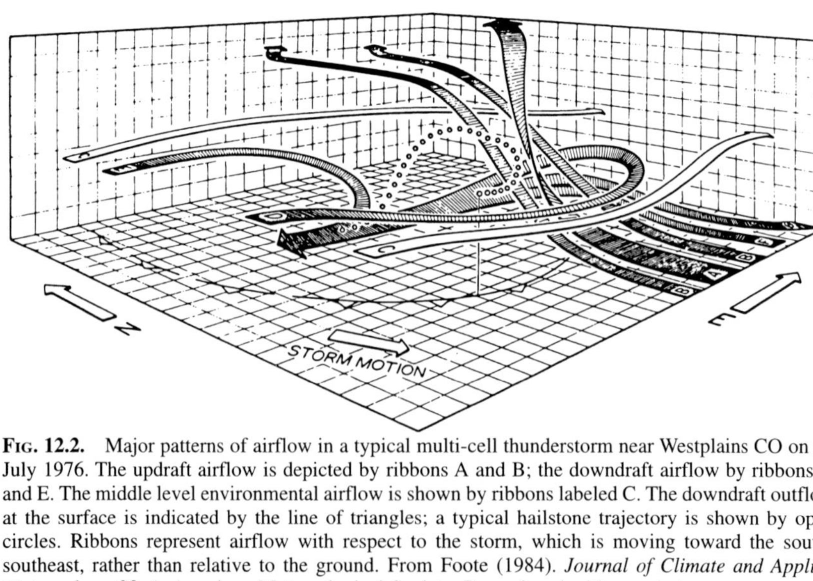

Fig. 12.2. Major patterns of airflow in a typical multi-cell thunderstorm near Westplains CO on 22 July 1976. The updraft airflow is depicted by ribbons A and B; the downdraft airflow by ribbons D and E. The middle level environmental airflow is shown by ribbons labeled C. The downdraft outflow at the surface is indicated by the line of triangles; a typical hailstone trajectory is shown by open circles. Ribbons represent airflow with respect to the storm, which is moving toward the south-southeast, rather than relative to the ground. From Foote (1984). *Journal of Climate and Applied Meteorology*, **23**, 1. American Meteorological Society. Reproduced with permission.

The maximum size of hailstones that can be produced by a given hailstorm is related to the maximum updraft velocity. Hailstones with fall speeds much greater than the maximum updraft velocity will fall out without growing substantially. Chisholm (1973) uses a moist adiabatic parcel model of an updraft with water loading to estimate the maximum updraft velocities for a variety of hail-producing thunderstorms in Alberta. He notes that the maximum observed hailstone size increases as the buoyant energy and maximum updraft velocity increase.

Maximum observed updraft velocities greater than $30\,\mathrm{m\,s^{-1}}$ have seldom been observed (Browning, 1977) although indirect evidence suggests that much higher updraft velocities may exist. Assuming a one-to-one correspondence between the fall velocity of the largest hailstone and the maximum updraft, vertical velocities in excess of $50\,\mathrm{m\,s^{-1}}$ should be expected. Chisholm (1973) calculates a maximum vertical velocity of $68.6\,\mathrm{m\,s^{-1}}$ in his studies of Alberta hailstorms. Observations of tropopause penetration by cloud tops would suggest vertical velocities as great as $100\,\mathrm{m\,s^{-1}}$. The lack of observational evidence for such strong updrafts argues against accepting these at face value.

12.2.2. *Supercell thunderstorms*

Under conditions of relatively large directional shear in the vertical in the storm environment, a multi-cell storm may evolve into a quasi-steady single-cell storm which may persist for several hours. The updraft velocities in supercell storms tend to be greater, leading to larger hail and continuous hailswaths, in contrast to smaller hail and discontinuous hailswaths from multi-cell storms. Supercell thunderstorms usually have a smoothly-shaped cloud dome at the top of the storm, as shown in Fig. 12.3, in contrast to the multiple turrets associated with a multi-cell storm.

The airflow associated with the Fleming supercell thunderstorm described by Browning and Foote (1976) is shown in Fig. 12.4. The low level inflow is from the south, rising in the updraft region, forming the upper level outflow. Large particles grow on the edges of the main updraft, circling around the main updraft on the right flank, forming an embryo curtain. The embryo curtain encloses the weak echo region, as shown in cross-section in Fig. 12.5, forming a bounded weak echo region or 'vault.'

The lack of a radar echo in the vault implies that virtually all of the condensate in the main updraft is in the form of small droplets, which will freeze as they reach the homogeneous freezing level. High concentrations of these small ice particles are injected into the anvil outflow, substantially reducing the precipitation efficiency of the supercell, compared to multi-cell thunderstorms.

12.3. Requirements for the growth of hail

The time required for hailstone growth processes and the fall speed of the growing particle define the requirements for the growth of large hail in a thunderstorm. The low fall speeds for the growing particles that will become hail

Fig. 12.3. Photograph of the Fleming supercell on 21 June 1972 showing the smoothly-shaped cloud dome at the top of the storm. From Browning and Foote (1976). Royal Meteorological Society. Reproduced with permission.

embryos requires relatively weak updraft velocities. The higher fall speeds for growing hailstones necessitates stronger updraft velocities. The transition from weaker to stronger updrafts must closely match the increase in fall speed as the particle grows.

12.3.1. *The embryo formation region*

The 10 to 15 minutes required to grow particles to millimeter sizes necessitates the location of the EFR in weaker updrafts. Since the particle fall speeds are less than $1\ \mathrm{m\,s^{-1}}$ during much of their growth, particles can be assumed to rise with the updraft. Graupel embryos grown by accretion from an ice crystal must be initiated above the 0°C level and cannot grow above the −40°C level since any liquid water present would freeze homogeneously.[2] This defines the upper and lower boundaries of the EFR for graupel embryos.

Embryos entering the HGZ at the −20°C level are generally those most likely to become hail. Figure 12.1 indicates that the −20°C level is roughly 3000 m above the 0°C level. Considering a 10 minute growth time for the embryos, updrafts in the EFR must be less than $5\ \mathrm{m\,s^{-1}}$. Natural ice nucleation processes such as deposition or contact freezing seldom produce ice crystals at temperatures much warmer than −10°C, reducing the available time for growth. Sedimentation of the particles as they grow larger serves to increase the available

[2] As noted in Chapter 4, the homogeneous freezing temperature is a function of the drop size, ranging from −35°C to −40°C for drops in the size range expected in the HGZ.

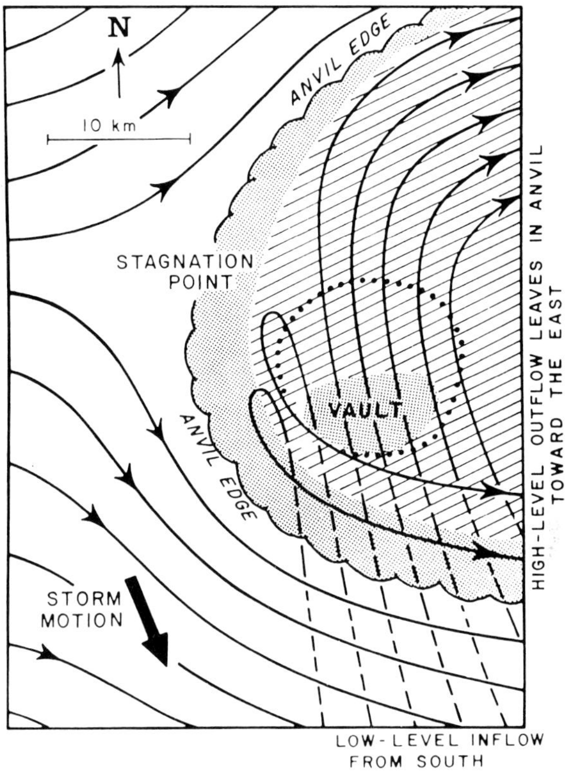

Fig. 12.4. Schematic representation of the airflow within and around the Fleming supercell. Areas devoid of radar echo are shown stippled; regions of radar echo are shown hatched. The dotted circle represents the core of the main updraft. Dashed arrows represent the low level inflow from the south, rising to form the upper level outflow. From Browning and Foote (1976). Royal Meteorological Society. Reproduced with permission.

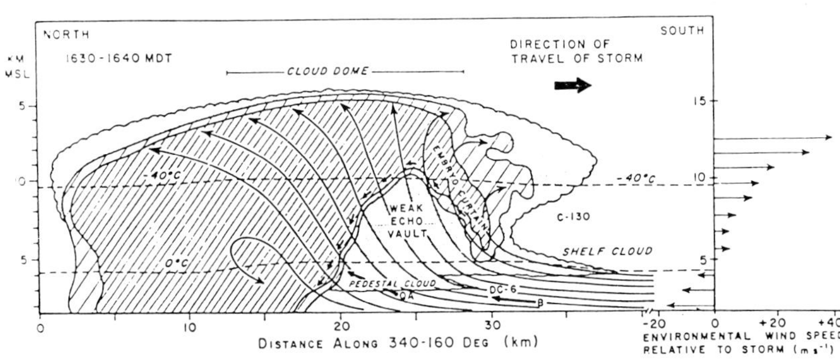

Fig. 12.5. Vertical cross-section of the Fleming supercell showing the visual boundaries of the cloud at 1630–1640 MDT superimposed on the radar echo. Volumes of cloud devoid of detectable echo are not shaded. Short arrows illustrate a hailstone trajectory. Arrows on the right ordinate represent a profile of the component of the environmental wind along the storm's direction of travel. Note the relatively smooth cloud dome and lack of daughter cells. The weak echo vault is bounded on the downshear side by a curtain of hail embryos formed on the edges of the main updraft. From Browning and Foote (1976). Royal Meteorological Society. Reproduced with permission.

Table 12.1. Hailstone embryo climatology. The fraction of hailstone embryos that were classified as graupel or frozen drops for six different geographical regions. The regions are in approximate order of decreasing cloud base temperature.

Region	Number of hailstones	Graupel (%)	Frozen drops (%)
South African Lowveld	1318	23	62
Oklahoma	1124	17	62
Switzerland	1220	37	63
South African Highveld[a]	1718	62	36
North Caucasus	210	65	35
Northeast Colorado	3660	87	9

[a] Percentages are an unweighted average of the four size categories provided.

Source: Knight et al. (1982).

time for growth. On the bases of these arguments, updraft velocities of 2 to $3\,\text{m}\,\text{s}^{-1}$ may be considered typical for the EFR.

Embryos grown from frozen drops may be initiated well below the 0°C level, depending on the cloud base temperature. Table 10.1 suggests that frozen drop embryos should dominate for cloud base temperatures of 10°C and warmer whereas graupel embryos should dominate for cloud base temperatures colder than 5°C. Table 12.1 presents a summary of hail embryo types for six different geographical regions (Knight et al., 1982) organized in approximate order of decreasing cloud base temperature. Less than 25% of the embryos of hailstones from the two regions with the warmest cloud bases were graupel; more than 60% of those from the three regions with the coldest cloud bases were graupel.

Embryos formed in the growing cells of multi-cell thunderstorms normally encounter increasing updraft velocities as the particles grow. Embryos formed early in the development of the growing cell may fall out of the cloud prematurely, before the updraft becomes strong enough to support large particles. Embryos formed late in the evolution of the growing cell may not grow to sizes large enough to be considered as potential hail embryos.

The EFR associated with supercell thunderstorms is restricted to a relatively narrow band of favorable updraft velocities on the edge of the main updraft. Browning and Foote (1976) suggest a three-stage-growth process, as shown in Fig. 12.6. During the first stage, the particle grows from the cloud base, rising in the weaker updrafts on the right flank of the main updraft, and eventually reaching millimeter sizes. At this level, most of the particles are carried into the anvil by the prevailing flow, with a relatively small fraction being carried around the updraft core to the south. During the second stage, the particles descend while being carried around the forward edge of the main updraft, forming the embryo curtain. Relatively little growth is expected in this region due to relatively low liquid water contents. Because of the updraft tilt, some of these particles fall into stronger updrafts and are carried up and over the main updraft (stage 3).

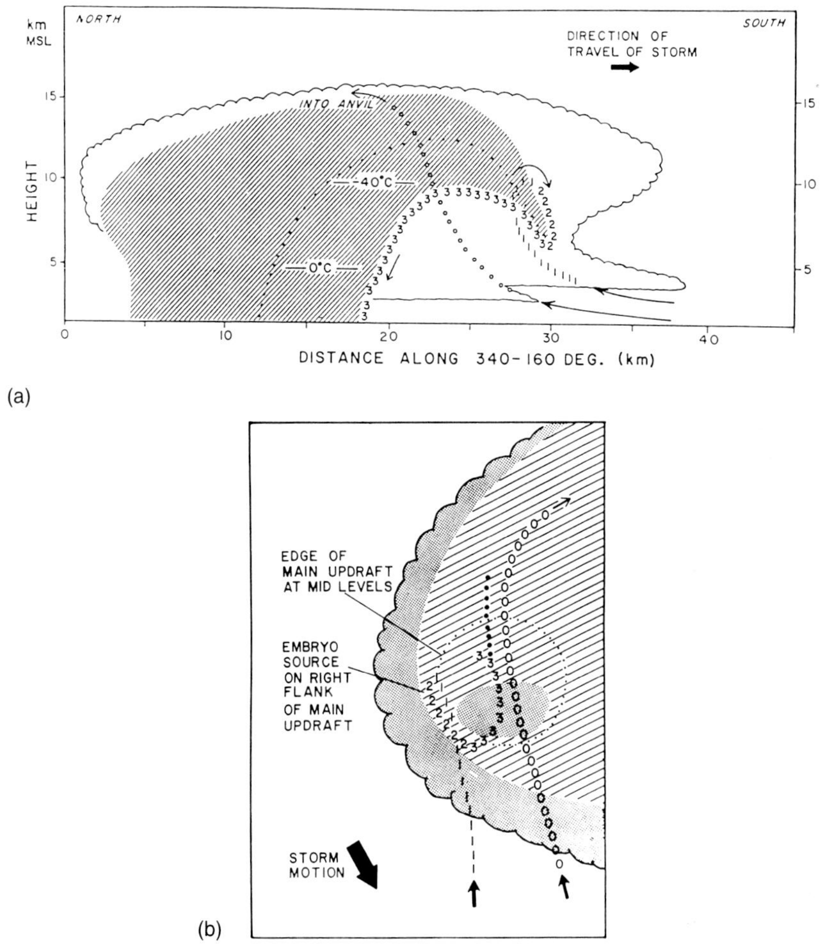

Fig. 12.6. Schematic model of hailstone trajectories within a supercell thunderstorm. (a) Vertical cross-section; (b) plan view. The trajectory defined by open circles represents particles formed in the core of the main updraft which are lost to the anvil. Trajectories 1, 2, and 3 represent the three stages in the growth of large hailstones as discussed in the text. The hatched regions indicate detectable radar echo. From Browning and Foote (1976). Royal Meteorological Society. Reproduced with permission.

12.3.2. *The hail growth zone*

A primary requirement of the HGZ is that there must be an ample supply of supercooled water to support particle grow by accretion. As noted above, this condition is met in strong updraft cores since there is insufficient time for significant ice particle growth which would deplete the liquid water. In the updraft core, liquid water contents may be expected to be close to adiabatic values. Near the −40°C level, homogeneous freezing reduces the liquid water content to zero, defining the upper boundary of the HGZ.

Temperatures within the HGZ must be cold enough to freeze a substantial portion of the accreted water. If ambient temperatures are too warm, the hailstone will be comprised of a spongy mixture of water and ice, which will either melt before reaching the ground, or 'splash' on impact. The $-20°C$ level is commonly assumed to define the lower boundary of the HGZ, implying the growth of solid ice particles.

The updraft velocities encountered by the growing particle must roughly match its increasing fall speed so that the particle remains within the HGZ. If the updraft velocities that the particles encounter increase faster than the fall speeds of the growing particles increase, the particles will follow a rising trajectory and may rise above the $-40°C$ level. This would preclude further growth until either the particles advect through the updraft or the cell begins to dissipate. In either event, growth of hail would be expected to be reduced. If the fall speeds of the growing particles increase faster than the updraft velocities that are encountered, the particles will follow a descending trajectory and fall out of the cloud before reaching the maximum size possible for the updraft.

The width of the updraft velocity profile defines the horizontal gradient of vertical velocities. A narrow updraft exhibits sharper horizontal gradients of vertical velocity than a broad updraft with the same maximum updraft velocity. Thus, a relatively narrow updraft would be more likely to produce a rising trajectory, whereas a relatively broad updraft would be more likely to produce a descending trajectory.

The tilt of the updraft defines the rate at which the growing particle is advected through the updraft. The more the updraft is tilted from the vertical, the faster the particles will advect through the updraft. This serves to accentuate the vertical velocity gradient, i.e., the greater the updraft tilt, the faster the growing particle will encounter stronger updrafts and would be more likely to follow a rising trajectory.

These considerations suggest that the width and tilt of the updraft define the likelihood that a given updraft will produce hail. Only a certain range of width/tilt combinations would be expected to produce hail; unfavorable combinations of width and tilt would not be expected to result in the formation of hail. These concepts will be pursued further in § 12.4.

12.4. Hailstone selection mechanisms

Typical concentrations of potential hail embryos are on the order of 10^3 to $10^4\,m^{-3}$, and yet hailstone concentrations are only the order of $0.1\,m^{-3}$ to $1\,m^{-3}$. Thus, only $1{:}10^4$ to $1{:}10^6$ of the potential hail embryos are actually selected to become hailstones. Clearly, mechanisms that select particular embryos to become hailstones must be operating. These mechanisms may be classified as deterministic or stochastic.

A consequence of the selection process is that the particles tend to be sorted by size. Typically, as a hailstorm passes overhead, the largest hailstones are encountered first, with particle sizes decreasing with time. The type of size sorting is determined by the concentration of large particles in the HGZ and reflects the degree of competition for the available supercooled water. This has important consequences for hail suppression, which will be discussed in the § 12.5.4.

12.4.1. *Deterministic versus stochastic selection*

A deterministic selection mechanism is one in which the **particular** embryos that **will** become hailstones can be identified, at least in principle. There are locations in time and space that favor selection of embryos to become hailstones. The size of the embryo determines its fall speed and thereby determines its position in the HGZ relative to the other embryonic hailstones and its ability to compete for the available supercooled water. Embryos that by virtue of their size and location are best able to compete for the available supercooled water will grow fastest. If the time they reside in the HGZ is sufficiently lengthy, then they will become hailstones.

A stochastic selection mechanism is one in which all embryos in the EFR have an equal chance for selection, regardless of their location or size, i.e., there are no distinguishing features of the embryos which determine their selection. Given a sufficient level of turbulence, the relative positions of the embryos are constantly and randomly changing. This means that each embryo has the same probability of being located in the zone of highest liquid water content and having the highest growth rate. Since the relative positions of the embryos are constantly changing, some of the embryos will be favored, spending an inordinate amount of time in the regions of highest liquid water content, thereby growing into hailstones. Those favored embryos are selected by chance.

In reality, both types of selection mechanisms are operating. Thus, while there is a tendency for larger embryos to become hailstones, there is enough indeterminacy in this process to allow much smaller embryos to become hailstones while some of the larger embryos do not. Similarly, even in the most favored locations for selection, not all of the potential hail embryos are selected to become hailstones.

The relative influences of location and size in determining the trajectories of the embryonic hailstones are illustrated in Fig. 12.7 using a Monte Carlo model to simulate the effects of turbulence. The updraft velocity is constant throughout the field at $15\,m\,s^{-1}$ with a tilt of 0.2 as shown. Turbulence was simulated using random perturbations in the x and z directions based on a normal distribution with a standard deviation of $1.17\,m\,s^{-1}$, scaled as the square root of the time step in seconds. Thus, each particle (packet) had upward and westward components of motion due to the mean flow field, a downward component due to the fall speed of the particle, and random components in the x and z directions due to turbulence.

Particles were followed as 'packets' containing 5 particles m^{-3} each. The grid size used in the model was 60 m, yielding 18,000 particles m^{-1} (two-dimensional model) in one packet. Embryos of three different sizes were followed; there were three packets of each size injected every 20 s, yielding a total embryo concentration of $45\,m^{-3}$. The liquid water content at the $-20°C$ level was assumed to be adiabatic from a cloud base at 5°C. Above the $-20°C$ level, the liquid water content was calculated based on the additional condensation due to cooling in the updraft and depletion resulting from accretional and diffusional growth of the embryonic hailstones.

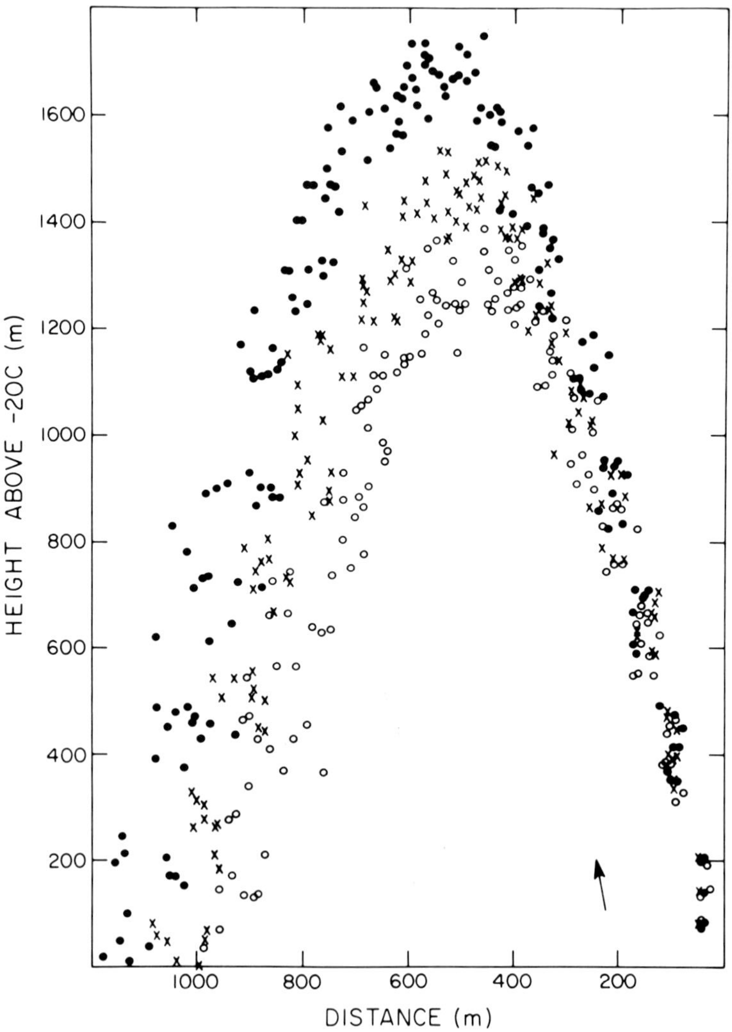

Fig. 12.7. The effects of particle size and turbulence on the trajectories for growing hailstones. Initial embryo sizes are (●), 200 μm; (×), 250 μm; and (○) 300 μm radius. The updraft velocity is $15\,m\,s^{-1}$ with a $3\,m\,s^{-1}$ component to the left (to the west in an eastward moving storm). The turbulence was assumed to superimpose a random perturbation determined using a Gaussian distribution with a standard deviation of $1.17\,m\,s^{-1}$. From Young (1978). *Journal of Atmospheric Sciences*, **35**, 11. American Meteorological Society. Reproduced with permission.

It is evident from Fig. 12.7 that the larger embryos tend to follow lower trajectories through the updraft while the initially smaller embryos tend to follow higher trajectories. There is considerable overlap in the exit positions of the hailstones with some of the largest and smallest embryos exiting at roughly the same position. The effects of the initial position of the embryos seems to be insignificant in this example.

12.4.2. *Size sorting*

Particles following lower trajectories will tend to encounter higher liquid water contents and thereby grow faster than particles following higher trajectories. However, particles following higher trajectories will spend more time in the HGZ than those following lower trajectories. If there are high concentrations of growing particles in the HGZ, the depletion of liquid water may be expected to reduce significantly the growth rates of particles following the higher trajectories. Hence, the most favored trajectory for the growth of hail is dependent on the concentration of growing particles in the HGZ.

If the concentration of embryos is relatively low, there will be relatively little depletion of the liquid water and the growth rates of particles following the lower and higher trajectories will be roughly the same. Thus, the particles following the higher trajectories will grow to the largest sizes in the HGZ by virtue of spending more time in the HGZ. They will also fall out of the HGZ further from their injection point, as shown in Fig. 12.8a. Thus, the hailstone sizes at the ground would increase with time as the storm passes overhead; this is termed 'positive size-sorting.'

Figure 12.8b illustrates negative size-sorting, resulting from relatively high concentrations of embryonic hailstones in the HGZ. Since the liquid water depletion is much greater than in the previous example, the growth rate of the particles following the higher trajectories is substantially lower than that of the particles following the lower trajectories. Thus, despite the longer time spent in the HGZ by particles following the higher trajectories, they do not grow as large as hailstones which followed the lower trajectories. This results in a decrease in hailstone sizes at the ground as time increases, i.e., negative size-sorting.

Positive size-sorting may be expected to be less pronounced than negative size-sorting since the advantage attained by following a higher trajectory is a small increase in the time available for growth, which must counterbalance the depletion of liquid water by the particles following lower trajectories. On the other hand, high concentrations of particles can severely deplete the liquid water, reducing the growth rates for particles following the higher trajectories by an order of magnitude. Thus, pronounced positive size-sorting may be expected to be rare compared to pronounced negative size-sorting.

Young (1977) suggests that the concentration of potential hail embryos resulting from the coalescence process would be much lower than the concentration resulting from the growth of graupel. Based on numerical model results, he finds the concentration of graupel larger than 100 μm radius at the

(a)

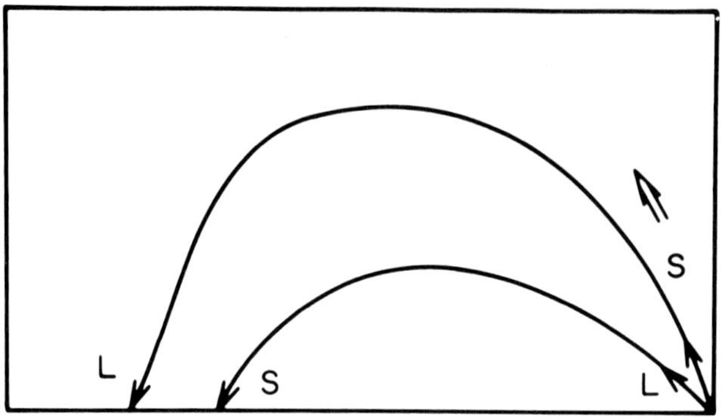

(b)

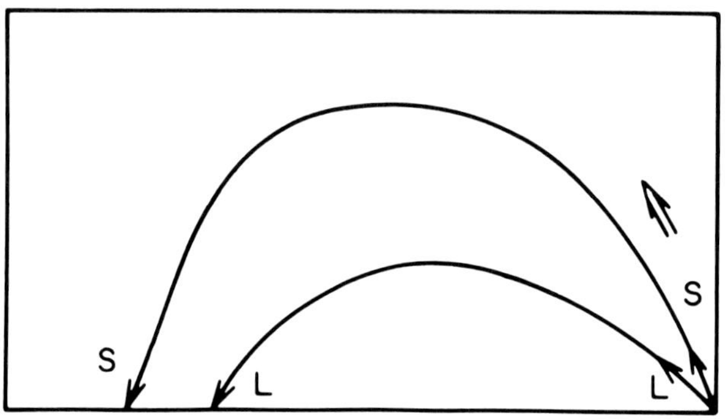

FIG. 12.8. Schematic diagram comparing the growth of hailstones following lower and higher trajectories. S and L indicate the relative sizes of the initial embryos (right side) and the resultant hailstones (left side). (a) The case with relatively low concentrations of embryos and relatively little depletion of the liquid water, resulting in positive size-sorting. (b) The case with relatively high concentrations of embryos and considerable depletion of the liquid water, resulting in negative size-sorting. From Young (1977). *Meteorological Monographs*, **16**, 38. American Meteorological Society. Reproduced with permission.

−30°C level in a typical updraft to be on the order of one particle per liter. This is comparable to typical ice particle concentrations as measured in northeast Colorado cumuli by Dye et al. (1974).

Young estimates the comparable concentration of large frozen drops to be on the order of $1.5\,m^{-3}$, roughly three orders of magnitude less than the concentrations of graupel. He notes that the largest drops are the most likely to freeze (see Problem 10.1) and, having frozen, will then remove smaller drops by accretion. The resultant frozen drop concentration depends on the relative rapidity of drops freezing *versus* drops being collected.

This suggests that the type of size sorting is related to the hail embryo type, with hailstorms producing hail from frozen drop embryos exhibiting a tendency for positive size-sorting, whereas those producing hail from graupel embryos would exhibit negative size-sorting. Examples of negative size-sorting are quite common; those of positive size-sorting are relatively rare. Held and van den Berg (1977) cite an example of positive size-sorting for a hailstorm on the South

African High Veld: 'Almost everywhere the hailstones fell after the rain had begun and the largest hailstones arrived on the ground either together with or after the smaller ones.'

12.4.3. *The effects of updraft tilt and width on hail production*

The Monte Carlo model described by Young (1977,1978) permits an examination of the effects of updraft tilt and width on the production of hail, as well as a definition of the EFR. In this two-dimensional, kinematic model, an updraft profile is specified as shown in Fig. 12.9. Embryos were injected in those streamlines east of the maximum updraft having vertical velocities less than $16\,\mathrm{m\,s^{-1}}$. For each streamline, seven packets with concentrations of 0.1, 0.3, 1.0, 3.0, 10, 30 and 100 particles $\mathrm{m^{-3}}$ were injected every 20 seconds. Embryo sizes were determined using a microphysical parcel model (Young, 1974), assuming a cloud base temperature of 5°C. Embryos ranged in size from 50 to 200 μm radius. The model was run to steady state, then the population of hailstones exiting the base of the model was sampled.

Various combinations of tilt and width can be examined by altering the horizontal component of the flow field and by scaling the horizontal axis to simulate different widths. The output of the model is displayed in Fig. 12.10 as $\mathrm{g\,m^{-2}\,s^{-1}}$ of hail; the total hail mass is determined assuming the storm is moving overhead at $10\,\mathrm{m\,s^{-1}}$. Increasing the tilt of the updraft from 0.2 to 0.4 decreased the total hail mass by 20% but, more significantly, eliminated virtually all hailstones larger than 1 cm radius.

An updraft of $25\,\mathrm{m\,s^{-1}}$ may be expected to support hailstones of 1.44 cm radius. This was achieved for the 0.2 tilt simulation with maximum hailstone radii greater than 1.5 cm. Clearly, increasing the tilt to 0.4 resulted in the growing

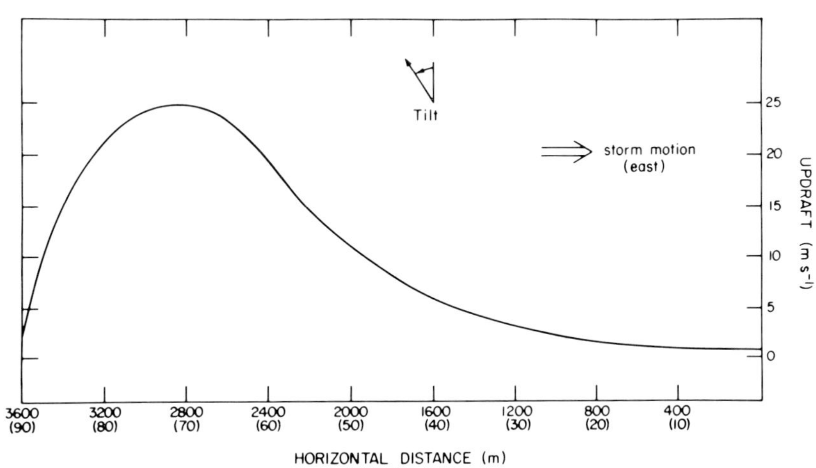

Fig. 12.9. Updraft profile used to simulate the effects of updraft tilt and width on hail production.

hailstones transiting the updraft too quickly to attain maximum growth. In this case, the hailstones followed a rising trajectory until reaching grids between 85 and 90, and fell out because the updrafts were too weak to support any particles larger than a centimeter. Increasing the tilt to 0.8 resulted in further reductions in the total hail mass and maximum hailstone sizes.

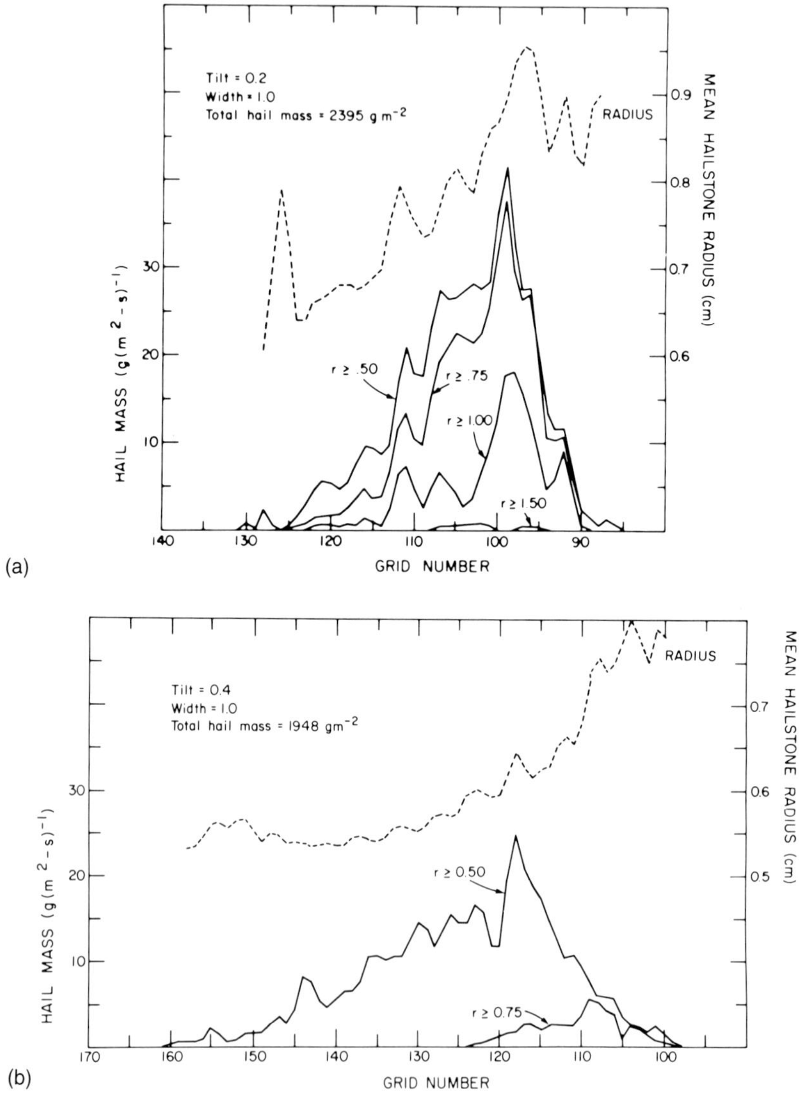

FIG. 12.10. Monte Carlo simulation of hail production in tilted updrafts. The mean hailstone radius (broken line) and hail mass production in hailstones of the specified radii (cm). Grid numbers increase to the west; the storm is moving to the east. (a) Tilt of 0.2; (b) tilt of 0.4. The standard updraft width as shown in Fig. 12.9 is used.

The maximum hailstone sizes scale linearly with width divided by tilt (W/T), as shown in Fig. 12.11. Although the maximum hailstone sizes were found to increase with increasing W/T for all the simulations included here, this trend could not continue indefinitely. Maximum hailstone radii greater than 2 cm were found for larger values of W/T, illustrating the continued growth of hail during its descent after achieving a balance level (at 1.44 cm radius) in the strongest updraft.

Based on his study of the Westplains storm depicted in Fig. 12.2, Foote (1984) concludes that the width of the updraft appears to be the main limiting factor determining the maximum size of the hailstones produced by a given storm. Maximum updrafts in the Westplains storm were slightly greater than 25 m s^{-1}, with maximum reported hailstone sizes on the order of 2 cm. These findings are remarkably similar to the results from the Monte Carlo simulations.

Nelson (1983) concludes that a long residence time in the HGZ is required to grow large hail. Long residence times would result from a broad updraft or slow advection (small tilt) through the updraft. The increases in fall speed associated with hailstone growth must nearly balance the increase in updraft velocity. Similar conclusions are reached by Ziegler et al. (1983).

Updraft widths reported by Marwitz (1972) and by Browning and Foote (1976) are somewhat larger than those shown here; reported updraft tilts ranged

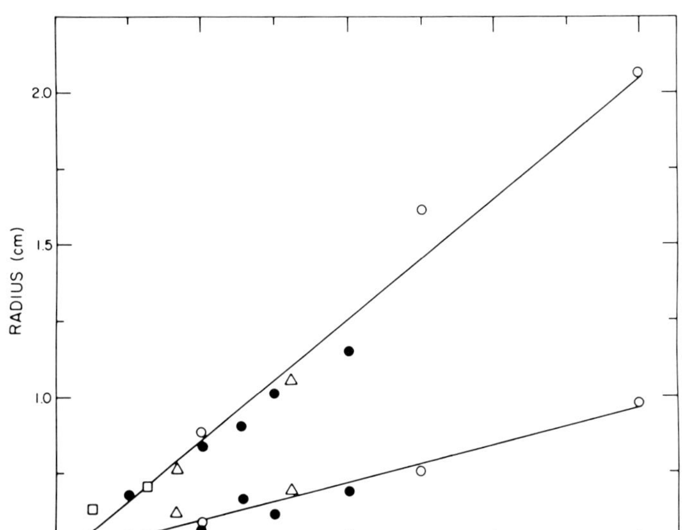

Fig. 12.11. Relation between maximum and average hailstone radius and the ratio of width to tilt based on Monte Carlo simulations of hail production. Maximum updraft velocity was 25 m s^{-1} in all cases. Least squares fits are shown with correlation coefficients of 0.97 and 0.95 for the maximum and average sizes, respectively. (○), T = 0.2; (●), T = 0.4; (△), T = 0.5; (▲), T = 0.6; (□), T = 0.8.

from 0.15 to 0.47, similar to the range studied here. Direct observations of the tilt and width of updrafts are not only infrequent and difficult to make, but the values obtained may be expected to depend on the level within the storm and the direction taken by the aircraft through the updraft. The tilt relevant to hail production is that between roughly the $-10°C$ and $-30°C$ levels. Methods such as that described by Warner (1972) may be used to estimate updraft tilt; triple-Doppler radar measurements may also provide a method for deducing updraft tilts and widths.

12.4.4. *Embryo source region*

The Monte Carlo model described in § 12.4.3 may be used to locate the initial particle location and size. This is accomplished by identifying each packet by the injection grid location, the initial size, and the injection time step, and then

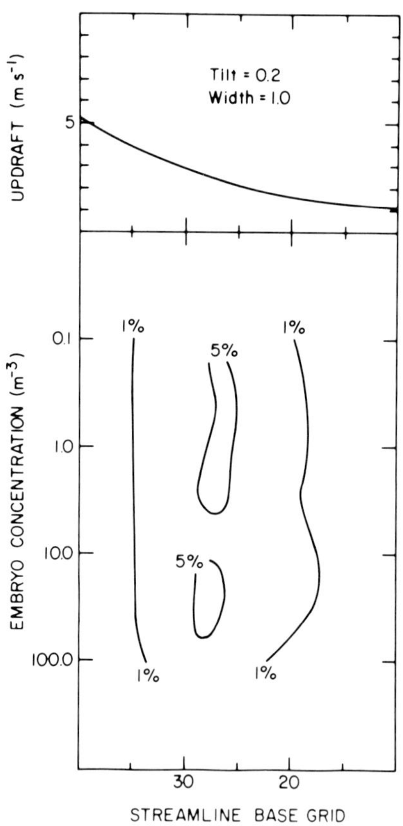

FIG. 12.12. Identification of the EFR using a Monte Carlo simulation of hail production. Isopleths reflect the fraction of embryos injected at a given location (streamline) of a given size that grew into hailstones (radii $\geq$ 0.5 cm). Relative initial sizes of the embryos may be inferred from the embryo concentration, with lower concentrations representing larger sizes. The corresponding updraft profile is shown at the top.

examining the identifications of the particles exiting the base of the model. Note that many packets exited the top of the model or the base of the model with particle sizes smaller than 0.5 cm radius and so were not classified as hailstones. Knowing the rate at which packets were injected provides the fraction of those packets injected at a given grid location that became hailstones.

Figure 12.12 suggests that even in the most preferred embryo location and for the most preferred embryo sizes, less than 10% of the potential embryos are selected to become hailstones. The embryo location appears to be more important than the embryo size, since the probability isopleths in Fig. 12.12 appear to be largely independent of the particle concentration (size). The preferred location corresponds to an updraft velocity of 2.5 m s^{-1}.

Foote (1984) used a particle tracing model in association with triple-Doppler observations of the airflow in the Westplains storm to determine the source region for large hailstones. Embryos of five different sizes ranging from 500 μm to 4 mm were injected at 7 km at 1 km intervals over the shaded region in Fig. 12.13. No attempt was made to tailor the embryo sizes to the updrafts they were injected into.

Most of the embryos forming hail greater than 0.5 cm radius were found to originate in a region a few kilometers wide along the southwest side of the main updraft, as shown in Fig. 12.13. The final hailstone size was found to be

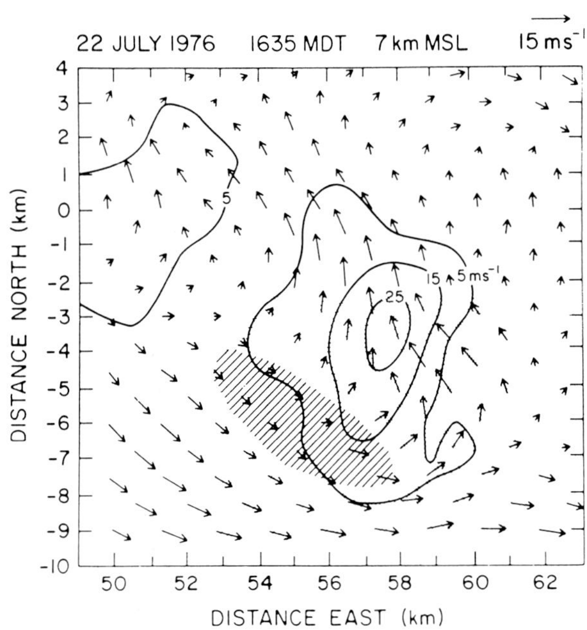

FIG. 12.13. Embryo starting positions relative to the airflow at 7 km MSL in the Westplains storm of 22 July 1976. Isopleths represent the updraft velocities in m s^{-1}; arrows represent wind vectors (scale in upper right). The stippled region represents the grid over which embryos were injected; the hatched region represents the starting locations of hail larger than 0.5 cm radius. From Foote (1984). *Journal of Climate and Applied Meteorology*, **23**, 1. American Meteorological Society. Reproduced with permission.

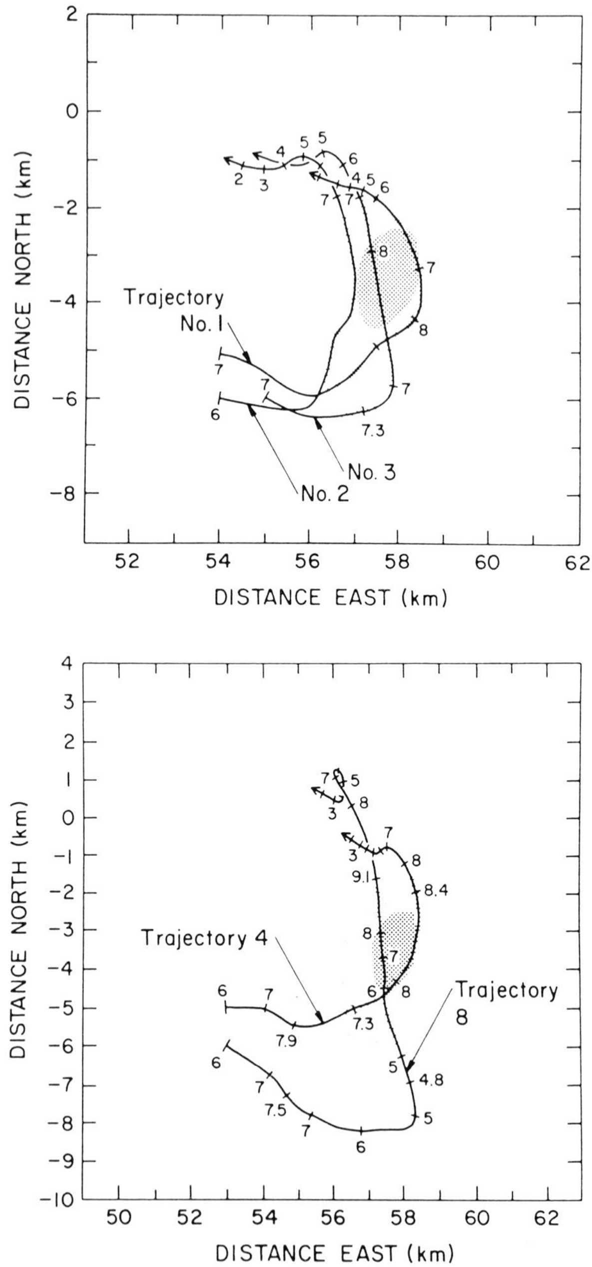

(a)

(b)

Fig. 12.14. Selected hailstone trajectories in plan view for embryos injected at 7 km (see Fig. 12.14). The 15 to 25 m s^{-1} (light shading) and greater than 25 m s^{-1} (dark shading) updraft regions are indicated. Heights in km MSL are indicated alongside the tracks. (a) The three largest hailstones, each growing to 0.95 cm radius; (b) smaller hailstones, (4) grew to 0.85 cm and (8) grew to 0.4 cm radius. From Foote (1984). *Journal of Climate and Applied Meteorology,* **23**, 1. American Meteorological Society. Reproduced with permission.

insensitive to the size of the initial embryo, and embryos of a single size could result in a spectrum of hailstone sizes due to the diversity of trajectories.

Hailstone trajectories for three embryos that grew into 0.95 cm radius hailstones are shown in Fig. 12.14a. Note that each of these trajectories passed near or through the 25 m s^{-1} updraft with relatively little variation in the height of the particles. This implies that the updraft width and tilt were suitably matched to the particle size, providing near optimal growth conditions for large hail. Trajectories that resulted in smaller hailstones shown in Fig. 12.14b exhibit rising trajectories while passing through the main updraft, due to their smaller size when entering the main updraft.

Ziegler et al. (1983) identify the EFR on the upwind side of the main updraft in an Oklahoma multicell hailstorm, with embryos coming from the adjacent growing cell and the debris between the mature and growing cells. Heymsfield (1983) also identifies the EFR as a narrow (1 km wide) region extending around the forward portion of the storm for a hailstorm in Colorado. He suggests that melted aggregates falling from the anvil contribute significantly to potential hail embryos in the EFR.

Knight (1990) presents another example of tracing particles to locate their origin. In this case, the origins of particles contributing to the large particle concentration in a particular target grid in a cumulus congestus are shown in Fig. 12.15. Note the diversity of the starting locations and times for particles that are

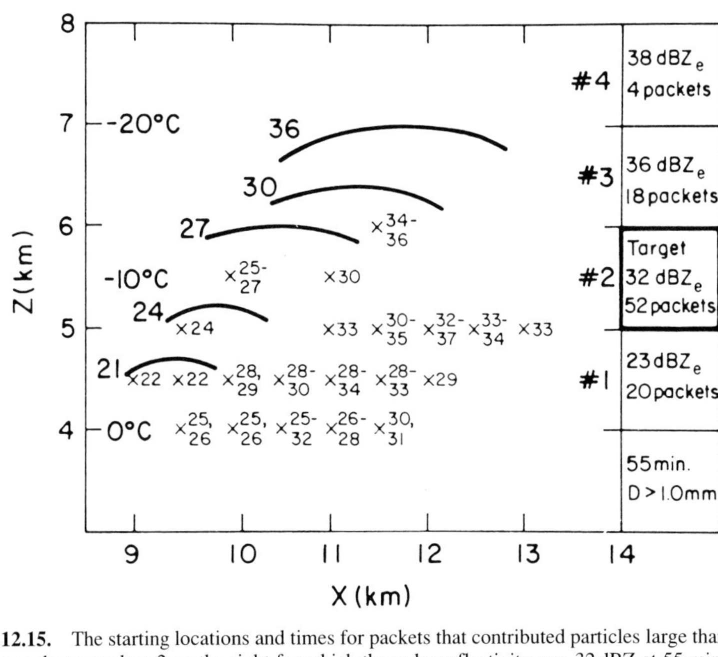

Fig. 12.15. The starting locations and times for packets that contributed particles large than 0.5 mm radius to the target box 2 on the right for which the radar reflectivity was 32 dBZ at 55 minutes. For each starting location, the earliest and latest times (minutes minus 0.5 min) are given. The cloud tops at various times are shown as arcs. From Knight (1990). *Journal of Applied Meteorology*, **29**, 5. American Meteorological Society. Reproduced with permission.

in the same grid at 55 minutes. The growth times range from 17 minutes to 33 minutes. These findings suggest that even the potential hail embryos entering the HGZ will arrive from a wide variety of initial locations.

12.5. Hail suppression concepts

Three hail suppression concepts have been advanced. Glaciation refers to the introduction of high concentrations of small ice particles in and below the HGZ which would be expected to reduce the amount of supercooled water available for hail growth. Beneficial competition refers to introducing more potential hail embryos in the EFR under the assumption that more embryos competing for the same amount of supercooled water in the HGZ will result in smaller particles. Trajectory lowering refers to introducing ultra-large embryos in the EFR such that they would pass through the main updraft below the HGZ, thereby reducing the supercooled water available in the HGZ.

12.5.1. *Glaciation*

If the supercooled water entering the base of the HGZ could be reduced, the accretional growth of all of the ice particles within the HGZ would be reduced. One way in which this could be accomplished is through massive seeding with an ice nucleant such as silver iodide or by direct injection of ice crystals using dry ice. By reducing the sizes of the hailstones, a larger fraction of their mass would be lost through melting and shedding while falling to the ground, thereby producing less damage.

Can the glaciation process proceed rapidly enough to reduce the liquid water content substantially before reaching the base of the HGZ? Young (1977) examines the seeding rates required to achieve given levels of liquid water depletion, as shown in Fig. 12.16 for an updraft which accelerates from $5\,\mathrm{m\,s^{-1}}$ at cloud base to $20\,\mathrm{m\,s^{-1}}$ at the $-40°\mathrm{C}$ level. Given that the largest hailstones tend to follow the lowest trajectories through the HGZ[3] and that the base of the HGZ is close to the $-20°\mathrm{C}$ level, seeding rates of roughly $4\,\mathrm{g\,AgI\,min^{-1}\,km^{-2}}$ are required to achieve a 10% reduction in the liquid water content relevant to the growth of large hail.

For a typical updraft area of 10 $\mathrm{km^2}$, seeding rates of $40\,\mathrm{g\,min^{-1}}$ are required.[4] Updrafts twice as strong would require seeding rates about an order of magnitude higher than this. To achieve a 50% reduction in the liquid water content would likewise require an order of magnitude increase in the seeding rates. Although direct injection of ice crystals using dry ice may initiate the glaciation process lower in the updraft, Young estimates seeding rates as high as $1000\,\mathrm{g\,min^{-1}\,km^{-2}}$ would be required to achieve a 50% reduction in liquid water for the strong updraft.

[3] This assumes negative size-sorting.
[4] The area enclosed by the $15\,\mathrm{m\,s^{-1}}$ isotach in Fig. 12.13 is roughly $12\,\mathrm{km^2}$.

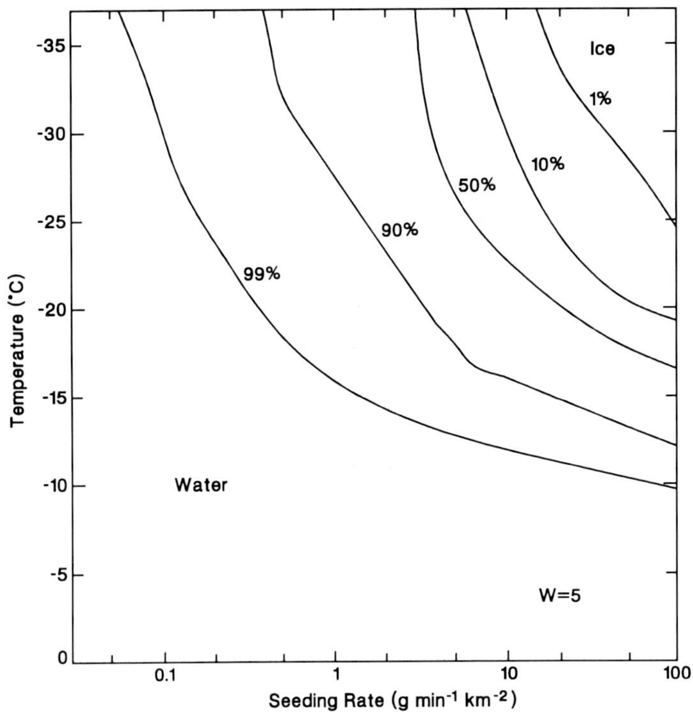

FIG. 12.16. The extent of glaciation expressed as the fraction of liquid water remaining versus height (temperature scale) and AgI seeding rate. The cloud base is at 9°C; the updraft undergoes a constant acceleration from 5 m s^{-1} at cloud base to 20 m s^{-1} at the −40°C level. From Young (1977). *Meteorological Monographs*, **16**, 38. American Meteorological Society. Reproduced with permission.

Dennis and Musil (1973) consider that hailstones undergoing growth in the moist or spongy growth regimes[5] may collect ice crystals through sintering. This would transfer mass to the growing hailstone as ice, allowing it to grow without being required to dissipate the latent heat of fusion associated with accretion of water drops. They assume a collection efficiency of unity for a moist hailstone collecting ice crystals and a collection efficiency of 0.25 for a hailstone growing in the dry growth regime. Thus, glaciation below the HGZ may simply result in a larger fraction of the mass collected by the growing hailstone to be in the form of ice crystals with relatively little affect on the overall growth rate of the hailstone.

Young suggests that glaciation of the EFR could reduce hailstone sizes by reducing the sizes of the potential hail embryos, and this could be achieved using much lower seeding rates (see Fig. 11.10). As shown in Fig. 12.14b, reducing the sizes of the embryos would result in rising trajectories through the main updraft and smaller hailstones at the ground, provided the largest naturally produced embryos would follow a nearly horizontal trajectory through the main updraft.

[5] See Fig 8.5.

Massive seeding of the EFR under the wrong conditions could lead to an increase in maximum hailstone sizes by reducing the 'screening' effect larger particles following lower trajectories would have.[6]

Regardless of seeding below the HGZ or seeding in the EFR, glaciation may be expected to increase the condensate flux into the upper portions of the storm, where a larger fraction of the condensate would be lost to the anvil outflow. This would serve to reduce the precipitation efficiency of the storm. Borland and Snyder (1975) show that a 5% decrease in precipitation would negate the benefits of a 20% reduction in hail damage in the Great Plains. Thus, even if the glaciation concept could be implemented, it is debatable whether there would be a net benefit.

12.5.2. *Beneficial competition*

The conventional expression of the beneficial competition concept is given by Sulakvelidze et al. (1974) who express the reduction in hailstone radius as

$$\frac{R_s}{R_n} = \left(\frac{N_n}{N_s}\right)^{1/3},$$

where the subscripts n and s refer to the natural and seeded cases, respectively, and N represents the number concentration of embryos in the HGZ. Thus, an order of magnitude increase in the concentration of embryos in the HGZ would be expected to reduce the hailstone sizes by $10^{1/3}$, or a factor of 2.15. However, this expression implies that the embryos compete for the liquid water on an *equal* basis.

It was noted above that equal access to the liquid water implies a level of turbulence such that the effects of particle size and position in the HGZ have a negligible influence on the growth of the particle, i.e., the selection mechanism is dominated by stochastic processes *and* there are no favored embryos. In order to assess the feasibility of the beneficial competition concept, it is necessary to ascertain whether the levels of turbulence expected within the HGZ are high enough to provide sufficient mixing of the particles so as to equalize the competition for the available supercooled water.

Young (1977) uses the Monte Carlo model described above to assess the effects of increasing the embryo concentration in the HGZ. In this simulation, only embryos of 0.25 mm radius were used, with each packet containing not more than 100 particles m^{-3}.[7] The resulting hail spectra are shown in Fig. 12.17. Although a reduction in the sizes of the largest hailstones occurs as a result of increasing the embryo concentration from 1 m^{-3} to 10 m^{-3} and to 100 m^{-3}, no further reduction in hailstone sizes results from increasing the concentration to 1000 m^{-3}.

[6] This concept is discussed further under the topic of trajectory lowering, § 12.5.3.

[7] Grouping the particles into packets assumes that all of the particles within a given packet compete for the liquid water on an equal basis. Thus, it is necessary to minimize the number of particles in a given packet to avoid this pseudo-competition.

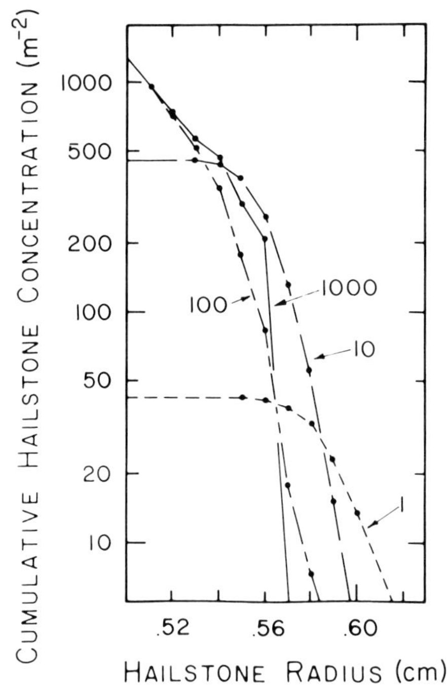

Fig. 12.17. Cumulative size spectra of hailstones as seen by a stationary observer at the ground as the HGZ passes overhead at 15 m s^{-1}. Only embryos of 0.25 mm radius were used with concentrations of 1, 10, 100, and 1000 embryos m^{-3}. From Young (1977). *Meteorological Monographs*, **16**, 38. American Meteorological Society. Reproduced with permission.

Figure 12.18 provides the basis for understanding this result. Low concentrations of embryos result in positive size-sorting, whereas high concentrations of embryos result in negative size-sorting, as discussed above. With positive size-sorting, the largest hailstones result from the higher trajectories. Increasing the total concentration of embryos increases the depletion of liquid water by the growing hailstones, thereby reducing the growth rates of the particles following the highest trajectories. Once the embryo concentration has increased to the point where the size-sorting is negative, further increases in the embryo concentration do not affect the liquid water content experienced by the particles following the lowest trajectories, which will become the largest hailstones.

Young (1977) suggests that the effects of turbulence serve to produce a set of 'equivalent' trajectories, that is, trajectories along which the growing particles compete equally for the liquid water. This zone of equivalent trajectories increases with an increasing level of turbulence, and decreases with increasing embryo concentration and with increasing embryo size disparity. Thus when the embryo concentrations are already relatively high, increasing the embryo concentration is counterbalanced by decreasing the zone of equivalent trajectories, resulting in the same production of hail.

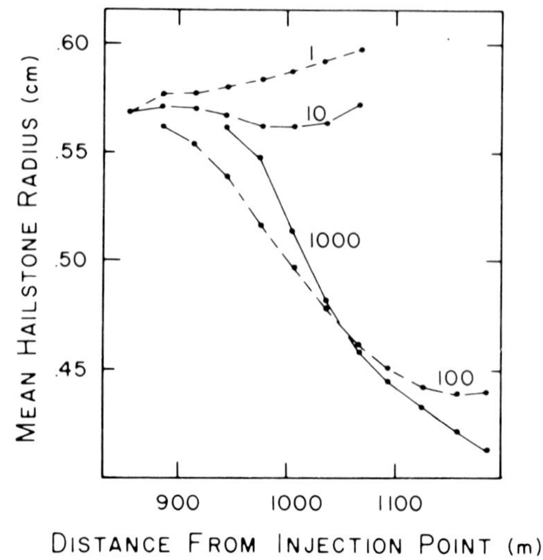

FIG. **12.18.** Mean hailstone radius as a function of distance from the injection point for total embryo concentrations of 1, 10, 100, and 1000 m^{-3}. Only embryos of 0.25 mm radius were used. From Young (1977). *Meteorological Monographs*, **16**, 38. American Meteorological Society. Reproduced with permission.

12.5.3. *Trajectory lowering*

Seeding with ultra-large particles may serve to provide a 'screen' of particles growing in the main updraft below the HGZ. This would reduce the amount of supercooled water reaching the HGZ, thereby reducing the growth of hailstones without the reduction in precipitation associated with the glaciation concept. Such ultra-large particles could be introduced into the EFR through large drop or hygroscopic particle seeding to initiate the coalescence process. Rokicki and Young (1978) show that such hygroscopic seeding would result in a substantial lowering of the radar first echo. This may be expected to transfer embryos into the main updraft below the HGZ.

Particles growing below the HGZ would be less able to freeze the water they accrete, either shedding the excess water[8] or incorporating it in a matrix of ice and water (spongy growth). Water that is shed may still be accreted by particles in the HGZ but the water that is retained by the spongy hailstones is removed as the particles fall out of the updraft. Since the particles are following lower trajectories, they spend less time above the 0°C level and the liquid water contents they experience are lower than if they followed trajectories through the HGZ. Thus, their sizes are smaller and, with less ice to melt, they are more likely to melt completely before reaching the ground.

Simulation of trajectory lowering using the Monte Carlo model is described by Young (1977). In this case, graupel embryos (radius = 0.25 mm, density = 0.5 g cm^{-3}) were injected in each of the four lowest levels of the model, and

[8] Shedding occurs when the ice fraction is less than 0.5, see § 8.3.3.

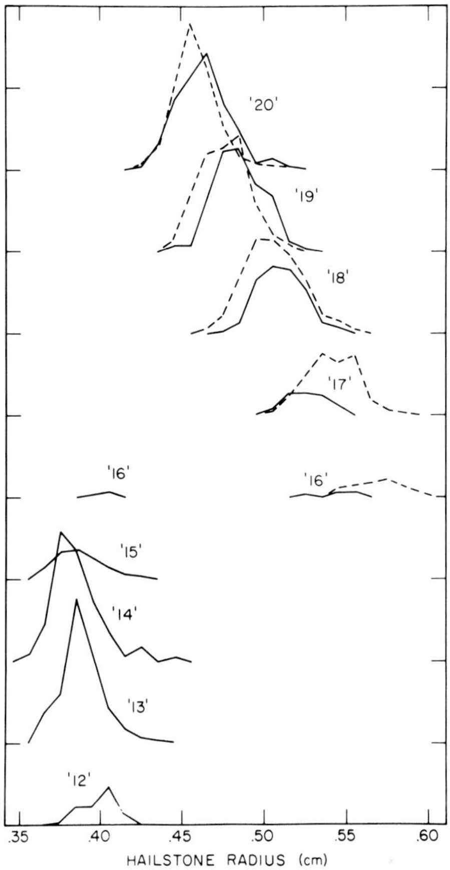

Fig. 12.19. Size spectra of hailstones by exit grid for simulations of trajectory lowering to suppress hail. Each grid is 60 m wide with grid 1 representing the injection point. Each division on the ordinate represents 5000 particles $m^{-2}\,s^{-1}\,cm^{-1}$. The natural case ($N = 100\,m^{-3}$, $r = 0.25$ mm) is shown by dashed lines; the seeded case (adding 100 frozen drops m^{-3}, $r = 0.5$ mm) is shown by solid lines. From Young (1977). *Meteorological Monographs*, **16**, 38. American Meteorological Society. Reproduced with permission.

frozen drop embryos (radius = 0.5 mm, density = $0.9\,g\,cm^{-3}$) were injected into the two lowest levels. The frozen drop concentration was $100\,m^{-3}$; two graupel concentrations of $10\,m^{-3}$ and $100\,m^{-3}$ were used.

The resulting hail spectra are shown in Fig. 12.19 for the case with $100\,m^{-3}$ graupel embryos. Note that the hailstones formed on the seed particles fall out first (grids 12 through 16) and are much smaller than the hailstones resulting from naturally occurring graupel. The graupel-based hailstones are shifted to smaller sizes and higher trajectories[9] as a result of the seeding. The resulting hail size spectra for both graupel concentrations are shown in Fig. 12.20, demonstrating a consistent reduction in hailstone sizes.

[9] Higher trajectories result in fallout further from the injection grid.

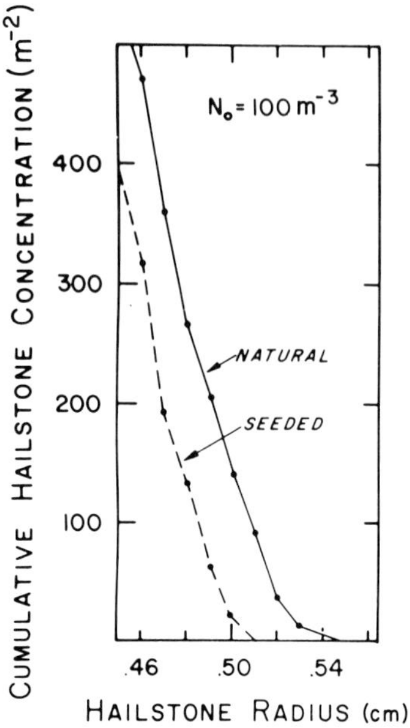

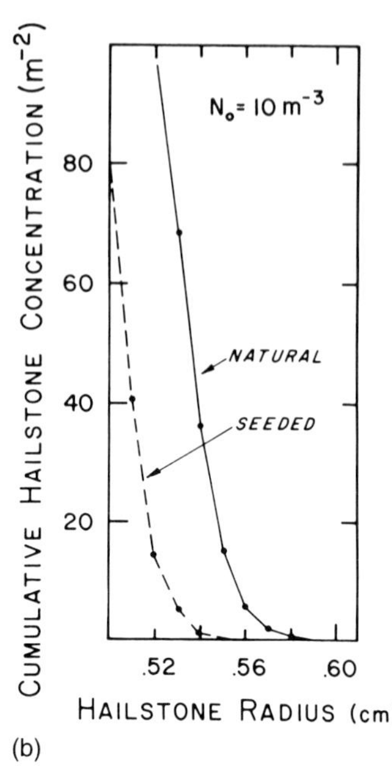

FIG. 12.20. Cumulative size spectra of hailstones, comparing the natural and frozen drop-seeded cases. Spectra are as would be observed by an observer at the ground as the HGZ passes overhead at $15\,m\,s^{-1}$. Seeding was simulated by injecting $100\,m^{-3}$ frozen drops (r = 0.5 mm, density = $0.9\,g\,cm^{-3}$) in the lowest two levels of the model. (a) The case with $10\,m^{-3}$ graupel embryos (r = 0.25 mm, density = $0.5\,g\,cm^{-3}$); (b) the case with $100\,m^{-3}$ graupel embryos. From Young (1977). *Meteorological Monographs*, **16**, 38. American Meteorological Society. Reproduced with permission.

The addition of the frozen drop embryos in a concentration of $100\,m^{-3}$ resulted in decreases in the liquid water content compared with the unseeded case of more than $1\,g\,m^{-3}$ in some portions of the main updraft. Simulations using smaller concentrations of frozen drops produced proportionately smaller reductions in the liquid water content and correspondingly resulted in a smaller reduction in hailstone sizes.

Screening of the HGZ by naturally occurring frozen drops may be expected for clouds with relatively warm cloud bases. Thus tropical cumulonimbi would be less likely to produce large hail despite quite high liquid water contents and relatively strong updrafts. Ludlam (1980) notes that large hail is rare in tropical regions. Appleman (1959) relates the occurrence of hail to the cloud base temperature and the positive area on a thermodynamic diagram which is directly related to the strength of the updraft. Figure 12.21 illustrates that these two parameters serve to delineate hail-producing thunderstorms from those that do not produce hail.

12.5.4. *Hail suppression projects*

Several hail suppression efforts were initiated in the late 1950s in the Soviet Union. These are described by Sulakvelidze et al. (1974) and are based on the

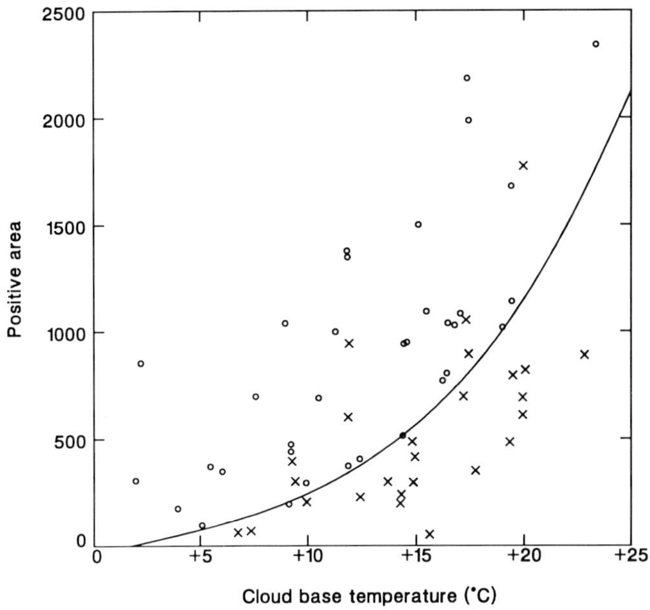

Fig. 12.21. The occurrence of hail as a function of cloud base temperature and available buoyant energy as determined from the positive area on a thermodynamic diagram. (○), hail; (×), no hail. (Based on data from Appleman (1959).

beneficial competition concept. Most of the Soviet hail suppression projects rely on direct injection by rockets of silver iodide or lead iodide into the HGZ, which was believed to contain high concentrations of large drops. One experiment in eastern Georgia relied on introducing both silver iodide and sodium chloride with the expressed intent of initiating precipitation earlier in the updraft and prematurely removing liquid water that would otherwise have entered the HGZ (trajectory lowering). Reductions in annual hail damage from 50% to 99% were reported although the statistical methods on which these results were based are subject to question.

The National Hail Research Experiment (NHRE) was initiated to explore hail suppression in northeast Colorado based on the reported successes from the Soviet Union. This included both a seeding experiment and extensive field observations designed to improve understanding of the physical processes leading to the development of hail. The design and conduct of the seeding experiment are described by Foote and Knight (1979). Seeding was conducted at cloud base below the main updraft using aircraft to disperse silver iodide. Although the seeding portion of the NHRE was scheduled to be conducted over a 5-year period starting in 1972, the seeding experiment was terminated following the 1974 season. Evidence gathered from the physical experiment suggested that the Soviet model for hail development based on the presence of an accumulation zone of large drops was not applicable to northeast Colorado.

Statistical evaluation of the effects of seeding on hail and precipitation in the NHRE are provided by Crow et al. (1979). Seeding was conducted on a total of 27 days with 30 days serving as controls. The null hypothesis that seeding had no effect on hail or precipitation could not be rejected at the 90% level, i.e., no effect of seeding could be shown.

The randomized hail suppression experiment *Grossversuch* IV was conducted in Switzerland between May 1977 and September 1981. This was intended to be a independent verification of the Soviet method of directly injecting massive amounts of silver iodide into the HGZ, based on the concept of beneficial competition. On the basis of hailpad observations at the ground and 10 cm radar observations, Federer et al. (1986) report that there was no significant difference between the 94 seeded and 122 not-seeded thunderstorm cells (on 33 and 43 days, respectively). In addition, no evidence for an accumulation zone containing large drops was found.

Problems

12.1* What is the radius of a hailstone that would just be balanced in a $40\,\mathrm{m\,s^{-1}}$ updraft at $-20°\mathrm{C}$ and 350 mbar? Assume the hailstone is spherical with a density of $0.9\,\mathrm{g\,cm^{-3}}$.

Computer problem

12.2** Test the beneficial competition and trajectory lowering concepts for hail suppression by constructing a simple Monte Carlo model to simulate the effects of turbulence on the growth of hailstones (see Young, 1978). Use a uniform updraft velocity and a constant tilt. Advect the particles with the mean flow, their fall speed, and a random component in both the x and z directions selected from a normal distribution with mean zero and specified standard deviation. Assume that the embryonic hailstones of radius r_i and fall speed v_i grow by accretion according to

$$\frac{dm}{dt} = \pi r_\mathrm{i}^2 \chi v_\mathrm{i} ,$$

where χ is the liquid water content and the collection efficiency is assumed to be unity. Assume that all of the accreted water is retained by the growing hailstone and that the hailstone grows as a sphere with a density of $0.9\,\mathrm{g\ cm^{-3}}$. Use the Reynolds number relationships in § 8.2.2 to determine the fall speeds for the hailstones.

Assume a pseudo-adiabatic lapse rate to determine pressure and temperature as a function of height. Assume the base of the model is at $-20°\mathrm{C}$ and 350 mbar with a liquid water content calculated assuming an adiabatic ascent from the cloud base at 5°C. Above the base of the model, determine the liquid water content within each grid from the pre-existing water content, the additional condensate due to the updraft, and the depletion due to the particles growing within that grid. A table of the additional condensate for each level can be derived from a thermodynamic diagram such as a skew T-log P diagram.

(a) Sample the model output for initial embryo concentrations of 1, 10, 100, and 1000 m^{-3}. Compare the total kinetic energy of the particles larger than 0.5 cm radius for each of the four cases. Initialize the model with 0.25 mm radius graupel particles with a density of 0.5 g cm^{-3}.

(b) Repeat (a) including 100 m^{-3} of 0.5 mm radius frozen drops with a density of 0.9 g cm^{-3} in addition to the graupel embryos.

REFERENCES

Appleman, H., 1959, 'An investigation into the formation of hail,' *Nubila* **2**, 28.

Borland, S.W., and Snyder, J.J., 1975, 'Effects of weather variables on the prices of Great Plains croplands,' *J. Appl. Meteorol.* **14**, 686.

Browning, K.A., 1977, 'The structure and mechanisms of hailstorms,' *Meteorol. Monogr.* **16**(38), 1.

Browning, K.A., and Foote, G.B., 1976, 'Airflow and hailgrowth in supercell storms and some implications for hail suppression,' *Quart. J. R. Meteorol. Soc.* **102**, 499.

Changnon, S.A., Jr., 1977, 'The climatology of hail in North America,' *Meteorol. Monogr.* **16**(38), 107.

Chisholm, A.J., 1973, 'Alberta hailstorms. Part I: Radar case studies and airflow models,' *Meteor Monogr.* **14**(36), 1.

Crow, E.L., Long, A.B., Dye, J.E., Heymsfield, A.J., and Mielke, P.W., Jr., 1979, 'Results of a randomized hail suppression experiment in northeast Colorado. Part II: Surface data base and primary statistical analysis,' *J. Appl. Meteorol.* **18**, 1538.

Dennis, A.S., and Musil, D.J., 1973, 'Calculations of hailstone growth and trajectories in a simple model,' *J. Atmos. Sci.* **30**, 278.

Dye, J.E., Knight, C.A., Tootenhoofd, V., and Cannon, T.W., 1974, 'The mechanism of precipitation formation in northeastern Colorado cumulus. III. Coordinated microphysical and radar observations and summary,' *J. Atmos. Sci.* **31**, 2152.

Federer, B., Waldvogel, A., Schmid, W., Schiesser, H.H., Hampel, F., Schweingruber, M., Stahel, W., Bader, J., Mezeix, J.F., Doras, N., d'Aubigny, G., derMegreditchian, G., and Vento, D., 1986, 'Main results from Grossversuch IV,' *J. Clim. Appl. Meteorol.* **25**, 917.

Foote, G.B., 1984, 'A study of hail growth utilizing observed storm conditions,' *J. Clim. Appl. Meteorol.* **23**, 84.

Foote, G.B., and Knight, C.A., 1979, 'Results of a randomized hail suppression experiment in northeast Colorado. Part I: Design and conduct of the experiment,' *J. Appl. Meteorol.* **18**, 1526.

Foote, G.B., Frank, H.W., Heymsfield, A.J., and Wade, C.G., 1982, 'The 22 July 1976 case study: Hail growth,' *Hailstorms of the Central High Plains*, Vol. II. Boulder, CO: Colorado Associated University Press, p. 181.

Goyer, G.G., 1977, 'Response to 'The climatology of hail in North America',' *Meteorol. Monogr.* **16**(38), 129.

Held, G., and van den Berg, H.J.C., 1977, 'A pre-frontal squall line on 14 November 1975,' *Arch. Met. Geoph. Biokl. Ser. A* **26**, 361.

Heymsfield, A.J., 1983, 'Case study of a hailstorm in Colorado. Part IV: Graupel and hail growth mechanisms deduced through particle trajectory calculations,' *J. Atmos. Sci.* **40**, 1482.

Knight, C.A., 1990, 'Lagrangian modeling of the ice process: A first echo case,' *J. Appl. Meteorol.* **29**, 418.

Knight, C.A., Cooper, W.A., Breed, D.W., Paluch, I.R., Smith, P.L., and Vali, G., 1982, 'Microphysics,' *Hailstorms of the Central High Plains*, Vol. I. Boulder, CO: Colorado Associated University Press, p. 151.

Ludlam, F.H., 1980, *Clouds and storms*. Pennsylvania State University Press.

Marwitz, J.D., 1972, 'The structure and motion of severe hailstorms. Part II. Multi-cell storms,' *J. Appl. Meteorol.* **11**, 180.

Nelson, S.P., 1983, 'The influence of storm flow structure on hail growth,' *J. Atmos. Sci.* **40**, 1965.

Rokicki, M.L., and Young, K.C., 1978, 'The initiation of precipitation in updrafts,' *J. Appl. Meteorol.* **17**, 745.

Sulakvelidze, G.K., Kiziriya, B.I., and Tsykunov, V.V., 1974, 'Progress of hail suppression work in the USSR,' *Weather and climate modification* (W.N. Hess, Ed.). New York: John Wiley and Sons, p. 410.

Warner, C., 1972, 'Calculations of updraft shapes in storms,' *J. Atmos. Sci.* **28**, 1516.

Young, K.C., 1974, 'A numerical simulation of wintertime, orographic precipitation. Part I. Description of model microphysics and numerical techniques,' *J. Atmos. Sci.* **31**, 1735.

Young, K.C., 1977, 'A numerical examination of some hail suppression concepts,' *Meteorol. Monogr.* **16**(38), 195.

Young, K.C., 1978, 'On the role of mixing in promoting competition between growing hailstones,' *J. Atmos. Sci.* **35**, 2190.

Ziegler, C.L., Ray, P.S., and Knight, N.C., 1983, 'Hail growth in an Oklahoma multicell storm,' *J. Atmos. Sci.* **40**, 1768.

13

CIRRIFORM CLOUDS

13.1. Introduction

Cirriform clouds are usually comprised entirely of ice crystals, typically bullet rosettes, bullets, and columns. Many of these crystals grow to surprisingly large sizes and are capable of surviving falls of several kilometers before subliming completely. In this manner, upper cirriform decks are capable of providing ice particles to lower clouds in which the concentrations of active ice nuclei are insufficient to develop an efficient precipitation process (Braham, 1967). This concept was discussed in § 10.3.2 with regard to the seeder–feeder mechanism for orographic clouds.

More recently, the role of cirriform clouds in modifying the global radiation budget has renewed interest in their macroscopic and microscopic structures. High clouds increase the planetary albedo while significantly reducing the outgoing infrared radiation. Stephens and Webster (1981) note that at low and mid-latitudes, thin cirriform clouds produce warming at the surface, in contrast to thicker low and middle level clouds, which produce cooling. Thus, global climate models need to consider cirriform clouds as distinct from low and middle level clouds in their effects on climate simulations.

Relatively few ice forming nuclei are available at typical cirriform altitudes, which are generally above 8 km MSL. Temperatures in cirriform clouds are typically -35°C and colder. Nucleation of cirrus particles appears to be predominantly through growth of haze particles[1], followed by homogeneous freezing nucleation as the solution droplet becomes more and more dilute.

Ice crystals in cirrus clouds at temperatures warmer than −50°C often have complex habits, with bullet rosettes being relatively common. Maximum crystal sizes tend to be fairly large; ice crystals with maximum dimensions up to 1 mm are not uncommon. At temperatures below −50°C, crystals tend to have simple habits and are much smaller. Figure 13.1 shows some examples of ice crystals collected at −80°C over the Marshall Islands. Maximum crystal dimensions are less than 50 μm; note the trigonal crystal in Fig. 13.1d.

[1] Haze particles are cloud condensation nuclei (CCN) below their critical size but above their deliquescent point.

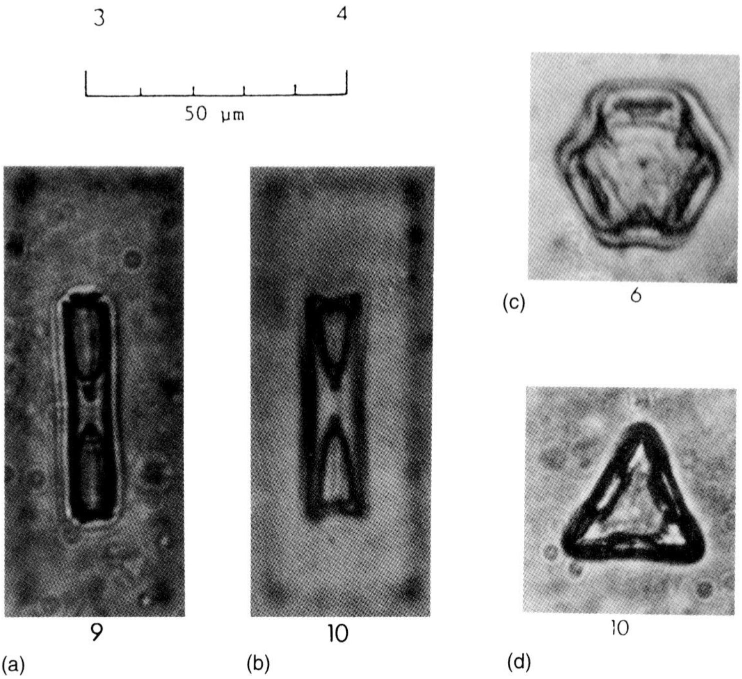

FIG. 13.1. Examples of ice crystals (replicas) collected at −80°C over the Marshall Islands. (a) and (b) are hollow columns (C1f); (c) is a simple plate (P1a); (d) is a trigonally symmetric plate crystal (P3b). Courtesy of A.J. Heymsfield.

13.2. Structure of cirriform clouds

Cirriform clouds consist of a generating zone in which ice particles are nucleated and grow to large sizes, and a trail as the particles exit the generating cell. A sublimation zone is located below the generating level in which the particles in the trail sublime as they fall. Generating zones typically are regions of weak updrafts, on the order of 1 to 2 $\mathrm{m\,s^{-1}}$; downdrafts may occur in the trail region due to particle loading or in the sublimation zone due to cooling as the ice particles sublime.

Cirrus uncinus clouds tend to be relatively short-lived, characterized by cells which have lifetimes on the order of 15 to 25 minutes. Cirrostratus clouds tend to be much more persistent and generally cover a much greater horizontal extent. Ice water contents tend to be lower in cirrostratus although particle sizes are comparable. Dominant crystal types tend to be bullets and bullet rosettes, suggesting a frozen drop origin for these ice particles.

13.2.1. *Cirrus uncinus*

The structure of a cirrus uncinus cloud is shown in Fig. 13.2. Nucleation occurs in a region of updraft (A), with the ice particles growing to form a visible cloud

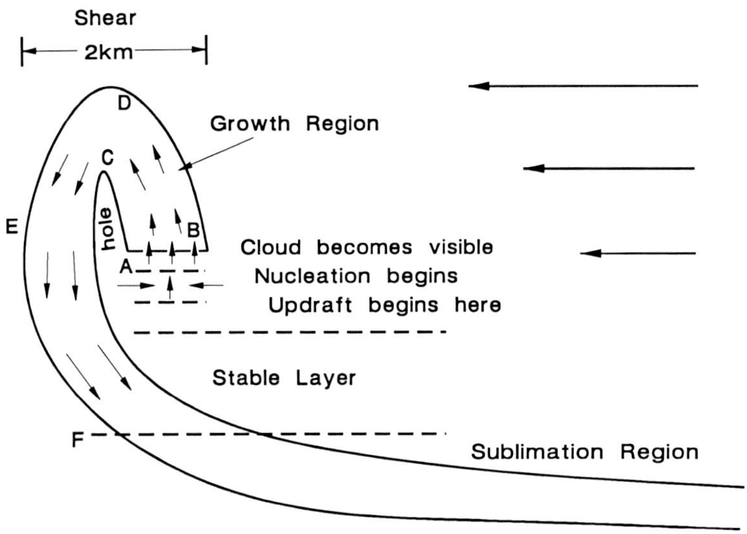

FIG. 13.2. Conceptual model of a cirrus uncinus cloud in conditions of positive wind shear. Ice particles are nucleated at A and grow as they rise (B,C,D). Particles continue growing as they fall (E), eventually falling into the sublimation region (F). After Heymsfield (1975b). *Journal of Atmospheric Sciences*, **32**, 4. American Meteorological Society. Reproduced with permission.

(B). The particles continue to grow as they rise in the updraft (C,D) but are carried downwind from the nucleation zone due to increasing wind speeds with height. Particles entering the downwind portion of the cirrus uncinus head may induce a downdraft or fall in an existing downdraft, making the slope of the trail region more nearly vertical. Downdrafts below the head region are damped out by the stable layer below. The particles fall into the sublimation region, which is subsaturated with respect to ice. Note that the trail extends upwind since the generating head is moving faster than the air in the trail.

Heymsfield (1975b) observes that the cirrus uncinus generating head is formed in a dry adiabatic layer with stable layers just above and below the generating head. If sufficient moisture is present, wave formation in the lower stable layer can trigger condensation and nucleation, in turn, releasing latent heat. This release of latent heat can produce positive buoyancy although some cirrus uncinus generating cells may not become positively buoyant and remain within the original wave.

Although individual cirrus uncinus generating cells may last only 15 to 25 minutes, the length of cirrus uncinus trails implies a more or less continuous supply of ice particles over a period of an hour or more. Heymsfield (1975b) suggests that cooling due to sublimation in the trail or convergence due to spreading of the downdraft in the trail region at the stable layer may result in new generating cells, which will continue to supply ice particles to the original trail.

Heymsfield (1975c) describes a numerical simulation of the growth (and sublimation) of particles in a cirrus uncinus cloud. The trajectories followed by

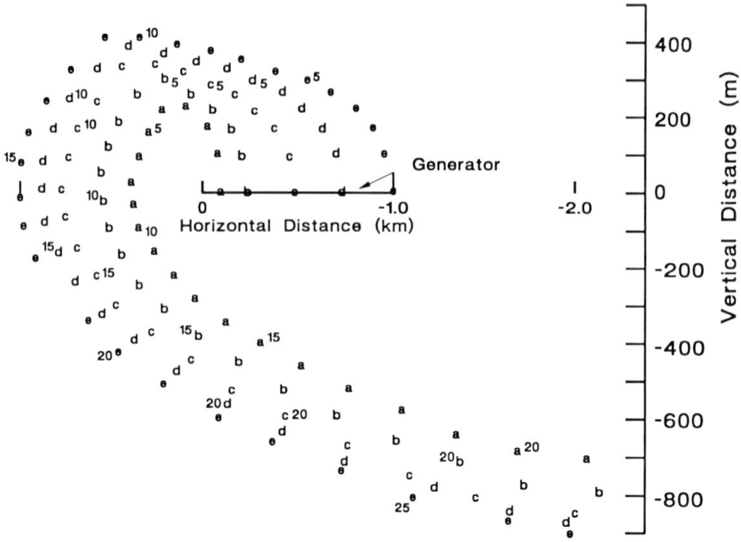

FIG. 13.3. Particle trajectories in a simulated cirrus uncinus cloud with an updraft velocity of $1\,m\,s^{-1}$ in a wind shear of $5\,m\,s^{-1}\,km^{-1}$. Nucleation is assumed to occur at $-40°C$ along a line between 0 and -1 km. Sublimation is assumed below -400 m. Numbers below particle positions indicate the time in 10^2 seconds. From Heymsfield (1975c). *Journal of Atmospheric Sciences*, **32**, 4. American Meteorological Society. Reproduced with permission.

the particles are shown in Fig. 13.3. Particles following trajectory a exit the updraft region most quickly and are the smallest size as they begin to fall, forming a 'hole' between the upwind generating region and the downwind trail region. These particles form the top of the trail by virtue of their lower fall speeds. The largest particles (e) are those nucleated furthest upwind since they remain in the updraft the longest.

Heymsfield (1975a) reports aircraft measurements of ice particles from four cirrus uncinus clouds between -36°C and -58°C. Ice water contents ranged from $0.15\,g\,m^{-3}$ to $0.43\,g\,m^{-3}$. Since the saturation vapor density over water at $-36°C$ is only $0.26\,g\,m^{-3}$, an accumulation mechanism must be operating in order to enhance the ice water content over its maximum possible adiabatic value. Particles larger than 150 μm in their maximum dimension were present in concentrations of $25\,\ell^{-1}$ to $40\,\ell^{-1}$, with mean ice crystal lengths of 500 μm to 750 μm reported. Total concentrations of ice particles of all sizes sampled were on the order of $10^2\,\ell^{-1}$.

13.2.2. *Cirrostratus*

Heymsfield (1975a) suggests that cirrostratus may evolve from cirrus uncinus under conditions where conditionally unstable lapse rates are present, such as near cold fronts. Under these conditions, embedded cirrus uncinus type trails and generating cells are evident. Cirrostratus may also be formed under stable conditions associated with overrunning, such as over a warm frontal boundary.

Under these conditions, the cirrostratus clouds tend to have a more uniform microstructure.

The smallest ice particles tend to be near the tops of cirrostratus decks, with particle sizes increasing toward the cloud base. Heymsfield (1975a) suggests this results from size-sorting when short-lived generating cells are present, with the larger particles falling faster. When generating cells were present, measured relative humidities were below ice saturation except near the cloud top. In more uniform cirrostratus, conditions of ice supersaturation are maintained throughout much of the depth of the cloud and particles continue to grow as they fall.

Ice water contents in cirrostratus clouds tend to be much lower than in cirrus uncinus. Heymsfield (1975a) reports ice water contents ranging from $0.025\,\mathrm{g\,m^{-3}}$ to $0.067\,\mathrm{g\,m^{-3}}$ for two cirrostratus clouds, based on aircraft measurements. Concentrations of ice particles larger than $150\,\mu\mathrm{m}$ in their maximum dimension were on the order of $4\,\ell^{-1}$ to $8\,\ell^{-1}$, with mean ice crystal lengths on the order of $350\,\mu\mathrm{m}$.

13.2.3 *Thunderstorm anvils*

Cirrus clouds formed as thunderstorn anvils are common in tropical regions. Temperatures in these clouds as cold as $-77°\mathrm{C}$ have been reported and clouds as thick as 10 km have been observed. High ice particle concentrations and the presence of aggregates of ice particles in thunderstorm anvils were noted in section 10.4.2. Thunderstorm anvils represent particles caried downwind of convective towers where they descend slowly and sublime. Heymsfield and Miller (1988) note that mass fluxes into the anvils of large thunderstorms are comparable to the precipitation flux.

13.3. Nucleation of ice particles

The formation of ice particles in cirriform clouds is generally believed to occur as a result of homogeneous freezing nucleation. As CCN are lifted above their deliquescent point, they form solution droplets. The likelihood of freezing homogeneously increases as the droplet temperature decreases and as the concentration of the solution decreases. As a particle rises, it becomes more dilute through condensation as well as colder, thereby enhancing the likelihood of homogeneous freezing. The following derivation of the homogeneous freezing nucleation rate for solution droplets follows that provided by Heymsfield and Sabin (1989).

The change in thermodynamic potential is similar to (4.6) with the saturation vapor pressure over water replaced by the saturation vapor pressure over the solution,

$$(\Delta Y)_{\mathrm{g}} = -\frac{4}{3}\pi r^3 N_{\mathrm{I}} kT \ln\left(\frac{e_{\mathrm{soln}}}{e_{\mathrm{s,i}})}\right) + 4\pi r^2 \sigma_{\mathrm{SL}}, \tag{13.1}$$

where σ_{SL} is the surface tension between the solution and ice, which will be assumed to be roughly equal to σ_{IL}. Following the treatment in § 3.4.2

$$\frac{e_{soln}}{e_{s,i}} = \left(\frac{e_{soln}}{e_{s,w}}\right)\left(\frac{e_{s,w}}{e_{s,i}}\right)$$

and

$$\frac{e_{soln}}{e_{s,i}} = \left(\frac{e_{s,w}}{e_{s,i}}\right)\frac{m}{m + im_s M_v/M_s} = X\left(\frac{e_{s,w}}{e_{s,i}}\right). \tag{13.2}$$

The radius of the critical embryo follows by substituting (13.2) in (13.1) as

$$r^* = \frac{2\sigma_{IL}}{N_I kT \ln\left(X\dfrac{e_{s,w}}{e_{s,i}}\right)}.$$

The change in thermodynamic potential at the critical radius is

$$(\Delta Y)_{g*} = \frac{16\pi\sigma_{IL}^3}{3\left[N_I kT \ln\left(X\dfrac{e_{s,w}}{e_{s,i}}\right)\right]^2},$$

or, in terms of the thermodynamic potential for pure water,

$$(\Delta Y)_{g*} = (\Delta Y_o)_{g*}\left[\frac{\ln\left(\dfrac{e_{s,w}}{e_{s,i}}\right)}{\ln X + \ln\left(\dfrac{e_{s,w}}{e_{s,i}}\right)}\right]^2. \tag{13.3}$$

Since

$$\ln X \ll \ln\left(\frac{e_{s,w}}{e_{s,i}}\right),$$

(13.3) can be approximated as

$$(\Delta Y)_{g*} \approx (\Delta Y_o)_{g*}\left\{1 - 2\frac{\ln X}{\ln\left(\dfrac{e_{s,w}}{e_{s,i}}\right)} + 3\left[\frac{\ln X}{\ln\left(\dfrac{e_{s,w}}{e_{s,i}}\right)}\right]^2\right\} \tag{13.4}$$

(see Problem 13.3).

The homogeneous freezing rate for a solution droplet then follows as

$$J \approx J_o \exp\left\{\frac{-(\Delta Y)_{g*}}{kT}\left[\frac{-2\ln X}{\ln\left(\dfrac{e_{s,w}}{e_{s,i}}\right)} + 3\left(\frac{\ln X}{\ln\left(\dfrac{e_{s,w}}{e_{s,i}}\right)}\right)^2\right]\right\} \tag{13.5}$$

where J_o is the homogeneous freezing nucleation rate for pure water. Since $\ln X$ is negative and $(e_{s,w}/e_{s,i})$ is positive, the effect of the dissolved salt is to reduce the nucleation rate. The decrease in temperature required to obtain the same nucleation rate as in pure water is roughly the same as the freezing point depression due to the solute.

The homogeneous freezing behavior for a range of CCN masses is shown in Fig. 13.4. As a parcel of air rises (broken lines), haze particles formed on the largest nuclei freeze first, followed by smaller and smaller CCN masses. For relatively warm temperatures, particles formed on the largest CCN freeze as haze particles, whereas for smaller CCN masses the particle must exceed its critical

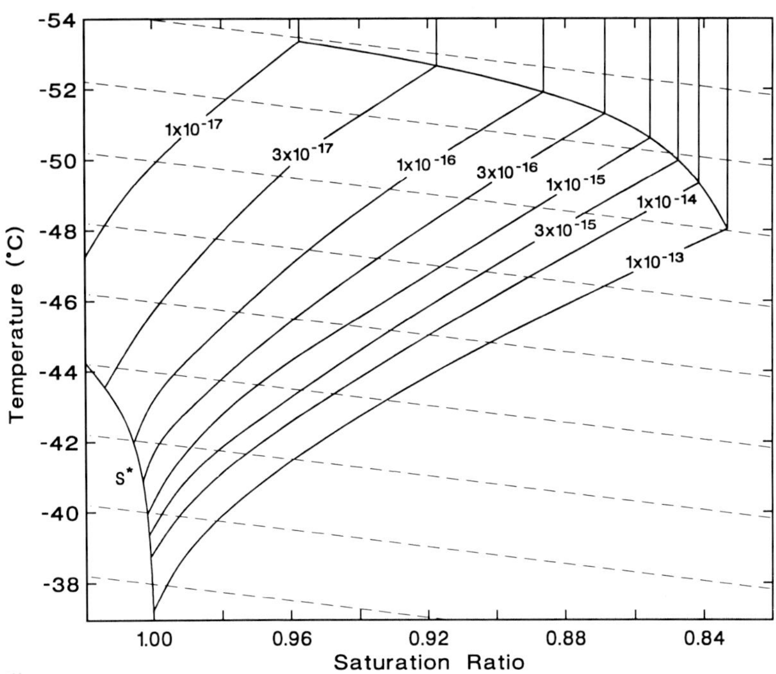

FIG. 13.4. Nucleation diagram for haze particles in an updraft. Homogeneous freezing nucleation thresholds ($J = 1\ s^{-1}$) are shown for eight CCN masses (NaCl) based on (13.5). The region to the left of the curve labeled S^* represents homogeneous freezing occuring only for particles larger than their critical radius. The broken lines represent adiabatic ascent from 350 mbar at 85% relative humidity. The vertical lines represent the transitions from a dry salt particle to a saturated solution.

radius before it has enough water to achieve a nucleation rate of 1 s^{-1}. Heymsfield and Sabin (1989) note that virtually all of the ice nucleation occurs in a narrow range of temperature since the growth of the ice particles rapidly depletes the water vapor in excess of ice saturation.

13.4. Radiative properties of cirriform clouds

Solar radiation received at the top of a cloud is largely scattered (reflected) by the cloud or transmitted through the cloud, the relative proportions depending on the optical depth. Relatively little of the visible radiation received from the sun is absorbed. Infrared radiation received from the surface below the cloud may be absorbed by the cloud or, if the cloud is optically thin enough, may pass through the cloud. The cloud emits infrared radiation in both the upward and downward directions.

For the atmosphere as a whole, the presence of a cloud increases the reflection of solar radiation, which is a cooling effect. Infrared radiation emitted from the surface is strongly absorbed by water and ice particles, which is a warming effect, i.e., the amount of radiation lost to space is decreased. For small to moderate optical depths, the warming due to absorption of infrared radiation dominates over the cooling due to reflection of solar radiation. Relatively thin clouds can absorb a large portion of the infrared radiation while reflecting a relatively small portion of the solar radiation.

For larger optical depths, the reflection of solar radiation continues to increase whereas virtually all of the infrared radiation is absorbed, i.e., the warming effect due to absorption of infrared radiation cannot increase further. Ackerman *et al.* (1988) estimate that an optical depth of >5 is required to produce net cooling by cirrus anvils in tropical regions although the cross-over point is dependent on the sizes, concentrations, and shapes of the particles, the location of the cloud and its temperature, and the solar zenith angle.

Ackerman *et al.* (1988) note that the exchange of infrared radiation has a net warming effect on cirriform clouds. For clouds with low to moderate IWC, there is warming at all levels; for clouds with high IWC, there is warming near the base of the cloud and cooling near its top. These effects are amplified for higher cirrus clouds due to lower air densities. They note that net warming of the cirrus anvil would be expected to result in several hundreds of meters of lifting which would be expected to result in additional particle growth. Starr and Cox (1985) note that net warming near the cloud base and net cooling near the cloud top would be expected to lead to greater instability and strong updrafts.

Ramaswamy and Detwiler (1986) show that smaller ice crystals have a larger effect on radiational cooling (heating) at the top (base) of a 1 km thick cloud than do larger crystals, as shown in Fig. 13.5. They attribute this to the greater optical depth attained by having higher concentrations of smaller ice crystals, assuming a fixed total ice water content. Thus, knowledge of the microstructure of cirriform clouds is crucial in evaluating their role in modifying the earth's radiation budget.

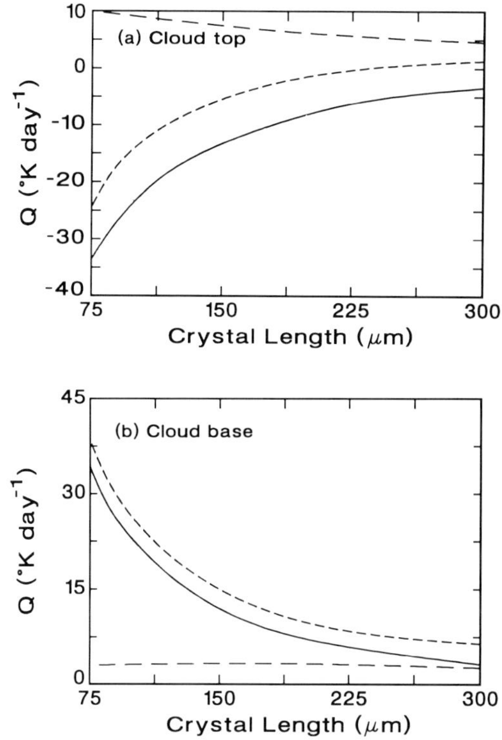

FIG. 13.5. Effect of ice crystal length on (----) solar, (——) longwave (IR), and (- - -) net heating rates at (a) the top and (b) the base of a 1 km thick cloud with a cloud ice water path fixed at 20 g m^{-2}. Cosine of the solar zenith angle is 0.5. Warming rates are in K day^{-1}. From Ramaswamy and Detwiler (1986). *Journal of Atmospheric Sciences*, **43**, 21. American Meteorological Society. Reproduced with permission.

13.5 Growth and survival of cirrus particles

Net loss of infrared radiation from the tops of cirriform clouds serves to enhance the growth of large ice particles and may similarly retard the sublimation of cirrus particles in the subsaturated air below the generating level. Barkstrom (1978) points out that a particle at the cloud top 'sees' outer space in the infrared window region (8 μm to 12 μm) and has perhaps half as much radiation to absorb as a particle deep within the cloud. Larger particles would be colder than smaller particles, enhancing their growth. These effects would in turn be expected to affect the radiative properties of the cloud.

13.5.1. Effects of radiative transfer

The loss (or gain) of energy through radiative processes is proportional to the surface area of the particle, whereas the diffusive transfer of sensible and latent heat is proportional to the particle radius. Thus, under conditions of net infrared radiation loss, the larger particles are colder than the smaller particles.

The net gain of radiant energy (Q_r) results in an additional heat term in (5.6),

$$L_x \frac{dm}{dt} + Q_r = -\frac{dQ}{dt}.$$

Roach (1976) suggests that for radiational cooling, Q_r may be as large as $-AQ_a$ where

$$Q_a \approx 0.00141\ [1 - \exp(-2800r)]$$

in units of cal s^{-1} cm^{-2}, and A is the particle surface area in cm^2.

The temperature for a particle of radius r then becomes

$$T_x = \frac{DL_x(\rho_\infty - b) + KT_\infty + rQ_a}{DL_x s + K}, \tag{13.6}$$

where s and b are the parameters for a linear approximation to the saturation vapor density. For an ice particle undergoing radiational cooling in an environment at ice saturation, the temperature depression below the ambient temperature is a linear function of the particle radius, i.e., larger particles are colder in direct proportion to their size.

The maximum temperature depression due to radiational cooling for a 100 μm radius ice sphere in an environment at ice saturation according to (13.6) is shown in Fig. 13.6 as a function of ambient temperature and pressure. As the ice particle loses infrared radiation and cools, the diffusion of water vapor to the ice particle increases, releasing latent heat and serving to warm the particle. This process is more efficient at higher altitudes due to the increase in the diffusivity of water vapor in air as pressure decreases, leading to smaller temperature depressions. At warmer temperatures, the flux of water vapor for a given temperature difference between the environment and the ice particle is greater due to an increase in the slope of the saturation vapor density versus temperature curve.

Heymsfield (1973) estimates that a column 1.0 mm in length in an environment at ice saturation at −35°C would grow to a length of 1.33 mm within 30 minutes due to radiational cooling. At −40°C, the resultant crystal length after 30 minutes would be 1.15 mm, and at −50°C, it would be 1.04 mm. The decrease in radiation-induced growth with decreasing temperature is a result of the decrease in saturation vapor density with temperature.

A population of ice (or water) particles of different sizes would tend to become more disparate in size due to the effects of radiational cooling. Consider a population of ice particles in an environment at ice saturation with net loss of infrared radiation. All of the particles would become colder, creating a vapor density difference that would result in diffusion of water vapor to the particles.

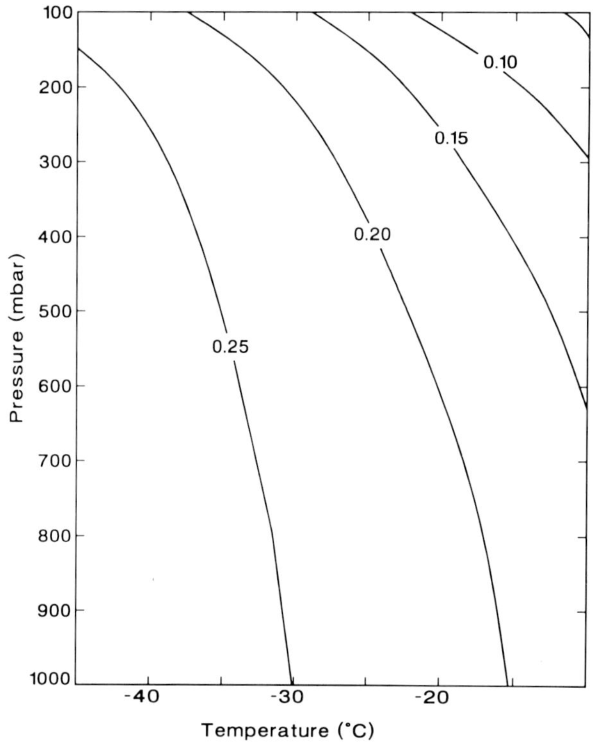

FIG. 13.6. Maximum temperature depression (°C) for a 100 μm radius ice sphere in an ice saturated environment as a function of ambient temperature and pressure. Values are calculated from (13.6).

This would decrease the ambient vapor density. Since the larger ice particles would be colder, they would be able to deposit water vapor on their surface at lower values of the ambient vapor density than would the smaller particles. Thus after some time, the ambient vapor density would have been reduced to a level in between the saturation vapor density over ice for the warmest particles and the coldest particles. At this point, the smaller ice particles are subliming while the larger particles are growing.

13.5.2. Cirrus particle survival

Braham and Spyers-Duran (1967) report the survival of cirrus crystals as much as 6000 m below the cirrus level despite ambient dew-point depressions up to 15°C. The relatively low values for the saturation vapor density over ice at temperatures colder than −30°C suggest that even in a completely dry environment, sublimation is a relatively slow process. Braham and Spyers-Duran estimate that most of the mass lost from the falling particles they observed had occurred in the last 2000 m of fall. They note that the particles at the cirrus level would have had to be considerably larger than would be expected to survive the

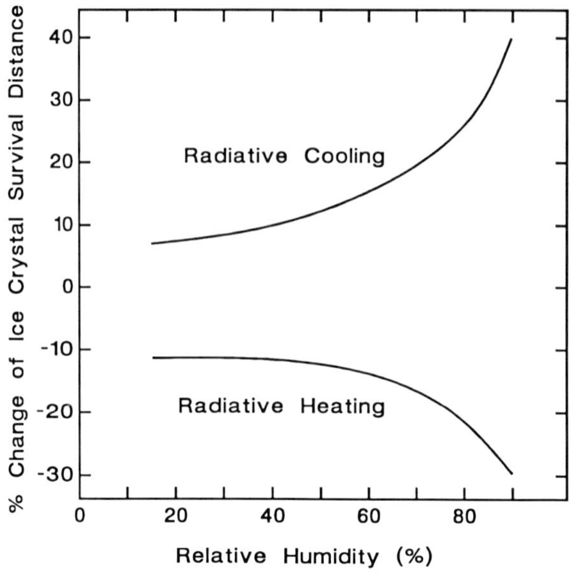

Fig. 13.7. The fractional changes in ice crystal survival distance as a function of the ambient relative humidity with respect to ice for a columnar crystal, with respect to the case without the effects of radiation. The initial crystal dimensions were $a = 82\,\mu\text{m}$ and $c = 400\,\mu\text{m}$ with an initial height of 9.15 km (US Standard Atmosphere). From Stephens (1983). *Journal of Atmospheric Sciences*, **40**, 7. American Meteorological Society. Reproduced with permission.

observed fall distance, and postulate that radiational cooling or downdrafts would enhance the survival of the falling particles.

Stephens (1983) compares the effects of radiational heating and cooling on the survival of ice crystals falling from cirrus levels and finds a substantial increase in the survival distance only for relative humidities greater than 80%, as shown in Fig. 13.7. For the 90% relative humidity case, the effects of radiational cooling increased the survival distance from 3550 m to 4930 m for ice crystals initially 800 μm in length, comparable to those estimated by Braham and Spyers-Duran. The survival of cirrus crystals and the possible seeding of lower cloud decks is dependent in part on radiational cooling reducing sublimation in the ice-subsaturated environment between the cirriform cloud and the lower clouds.

Problems

13.1* Calculate typical values for $\ln X$ and $\ln(e_{s,w}/e_{s,i})$ for temperatures near $-40°$C and saturation ratios between 0.95 and 1.0.

13.2* Plot the maximum temperature depression $(T_x - T_\infty)$ of a spherical ice particle as a function of radius for ambient conditions of $-40°$C and 350 mbar. Assume maximum radiational cooling.

13.3** Derive (13.4) from (13.3).

REFERENCES

Ackerman, T.P., Liou, K-N, Valero, F.P.J., and Pfister, L. 1988; 'Heating rates in tropical anvils,' *J. Atmos. Sci,* **45**, 1606.

Barkstrom, B.R., 1978, 'Some effects of 8–12 μm radiant energy transfer on the mass and heat budgets of cloud droplets,' *J. Atmos. Sci.* **35**, 665.

Braham, R.R., Jr., 1967, 'Cirrus cloud seeding as a trigger for storm development,' *J. Atmos. Sci.* **24**, 311.

Braham, R.R., Jr., and Spyers-Duran, P., 1967, 'Survival of cirrus crystals in clear air,' *J. Appl. Meteor.* **6**, 1053.

Heymsfield, A.J., 1973, *Cirrus uncinus generating cells and the evolution of cirriform clouds*. Ph.D. Thesis, The University of Chicago.

Heymsfield, A.J., 1975a, 'Cirrus uncinus generating cells and the evolution of cirriform clouds. Part I: Aircraft observations of the growth of the ice phase,' *J. Atmos. Sci.* **32**, 799.

Heymsfield, A.J., 1975b, 'Cirrus uncinus generating cells and the evolution of cirriform clouds. Part II: The structure and circulations of the cirrus uncinus generating head,' *J. Atmos. Sci.* **32**, 809.

Heymsfield, A.J., 1975c, 'Cirrus uncinus generating cells and the evolution of cirriform clouds. Part III: Numerical computations of the growth of the ice phase,' *J. Atmos. Sci.* **32**, 820.

Heymsfield, A.J., and Miller, K.M., 1988; 'Water vapor and ice mass transported into the anvils of CCOPE thunderstorms: comparison with storm influx and rainout,' *J. Atmos. Sci,* **45**, 3501.

Heymsfield, A.J., and Sabin, R.M., 1989, 'Cirrus crystal nucleation by homogeneous freezing of solution droplets,' *J. Atmos. Sci.* **46**, 2252.

Ramaswamy, V., and Detwiler, A., 1986, 'Interdependence of radiation and microphysics in cirrus clouds,' *J. Atmos. Sci.* **43**, 2289.

Roach, W.T., 1976, 'On the effect of radiative exchange on the growth by condensation of a cloud or fog droplet,' *Quart. J. R. Meteorol. Soc.* **102**, 361.

Starr, D. O'C., and Cox, S.K., 1985, 'Cirrus clouds. Part II: Numerical experiments on the formation and maintenance of cirrus,' *J. Atmos. Sci.* **42**, 2682.

Stephens, G.L., 1983, 'The influence of radiative transfer on the mass and heat budgets of ice crystals falling in the atmosphere,' *J. Atmos. Sci.* **40**, 1729.

Stephens, G.L., and Webster, P.J., 1981, 'Clouds and climate: sensitivity of simple systems,' *J. Atmos. Sci.* **38**, 235.

Appendix 1

REVIEW OF THERMODYNAMICS

Clouds form when air reaches and exceeds water saturation. In the free atmosphere, this almost always occurs as a result of an adiabatic pressure change (lifting) which produces cooling and supersaturated conditions, thereby allowing nucleation of a condensed water phase. Thus, any quantitative description of clouds requires formulation of 'dry' and 'wet' adiabatic processes and a description of phase changes in the water system. This appendix will review these thermodynamic processes on which the formation of clouds in the free atmosphere depends.

A.1.1. Dry adiabatic processes

A dry adiabatic process is a process during which a parcel of air does not exchange heat with its environment and no phase changes occur within the parcel. Such a parcel may contain water vapor and thus be considered to be 'moist,' but as long as there is no change of phase, the process is considered to be 'dry.'[1] Since the fraction of water vapor in air may vary from essentially zero to 3% whereas the relative fractions of the gases comprising 'dry' air remains quite constant, it is customary to consider moist air as a two-component system. The main gaseous components of the atmosphere which do not condense under the range of temperatures and pressures normally encountered within the earth's atmosphere are nitrogen (78.1% by volume), oxygen (21.0%) and argon (0.9%). By applying Dalton's law of partial pressures, dry air can be considered to be a single-component gas with an effective molecular weight of 28.967.

A.1.1.1. *Equation of state for moist air*

Considering moist air to be an ideal gas, the density of moist air ρ_a can be expressed in terms of the dry air pressure (P_d) and water vapor pressure (e)[2] as

$$\rho_a = \rho_d + \rho_v = \frac{P_d M_d}{RT} + \frac{e M_v}{RT}. \qquad \text{(A.1.1)}$$

[1] This definition applies even under conditions when water saturation may be exceeded.

[2] The symbol e for water vapor pressure here is in keeping with standard meteorological practice. In thermodynamical treatments of nucleation, the symbol P is commonly employed. In this text, the initial development of the nucleation equations use P but reverts to the use of e for most of the development of the nucleation equations.

The subscripts d, v, and a refer to dry air, water vapor, and their mixture, respectively, with M referring to the molecular weight. R is the ideal gas constant, and T is the temperature. Applying Dalton's law of partial pressures, the total pressure P is $P_d + e$ and (A.1.1) reduces to

$$\rho_a = \frac{PM_d}{RT}\left[1 - \frac{e}{P}\left(1 - \frac{M_v}{M_d}\right)\right]. \qquad \text{(A.1.2)}$$

It is useful to define a 'virtual' temperature T^* as the temperature dry air would have if its density and pressure were the same as the moist air under consideration. The equation of state then becomes

$$\rho_a = \frac{PM_d}{RT^*} \qquad \text{(A.1.3)}$$

and the definition of virtual temperature then follows from (A.1.2) and (A.1.3) as

$$T^* = T\left[1 - \frac{e}{P}\left(1 - \frac{M_v}{M_d}\right)\right]^{-1} = T\left[1 - 0.378\,\frac{e}{P}\right]^{-1}. \qquad \text{(A1.1.4)}$$

The virtual temperature is never less than the actual temperature since $M_v < M_d$ and may be as much as 4°C or 5°C warmer, although differences greater than 3°C are rare.

A.1.1.2. *Poisson's equation and potential temperature*

The change of temperature with changing pressure in a reversible adiabatic process may be derived from the first law of thermodynamics. For an open system such as the atmosphere, the first law may be expressed as

$$dE = dQ - P\,d\alpha = c_v T, \qquad \text{(A.1.5)}$$

where E is the internal energy, Q is the heat, α is the specific volume ($1/\rho$) and c_v is the specific heat of air at constant volume. Note that in an open system, E and Q are expressed per unit mass.

Since we wish to express temperature changes in terms of pressure changes, the ideal gas law needs to be employed in differential form to express $d\alpha$ in terms of dP and dT. Recognizing that $\alpha = R_d T/P$, (A.1.5) may be expressed as

$$dQ = \frac{-R_d T}{P}\,dP + (c_v + R_d)\,dT = \frac{-R_d T}{P}\,dP + c_p dT, \qquad \text{(A.1.6)}$$

where $R_d = R/M_d$ and c_p is the specific heat of air at constant pressure.

For an adiabatic process, $dQ = 0$ and (A.1.6) becomes

$$\frac{dP}{dT} = \frac{c_p P}{R_d T}. \tag{A.1.7}$$

Integrating from some initial state (P_o,T_o) to some final state (P,T), Poisson's equation follows as

$$T = T_o\left(\frac{P}{P_o}\right)^{R_d/c_p} = T_o\left(\frac{P}{P_o}\right)^{0.286}. \tag{A.1.8}$$

The potential temperature θ is defined as that temperature an air parcel would have if it were brought dry adiabatically to a pressure of 1000 mbar. From (A.1.8), the potential temperature is

$$\theta = T\left(\frac{1000}{P}\right)^{0.286},$$

where P is expressed in mbar and T in Kelvin.

A.1.1.3. *The dry adiabatic lapse rate*

The dry adiabatic lapse rate may be derived from (A.1.7) and the hydrostatic equation

$$\frac{dP}{dz} = -\rho_a g .$$

Taking the differential of (A.1.7) with respect to z and expressing the air density ρ_a using the ideal gas law,

$$\frac{1}{P}\frac{dP}{dz} = \frac{c_p}{R_d T}\frac{dT}{dz} = \frac{-g}{R_d T}. \tag{A.1.9}$$

The dry adiabatic lapse rate follows as

$$\frac{dT}{dz} = \frac{-g}{c_p}. \tag{A.1.10}$$

The dry adiabatic lapse rate is virtually constant in the troposphere, with a value of 9.76°C km^{-1}. If the actual lapse rate exceeds this value, an ascending parcel would become warmer and less dense than its environment and would accelerate upwards. Similarly, a descending parcel would become colder than its environment and accelerate downwards. Such a superadiabatic condition is obviously unstable and is rare in the atmosphere.

A.1.1.4. *Water vapor correction to Poisson's equation*

Since the exponent in (A.1.8) is not the same for both dry air and water vapor and the ratio of these components may vary, Poisson's equation should be developed for this two-component system. It will be shown that this correction is quite small and for most purposes, can be ignored.

Expressing (A.1.6) for the two component system and setting $dQ = 0$,

$$\frac{dP}{\rho_a} = [c_p\,(1-q) + c_{wv}q]\;dT, \tag{A.1.11}$$

where $\rho_a = \rho_d + \rho_v$, c_{wv} is the specific heat of water vapor at constant pressure, and q is the specific humidity, ρ_v/ρ_a. Expressing ρ_a in terms of the total pressure and virtual temperature

$$\frac{dP}{P}\,R_d T^* = [c_p\,(1-q) + c_{wv}q]\;dT. \tag{A.1.12}$$

The virtual temperature may be expressed in terms of the specific humidity (see Problem A.1.1) as

$$T^* = T\left[1 + q\left(\frac{M_d}{M_v} - 1\right)\right]. \tag{A.1.13}$$

Recognizing that the specific humidity is conserved during dry adiabatic processes, (A.1.12) and (A.1.13) can be combined and integrated as before. The exponent in the Poisson equation then becomes

$$\frac{R_d\left[1 + q\left(\dfrac{M_d}{M_v} - 1\right)\right]}{c_p\,(1-q) + c_{wv}q}. \tag{A.1.14}$$

Employing the definitions of mixing ratio ($w = \rho_v/\rho_d$) and specific humidity, (A.1.14) may be simplified and Poisson's equation becomes

$$T = T_o\left(\frac{P}{P_o}\right)^{\frac{R_d + wR_v}{c_p + wc_{wv}}}. \tag{A.1.15}$$

Since $c_{wv} > R_v$, the simplified form of the Poisson equation overestimates the temperature change due to dry adiabatic pressure changes. However, the errors even under the most extreme conditions expected in the troposphere are less than 0.08°C.

1.1.2. Phase change and latent heat

The study of cloud microphysical processes is intimately related to phase changes in the water system. The initial formation (nucleation) of water drops and their growth by diffusion represents a vapor to liquid transition. The diffusional growth of ice crystals represents a vapor to solid transition. The freezing of water drops and the collection of water drops by ice particles represents a liquid to solid conversion. Since these processes constitute the bulk of the subject matter of this book, a review of phase changes with an emphasis on the water system is in order.

Three phases of water which are of importance in the troposphere can be identified, namely, water vapor, liquid water, and solid water or ice (Ice Ih). In keeping with common usage, these three phases will be referred to as water vapor, water, and ice, respectively.

Three possible pathways between these three phases can be identified, each having two directions, as shown in Fig. A.1.1. The water vapor–water transitions are termed condensation and evaporation; the water vapor–ice transitions are termed deposition and sublimation; the water–ice transitions are termed freezing and melting.

A.1.2.1. *PαT diagrams for the water system*

Pα and PT diagrams are convenient two-dimensional representations of the three dimensional (P,α,T) behavior of a one-component system. These are symmetric in the sense that equivalent information may be obtained from either a Pα diagram with isotherms defined or a PT diagram with isosteres defined.[3] PT diagrams are commonly represented without isostere mapping and are used to delineate the stable phases, although they may be used to depict metastable phases as well. Such diagrams are termed phase diagrams.

To begin, consider the Pα and PT phase diagrams, i.e., without isopleths and showing only the phase boundaries. The PT diagram shown in Fig. A.1.2 delineates the regimes of P and T for which each phase is stable. The phase boundary between water vapor and metastable (supercooled) water is indicated by the broken line. The triple point for the water system occurs at +0.0075°C and 6.11 mbar. The critical point occurs at 374.0°C and 217.72 atmospheres (2.206 $\times$ 10^5 mbar). The apparent negative curvature of the water vapor–water and water vapor–ice phase transition lines in Fig. A.1.2 is a result of the semilogarithmic scales used. On a linear scale, the curvature is positive, as shown in Fig. A.1.4a. This has important implications in that supersaturated conditions may be obtained by mixing two saturated (or nearly saturated) parcels having different temperatures.

The Pα diagram shown in Fig. A.1.3 exhibits one of the complexities of the water system, namely that the solid phase is less dense than the liquid phase. This

[3] The equivalent information can also be obtained from a Tα diagram with isobars plotted (see Problem A.1.8).

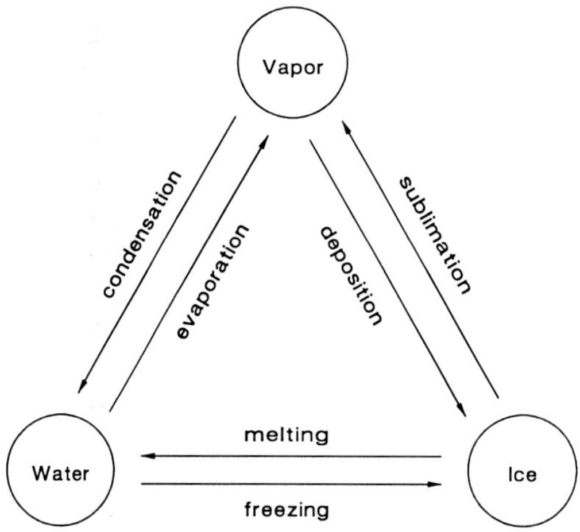

FIG. A.1.1. Nomenclature for phase transitions in the water system.

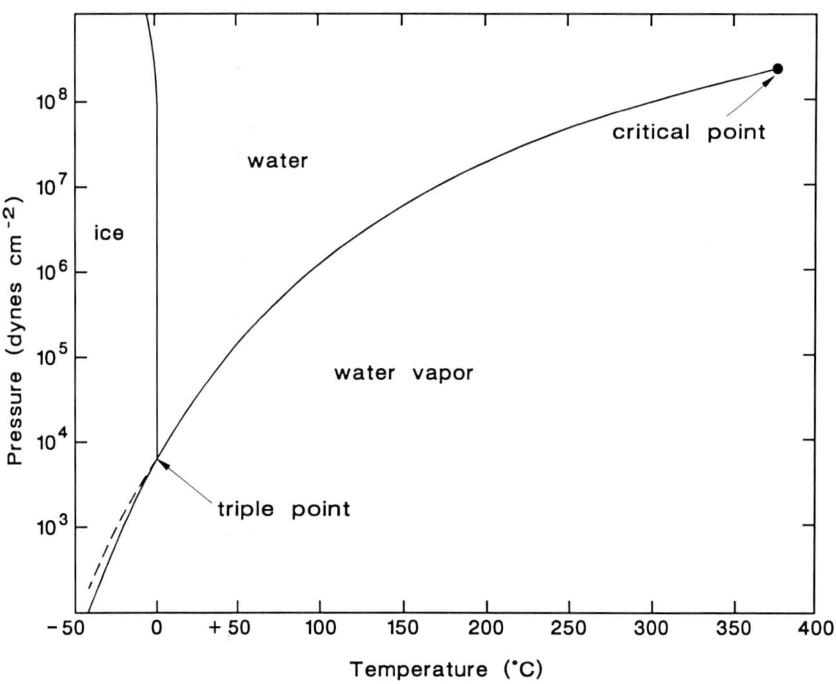

FIG. A.1.2. PT phase diagram for water. Stable phase boundaries are shown by solid lines; metastable phase boundaries are indicated by broken lines.

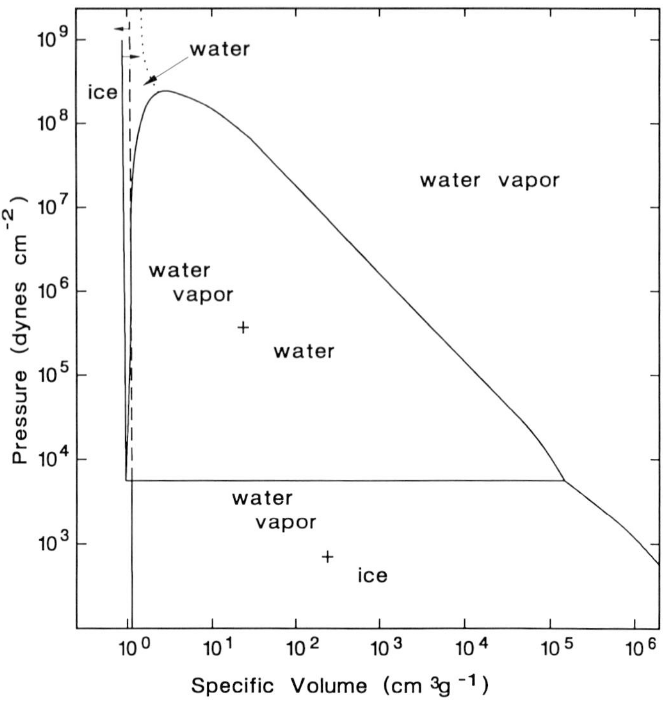

FIG. A.1.3. Pα phase diagram for water. Regions where pure and mixed phases are stable are shown. A section of the critical isotherm (dotted line) is used to bound the water region. Ice, water, or their mixture may exist between the parallel solid and broken lines.

anomalous behavior of water accounts for the negative slope of the water–ice phase transition line in Fig. A.1.2 and does not permit a unique specification of temperature and phase given only pressure and specific volume.

For the purposes of better understanding these diagrams, Figs. A.1.2 and A.1.3 may be presented in schematic form, exaggerating the more important features of the diagrams. Four illustrative isotherms are drawn on the PT and Pα phase diagrams shown in Figs. A.1.4a and A.1.4b, respectively. The isotherm through the critical point (T_{CP}) separates those temperatures at which water vapor and water are distinguishable from those temperatures at which they are not.

Following the isotherm between the triple point isotherm (T_{TP}) and the critical point isotherm, as the water vapor pressure is increased, the water vapor density increases until the water vapor–water phase transition line is reached. Without any further increase in pressure,[4] the density of the system can continue to increase as the ratio of water vapor to water decreases. At the pressure where the isotherm intersects the phase transition line in the PT diagram, the density may range from that of pure water at (P,T) to that of pure water vapor at (P,T).

[4] Latent heat must be extracted in order to maintain the constant temperature.

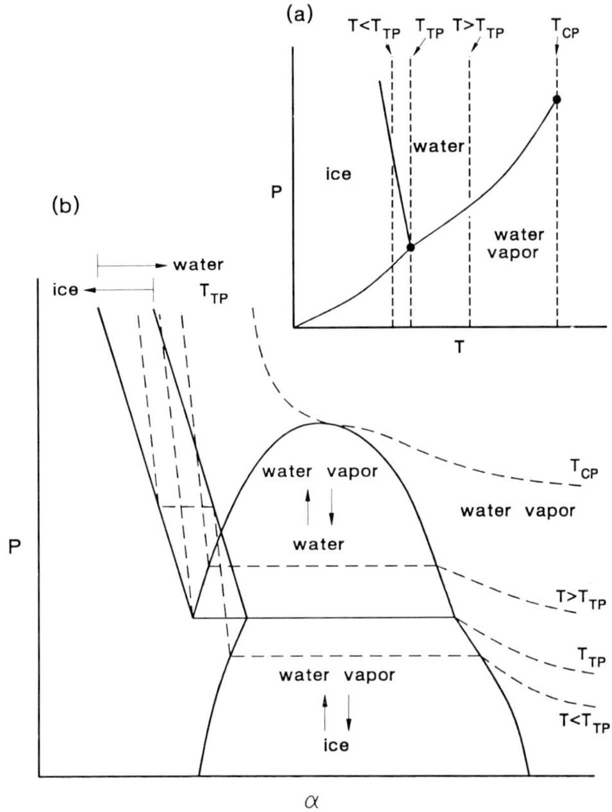

Fig. A.1.4. Schematic PT and Pα diagrams for water, emphasizing the important aspects of the water system. Phase transition lines are shown as solid lines; selected isotherms as broken lines.

Densities in between these extremes are attained by the proper mixture of water vapor and water, as indicated on the Pα diagram. If one were to plot isosteres on a PT diagram, the phase transition lines would be evident as a number of coincident isosteres (see Problem A.1.6).

The isotherm through the triple point exhibits behavior similar to the isotherm $T > T_{TP}$. Note that once the system has condensed completely to water, this isotherm remains entirely within the water region.

The isotherm $T < T_{TP}$ exhibits the most complicated behavior since it must pass through successive regions of water vapor, ice, and then water as the pressure is increased. With increasing pressure, the water vapor–ice phase transition is reached, whereupon the density changes by varying the relative fractions of water vapor and ice at constant pressure. Once the system has been transformed completely to ice, its density increases as pressure increases until the water–ice phase transition is reached. Once again, the density may increase at constant pressure, this time by increasing the ratio of water to ice. Once the system has completely melted, the density once again rises in response to increasing pressure within the water region.[5]

In the Pα diagram for the water system, a given combination of P and α may not uniquely define the phase or phases of the system, i.e., at one temperature, the system may consist entirely of ice while at a warmer temperature, for the same pressure and density, the system may consist entirely of water.[6] Definition of the water vapor regions are fairly straightforward. To the right of the 'double dome', water vapor is the stable phase. Within the full dome, water vapor and water are in stable equilibrium; within the truncated dome, water vapor and ice are in stable equilibrium.

The water region may be traced out by describing the isotherms which pass through the water region. The water region extends to the right from the line as shown in Fig. A.1.4b. Similarly, the ice region lies to the left of the line as shown. The region between these two lines represents three possible situations, namely, pure water, pure ice, or a mixture of water and ice.[7] In other words, for a given pressure, it is possible to define a stable pure water system, a stable pure ice system, and a system consisting of both water and ice in equilibrium such that they all have the same density, although they would have different temperatures.

A.1.2.2. *Latent heat and the Clausius–Clapeyron relation*

Although equilibrium phase changes occur without concomitant changes in the pressure or temperature, i.e, there is no change in the Gibbs free energy, there is a marked change in the order or entropy of the system, and hence there is an exchange of heat with the environment. As a system progresses from a less ordered to a more ordered state, the entropy decreases, implying that the system loses heat to the environment. This heat is termed 'latent heat' and is named according to the particular phase change involved, e.g., sublimation, fusion, etc.

Along any phase transition line separating two phases 1 and 2 in a PT diagram, the two phases are in equilibrium. Thus, following a given phase transition line, dG_1 must equal dG_2. From the definition of Gibbs free energy,

$$dG_1 = dG_2 = \alpha_1 dP - S_1 dT = \alpha_2 dP - S_2 dT$$

or

$$\frac{dP}{dT} = \frac{S_2 - S_1}{\alpha_2 - \alpha_1}.$$

[5] If the isotherm $T < T_{TP}$ is warmer than $-17°C$, increasing pressure will result in successive transitions to Ices V, VI, and VIII. The reader is referred to Fletcher (1970) for a more complete description of these phase changes.

[6] This would not occur for a system in which the solid phase is more dense than the liquid phase.

[7] Due to the relative incompressibilities of ice and water, much of the regions attributed to ice or water are not accessible.

The entropy difference is simply the latent heat (L_{12}) divided by the temperature. This yields the Clausius–Clapeyron relationship

$$\frac{dP}{dT} = \frac{L_{12}}{T(\alpha_2 - \alpha_1)}. \quad \text{(A.1.16)}$$

In the freezing of water, for example, $L_f = 333.7\,\mathrm{J\,g^{-1}}$ (at 0°C), indicating that heat is taken up (going from phase 1, ice, to phase 2, water). Since the density of ice is less than that of water, $\alpha_2 - \alpha_1$ is negative and dP/dT is negative, indicating the water–ice phase transition line in the PT diagram slopes toward colder temperatures with increasing pressure.

For vapor transitions far from the critical temperature, α_1 may be neglected compared to α_2. Using the ideal gas law and replacing P with the saturation vapor pressure (e_s), (A.1.16) reduces to

$$\frac{d(\ln\ e_s)}{dT} = \frac{L_v}{R_v T^2}. \quad \text{(A.1.17)}$$

Here, L_v is the latent heat of vaporization and the slope de_s/dT is positive. Note that the curvature (d^2e_s/dT^2) is positive for conditions normally encountered in the atmosphere (see Problem A.1.2).

Often, the latent heat is desired for temperatures not provided in the available tables or for a phase transition in which the initial and final temperatures are not the same. In this case it is necessary to calculate the heat required to warm or cool the two phases to some temperature for which the latent heat is provided. Figure A.1.5 provides an example of calculating the latent heat of fusion for an ice particle at −10°C collecting water drops at −15°C. The water

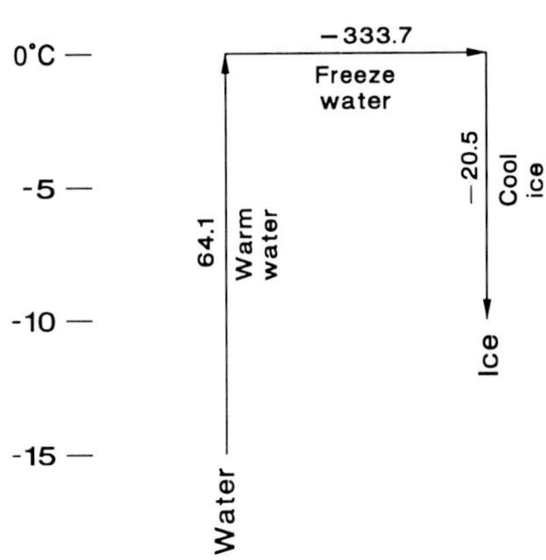

Fig. A.1.5. Example of the calculation of latent heat when the initial and final temperatures are not the same. The water is warmed from −15°C to 0°C, absorbing $64.1\,\mathrm{J\,g^{-1}}$; water freezes at 0°C, releasing $333.7\,\mathrm{J\,g^{-1}}$; the ice is cooled from 0°C to −10°C, releasing $20.5\,\mathrm{J\,g^{-1}}$. The latent heat for the overall transition is $-290.1\,\mathrm{J\,g^{-1}}$.

is warmed by 15°C at an average specific heat of 4.266 J g^{-1} °C^{-1}, absorbing 64.1 J g^{-1}. The water is frozen at 0°C, releasing 333.7 J g^{-1}. The resulting ice is then cooled to −10°C at an average specific heat of 2.07 J g^{-1} °C^{-1}, releasing 20.5 J g^{-1}. The latent heat of fusion for the transition is the sum of the heats involved in the three steps, or −290.1 J g^{-1}, indicating heat is given off.

A.1.3. Expressions of water vapor content

In addition to expressing the water vapor content in terms of its density (ρ_v) and its pressure (e), various other useful definitions are in common use. These include the mixing ratio, specific humidity, wet-bulb and dewpoint temperatures, and relative humidity.

A.1.3.1. *Mixing ratio and specific humidity*

The mixing ratio (w) and specific humidity (q) are both conserved during volume changes not involving condensation, and are useful when considering vertical motions in the atmosphere. The mixing ratio is defined as the mass ratio of water vapor to dry air,

$$w = \frac{\rho_v}{\rho_d} = \frac{eM_v}{P_dM_d} = 0.622\left(\frac{e}{P - e}\right).$$

The specific humidity is defined as the mass ratio of water vapor to moist air,

$$q = \frac{\rho_v}{\rho_v + \rho_d} = \frac{eM_v}{eM_v + P_dM_d} = \frac{0.622\,e}{P - 0.378\,e}. \qquad \text{(A.1.18)}$$

By dividing both numerator and denominator of (A.1.18) by ρ_d, the conversion between specific humidity and mixing ratio becomes

$$q = \frac{\rho_v/\rho_d}{1 + \rho_v/\rho_d} = \frac{w}{1 + w}.$$

A.1.3.2. *Dewpoint temperature and relative humidity*

The dewpoint temperature and relative humidity are measures of water vapor content related to saturation values. The equilibrium vapor pressure over a plane water surface increases with increasing temperature and is essentially independent of the ambient (total) pressure (McDonald, 1963). The dewpoint temperature is the temperature to which air must be cooled isobarically for it to reach saturation. Introducing the subscript s,w to indicate values taken at water saturation (and s,i to indicate ice saturation), the dewpoint temperature (T_d) may be defined as

$$e_{s,w}(T_d) = e(T),$$

where $e(T)$ refers to the actual vapor pressure at temperature T and $e_{s,w}(T_d)$ refers to the saturation vapor pressure at T_d. Similarly, a 'frostpoint' temperature (T_f) can be defined with respect to ice saturation for $T_f < 0°C$,

$$e_{s,i}\,(T_f) = e(T)\,.$$

The relative humidity expresses the fraction of saturation represented by the water vapor actually present, i.e.,

$$RH = \frac{\rho_v}{\rho_{s,w}(T)} = \frac{e(T)}{e_{s,w}(T)}\,.$$

A.1.3.3. *Wet-bulb temperature*

The wet-bulb temperature, as defined by Huschke (1959), represents the temperature an air parcel would have if it were cooled *isobarically* by evaporation to the point of water saturation. All the heat required to evaporate the water is assumed to come from the air parcel and the water vapor resulting from the evaporation is added to the parcel.

The most direct method of solution employs a graphical or iterative technique, as in the following example. Let us determine the wet-bulb temperature for an air parcel initially at 20°C and 1000 mbar with 30% relative humidity. Consider what happens to 1 cm^3 of moist air. Values for $\rho_{s,w}$ and L_v will be taken from List (1958).

Initially, the 1 cm^3 of air contains 5.190×10^{-6} g of water vapor which is 30% of the saturation value of $17.30 \times 10^{-6}\,g\,cm^3$, and 1.1800×10^{-3} g of dry air. In order to cool this air by 1°C, 2.8549×10^{-4} cal of heat ($m_d c_p + m_v c_{wv}$) must be extracted. By estimating the final (wet-bulb) temperature and using the value of latent heat corresponding to that temperature, it is not necessary to consider the heat exchange required to cool the additional water vapor; only the water vapor and air need be considered.

With a relative humidity of 30%, it will be necessary to roughly double the amount of water vapor present. As a first estimate, evaporate 5×10^{-6} g of water and calculate the resultant temperature and vapor density in the parcel, bearing in mind that the final vapor density should equal the saturation vapor density at the final temperature for the solution to be complete. From the saturation vapor density tables, the wet-bulb depression may be estimated as roughly 10°C and so the latent heat of vaporization at +10°C will be used (2.477 kJ g^{-1}). This gives a heat change of 12.39×10^{-3} J and a final temperature of 9.637°C.

Now that water vapor has been added to the parcel and the parcel has been isobarically cooled, its volume has changed. The actual vapor density in the parcel is 10.190×10^{-6} divided by the new volume V. The volume V may be expressed in terms of m_d and m_v by using the equation of state to set up two equations in two unknowns, V and e. Eliminating the vapor pressure e,

$$V = \frac{RT}{P}\left(\frac{m_v}{M_v} + \frac{m_d}{M_d}\right).$$

Now, V is 0.97111 cm^3 and the actual vapor density is 10.493×10^{-6} g cm^{-3} compared to the saturation value at 9.637°C, which is 9.184×10^{-6} g cm^{-3}.

The estimate of adding 5×10^{-6} g of water vapor is too high, resulting in a supersaturated condition. Now repeat these calculations with 4×10^{-6} g of evaporated water. This yields an actual vapor density of 9.407×10^{-6} g cm^{-3} compared to the saturation value of 10.467×10^{-6} g cm^{-3} at the final temperature of 11.710°C.

Note that these two points bracket the desired solution. By plotting these two sets of vapor densities as a function of the final temperature, as shown in Fig. A.1.6, the wet-bulb temperature can be estimated as 10.79°C at the intersection of the two lines. Alternatively, the vapor densities can be plotted as functions of the added water vapor mass, obtaining a value of 4.45×10^{-6} g for the proper water mass to evaporate. By repeating the calculations once again, the wet-bulb temperature is found to be 10.80°C.

Although the above definition of wet-bulb temperature is the preferred definition, several other similar but not identical definitions are in common use. The adiabatic or pseudo wet-bulb temperature is the temperature an air parcel would have if it were cooled *adiabatically* to saturation and then compressed to its original pressure while saturation is maintained by evaporation into the parcel. This method is commonly employed using a thermodynamic diagram and the

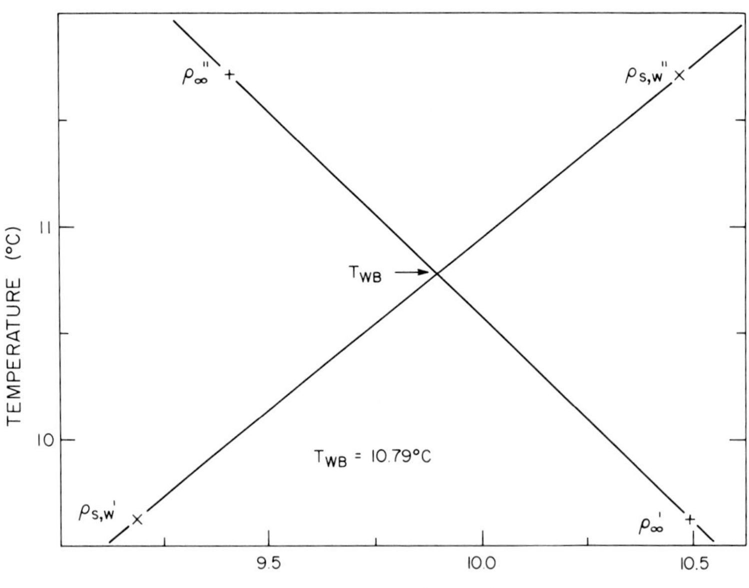

Fig. A.1.6. Graphical solution for the wet-bulb temperature. (×), The actual saturation vapor density at the calculated temperature; (+), the calculated vapor density for that temperature. The solution is the intersection of the two lines, which indicates the temperature at which the calculated vapor density corresponds to water saturation.

resultant temperature is always less than the isobaric wet-bulb temperature as may be deduced from (A.1.5) when one considers the effect of the work involved in this process.

The wet-bulb temperature may also be defined as the temperature a well-ventilated thermometer with a wet muslin covering would attain under steady state conditions. This is similar, but not identical, to the temperature a freely falling water drop would achieve. The latter definition is dealt with in more detail in Chapter 5.

A.1.4. Pseudo-adiabatic processes

Once phase changes occur within a parcel, latent heat is exchanged with the air and the Poisson relationship no longer holds. Two types of 'wet' adiabatic processes may be envisaged. One of these specifies that the condensed water is carried along with the parcel and exchanges heat with the air as the condensed water changes temperature. The second assumes that the condensed water has no effect on the parcel, except for the exchange of latent heat as it condenses or evaporates.

The first 'wet' adiabatic process is dependent on the amount of condensed water present, and thus dT/dP is not uniquely defined if only temperature and pressure are specified, since the liquid water mixing ratio χ must also be defined. The second process is preferred since it allows dT/dP to be uniquely defined given the temperature and pressure. However, one should recognize that the process is not realistic in the sense that water does have a finite specific heat.

The second process has been visualized as condensation followed by immediate precipitation or removal of the condensed water from the parcel, and for this reason has been termed a 'pseudo-adiabatic' process. However, this description implies that a pseudo-adiabatic process is not reversible whereas one commonly employs the relationship in a reversible manner, e.g., in the calculation of a pseudo-wet-bulb temperature. For this reason, the description of the pseudo-adiabatic process which implies that the specific heat of the condensed water is zero is preferred.

A.1.4.1. *Derivation of the pseudo-adiabatic equation*

The first law of thermodynamics for a two-component system (water and dry air) with a condensing phase may be written as

$$-c_w \chi dT - d(L_v q_s) = [c_p (1 - q_s) + c_{wv} q_s] dT - \frac{dP}{\rho_a}, \qquad \text{(A.1.19)}$$

where q_s refers to the saturation value of specific humidity and χ refers to the weight of condensed water per gram of air.[8] Note that (A.1.19) is expressed per gram of air.

[8] This is not quite the same as the liquid water mixing ratio since it is in reference to moist air, although the distinction is not made in practice.

The right hand side of (A.1.19) represents the changes due to the adiabatic process without condensation; the left-hand side represents the changes due to condensation. Since moist air and dry air behave nearly identically for a dry adiabatic process, the right-hand side may be simplified by assuming $c_{wv} \approx c_p$ and $T \approx T^*$. In addition, L_v will be assumed to be constant and, under our definition of a pseudo-adiabatic process, the specific heat of water c_w will be assumed to be zero. Under these assumptions, (A.1.19) reduces to

$$c_p dT - \frac{R_d T}{P}\, dP = -L_v dq_s \,. \tag{A.1.20}$$

Now, dq_s must be expressed in terms of dP and dT so that (A.1.20) can be integrated. Differentiating the definition of specific humidity and taking $\rho_d + \rho_v = \rho_a$,

$$dq_s = \frac{\rho_a d\rho_{s,w} - \rho_{s,w} d\rho_a}{\rho_a^2} \,. \tag{A.1.21}$$

Using the differential forms of the equation of state for air and water vapor, dq_s can be expressed in terms of the total pressure and saturation vapor pressure as

$$dq_s = \frac{M_v}{M_d}\left[\frac{T de_s - e_s dT}{PT} - \frac{e_s}{P^2 T}(T dP - P dT)\right]. \tag{A.1.22}$$

The Clausius–Clapeyron equation allows de_s to be expressed in terms of dT, reducing (A.1.22) to

$$dq_s = \frac{M_v}{M_d}\left[\frac{L_v e_s}{R_v T^2 P}\, dT - \frac{e_s}{P^2}\, dP\right]. \tag{A.1.23}$$

Replacing dq_s in (A.1.20) with (A.1.23) and solving for dT/dP,

$$\frac{dT}{dP} = \frac{\dfrac{R_d T}{P} + \dfrac{M_v e_s L_v}{P^2 M_d}}{c_p + \dfrac{L_v^2 M_v e_s}{M_d R_v P T^2}} \tag{A.1.24}$$

which expresses the change of temperature with changing pressure for a pseudo-adiabatic process.

Integrating (A.1.24) to a form analogous to Poisson's equation (A.1.12) is not trivial and is usually done numerically (see Problem A.1.9).

A.1.4.2. *Pseudo-adiabatic lapse rate*

The pseudo-adiabatic lapse rate follows directly from (A.1.24) by employing the hydrostatic relationship in the same manner as for the dry adiabatic lapse process,

$$\frac{dT}{dz} = \frac{-g\left(1 + \dfrac{L_v \rho_{s,w}}{P}\right)}{c_p + \dfrac{L_v^2 M_v \rho_{s,w}}{M_d P T}} . \tag{A.1.25}$$

This reduces to the dry adiabatic lapse rate, (A.1.10), with ρ_s set equal to 0. The pseudo-adiabatic lapse rate in the lower troposphere in temperate regions is on the order of $5°C\,km^{-1}$ although it may range from as small as $3.5°C\,km^{-1}$ to the dry adiabatic lapse rate of $9.76°C\,km^{-1}$.

A.1.4.3. *Equivalent temperatures*

The equivalent temperature is defined as that temperature a parcel would have if all the water vapor were condensed out and the resulting latent heat used to warm the parcel. This may be accomplished as an isobaric process or via adiabatic processes.

The isobaric equivalent temperature may be readily calculated as

$$T_e = T\left(1 + \frac{L_v w}{c_p T}\right).$$

Since all of the water vapor is being condensed out, it is only necessary to warm the air.

The adiabatic equivalent (or pseudo-equivalent) temperature is that temperature which would result if the parcel were lifted dry adiabatically to saturation, pseudo-adiabatically until all moisture is condensed out, and then compressed dry adiabatically to its original pressure. The second step is physically unrealizable since it implies an expansion to zero pressure. In practice, expansion to -60°C is adequate since the dry and pseudo-adiabats are essentially parallel at this temperature and colder.

The pseudo-equivalent potential temperature is that temperature a parcel would attain if it were lifted dry adiabatically to saturation, pseudo-adiabatically until all the moisture is condensed out and then compressed dry adiabatically to 1000 mbar. This is noted to be the potential temperature corresponding to the adiabatic equivalent temperature. The equivalent potential temperature has the advantage that it is conserved for both dry and pseudo-adiabatic processes, whereas the potential temperature is conserved only for dry adiabatic processes.

Problems

A.1.1* Show that

$$T^* = T\left[1 + q\left(\frac{M_d}{M_v} - 1\right)\right].$$

A.1.2* Show that the plot of saturation vapor pressure (e_s) versus temperature has positive curvature for temperatures normally encountered in the troposphere.

A.1.3* What is the maximum wet-bulb depression possible at 0°C? Assume P = 1000 mbar.

A.1.4* Given ambient conditions of 10°C and 800 mbar with 40% relative humidity, calculate the following:
(a) density of water vapor.
(b) density of dry air.
(c) water vapor mixing ratio.
(d) specific humidity.
(e) virtual temperature.
(f) potential temperature.
(g) isobaric equivalent temperature.
(h) equivalent potential temperature.

A.1.5* What is the boiling point temperature for water at 700 mbar?

A.1.6** Plot representative isosteres for the PT diagram for the water system.

A.1.7** How does the curvature of the plot of saturation vapor pressure versus temperature change as you approach the critical temperature?

A.1.8*** Construct a Tα diagram for the water system. Use log α and linear T scales. Sketch in some illustrative isobars.

Computer problems

A.1.9** Compute the change in temperature for a parcel lifted pseudo-adiabatically from 900 mbar to 700 mbar by numerically integrating (A.1.24) using ΔP steps of 10 mbar.

A.1.10*** Compute the change in temperature for a pseudo-adiabatic process as in Problem A.1.9 but use the following stepwise approach. Start with a parcel of 1 cm^3.
(a) Lift the parcel dry adiabatically ΔP and compute the temperature.
(b) Compute the volume of the expanded parcel.
(c) Determine the saturation vapor density at the new temperature.

(d) Compute the actual vapor density and the amount of condensate, assuming water saturation is maintained.
(e) Compute the latent heat released by the additional condensate and the change in sensible heat for the pre-existing condensate.
(f) Recompute the temperature of the parcel and repeat steps (b), (c), (d), and (e).
(g) Go to step (a) until 700 mbar is reached.

Try different steps ΔP until the integration produces stable results. You may need to use double precision arithmetic.

REFERENCES

Fletcher, N.H., 1970, *The chemical physics of ice*. Cambridge: Cambridge University Press.

Huschke, R.E. (Ed.), 1959, *Glossary of meteorology*. Boston, MA: American Meteorological Society.

List, R.J. (Ed.), 1958, *Smithsonian meteorological tables*. Washington, DC: Smithsonian Institute.

McDonald, J.E., 1963, 'Intermolecular attractions and saturation vapor pressure,' *J. Atmos. Sci. 20, 178.*

Appendix 2

A BRIEF OVERVIEW OF NUMERICAL MODELS FOR MICROPHYSICAL PROCESSES

The interactions between nucleation, diffusional growth, and growth by collection are sufficiently complex so as to require some type of numerical solution to understand these interactions more fully. The formulation of a numerical solution constitutes a 'numerical model' which is a set of equations describing the formation and growth (evaporation) of the particles. Equations may be used in their complete form, or simplifying assumptions can be made to reduce the programming complexity and/or the computer time required for a solution. Equations which have been simplified to an extreme degree are often referred to as 'parameterized' equations.

Some examples of simple microphysical models have been introduced in the text. Results from more sophisticated models are discussed throughout Chapters 9 through 13. These models may be classified as (a) particle-tracing models, (b) bin models, or (c) spectral models, depending on the manner in which the particles are handled.

A.2.1 *Particle-Tracing Models*

Particle-tracing models follow individual particles or groups of particles. The model of the activation of CCN described in section 9.2.1(ii) (Young, 1993) follows 37 groups of particles distinguished by the mass of ammonium sulphate. The number of groups in the model remains constant with time. All of the particles within a group are assumed to have identical growth histories. The model is also steady-state in that it tacitly assumes that particles with the same ammonium sulphate mass injected in the updraft at different times will experience the same growth history.

The hail growth model discussed in section 12.4.1 (Young, 1978) is another example of a particle-tracing model. This model relies on Monte Carlo techniques to simulate the effects of turbulence on the growth of particles which means identical particles injected into the updraft at different times will follow different growth histories. Here the advantages of a particle-tracing model become evident since it is possible to 'trace' the growth history of a particular group of particles. Thus, one can deduce the origin of particles that become hailstones and determine the length of time the particle resided in the updraft.

A2.2 *Bin Models*

Bin models are used when the particle interactions become too complex for a particle-tracing model to handle. This commonly is a result of modeling collection processes as described in section 7.3.1 (Young, 1975). The quasi-stochastic collection model specifies that only a fraction of the drops in a particular group will undergo a collection event in a given time step. Thus, at each step, each group must be divided into two groups, one which undergoes a collection and one which does not. Thus, the number of groups to be followed progressively increases. The number of drops in each group becomes progressively smaller as the number of groups increases.

The bin approach is a way in which particles of similar size and location are grouped together.[1] One could visualize this as a 'consolidation' process in which a periodic examination of the groups results in combining groups whose drop radii and location agree to within preset limits. This serves to reduce the number of groups being followed. However, once groups are consolidated, their growth histories become obscure since drops with a variety of growth histories are included in the same group.

The bin model defines ranges of particle sizes and groups all particles having radii within a set range together. This may be accomplished in either of two ways. The discrete bin model assumes all of the particles within a given bin are of the same size. Due to the requirement of conserving both total number and total mass, the group radius varies as the composition of the group changes. The continuous bin model assumes that the particles are distributed throughout the range of sizes for that particular bin. Young (1974) introduces a continuous bin approach assuming a linear distribution of particles within a bin as shown in Fig. A.2.1.

The continuous bin model is preferred to the discrete bin model since particles in the continuous bin model are promoted from bin-to-bin in a more realistic manner. In the discrete model, particles growing by diffusion would remain within a given bin until **all** of the particles grew large enough to be transferred into the next larger bin. In the continuous bin model, some of the particle would be transferred into the next larger bin at each time step as shown in Fig. A.2.2. Thus, for a simulation of collection processes, the effect of the range of sizes within a given bin on the collection process is modeled more accurately since the larger drops within a bin at a given time are advanced to the next larger bin before the smaller drops.

The inability of bin models to provide drop growth histories can be partially alleviated by employing Markov chain techniques as described by Valdez and Young (1985). Output from the bin model is used to develop a transition probability matrix in which transitions between any combination of bins is determined. In addition, 'birth' and 'death' probabilities are determined for each bin. These transition probabilities may then be used to describe the probability of any given drop history.

[1] In a multi-dimensional model, particles are also grouped in location bins defined by pre-set ranges of height and/or distance.

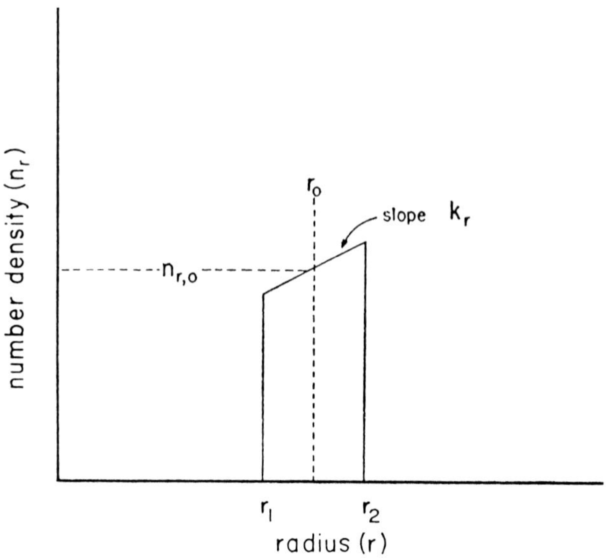

FIG. A.2.1. Distribution of particles for the continuous bin technique. Bin limits are r_1 and r_2 with mid-point r_o. Particles are distributed linearly within this bin with slope k_r and a mean concentration $n_{r,o}$.

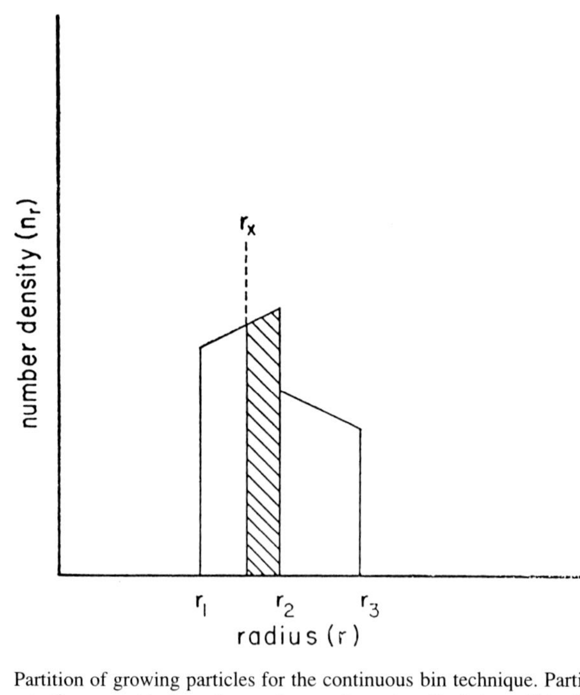

FIG. A.2.2. Partition of growing particles for the continuous bin technique. Particles with radius r_x will just grow to the upper bin boundary r_2 in one time step. Particles within the shaded area are transferred to bin (r_2, r_3). From Young (1974a).

A.2.3. *Spectral Models*

Spectral models assume a particle size distribution and derive equations which describe the evolution of the parameters of the distribution as the particles evolve. Srivastava (1978) derived expressions for the evolution of the parameters n_o and Λ of the Marshall-Palmer distribution given by (9–2), allowing a relatively simple simulation of the evolution of a raindrop spectrum within a cloud. Clark (1974) used the gamma distribution to model coalescence processes; Clark and Hall (1983) used several log-normal distributions for the same purpose.

Brown (1991) combines features of the spectral and bin models to simulate detailed features of the evolution of the drop spectrum as it approaches equilibrium. Brown defines four bins defined by drop radius. A spectral model is employed within each bin.

A.2.4. *Dimensionality of Models*

The dimensionality of a model defines the spatial and temporal context of the microphysical model. The simplest models are 'parcel' models in which location is not considered. The rationale for parcel models is that whatever particles fall out of the bottom of the modeled volume are matched by identical particles falling into the top of the volume. Parcel models are useful for modeling microphysical processes leading up to precipitation, i.e., conditions under which sedimentation is unimportant.

Steady-state models assume time invariance, growing particles by advecting particles through the model. Young (1974b) provides an example of a two-dimensional, steady-state model in which particles grow as they are advected over an orographic barrier. An assumed variation of wind speed with height and distance results in a time step dependent on height and distance, as described briefly in section 11.5.1. The model is in essence, a one-dimensional (vertical), time-dependent model where the horizontal distance axis is a quasi-time axis.

Two-dimensional, time-dependent models have been developed for both vertical slab (distance versus height) and axisymmetric (radial distance versus height) configurations. Hall (1980) describes a two-dimensional, slab symmetric, anelastic cloud model that incorporates detailed microphysical descriptions for both water and ice processes.

A.2.5. *Kinematic versus Dynamic Models*

Microphysical models may be decoupled from their dynamical structure, assuming the interactions between the microphysical and dynamical processes are negligible. Such decoupled models are termed 'kinematic' models. The two-dimensional, steady-state model described by Young (1974a) is such a kinematic model in which the flow field was adopted from a dynamical model. The flow was assumed to be stable and interactions between the microphysical and dynamical processes was neglected.

Dynamical models incorporating detailed microphysical processes allow simulation of interactions between the microphysical and dynamical processes but are much more costly in terms of the required computer resources. Such interactions are important in simulations of convective clouds. Clark (1973) and Hall (1980) describe models that permit such interactions between the microphysical and dynamical processes.

REFERENCES

Brown, P.S., Jr., 1991; 'Parameterization of the evolving drop-size distribution based on analytic solution of the linearized coalescence-breakup equation,' *J. Atmos. Sci.,* **35**, 108-117.

Clark, T.L., 1973; 'Numerical modeling of the dynamics and micro-physics of warm cumulus convection,' *J. Atmos. Sci.,* **30**, 857–878.

Clark, T.L., 1974; 'A study in cloud phase parameterization using the gamma distribution,' *J. Atmos. Sci.,* **31**, 142–155.

Clark, T.L. and Hall, W.D., 1983; 'A cloud physical parameteriza-tion using movable basis functions: Stochastic coalescence parcel calculations,' *J. Atmos. Sci.,* **40**, 1709–1728.

Hall, W.D., 1980; 'A detailed microphysical model within a two-dimensional dynamic framework: Model description and preliminary results,' *J. Atmos. Sci.,* **37**, 2486–2507.

Srivastava, R.C., 1978; 'Parameterization of rain-drop size distributions. *J. Atmos. Sci.,* **35**, 108–117.

Valdez, M.P., and K.C. Young, 1985; 'Number fluxes in equilibrium raindrop populations: A Markov chain analysis,' *J. Atmos. Sci.,* **42**, 1024–1036.

Young, K.C., 1974a; 'A numerical simulation of wintertime, orographic precipitation: Part I. Description of model microphysics and numerical techniques,' *J. Atmos. Sci.,* **32**, 1735-1748.

Young, K.C., 1974b; 'A numerical simulation of wintertime, orographic precipitation: Part II. Comparison of natural and AgI-seeded conditions,' *J. Atmos. Sci.,* **32**, 1749–1767.

Young, K.C., 1975; 'The evolution of drop spectra due to condensation, coalescence and breakup,' *J. Atmos. Sci.,* **32**, 965–973.

Young, K.C., 1978; 'On the role of mixing in promoting competition between growing hailstones,' *J. Atmos. Sci.,* **35**, 2190-2193.

Young, K.C., 1993; 'Effects of simplifications of the Köhler equation on the activation of CCN in an updraft,' J. Atmos. Sci. (in press)

Appendix 3

USEFUL CONSTANTS AND DEFINITIONS

Constants

c_p	$1.005\ J\ g^{-1}\ K^{-1}$	specific heat of air at constant pressure
c_{wv}	$1.846\ J\ g^{-1}\ K^{-1}$	specific heat of water vapor at constant pressure
g	$980.629\ cm\ s^{-2}$	acceleration due to gravity at mean sea level and 45° latitude
h	$6.626\ 176 \times 10^{-34}\ J\ s$	Planck's constant
k	$1.380\ 662 \times 10^{-23}\ J\ K^{-1}$	Boltzmann's constant
$L_{f,o}$	$333.69\ J\ g^{-1}$	latent heat of fusion at 0°C
$L_{s,o}$	$2834.46\ J\ g^{-1}$	latent heat of sublimation at 0°C
$L_{v,o}$	$2500.78\ J\ g^{-1}$	latent heat of vaporization at 0°C
M_d	$28.9644\ g\ mol^{-1}$	apparent molecular weight for dry air
M_v	$18.0160\ g\ mol^{-1}$	molecular weight for water vapor
N_A	6.0228×10^{23} molecules mol^{-1}	Avogadro's number
N_I	3.0656×10^{22} molecules g^{-1}	number density for ice at 0°C
N_L	3.3425×10^{22} molecules g^{-1}	number density for water at 0°C
R	$8.31436 \times 10^{7}\ erg\ mol^{-1}\ K^{-1}$	gas constant
R_d	$2.8704 \times 10^{6}\ erg\ g^{-1}\ K^{-1}$	gas constant for dry air
R_v	$4.615 \times 10^{6}\ erg\ g^{-1}\ K^{-1}$	gas constant for water vapor
σ_{VS}	$109 \pm 3\ erg\ cm^{-2}$	surface energy for water vapor–ice
σ_{LS}	$33 \pm 3\ erg\ cm^{-2}$	interfacial energy for water–ice

Conversion factors

1 inch	= 2.54 cm
1 liter	= $1000.028\ cm^3$
1 international knot	= $0.514444\ m\ s^{-1}$
1 pound av	= 453.5923 g
1 standard atmosphere	= 1013.250 mb
	= 760 mmHg
	= 29.9213 in Hg
	= $14.6960\ lb\ in^{-2}$
1 cal	= 4.18684×10^{7} erg
1 joule	= 0.238844 cal

Dimensionless numbers

N_{Bo} Bond number: ratio of gravitational force to surface force

$$N_{Bo} = \frac{16\ (\rho_x - \rho_a)\ gr^2}{3\sigma}$$

where σ is the surface tension, ρ is the density, and the subscripts x and a refer to the particle and air, respectively.

N_{Da} Davies number: ratio of gravitational force to drag force

$$N_{Da} = \frac{32\rho_a\ (\rho_x - \rho_a)\ gr^3}{3\eta^2}$$

where η is the dynamic viscosity, ρ is the density, and the subscripts x and a refer to the particle and air, respectively.

N_{Kn} Knudsen number: ratio of mean free path for air molecules to radius of the aerosol

$$N_{Kn} = \frac{\lambda}{r}$$

where λ is the mean free path for air molecules.

N_{Pr} Prandtl number: ratio of viscosity to conductivity

$$N_{Pr} = \frac{\eta_a c_p}{K_a}$$

where c_p is the specific heat of air at constant pressure, η is the dynamic viscosity, K is the thermal conductivity, and the subscript a refers to air.

N_{Re} Reynolds number: ratio of inertial force to viscous force

$$N_{Re} = \frac{2v_t r_c \rho_a}{\eta_a}$$

where r_c is the characteristic dimension of the particle, v_t is its terminal fall velocity, ρ and η refer to density and dynamic viscosity, and the subscript a refers to air.

N_{Sc} Schmidt number: ratio of viscosity to diffusivity

$$N_{Sc} = \frac{\eta_a}{D\rho_a}$$

where D is the diffusivity, ρ and η refer to density and dynamic viscosity, and the subscript a refers to air.

N_{We} Weber number: ratio of dynamic pressure to pressure across a curved interface

$$N_{We} = \frac{\rho_w r_L}{\sigma_{LV}} (v_L - v_S)^2$$

where v is the terminal fall velocity, ρ is the density, and the subscripts L and S refer to the large and small drops, respectively.

INDEX